Biocatalysis
and Agricultural
Biotechnology

Biocatalysis and Agricultural Biotechnology

Edited by

Ching T. Hou
Jei-Fu Shaw

International Society of Biocatalysis
and Biotechnology

National Chung Hsing University

CRC Press
Taylor & Francis Group
Boca Raton London New York

CRC Press is an imprint of the
Taylor & Francis Group, an **informa** business

CRC Press
Taylor & Francis Group
6000 Broken Sound Parkway NW, Suite 300
Boca Raton, FL 33487-2742

First issued in paperback 2019

ISBN-13: 978-1-4200-7703-2 (hbk)
ISBN-13: 978-0-367-38569-9 (pbk)

Library of Congress Cataloging-in-Publication Data

Biocatalysis and agricultural biotechnology / editors, Ching T. Hou, Jei-Fu Shaw.
    p. cm.
   "Selected papers presented at the third International Symposium on
Biocatalysis and Biotechnology held at the National Chung Hsing University
(NCHU), Taichung, Taiwan, on November 28-30, 2007."
   Includes bibliographical references and index.
   ISBN 978-1-4200-7703-2 (hardcover : alk. paper) 1. Agricultural
biotechnology--Congresses. 2. Enzymes--Biotechnology--Congresses. I. Hou,
Ching T. (Ching-Tsang), 1935- II. Shaw, Jei-Fu. III. International Symposium
on Biocatalysis and Biotechnology 2007 : Taichung, Taiwan

S494.5.B563B48 2009
631--dc22
    2009002055

Visit the Taylor & Francis Web site at
http://www.taylorandfrancis.com

and the CRC Press Web site at
http://www.crcpress.com

Contents

SECTION 1 Improvement of Agronomic and Microbial Traits

SECTION 2   Bio-Based Industry Products

Preface

This book was assembled with the intent of bringing together current advances and in-depth reviews of biocatalysis and agricultural biotechnology with emphasis on bio-based products and agricultural biotechnology. The book consists of selected papers presented at the third International Symposium on Biocatalysis and Biotechnology held at the National Chung Hsing University (NCHU), Taichung, Taiwan, on November 28–30, 2007. At this symposium, 55 distinguished international scientists from the United States, Japan, Korea, Canada, Brazil, Belgium, Slovak Republic, France, and Taiwan shared their valuable research results. In addition, there were 13 selected posters, 12 bioenergy exhibitions, and over 400 attendees. A few chapters contained in this book were contributed by distinguished scientists who could not attend this meeting. This meeting was a great success, and we greatly appreciate the contribution of local organization committee members at NCHU, including Dr. C. H. Yang, director of NCHU Biotechnology Center; Dr. T. J. Fang; Dr. C. S. Wang; Dr. C. C. Huang; Dr. S. W. Tsai; Dr. F.-J. Jan; Dr. Y.-T. Chen; Dr. C. Chang; and their colleagues and students.

Recent energy and food crises point out the importance of bio-based products from renewable resources and agricultural biotechnology. Use of modern tools of molecular engineering on plants, animals, and microorganisms to solve these crises and improve the wellness of humankind is inevitable. This is the most current book on agricultural biotechnology and bio-based industrial products. The authors are internationally recognized experts from all sectors of academia, industry, and government research institutes.

The book consists of 28 chapters divided into two sections. The first 11 chapters describe the world's newest research on improvement of agronomic and microbial traits. Included are molecular strategies for enhancing seed oil content, engineering biosynthesis of α-eleostearic acid in transgenic systems, cloning and characterization of Kennedy pathway acyltransferases from developing seeds of tung, accumulation of epoxy fatty acids in plant oils, high-erucic acid rapeseed cultivar development in Canada, new approaches for acquired tolerance to biotic and abiotic stress of economically important crops, engineering plants for industrial uses, improvement of plant transformation efficiency, bifidobacterial lacto-N-biose/galacto-N-biose pathway involved in intestinal growth, improvement of agronomic traits by using different isoforms of ferredoxin for plant development and disease resistance, biosynthesis of unusual fatty acids in microorganisms and their production in plants, and evaluation of gene flow in a minor crop (safflower for the production of plant-made pharmaceuticals in Canada).

The next 17 chapters describe bio-based industrial products. Included are production of functional γ-linolenic acid by expression of fungal Δ12 and Δ6 desaturase genes in the oleaginous yeast *Yarrowia lipolytica*, production of monoacylglycerols through lipase-catalyzed reactions, physiological function of a Japanese traditional fish source (*ishiru*), modifying enzyme character by gene manipulation (preparation of chimeric genes), biocatalytic synthesis of chiral intermediates for development of drugs, selective removal of free fatty acids from fats and oils using biological systems, branched chain fatty acid as a functional lipid, retrobiosynthetic production of 2′-deoxyribonucleoside from glucose, acetaldehyde, and nucleobase through multistep enzymatic reactions, omega-3 phospholipid nanocapsules as effective forms for transporting immune response modifier, regiospecific quantification of triacylglycerols by mass spectrometry and its use in olive oil analysis, utilization of membrane separation technology for purification and concentration of anserine and carnosine extracted from chicken meat, preparation and characterization of novel phospholipids containing terpene mediated by phospholipase D, synthesis of useful glycosides by cyclodextrin glucanotransferases and glycosidases, biotechnological valorization of agroindustrial materials to value-added bioproducts, marine functional ingredients and advanced technology for health food development, a new method for in vitro glycogen synthesis, immunostimulating activity of glycogen, and biological synthesis of metal nanoparticles.

This book serves as reference for teachers, graduate students, and industrial scientists who conduct research in biosciences and biotechnology.

Ching T. Hou
Peoria, Illinois

Jei-Fu Shaw
Taichung, Taiwan

Contributors

Yoshiyuki Adachi
Laboratory for Immunopharmacology
  of Microbial Products
Tokyo University of Pharmacy and Life
  Science
Tokyo, Japan

Tsunehisa Akiyama
Biochemical Research Laboratory
Ezaki Glico Company, Ltd.
Osaka, Japan

Joosh Baljinnyam
National Food Research Institute
Tsukuba, Japan
Mongolian University of Science and
  Technology
Bagatoiruu, Ulaanbaatar, Mongolia

Alex Boyko
University of Lethbridge
Lethbridge, Alberta, Canada

Emília Breierová
Slovak Academy of Sciences
Bratislava, Slovak Republic

Milan Čertík
Faculty of Chemical and Food
  Technology
Slovak University of Technology
Bratislava, Slovak Republic

Jeng-Sheng Chang
Food Science Department
National Taiwan Ocean University
Keelung, Taiwan

Ke Liang B. Chang
Food Science Department
National Taiwan Ocean University
Keelung, Taiwan

Pi-Fang Linda Chang
National Chung Hsing University
Taichung City, Taiwan

Dzi-Chi Chen
Yeastern Biotech Company, Ltd.
Taipei, Taiwan

Ying-Hsuan Chen
Yeastern Biotech Company, Ltd.
Taipei, Taiwan

Lu-Te Chuang
Yuanpei University
Hsin-Chu, Taiwan

Toshiki Enomoto
Ishikawa Prefectural University
Ishikawa, Japan

Teng-Yung Feng
Academia Sinica
Taipei, Taiwan

W. G. Dilantha Fernando
University of Manitoba
Winnipeg, Manitoba, Canada

Takashi Furuyashiki
Biochemical Research Laboratory
Ezaki Glico Company, Ltd.
Osaka, Japan

Shoji Hagiwara
National Food Research Institute
Tsukuba, Japan

Linda M. Hall
Agricultural Food and Nutritional
    Science
University of Alberta
Edmonton, Alberta, Canada

Vladimíra Hanusová
Slovak University of Technology
Bratislava, Slovak Republic

John L. Harwood
Cardiff University
Cardiff, Wales, United Kingdom

Tomoko Hatanaka
Kobe University
Kobe, Japan

Kiyoshi Hayashi
National Food Research Institute
Tsukuba, Japan

David Hildebrand
University of Kentucky
Lexington, Kentucky

Nobuyuki Horinouchi
Division of Applied Life Sciences
Graduate School of Agriculture
Kyoto University
Kyoto, Japan

Masashi Hosokawa
Hokkaido University
Hakodate, Japan

Hsiang-En Huang
National Taitung University
Taitung, Taiwan

Yung-Sheng Huang
National Chung-Hsing University
Taichung, Taiwan

Hideki Kajiura
Food Material Department
Glico Foods Company, Ltd.
Takatsuki, Japan

Ryo Kakutani
Biochemical Research Laboratory
Ezaki Glico Company, Ltd.
Osaka, Japan

Beom Soo Kim
Department of Chemical Engineering
Chungbuk National University
Cheongju, Chungbuk, Korea

Hyun-Jin Kim
Korea Food Research Institute
Kyonggi, South Korea

Taro Kiso
Osaka Municipal Technical Research
    Institute
Osaka, Japan

Motomitsu Kitaoka
National Food Research Institute
Tsukuba, Japan

Zwe-ling Kong
Food Science Department
National Taiwan Ocean University
Keelung, Taiwan

Igor Kovalchuk
University of Lethbridge
Lethbridge, Alberta, Canada

Takashi Kuriki
Biochemical Research Laboratory
Ezaki Glico Company, Ltd.
Osaka, Japan

Dae Young Kwon
Korea Food Research Institute
Kyonggi, South Korea

André Laroche
Agriculture and Agri-Food Canada
Lethbridge, Alberta, Canada

Genyi Li
University of Manitoba
Winnipeg, Manitoba, Canada

Runzhi Li
University of Kentucky
Lexington, Kentucky

Jiann-Tsyh Lin
USDA-ARS, Western Regional
    Research Center
Albany, California

Wei-Chin (Wayne) Lin
Agriculture and Agri-Food Canada
Pacific Agri-Food Research Centre
Agassiz, British Columbia, Canada

Yi-Hsien Lin
Academia Sinica
Taipei, Taiwan

Catherine Madzak
Laboratoire de Microbiologie et
    Génétique Moléculaire
AgroParisTech
Thiverval-Grignon, France

Ivana Márová
Brno University of Technology
Brno, Czech Republic

Thomas A. McKeon
USDA-ARS Western Regional Research
    Center
Albany, California

Marc A. McPherson
Agricultural Food and Nutritional
    Science
University of Alberta
Edmonton, Alberta, Canada

Peter B. E. McVetty
University of Manitoba
Winnipeg, Manitoba, Canada

Dauenpen Meesapyodsuk
University of Saskatchewan
Saskatoon, Saskatchewan, Canada

Toshihide Michihata
Industrial Research Institute of
    Ishikawa
Ishikawa, Japan

Kazuo Miyashita
Hokkaido University
Hakodate, Japan

Hiroshi Nabetani
National Food Research Institute
Tsukuba, Japan

Hirohisa Naito
Hokkaido University
Hakodate, Japan

Mitsutoshi Nakajima
National Food Research Institute
Tsukuba, Japan

Hirofumi Nakano
Osaka Municipal Technical Research
    Institute
Osaka, Japan

Jean-Marc Nicaud
Laboratoire de Microbiologie et
    Génétique Moléculaire
AgroParisTech
Thiverval-Grignon, France

Mamoru Nishimoto
National Food Research Institute
Tsukuba, Japan

Jun Ogawa
Division of Applied Life Sciences
Graduate School of Agriculture
Kyoto University
Kyoto, Japan

Takeshi Ohkubo
NOF Corporation
Tokyo, Japan

Naohito Ohno
Laboratory for Immunopharmacology
of Microbial Products
Tokyo University of Pharmacy and Life
Science
Tokyo, Japan

Hirosuke Oku
University of the Ryukyus
Okinawa, Japan

Bonnie Sun Pan
Food Science Department
National Taiwan Ocean University
Keelung, Taiwan

Min-Hsiung Pan
Department of Seafood Science
National Kaohsiung Marine University
Kaohsiung, Taiwan

Ramesh N. Patel
Bristol-Myers Squibb
New Brunswick, New Jersey

Xiao Qiu
Department of Applied Microbiology
and Food Science
University of Saskatchewan
Saskatoon, Saskatchewan, Canada

Peter Rapta
Slovak University of Technology
Bratislava, Slovak Republic

Saleh Shah
Alberta Research Council
Vegreville, Alberta, Canada

Yuji Shimada
Osaka Municipal Technical Research
Institute
Osaka, Japan

Sakayu Shimizu
Division of Applied Life Sciences
Graduate School of Agriculture
Kyoto University
Kyoto, Japan

Shigenobu Shiotani
National Food Research Institute
Tsukuba, Japan
Tokai Bussan Company Ltd.
Tokyo, Japan

Jay Shockey
USDA-ARS Southern Regional
Research Center
New Orleans, Lousiana

Jae Yung Song
Chungbuk National University
Cheongju, Chungbuk, Korea

Muhammad Tahir
University of Manitoba
Winnipeg, Manitoba, Canada

Koretaro Takahashi
Hokkaido University
Hakodate, Japan

Mikako Takasugi
Kyushu Sangyo University
Fukuoka City, Japan

Hiroki Takata
Biochemical Research Laboratory
Ezaki Glico Company, Ltd.
Osaka, Japan

Hajime Taniguchi
Department of Food Science
Ishikawa Prefectural University
Ishikawa, Japan

David C. Taylor
Plant Biotechnology Institute
National Research Council
Saskatoon, Saskatchewan, Canada

Yomi Watanabe
Osaka Municipal Technical Research
    Institute
Osaka, Japan

Randall J. Weselake
University of Alberta
Edmonton, Alberta, Canada

Li-Cyuan Wu
Food Science Department
National Taiwan Ocean University
Keelung, Taiwan

Yukihiro Yamamoto
Hokkaido University
Hakodate, Japan

Teruyoshi Yanagita
Department of Applied Biochemistry
    and Food Science
Saga University
Saga, Japan

Nobuya Yanai
National Food Research Institute
Tsukuba, Japan
Tokai Bussan Company Ltd.
Tokyo, Japan

Suk Hoo Yoon
Korea Food Research Institute
Kyonggi, South Korea

Keshun Yu
University of Kentucky
Lexington, Kentucky

Carla Zelmer
University of Manitoba
Winnipeg, Manitoba, Canada

Editors

Ching T. Hou, Ph.D. is lead scientist at the U.S. Department of Agriculture (USDA) National Center for Agricultural Utilization Research in Peoria, Illinois. He has also worked as director of the Department of Microbial Biochemistry and Genetics at the Squibb Institute for Medical Research in Princeton, New Jersey, and was a principal investigator and group leader at Exxon Research Center and a National Research Council research associate at the USDA. He is a Fellow of the American Academy of Microbiology 1994, the Society for Industrial Microbiology (SIM, 1996), and the American Oil Chemists' Society (AOCS, 2006). Dr. Hou holds 17 patents, has published 5 books, and has authored or coauthored over 210 peer-reviewed scientific papers. He organized and chaired 50 symposia in both national and international conferences and has written 25 invited review chapters. Dr. Hou was a consultant on petroleum microbiology, biochemistry, and environmental issues to Exxon Research Center (1990–1992). He served as an advisor to the government of Taiwan on biotechnology research planning in 1983, to the Institute of AgroBiotechnology, Academia Sinica (1997–2001), and to the Institute of Botany, Academia Sinica (2001–2004) in Taiwan. Recently, he was appointed a member of the advisory board of the National Chung Hsing University, Taiwan (2005–2008). Dr. Hou is a longtime member of SIM, ASM (American Society for Microbiology), ACS (American Chemical Society), and AOCS.

He was a founding member of the Biotechnology Division, AOCS. He has organized two biocatalysis symposia sessions each year for the past 19 years for the AOCS annual meeting. Dr. Hou established a new award, the AOCS Biotechnology Lifetime Achievement Award, in 1998. He is an associate editor of the *Journal of American Oil Chemists' Society* and an editorial board member of the *Journal of Industrial Microbiology*. He is past president of the Chinese-American Microbiology Society (1995–1996) and was chairperson (1998–2001) of the Biotechnology Division, American Oil Chemists' Society. He is currently vice chairperson of the AOCS Foundation and is U.S. side chairperson of the biocatalysis and biotechnology area of the United States–Japan Natural Resources Utilization Collaboration Panel.

Jei-Fu Shaw, Ph.D. is chair professor and president of National Chung Hsing University, Taiwan, a renowned agricultural biotechnology research university. He received his Ph.D. from the Department of Biochemistry, University of Arkansas, in 1977 and was a visiting scientist at the Department of Applied Biological Sciences, Massachusetts Institute of Technology, from 1985 to 1986. He is a distinguished scientist who has been devoted to research, education, and promotion of biotechnology in Taiwan for nearly 30 years. He is the author or coauthor of over 290 publications and holds six patents, with three patents pending and three technical transfers. He has received over 1,000 SCI (Science Citation Index) citations of his original research papers. He has been the invited speaker or chairperson at many national and international scientific conferences. His major research contributions include novel discoveries on the structure, function, and biotechnological applications of several important enzymes and outstanding contributions in plant science and biotechnology, including discovery of important plant senescence-associated genes, application of genetic engineering approaches to prolonging the shelf lives of agricultural produce, and invention of enzymatic methods and utilizing transgenic plant technology to convert low agricultural staples into valuable industrial products, thus improving nutrition by producing high-protein foods.

Dr. Shaw has served as director of the Institute of Botany, Academia Sinica; director of the Life Science Promotion Center, National Science Council; president of the Taiwan Society for Biochemistry and Molecular Biology; president of Academia-Industry consortium of central Taiwan; and vice president of the International Society of Biocatalysis and Biotechnology. He helped establish the National Agricultural Biochemistry Program and serves as the coordinator of the Plant Gene Utilization Division. He led the Taiwan Research Team in the International Rice Genome Sequence Project to complete the sequencing of rice chromosome 5. His leadership has had an important impact on the development of bioindustry and agriculture. His career has been marked by many national and international honors and awards, including the Distinguished Research Award of the National Science Council, the National Agronomy Award, AAAS (American Association for the Advancement of Science) Fellow, and the Biotechnology Lifetime Achievement Award of AOCS.

Section 1

Improvement of Agronomic
and Microbial Traits

1 Molecular Strategies for Increasing Seed Oil Content

Randall J. Weselake, David C. Taylor, Saleh Shah, André Laroche, and John L. Harwood

CONTENTS

Key Words: Acyltransferase; *Arabidopsis*; *Brassica napus*; carbon flux; fatty acid and triacylglycerol biosynthesis; overexpression; transcription factor.

1.1 INTRODUCTION

The rising demand for vegetable oils to satisfy the food and industrial needs of a growing global population will place increasing pressure on the available supply of oil, which will in turn result in substantial increases in the price of seed oils. In the case of Canadian canola (*Brassica napus* L. and *B. rapa* L.), it has been estimated that an absolute increase of 1% of oil in the seed presently has an economic value of $60 million per annum for the Canadian oilseed crushing and processing industry (Canola Council of Canada). In addition to increasing acreage for production of oilseed crops and recruiting nonconventional crops as sources of industrial oil, there is considerable interest in developing molecular strategies for increasing seed oil content.

Plant breeding initiatives have developed oleaginous crops with enhanced seed oil content. The use of molecular markers and the development of quantitative trait loci (QTL) mapping methods have facilitated the identification of genes controlling lipid accumulation, and in turn, marker-assisted selection has been used to move favorable QTL alleles into genotypes for oilseed breeding programs (Zhao et al. 2006). Although genetic studies and breeding have revealed that seed oil content is controlled by a number of QTL (Hobbs et al. 2004, Delourme et al. 2006, Wu et al. 2006, Zhao et al. 2006), several studies with single-gene manipulations have led to substantial increases in seed oil content.

The current review focuses mainly on recent advances in the development of molecular strategies to enhance seed oil content. So far, this type of research has mostly been limited to the use of the model plant *Arabidopsis*, *B. napus*, and tobacco. The reviews of Thelen and Ohlrogge (2002) and Hills (2004) provided earlier coverage of some of the topics in the current review. Most of the discussion in the current review focuses on strategies involving the overexpression of genes encoding enzymes involved in lipid biosynthesis. Therefore, the review begins by providing some background information on seed oil biosynthesis. Strategies involving modification of other reactions in cellular metabolism that may alter the flow of carbon from nonlipid molecules into storage lipid are also discussed. Finally, transcription factors, which can affect the expression of several enzymes in a pathway, are discussed in relation to the regulation of seed oil accumulation. Thus, overproduction of specific transcription factors during seed development represents a way of attenuating the activity of an entire pathway to increase seed oil accumulation.

1.2 BIOSYNTHESIS OF SEED OIL

Genetic engineering has created numerous possibilities for enhancing seed oil content. Prior to entering a discussion of various genetic engineering approaches for enhancing seed oil content, we provide a general overview of the biosynthesis of seed oil based on information from the following detailed reviews on the topic: Stymne and Stobart (1987), Ohlrogge and Browse (1995), Harwood (2005), Weselake (2005), Lung and Weselake (2006), and Cahoon et al. (2007).

Seed oil is predominantly composed of triacylglycerol (TAG), which consists of a glycerol backbone with three esterified fatty acyl groups. A generalized scheme for fatty acid (FA) and TAG biosynthesis in developing seeds of oil crops is shown in Figure 1.1. Enzymes identified as having a key role in enhancing carbon flow into seed oil are indicated with an asterisk. De novo FA biosynthesis occurs in the subcellular organelle known as the plastid through the catalytic action of acetyl-coenzyme A carboxylase (ACCase) and the FA synthase complex. The ACCase-catalyzed reaction provides the malonyl-CoA (coenzyme A) for addition of two-carbon units to the growing FA chain, which is covalently linked to acyl carrier protein (ACP) in the FA synthase complex. A soluble acyl-ACP desaturase catalyzes the formation of monounsaturated FA such as oleate. Acyl-ACP hydrolase (or thioesterase) catalyzes the release of FAs, which in turn move to the outer plastidial membrane, where acyl-CoA synthetase catalyzes their conversion to acyl-CoAs. Molecular species of acyl-CoA derived from plastidial FA can be further elongated and desaturated in the endoplasmic reticulum (ER).

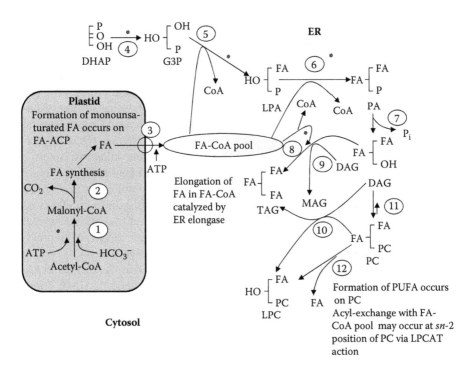

FIGURE 1.1 Generalized scheme for fatty acid (FA) and triacylglycerol (TAG) biosynthesis in developing seeds of oil crops. Reactions having a substantial effect on the flow of carbon into seed oil are indicated with an asterisk. The formation of TAG can occur through acyl-coenzyme A (acyl-CoA)-dependent processes (enzyme 8) and acyl-CoA-independent processes (enzymes 9 and 10). Enzymes 1–12 are 1, acetyl-CoA carboxylase; 2, fatty acid synthase complex; 3, acyl-CoA synthetase; 4, sn-glycerol-3-phosphate dehydrogenase; 5, sn-glycerol-3-phosphate acyltransferase; 6, lysophosphatidic acid acyltransferase; 7, phosphatidic acid phosphatase; 8, diacylglycerol acyltransferase; 9, diacylglycerol transacylase; 10, phospholipid:diacylglycerol acyltransferase; 11, choline phosphotransferase; 12, phospholipase A$_2$. Additional abbreviations: ATP, adenosine triphosphate; CoA, coenzyme A; DAG, sn-1, 2-diacylglycerol; DHAP, dihydroxyacetone phosphate; ER, endoplasmic reticulum; FA, fatty acid; FA-ACP; fatty acyl-acyl carrier protein; FA-CoA, fatty acyl-coenzyme A; G3P, sn-glycerol-3-phosphate; LPA, lysophosphatidic acid; LPC, lysophosphatidylcholine; LPCAT, lysophosphatidylcholine acyltransferase; MAG, monoacylglycerol; PA, phosphatidic acid; PC, phosphatidylcholine; PUFA, polyunsaturated fatty acid. The depicted scheme is based on information from Stymne and Stobart (1987), Ohlrogge and Browse (1995), Harwood (2005), and Weselake (2005).

The cytoplasmic pool of plastidially derived acyl-CoAs can in turn serve as substrates for the reactions of TAG assembly, which occur through the catalytic action of membrane-bound enzymes in the ER in a process that involves membrane metabolism (Weselake 2005). The glycerol backbone used for TAG bioassembly is derived from sn-glycerol-3-phosphate (G3P), which is produced from dihydroxyacetone phosphate (DHAP) and reduced nicotinamide adenine dinucleotide (NADH) through the catalytic action of *sn*-glycerol-3-phosphate dehydrogenase. *sn*-Glycerol-3-phosphate acyltransferase (GPAT) catalyzes the acyl-CoA-dependent acylation of the *sn*-1 position of G3P to produce lysophosphatidic acid (LPA), which is further acylated at the *sn*-2 position by the catalytic action of acyl-CoA-dependent lyso-phosphatidic acid acyltransferase (LPAAT) to produce phosphatidic acid (PA). The removal of the phosphate group from PA is catalyzed by PA phosphatase. The resulting *sn*-1,2-diacylglycerol (DAG) can be converted to TAG by acyl-CoA-dependent and acyl-CoA-independent processes. Diacyglycerol acyltransferase (DGAT) catalyzes the acyl-CoA-dependent acylation of DAG. Two genes encoding two isoforms of membrane-bound DGAT (type 1 and type 2) have been identified in a number of organisms, including plants. *DGAT1* does not share homology with *DGAT2* (Lardizabal et al. 2001). In addition, Saha et al. (2006) described the molecular cloning and expression of a soluble DGAT from developing peanut (*Arachis hypogaea*) cotyledons. This third isoform of DGAT shared less than 10% identity with previously identified forms of DGAT1 or DGAT2 from other organisms.

Acyl-CoA-independent processes have been identified that depend on either phosphatidylcholine (PC) or DAG as donors of fatty acyl moieties to DAG. These reactions are catalyzed by phospholipid:diacylglycerol acyltransferase (PDAT) or diacylglycerol transacylase (DGTA), respectively. Complementary DNAs (cDNAs) encoding PDAT have been isolated (Dahlqvist et al. 2000, Ståhl et al. 2004), whereas a cDNA encoding DGTA has not been identified. A study with a knockout for PDAT in *Arabidopsis*, however, has suggested that the enzyme does not play a major role in TAG biosynthesis during seed development (Mhaske et al. 2005).

TAG produced through acyltransferase action accumulates as droplets between the leaflets of the ER. Oil bodies surrounded by a half-unit membrane and ranging in size from 0.2 to 2 micrometers eventually bud off the surface of the ER.

DAG produced via the action of PA phosphatase can also be converted to PC by the catalytic action of choline phosphotransferase (CPT). Production of polyunsaturated FAs via the catalytic action of membrane-bound desaturases occurs while monounsaturated FAs are esterified to the glycerol backbone of PC. The combined forward and reverse reactions catalyzed by lysophosphatidylcholine acyltransferase (LPCAT) are believed to facilitate acyl exchange at the *sn*-2 position of PC, with the acyl-CoA pool thereby allowing incorporation of polyunsaturated FA into TAG via the acyl-CoA-dependent acyltransferase reactions in the mainstream of TAG biosynthesis. The reverse reaction of CPT to produce DAG provides a second opportunity to introduce polyunsaturated FA into TAG. In addition, phospholipase A_2 may catalyze the release of free FA from PC. The liberated FA could be reesterified as acyl-CoA for use in the mainstream of TAG biosynthesis.

1.3   OVEREXPRESSION OF LIPID BIOSYNTHETIC ENZYMES

1.3.1   OVEREXPRESSION OF ENZYMES PRODUCING TAG BUILDING BLOCKS

Lipid biosynthetic enzymes catalyzing the production of TAG building blocks (FA and the glycerol backbone) and acyltransferases catalyzing TAG bioassembly have been explored as targets for enhancing seed oil content. ACCase has been investigated as an enzyme regulating the availability of FA. Roesler et al. (1997) examined the expression of *Arabidopsis* cytosolic homomeric ACCase in the plastids of developing seeds of *Brassica napus* (Roesler et al. 1997). The underlying rationale for this investigation was that the cytosolic homomeric form of ACCase was not subject to the same metabolic regulatory controls as the heteromeric plastidial isozyme of ACCase. This strategy resulted in relative increases in seed oil content of about 5% in T_2 generation seeds.

The glycerol backbone has also been examined as a limiting factor in the production of TAG. Analyses of the G3P content of developing seeds of *B. napus* and *Arabidopsis* have shown that the rate of provision of G3P is not adequate to sustain high G3P levels during the rapid phase of oil accumulation (Vigeolas and Geigenberger 2004). Subsequently, Vigeolas et al. (2007) overexpressed a yeast G3P dehydrogenase during seed development in *B. napus* under greenhouse conditions as a means of increasing the abundance of G3P. The level of G3P increased by three- to four-fold in developing seeds of the T_4 generation, and the lipid content of mature seed increased significantly by up to 40% on a relative basis.

1.3.2 EXPRESSION OF PLASTIDIAL OR BACTERIAL *GPAT*

Jain et al. (2000) investigated the GPAT-catalyzed reaction as a target for enhancing seed oil content. The investigators expressed a cDNA encoding a plastidial safflower GPAT or the gene encoding *Escherichia coli* GPAT during seed development in *Arabidopsis* under greenhouse conditions. The construct for plastidial safflower GPAT was designed to remove the plastidial targeting sequence. Analyses of T_2 seeds revealed average oil content increases, on a relative basis, of 22% and 15% for expression of the safflower GPAT cDNA and the *E. coli* GPAT gene, respectively.

1.3.3 EXPRESSION OF A YEAST *SLC1-1* GENE

Zou et al. (1997) enhanced seed oil content in *Arabidopsis* and high-erucic acid *B. napus* L. cv Hero through expression of a mutated yeast LPAAT gene (*SLC1-1*) using a constitutive promoter. The various transgenic lines, grown under greenhouse conditions, exhibited relative increases in seed oil content ranging from 8% to 48%. Field studies from two successive years with T_4 and T_5 generations of *B. napus* L. cv Hero expressing *SLC1-1* resulted in relative increases in seed oil content of up to 13.5% (Taylor et al. 2001).

1.3.4 OVEREXPRESSION OF *DGAT*

Bouvier-Navé et al. (2000) demonstrated that transformation of tobacco with cDNA encoding type 1 *Arabidopsis* DGAT1 (*AtDGAT1*), using a constitutive promoter, resulted in as much as a seven-fold increase in the TAG content of leaves compared to the wild-type plant. Jako et al. (2001) further showed that seed-specific expression of *AtDGAT1* in *Arabidopsis* using a napin promoter resulted in 11% to 28% relative increases in seed oil content in homozygous *napin:AtDGAT1* lines. All transgenic

lines with increased *AtDGAT1* transcript and DGAT activity during seed development (compared to the wild type) displayed enhanced seed oil content at maturity.
More recently, seed-specific overexpression of *DGAT1* in a major oilseed crop has
also been shown to lead to enhanced seed oil content. Overexpression of *DGAT1* from
either *Arabidopsis* (*AtDGAT1*) or *B. napus* L. (*BnDGAT1*) during seed development
in two cultivars of oilseed rape (*B. napus* L.) has been shown to result in enhanced
seed oil content under both greenhouse and field conditions (Weselake et al. 2007).
A high oil transgenic line overexpressing *BnDGAT1*, grown under greenhouse conditions, displayed a decrease in the DAG:TAG ratio and fourfold enhanced DGAT
activity in developing seeds obtained about 4 weeks after flowering. Top-down control analysis (see Ramli et al. 2002) of this transgenic line indicated a decrease in
flux control from about 70% to 50% in the TAG assembly block (block B) (Weselake
et al. 2007). Production of FA is represented by reactions in block A with acyl-CoA
linking block A to the TAG assembly reactions of block B (see Ramli et al. 2002).
The results validated the use of control analysis for identifying molecular targets for
modification of seed traits (Weselake et al. 2007).

DGAT2 also has been overexpressed during seed development. Lardizabal et al.
(2006) achieved a small but statistically significant increase in seed oil content in
soybean (*Glycine max* L.) through the overexpression of an *Umbelopsis Mortierella*
(formerly) *ramanniana DGAT2* during seed development.

Previously, several lines of evidence suggested that overexpression of *DGAT* was
a promising target for enhancing seed oil content. Harwood and colleagues demonstrated that DAG was the most abundant Kennedy pathway intermediate in developing seeds of *B. napus* undergoing rapid oil accumulation, suggesting that the level
of DGAT activity was limiting the conversion of this intermediate into TAG (Perry
and Harwood 1993a, 1993b, Perry et al. 1999). In addition, in studies with developing safflower (*Carthamus tinctorius* L.) and *B. napus* seed, DGAT activity was
shown to be relatively low in comparison to other enzyme activities in the Kennedy
pathway leading to TAG (Ichihara et al. 1988, Perry et al. 1999). Furthermore, in
B. napus and safflower, it has been shown that DGAT activity is coordinated with
oil accumulation, with enzyme activity reaching a maximum during the rapid phase
of lipid accumulation and declining thereafter with seed maturity (Weselake et al.
1993). In addition, an *Arabidopsis* mutant (AS11) with decreased seed oil content
displayed reduced DGAT activity and an increased DAG:TAG ratio (Katavic et al.
1995). Somewhat later, it was shown that the AS11 had a mutation in the gene encoding DGAT1 (Zou et al. 1999b). *BnDGAT1* was the first DGAT cloned from a major
oilseed crop (*B. napus*) (Nykiforuk et al. 2002). Zhang et al. (2005) silenced *DGAT1*
in tobacco (*Nicotiana tabacum*), showing reductions of 9% to 49% in seed oil content on a relative basis, with concomitant increases in protein and sugar content.
Interestingly, a very recent study by Zheng et al. (2008) showed that a high-oil QTL
affecting corn (*Zea mays*) seed oil and oleic acid contents encodes a DGAT1. This
gene encoding the DGAT1 represented a specific allele (*DGAT1-2*), which resulted
in a phenylalanine insertion at position 469 in the amino acid sequence representing
the enzyme. Ectopic expression of the high-oil *DGAT1-2* allele resulted in relative
increases in seed oil content of up to 41%.

There appear to be species-based differences in the relative contributions of type 1 versus type 2 DGAT to TAG accumulation during seed development. DGAT2 has been shown to play a major role in TAG accumulation in the seeds of the tung tree (*Vernicia fordii*) and castor bean (*Ricinus communis*), in which the enzyme is involved in the incorporation of unusual fatty acyl moieties (Kroon et al. 2006, Shockey et al. 2006). The in vivo contribution of DGAT2 to oil biosynthesis in developing seeds of *B. napus*, however, remains to be elucidated (Cahoon et al. 2007).

1.4 OVEREXPRESSION/REPRESSION OF ENZYMES REGULATING CARBON FLUX

The diversion of carbon from carbohydrate and other nonlipid molecules to lipid represents another possible strategy for enhancing seed oil content. The relationship of carbohydrate catabolism to lipid biosynthesis in photosynthetic developing seeds of *B. napus* is outlined in Figure 1.2. Reactions that have been shown to have a substantial effect on seed oil accumulation are identified with an asterisk.

Tomlinson et al. (2004) hypothesized that a significant degree of control over the flow of carbon into seed oil was associated with pathways of sucrose catabolism and entry into metabolism. Invertase and hexokinase represent one of two routes through which sucrose can enter metabolism, with activities of these enzymes also having possible effects on the status of sugar-sensing pathways. Using constructs with seed-specific promoters and yeast genes encoding invertase or hexokinase, the investigators introduced additional invertase and hexokinase activity into the apoplast and cytosol of developing tobacco (*N. tabacum* L.) seeds, respectively. The yeast enzymes were expressed alone in tobacco seed or in combination. Despite enormous increases in the activities of these enzymes during seed development, there was essentially no effect on seed oil accumulation, indicating that control over oil accumulation is associated with other levels of metabolism or metabolite transport.

Mitochondrial pyruvate dehydrogenase (PDH) complex catalyzes the first committed step in respiratory carbon metabolism and represents a link between glycolytic carbon metabolism and the tricarboxylic acid cycle. Seed-specific antisense repression of the gene encoding mitochondrial pyruvate dehydrogenase kinase (PDHK), a negative regulator of the mitochondrial PDH complex, has been shown to result in increased seed oil content and average seed weight in *Arabidopsis* (Zou et al. 1999a, Marillia et al. 2003). Identical experiments conducted with *B. napus* produced similar results, and in preliminary field trials, the seed oil contents of *B. napus* antisense mitochondrial PDHK transgenics were increased by 3%–5% (Marillia and Taylor, unpublished data). Feeding studies with [3–^{14}C]pyruvate, using siliques of transgenic *Arabidopsis*, supported the hypothesis that the observed increase in seed oil accumulation was attributable to an increased supply of acetyl-CoA from the mitochondria due to the enhanced action of the PDH complex in the decarboxylation of pyruvate (Marillia et al. 2003). It was suggested that mitochondrial-generated acetyl-CoA was probably hydrolyzed to free acetate in the mitochondria. It was further suggested that the acetate may in turn move into the plastid for conversion into acetyl-CoA by the catalytic action of plastidial acetyl-CoA synthetase. Li et al. (2006) used subtractive

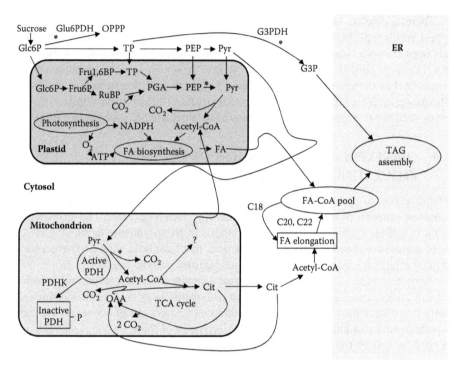

FIGURE 1.2 Carbohydrate catabolism in relation to lipid biosynthesis in photosynthetic developing seeds of *B. napus* L. Reactions having a substantial effect on the flow of carbon into seed oil are indicated with an asterisk. Glycolysis operates in the cytosol and plastid. Cytosolic triose phosphate (TP) in the form of dihydroxyacetone phosphate provides substrate for cytosolic *sn*-glycerol-3-phosphate dehydrogenase (G3PDH) to produce the glycerol backbone for triacylglycerol (TAG) bioassembly in the endoplasmic reticulum (ER). The oxidative pentose phosphate pathway (OPPP) also operates in the cytosol and plastid. *Arabidopsis* knockouts for cytosolic glucose-6-phosphate dehydrogenase (Glu6DH) have been shown to accumulate more seed oil, suggesting that more glucose-6-phosphate (Glu6P) is available for lipid production (Waako et al. 2007). The nonoxidative reactions of the OPPP combined with ribulose 1,5-bisphosphate carboxylase/oxygenase (Rubisco) action (without the Calvin cycle) drive the conversion of fructose-6-phosphate (Fru6P) into 3-phosphoglyceric acid (PGA) (Schwender et al. 2004a). This process bypasses the glycolytic enzymes glyceraldehyde-3-phosphate dehydrogenase and phosphoglycerate kinase, recycling half of the CO_2 produced by plastidial pyruvate dehydrogenase action. As a result, 20% more acetyl-CoA is available for fatty acid (FA) synthesis with 40% less loss of carbon as CO_2. Photosynthesis provides reducing power (NADPH, nicotinamide adenine dinucleotide phosphate) and O_2 for adenosine triphosphate (ATP) production for FA biosynthesis. Acetyl-CoA is produced in both the plastid and mitochondrion through the catalytic action of the pyruvate dehydrogenase (PDH) complex. Flux analysis has demonstrated that mitochondrial metabolism provides acetyl-CoA for FA elongation and not FA biosynthesis in the plastid (Schwender et al. 2006). Citrate (Cit) derived from acetyl-CoA through the TCA (tricarboxylic acid) cycle reactions is transported out of the mitochondrion and is converted to cytosolic acetyl-CoA through the catalytic action of ATP:citrate lyase. Antisense repression of mitochondrial pyruvate dehydrogenase kinase (PDHK), however, has been shown to result in more mitochondrial *(continued on page 11)*

suppression hybridization to identify differentially expressed genes in seeds of two near-isogenic *B. napus* L. lines differing in seed oil content. Interestingly, *PDHK* was identified as one of the few down-regulated genes in the high-oil line. Thus, the decrease in *PDHK* expression is consistent with results on antisense repression of the gene.

In plants, glycolysis occurs in both the cytosol and plastid, with both compartments linked via membrane transporters in the plastidial envelope (Plaxton and Podesta 2006). Disruption of the gene encoding the β_1-subunit of heteromeric plastidic pyruvate kinase complex in *Arabidopsis* has been shown to lead to reduced plastidial pyruvate kinase activity and a 60% reduction in seed oil content (Andre et al. 2007). The seed oil phenotype, however, was completely restored through expression of the cDNA encoding the β_1-subunit of the complex. Thus, pyruvate kinase was critical in the formation of seed oil, with the enzyme of the plastidial pathway representing the preferred route for producing precursors of FA biosynthesis. Analysis of differential gene expression in two near-isogenic lines of *B. napus* differing in seed oil content indicated that expression of pyruvate kinase along with the activity of the enzyme was increased in the high-oil line (Li et al. 2006).

Wakao et al. (2008) suggested that cytosolic glucose-6-phosphate dehydrogenase had a role in supplying NADPH (nicotinamide adenine dinucleotide phosphate), via the oxidative pentose phosphate pathway, for oil accumulation in developing seeds in which photosynthesis may be limited. Studies with single and double mutants disrupted in two cytosolic forms of the enzyme indicated that loss of cytosolic glucose-6-phosphate dehydrogenase (G6PDH) activity affects developing seed metabolism by increasing carbon substrates for synthesis of oil instead of decreasing the supply of NADPH for FA synthesis. Seeds of the double mutant had higher seed oil content and increased seed weight compared to the wild type. It was concluded that a highly dynamic metabolic network compensates for the inactivation of one or both G6PDHs.

Given that inactivation of PDHK or G6PDH has been shown to result in enhanced seed oil content, these enzymes may thus represent promising targets for enhancing seed oil content using a targeting-induced local lesions in genomes (TILLING) approach. TILLING involves screening for mutations in a known target of interest (McCallum et al. 2000). In the case of an enzyme, a single-base mutation in the encoding gene could potentially lead to reduction or even elimination of enzyme activity. The application of TILLING to a major oilseed crop (Slade and Knauf 2005), such as *B. napus*, could potentially constitute a nongenetic engineering approach to enhance seed oil content through inactivation or partial inactivation of PDHK or G6PDH.

FIGURE 1.2 *(continued from page 10)* PDH complex activity, leading to the production of more acetyl-CoA, resulting in enhanced accumulation of seed oil (Zou et al. 1999, Marillia et al. 2003). The precise mechanism by which mitochondrial acetyl-CoA promotes enhanced seed oil accumulation remains to be elucidated. Additional metabolites: Fru 1, 6BP, fructose-1, 6-bisphosphate (Fru1,6BP); OAA, oxaloacetate; PEP, phosphoenolpyruvate. Additional information for the depicted scheme is available in the work of Ruuska et al. (2004).

Conversely, mutations might also be identified that lead to enzyme activation, such as the one associated with the phenylalanine insertion at position 469 of DGAT1-2 from Z. *mays* (Zheng et al. 2008). Therefore, a TILLING strategy could potentially be useful for activation of DGAT1 to produce seed with enhanced oil content.

Recent advances in metabolite analysis and pathway flux are likely to lead to novel strategies for enhancing seed oil content. Stable isotope labeling methods have been developed to probe in vivo intermediary metabolism (Schwender et al. 2004b). The flow of carbon in developing seeds of B. *napus* has been studied extensively through ^{13}C-labeling experiments on cultured embryos (Schwender and Ohlrogge 2002, Schwender et al. 2003, 2004a, 2006). The experiments provided additional insights into seed metabolism, such as the involvement of amino acids in the production of cytosolic acetyl-CoA for FA elongation. In addition, it was demonstrated that mitochondrial metabolism did not provide acetyl-CoA for plastidial FA biosynthesis (Schwender and Ohlrogge 2002, Schwender et al. 2006). This observation, however, makes it difficult to interpret the proposed role of the mitochondrial PDH complex in providing acetyl moieties for use in plastidial FA biosynthesis (Marillia et al. 2003). Thus, it appears that the contribution of mitochondrial respiration to lipid biosynthesis is more complex than outlined in some metabolic flux models.

1.5 TRANSCRIPTION FACTORS GOVERNING SEED OIL ACCUMULATION

It is not well understood how the oil formation pathway fits into the seed development program (Wang et al. 2007). Genetic studies of the regulation of seed development and maturation, however, point to the involvement of several key transcription factors regulating seed oil accumulation. Thus, overexpression of key transcription factors may represent yet another approach for enhancing seed oil content. Both knockout and overexpression studies are typically used to probe the roles of transcription factors in attenuating cellular metabolism (Zhang 2003).

Arabidopsis *mutant wrinkled1* (*wri1*) has been shown to have an 80% reduction in seed oil content (Focks and Benning 1998). The activities of several glycolytic enzymes were reduced in developing homozygous *wri1* seeds, suggesting that *WRI1* was involved in regulation of carbohydrate metabolism during seed development. The suggestion that *WRI1* encodes a transcription factor regulating genes involved in seed oil accumulation was based on cDNA microarray analysis of gene expression in wild-type versus mutant *wri1* seeds (Ruuska et al. 2002). The expression of only a few genes encoding glycolytic enzymes, however, was decreased in *wri1* mutant seeds, suggesting that general downregulation of the glycolytic enzymes probably occurred at the biochemical rather than transcript level. Somewhat later, *WRI1* was shown to encode an APETALA2-ethylene-responsive element-binding protein (AP2/EREBP) and expression of *WRI1* cDNA under the control of a constitutive promoter was shown to result in enhanced seed oil accumulation (Cernac and Benning 2004). In addition to having a role in seed oil accumulation, *WRI1* has also been shown to affect the regulation of sugar metabolism during seed germination and early-stage seedling establishment (Cernac et al. 2006).

Through screening of high and low mutant oil lines of *Arabidopsis*, Shen et al. (2006) discovered that a knockout of homeobox gene *GLABRA2* resulted in a high-oil phenotype. *GLABRA2* encodes a transcription factor required for normal trichome and root hair development. The mechanism by which this transcription factor affects seed oil accumulation is not known.

Wang et al. (2007a) ascribed a central regulatory role to *FUS3* in oil deposition in *Arabidopsis*. Analysis of published seed transcriptome data indicated that the *FUS3* transcript increased in abundance together with almost all of the plastidial FA biosynthetic transcripts. In addition, transgenic *Arabidopsis* seedlings expressing a dexamethasone-inducible *FUS3*, and *Arabidopsis* mesophyll protoplasts transiently expressing the *FUS3* displayed rapid induction of FA biosynthetic gene expression.

In a study with soybean, Wang et al. (2007b) demonstrated that overexpression of *GmDof4* and *GmDof11* genes (soybean Dof-type transcription factor genes) resulted in enhanced expression of the gene encoding the β-subunit of ACCase and the gene encoding long-chain acyl-CoA synthetase, respectively. In contrast, expression of the *CRA1* gene, which encodes 12S storage protein, was decreased. DNA-binding assays indicated that GmDof4 and GmDof11 interacted with *cis*-DNA elements in the promoter regions of the ACCase gene and long-chain acyl-CoA synthetase gene, and both transcription factors also downregulated *CRA1* expression through direct binding. Observations on relative lipid content increases in transgenic lines displaying the greatest level of *GmDof4* or *GmDof11* expression ranged from 11% to 24%.

1.6 CONCLUSIONS AND FUTURE DIRECTIONS

This review examined the effects of manipulating specific enzyme-catalyzed reactions and the expression of transcription factors on seed oil accumulation. The substantial increases in seed oil content achieved through manipulation of a variety of single genes are probably caused by complex changes in metabolic networks resulting from changes in gene expression, protein–protein interactions, and modulation of cellular enzyme activity. Our growing understanding of metabolic networks is driven largely by functional genomics, which is providing a detailed picture of global molecular events during seed development (Ruuska et al. 2002, Gutierrez et al. 2007, Hajduch et al. 2006, Joosen et al. 2007, Xiang et al. 2008). Investigation of the transcriptome, proteome, metabolome (Kusano et al. 2007), and fluxome (Sauer 2004) of oil-forming transgenic cells produced through single-gene manipulations could potentially reveal other cellular factors that may represent additional targets for enhancing seed oil content. In addition, this information would be valuable for genomics-assisted breeding for enhancing seed oil content. Further research into the structure/function of lipid biosynthetic enzymes could lead to strategies for modifying the catalytic efficiency, selectivity, and sensitivity of enzymes to allosteric effectors using protein engineering. Future strategies for enhancing oil accumulation in oleaginous crops will likely involve the production of transformed plants that have a number of modifications introduced at different points in the metabolic network to maximize seed oil content. From an economic perspective, maximizing the quantity of oil obtained per unit land mass represents the ultimate goal for oil production. Therefore, it is necessary to complement research on increasing seed oil content with investigations of yield.

ACKNOWLEDGMENTS

We are grateful for the research support provided by the Alberta Agricultural Research Institute, Alberta Crop Industry Development Fund, BBSRC (United Kingdom), Canada Research Chairs Program, Canada Foundation for Innovation, Natural Sciences and Engineering Research Council of Canada, Saskatchewan Agriculture Development Fund, and University of Alberta.

REFERENCES

Andre, C., J.E. Froehlich, M.R. Moll, and C. Benning. 2007. A heteromeric plastidic pyruvate kinase complex involved in seed oil biosynthesis in *Arabidopsis*. *Plant Cell* 19:2006–2022.

Bouvier-Navé, P., P. Benveniste, P. Oelkers, S.L. Sturley, S.L., and H. Schaller. 2000. Expression in yeast and tobacco of plant cDNAs encoding acyl CoA:diacylglycerol acyltransferase. *Eur. J. Biochem.* 267:85–96.

Cahoon, E.B., J.M. Shockey, C.R. Dietrich, S.T. Gidda, R.T. Mullen, and J.M. Dyer. 2007. Engineering oilseeds for sustainable production of industrial and nutritional feedstocks: solving bottlenecks in fatty acid flux. *Curr. Opin. Plant Biol.* 10:236–244.

Cernac, A., C. Andre, S. Hoffmann-Benning, and C. Benning. 2006. WRI1 is required for seed germination and seedling establishment. *Plant Physiol.* 141:745–757.

Cernac, A., and C. Benning. 2004. *WRINKLED1* encodes an AP2/EREB domain protein involved in the control of storage compound biosynthesis in *Arabidopsis*. *Plant J.* 40: 575–585.

Dahlqvist, A., U. Ståhl, M. Lenman, et al. 2000. Phospholipid:diacylglycerol acyltransferase: an enzyme that catalyzes the acyl-CoA-independent formation of triacylglycerol in yeast and plants. *Proc. Natl. Acad. Sci. U.S.A.* 97:6487–6492.

Delourme, R., C. Falentin, V. Huteau, et al. 2006. Genetic control of oil content in oilseed rape (*Brassica napus* L.). *Theor. Appl. Genet.* 113:1331–1345.

Focks, N., and C. Benning. 1998. *wrinkled1*: A novel, low-seed-oil mutant of *Arabidopsis* with a deficiency in the seed-specific regulation of carbohydrate metabolism. *Plant Physiol.* 118:91–101.

Gutierrez, L., O.V. Van Wuytswinkel, M. Castelain, and C. Bellini. 2007. Combined networks regulating seed maturation. *Trends Plant Sci.* 12:294–300.

Hajduch, M., J.E. Casteel, K.E. Hurrelmeyer, Z. Song, G.K. Agrawal, and J.J. Thelen. 2006. Proteomic analysis of seed filling in *Brassica napus*. Developmental characterization of metabolic isozymes using high-resolution two-dimensional gel electrophoresis. *Plant Physiol.* 141:32–46.

Harwood, J.L. 2005. Fatty acid biosynthesis. In: *Plant lipids: biology, utilization and manipulation*, ed. D.J. Murphy, pp. 27–66. Oxford, U.K.: Blackwell.

Hills, M.J. 2004. Control of storage-product synthesis in seeds. *Curr. Opin. Plant Biol.* 7:302–308.

Hobbs, D.H., J.E. Flintham, and M.J. Hills. 2004. Genetic control of storage oil synthesis in seeds of *Arabidopsis*. *Plant Physiol.* 136:3341–3349.

Ichihara, K., T. Takahashi, and S. Fujii. 1988. Diacylglycerol acyltransferase in maturing safflower seeds: its influences on the fatty acid composition of triacyglycerol and on the rate of triacylglycerol synthesis. *Biochim. Biophys. Acta* 958:125–129.

Jain, R.K., M. Coffey, K. Lai, A. Kumar, and S.L. MacKenzie. 2000. Enhancement of seed oil content by expression of glycerol-3-phosphate acyltransferase genes. *Biochem. Soc. Trans.* 28:958–961.

Jako, C., A. Kumar, Y. Wei, et al. 2001. Seed-specific over-expression of an *Arabidopsis* cDNA encoding a diacylglycerol acyltransferase enhances seed oil content and seed weight. *Plant Physiol.* 126:861–874.

Joosen, R., J. Cordewener, E.D.J. Supena, et al. 2007. Combined transcriptome and proteome analysis identifies pathways and markers associated with the establishment of rapeseed microspore-derived embryo development. *Plant Physiol.* 144:155–172.

Katavic, V., D.W. Reed, D.C. Taylor, et al. 1995. Alteration of seed fatty acid composition by an ethyl methanesulfonate-induced mutation in *Arabidopsis thaliana* affecting diacylglycerol acyltransferase activity. *Plant Physiol.* 108:399–409.

Kroon, J.T., W. Wei, W.J. Simon, and A.R. Slabas. 2006. Identification and functional expression of a *type 2 acyl-CoA:diacylglycerol acyltransferase (DGAT2)* in developing castor bean seeds which has high homology to the major triglyceride biosynthetic enzyme of fungi and animals. *Phytochemistry* 67:2541–2549.

Kusano, M., A. Fukushima, M. Arita, et al. 2007. Unbiased characterization of genotype-dependent metabolic regulations by metabolomic approach in *Arabidopsis thaliana*. *BMC Syst. Biol.* 1:53 (doi:10.1186/1752-0509-1-53).

Lardizabal, K.D., J.T. Mai, N.W. Wagner, A. Wyrick, T. Voelker, and D.J. Hawkins. 2001. DGAT2 is a new diacylglycerol acyltransferase gene family: purification, cloning, and expression in insect cells of two polypeptides from *Mortierella ramanniana* with diacylglycerol acyltransferase activity. *J. Biol. Chem.* 276:38862–38869.

Lardizabal, K.D., G.A. Thompson, and D. Hawkins. 2006. Diacylglycerol acyltransferase proteins. U.S. Patent 7135617; issued November 14.

Li, R.-J., H.-Z.Wang, H. Mao, Y.-T. Lu, and W. Hua. 2006. Identification of differentially expressed genes in seeds of two near-isogenic *Brassica napus* lines with different oil content. *Planta* 224:952–962.

Lung, S.-C., and R.J. Weselake. 2006. Diacylglycerol acyltransferase: a key mediator of plant triacylglycerol synthesis. *Lipids* 41:1073–1088.

Marillia, E.-F., B.J. Micallef, M. Micallef, et al. 2003. Biochemical and physiological studies of *Arabidopsis thaliana* transgenic lines with repressed expression of the mitochondrial pyruvate dehydrogenase kinase. *J. Exp. Bot.* 54:259–270.

McCallum, C.M., L. Comai, E.A. Greene, and S. Henikoff. 2000. Targeting induced local lesions in genomes (TILLING) for plant functional genomics. *Plant Physiol.* 123:439–442.

Mhaske, V., K. Beldjilali, J. Ohlrogge, and M. Pollard. 2005. Isolation and characterization of an *Arabidopsis thaliana* knockout line for phospholipids:diacylglycerol transacylase gene (At5g13640). *Plant Physiol. Biochem.* 43:413–417.

Nykiforuk, C.L., T.L. Furukawa-Stoffer, P.W. Huff, et al. 2002. Characterization of cDNAs encoding diacyglycerol acyltransferase from cultures of *Brassica napus* and sucrose-mediated induction of enzyme biosynthesis. *Biochim. Biophys. Acta* 1580:95–109.

Ohlrogge, J.B., and J. Browse. 1995. Lipid biosynthesis. *Plant Cell* 7:957–970.

Perry, H.J., R. Bligny, E. Gout, and J.L. Harwood. 1999. Changes in Kennedy pathway intermediates associated with increased triacylglycerol synthesis in oil-seed rape. *Phytochemistry* 52:799–804.

Perry, H.J., and J.L. Harwood. 1993a. Changes in lipid content of developing seeds of *Brassica napus*. *Phytochemistry* 32:1411–1415.

Perry, H.J., and J.L. Harwood. 1993b. Radiolabelling studies of acyl lipids in developing seeds of *Brassica napus*: use of [1–^{14}C]acetate precursor. *Phytochemistry* 33:329–333.

Plaxton, W.C., and F.E. Podesta. 2006. The functional organization and control of plant respiration. *CRC Crit. Rev. Plant Sci.* 25:159–198.

Ramli, U.S., D.S. Baker, P.A. Quant, and J.L. Harwood. 2002. Control analysis of lipid biosynthesis in tissue cultures from oil crops shows that flux control is shared between fatty acid synthesis and lipid assembly. *Biochem. J.* 364:393–401.

Roesler, K., D. Shintani, L. Savage, S. Boddupalli, and J. Ohlrogge. 1997. Targeting of the *Arabidopsis* homomeric acetyl-coenzyme A carboxylase to plastids of rapeseeds. *Plant Physiol.* 113:75–81.

Ruuska, S.A., T. Girke, C. Benning, and J.B. Ohlrogge. 2002. Contrapuntal networks of gene expression during *Arabidopsis* seed filling. *Plant Cell* 14:1191–1206.

Ruuska, S.A., J. Schwender, and J.B. Ohlrogge. 2004. The capacity of green oilseeds to utilize photosynthesis to drive biosynthetic processes. *Plant Physiol.* 136:2700–2709.

Saha, S., B. Enugutti, S. Rajakumari, and R. Rajasekharan. 2006. Cytosolic triacylglycerol biosynthetic pathway in oilseeds. Molecular cloning and expression of peanut cytosolic diacylglycerol acyltransferase. *Plant Physiol.* 141:1533–1543.

Sauer, U. 2004. High-throughput phenomics: experimental methods for mapping fluxomes. *Curr. Opin. Biotechnol.* 15:58–63.

Schwender, J., F. Goffman, J.B. Ohlrogge, and Y. Schachar-Hill. 2004a. Rubisco without the Calvin cycle improves the carbon efficiency of developing green seeds. *Nature* 432:779–782.

Schwender, J., and J.B. Ohlrogge. 2002. Probing in vivo metabolism by stable isotope labeling of storage lipids and proteins in developing *Brassica napus* embryos. *Plant Physiol.* 130:347–361.

Schwender, J., J.B. Ohlrogge, and Y. Shachar-Hill. 2003. A flux model of glycolysis and the oxidative pentosephosphate pathway in developing *Brassica napus* embryos. *J. Biol. Chem.* 278:29442–29453.

Schwender, J., J. Ohlrogge, and Y. Shachar-Hill. 2004b. Understanding flux in plant metabolic networks. *Curr. Opin. Plant Biol.* 7:309–317.

Schwender, J., Y. Shachar-Hill, and J.B. Ohlrogge. 2006. Mitochondrial metabolism in developing embryos of *Brassica napus*. *J. Biol. Chem.* 281:34040–34047.

Shen, B., K.W. Sinkevicius, D.A. Selinger, and M.C. Tarczynski. 2006. The homeobox gene GLABRA2 affects seed oil content in *Arabidopsis*. *Plant Mol. Biol.* 60:377–387.

Shockey, J.M., S.K. Gidda, D.C. Chapital, et al. 2006. Tung tree *DGAT1* and *DGAT2* have nonredundant functions in triacylglycerol biosynthesis and are localized to different subdomains of the endoplasmic reticulum. *Plant Cell* 18:2294–2313.

Slade, A.J., and V.C. Knauf. 2005. TILLING moves beyond functional genomics into crop improvement. *Transgenic Res.* 14:109–115.

Ståhl, U., A.S. Carlsson, M. Lenman, et al. 2004. Cloning and functional characterization of a phospholipids:diacylglycerol acyltransferase from Arabidopsis. *Plant Physiol.* 135:1324–1335.

Stymne, S., and A.K. Stobart. 1987. Triacylglycerol biosynthesis. In: *The biochemistry of plants, vol. 9, lipids: structure and function*, ed. P.K. Stumpf, pp. 175–214. New York: Academic Press.

Taylor, D.C., V. Katavic, J.-T. Zou, et al. 2001. Field-testing of transgenic rapeseed cv. Hero transformed with a yeast sn-2 acyltransferase results in increased oil content, erucic acid content and seed yield. *Mol. Breeding* 8:317–322.

Thelen, J.J., and J.B. Ohlrogge. 2002. Metabolic engineering of fatty acid biosynthesis in plants. *Metab. Eng.* 4:12–21.

Tomlinson, K.L., S. McHugh, H. Labbe, et al. 2004. Evidence that the hexose-to-sucrose ratio does not control the switch to storage product accumulation in oilseeds: analysis of tobacco seed development and effects of overexpressing apoplastic invertase. *J. Exp. Bot.* 554:2291–2303.

Vigeolas, H., and P. Geigenberger. 2004. Increased levels of glycerol-3-phosphate lead to a stimulation of flux into triacylglycerol synthesis after supplying glycerol to developing seeds of *Brassica napus* L. in planta. *Planta* 219:827–835.

Vigeolas, H., P. Waldeck, T. Zank, and P. Geigenberger. 2007. Increasing seed oil content in oilseed rape (*Brassica napus* L.) by over-expression of a yeast glycerol-3-phosphate dehydrogenase under the control of a seed-specific promoter. *Plant Biotech. J.* 5:431–441.

Wakao, S., C. Andre, and C. Benning. 2008. Functional analyses of cytosolic glucose-6-phosphate dehydrogenases and their contribution to seed oil accumulation in *Arabidopsis*. *Plant Physiol.* 146:277–288.

Wang, H., J. Guo, K.N. Lambert, and Y. Lin. 2007a. Developmental control of *Arabidopsis* seed oil biosynthesis. *Planta* 226:773–783.

Wang, H.-W., B. Zhang, Y.-J. Hao, et al. 2007b. The soybean Dof-type transcription factor genes, *GmDof4* and *GmDof11*, enhance lipid content in the seeds of transgenic *Arabidopsis* plants. *Plant J.* 52:716–729.

Weselake, R.J. 2005. Storage lipids. In: *Plant lipids: biology, utilization and manipulation*, ed. D.J. Murphy, pp. 162–221. Oxford, U.K.: Blackwell.

Weselake, R.J., M.K. Pomeroy, T.L. Furukawa, J.L. Golden, D.B. Little, and A. Laroche. 1993. Developmental profile of diacylglycerol acyltransferase in maturing seeds of oilseed rape and safflower and microspore-derived cultures of oilseed rape. *Plant Physiol.* 102:565–571.

Weselake, R.J., S. Shah, D.C. Taylor, et al. 2007. Transformation of *Brassica napus* with diacylglycerol acyltransferase-1 results in increased seed oil content. In: *Current advances in the biochemistry and cell biology of plant lipids*, eds. C. Benning and J. Ohlrogge, pp. 232–234. Salt Lake City, UT: Aardvark Global.

Wu, J.-G., C.-H. Shi, and H.-Z. Zhang. 2006. Partitioning genetic effects due to embryo, cytoplasm and maternal parent for oil content in oilseed rape (*Brassica napus* L.). *Gen. Mol. Biol.* 29:533–538.

Xiang, D., R. Datla, F. Li, et al. 2008. Development of a *Brassica* seed cDNA microarray. *Genome* 51:236–242.

Zhang, F.-Y., M.-F. Yang, and Y.-N. Xu. 2005. Silencing of *DGAT1* in tobacco causes a reduction in seed oil content. *Plant Sci.* 169:689–694.

Zhang, Z.Z. 2003. Overexpression analysis of plant transcription factors. *Cur. Opin. Plant Biol.* 6:430–440.

Zhao, J., H.C. Becker, D. Zhang, Y. Zhang, and W. Ecke. 2006. Conditional QTL mapping of oil content in rapeseed with respect to protein content and traits related to plant development and grain yield. *Theor. Appl. Genet.* 113:33–38.

Zheng, P., W.B. Allen, K. Roesler, et al. 2008. A phenylalanine in DGAT is a key determinant of oil content and composition in maize. *Nat. Genet.* 40:367–372.

Zou, J.-T., V. Katavic, E.M. Giblin, et al. 1997. Modification of seed oil content and acyl composition in *Brassicaceae* by expression of a yeast *sn*-2 acyltransferase gene. *Plant Cell* 9:909–923.

Zou, J.T., Q. Qi, V. Katavic, E.-F. Marillia, and D.C. Taylor. 1999a. Effects of antisense repression of an *Arabidopsis thaliana* pyruvate dehydrogenase kinase cDNA on plant development. *Plant Mol. Biol.* 41:837–849.

Zou, J., Y. Wei, C. Jako, A. Kumar, G. Selvaraj, and D.C. Taylor. 1999b. The *Arabidopsis thaliana* TAG1 mutant has a mutation in a diacylglycerol acyltransferase gene. *Plant J.* 19:645–653.

2 Engineering Industrial Oil Biosynthesis
Cloning and Characterization of Kennedy Pathway Acyltransferases from Novel Oilseed Species

Jay Shockey

CONTENTS

Key Words: Acyltransferase; diacylglycerol; glycerol-3-phosphate; Kennedy pathway; Land's cycle; lysophosphatidic acid; triacylglycerol.

2.1 INTRODUCTION

For more than 20 years, various industrial, governmental, and academic laboratories have developed and refined genetic engineering strategies aimed at manipulating lipid metabolism in plants and microbes. The goal of these projects is to produce renewable specialized oils that can effectively compete with or completely replace traditional nonrenewable feedstocks. The target oils produced in most of these projects are used in either nutritional or industrial applications. Some nutritional breeding programs seek to improve the physical properties and shelf life of oils used in cooking and frying applications. These objectives are generally accomplished by reductions in the levels of saturated and polyunsaturated fatty acids accompanied by relative increases in monounsaturated fatty acids [1].

Recently, much focus has been directed toward creating "fish oils" in plants. Fish oils positively affect eye and heart health, and many aspects of neurological development in infants, due to the high levels of polyunsaturated very long-chain (C20–C22) fatty acids. Some consumers have become reluctant to consume many types of fish or the oils derived from them due to the overfishing of certain species and evidence of increased levels of heavy metals and other toxins found in some types of fish and other seafood. Therefore, alternative sources of these oils must be found or created, and engineering of plants or microbes to produce fishlike oils offers an attractive, low-impact, renewable source of these valuable commodities. Much progress has been made in this exciting field in recent years, and excellent reviews summarizing these results have recently been published [2–4]; therefore, these topics are not discussed in detail here. This chapter focuses on the efforts to engineer plants and microbes to produce industrially useful oils.

Modern society has become increasingly dependent on crude oil in recent decades. The demand for, and cost of, fossil fuels has continued to climb precipitously. In addition to the rapid increases in demand, the supply of fossil fuels is also of major concern to governments and consumers around the world. Although long thought to be a limitless resource, the amount of available crude oil is finite. As this resource becomes limited, the laws of supply and demand state that oil prices will climb to levels never before seen or imagined. The resulting negative impacts on society and the global economy will be severe unless alternative, sustainable sources of energy and industrial feedstocks are developed.

The massive success of petroleum oil as a commercial commodity is due in large part to its consistently high quality and its chemical complexity, which lends itself to production of a large number of products and coproducts. Gasoline, diesel fuel, and automotive lubricant oil are the most obvious derivatives, but various other hydrocarbons, such as tar, kerosene, and volatile short-chain hydrocarbons like methane, propane, and butane, are also produced during the distillation of crude oil. Importantly, approximately 10% of this vast resource is used to create feedstocks for the synthesis of a large number of different consumer products, such as plastics, detergents, synthetic rubber, and chemical fertilizers. Plant seed oils are structurally similar to many long-chain hydrocarbons derived from crude oil and thus hold great potential as petrochemical replacements in the synthesis of both fuels (in the form of biodiesel) and many consumer products.

Nature contains plant species that produce several hundred different fatty acid structures [5]. Most of these fatty acids differ from the normal suite of fatty acids found in all plants due to changes in acyl chain length, the presence of unusual double-bond arrangements, or other unusual chemical functionalities, such as triple bonds or epoxy, hydroxyl, or furan groups. Some of these fatty acids provide chemical versatility not found in crude oil and its derivatives, thus providing unique possibilities for product development. Many plant oils provide the additional benefits of stable supply and price. Processed soybean oil is consistently produced at approximately US$80/barrel, a price that compares very well to the volatile market for crude oil, which was priced above US$138/barrel as of June 6, 2008, and is expected to climb even higher as the demand from rapidly expanding economies in countries like China and India increases.

As is the case for nutritional oil engineering projects, the field of industrial oil engineering has seen much recent progress and yet still faces many obstacles. This chapter focuses on our recent efforts to discover and characterize the genes and enzymes needed to produce novel, industrially useful oils in transgenic systems. Our model systems are castor bean (*Ricinus communis*) and tung tree (*Vernicia fordii*). Our early findings with these two species suggest that production designs for many types of novel oils face common bottlenecks and obstacles, and therefore we hope that many of the findings in this project will be broadly applicable to other oilseed engineering strategies.

2.2 CASTOR AND TUNG: TEXTBOOK EXAMPLES OF THE ADVANTAGES AND DISADVANTAGES OF EXOTIC OILSEED SPECIES

Ricinoleic acid (12-D-hydroxy-octadeca-*cis*-9-enoic acid) is an important natural product that can be used as a petrochemical replacement in the synthesis of products such as lubricants, nylon, soaps, dyes, and adhesives, primarily due to the presence of the hydroxyl group at the C-12 position [6]. Castor oil contains approximately 90% ricinoleic acid. The seed oil of the tung tree is used as a drying agent in the formulations of inks, dyes, coatings, and resins [7]. The unique drying qualities of tung oil are based on the enrichment of α-eleostearic acid ($18:3\Delta9^{cis},11^{trans},13^{trans}$), which makes up 80% of total fatty acid composition of tung seed oil. The conjugated double-bond arrangement in eleostearic acid facilitates rapid oxidation and polymerization, thus imparting the excellent drying properties inherent to tung oil.

However, both tung tree and castor bean have a number of unfavorable agronomic characteristics (limited growing areas, toxic by-products, etc.) and are not viable oilseed crops in the United States and many other countries. Therefore, most tung and castor oils must be imported, and the quality, availability, and price (e.g., $0.50–$1.00 per pound for tung oil compared to approximately $0.15–$0.22/pound for soybean oil) can fluctuate dramatically from one year to the next. Our major goal is to provide cheaper, more reliable sources of industrial oils to industries through production of tunglike drying oils and oils containing high levels of hydroxy fatty acids (HFAs) in transgenic crops. These goals cannot be met until we learn more about how oils are synthesized and packaged in developing oilseeds.

2.3   PRIMARY BIOSYNTHETIC ENZYMES ARE INSUFFICIENT FOR HIGH-LEVEL PRODUCTION OF UNUSUAL FATTY ACIDS

The pace of progress in the area of engineering industrial fatty acids accelerated considerably with the discovery of several genes responsible for the synthesis of unusual fatty acids. Depending on the type of fatty acid in question, these enzymes generally fall into one of two main categories: acyl-acyl carrier protein thioesterases (for short- and medium-chain fatty acids, such as lauric acid) [8] or diverged fatty acid desaturases. The castor bean fatty acid hydroxylase enzyme (FAH12) [9] (which produces ricinoleic acid from oleic acid), and tung tree fatty acid conjugase (FADX) [10], which converts linoleic acid to α-eleostearic acid, are just two examples of the latter category. While the cloning of the genes responsible for the biosynthesis of these fatty acids represented major advances in this area, practical application of them to engineering oilseed fatty acid content was met with limitations. As shown in Table 2.1, many different genes capable of unusual industrial fatty acid biosynthesis have been introduced into several different oilseed host species, including *Arabidopsis thaliana*, soybean, and canola. In each case, a significant amount of the target fatty acid is produced, but at a much lower level than seen in the exotic species from which the biosynthetic enzyme was identified. In many cases, much of the target fatty acid remains in the phospholipid pool, inaccessible to the downstream enzymatic reactions of triacylglycerol (TAG) synthesis [14], or otherwise accumulates as substrates that are poorly utilized in these reactions [16]. The consistency of this phenomenon strongly suggests that other TAG biosynthetic enzymes in the native species may have coevolved with the primary biosynthetic enzymes to help remove the novel fatty acids from the phospholipid pool or channel the novel fatty acids into neutral storage lipids.

TABLE 2.1

Comparison of Novel Fatty Acid Content in Seed Oils of Native Species and Transgenic Species Expressing Primary Biosynthetic Genes from Native Species

Exotic Fatty Acid Type	Chemical Functionality	Source Plant		Target Plant		
		Name	Maximum % in Oil	Name	Maximum % in Oil	Reference
Ricinoleic	Hydroxy	*Ricinus communis*	~90	*Arabidopsis thaliana*	16–18	11, 12
Eleostearic	Conjugated	*Momordica charantia*	70–80	*Glycine max*	17	13
		Vernicia fordii	~80	*A. thaliana*	6	14
Crepenynic	Triple bond	*Crepis alpina*	70	*A. thaliana*	25	15
Vernolic	Epoxy	*Vernonia galamensis*	80	*A. thaliana*	15	15

2.4 VARIOUS ENZYMES OF THE KENNEDY PATHWAY AND LAND'S CYCLE MAY BE NECESSARY FOR EFFICIENT PACKAGING OF UNUSUAL FATTY ACIDS IN STORAGE LIPIDS

The major pathways that establish bulk fatty acid flux into membrane and storage lipid pools have been established for almost 50 years. Land's cycle [17] is responsible for membrane remodeling of phospholipids. In this pathway, fatty acids are exchanged between the acyl-coenzyme A (CoA) and phospholipid pools, most likely in both directions [18]. Such a pathway is likely used in some mammalian tissues for production of surfactants and in oilseeds as a method for selective channeling of polyunsaturated fatty acids into neutral lipids. This acyl-CoA pool is used in part by the acyltransferase enzymes of the Kennedy pathway [19], which successively acylate each of the three hydroxyl groups of glycerol, starting with glycerol-3-phosphate. These two pathways are combined in the proposed synthetic route of a molecule of tung oil containing eleostearic acid as shown in Figure 2.1, complete with each of the possible relevant enzymatic reactions. The biosynthetic route for ricinoleic acid in castor is likely very similar, except that this fatty acid is synthesized from oleic acid (18:1) rather than linoleic acid (18:2).

2.4.1 DIACYLGLYCEROL ACYLTRANSFERASE

As stated, the inability of individual biosynthetic enzymes to produce levels of novel fatty acids comparable to that of their native species indicates that one or more enzymes of Land's cycle or Kennedy pathway may have coevolved to aid in the

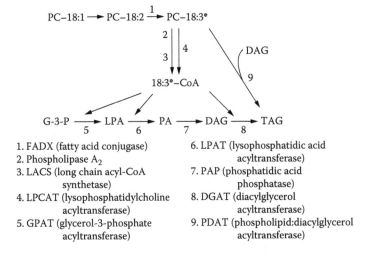

1. FADX (fatty acid conjugase)
2. Phospholipase A$_2$
3. LACS (long chain acyl-CoA synthetase)
4. LPCAT (lysophosphatidylcholine acyltransferase)
5. GPAT (glycerol-3-phosphate acyltransferase)
6. LPAT (lysophosphatidic acid acyltransferase)
7. PAP (phosphatidic acid phosphatase)
8. DGAT (diacylglycerol acyltransferase)
9. PDAT (phospholipid:diacylglycerol acyltransferase)

FIGURE 2.1 Schematic diagram of α-eleostearic acid and triacylglycerol biosynthesis in developing tung (*Vernicia fordii*) seeds. Eleostearic acid is synthesized from linoleic acid attached to endoplasmic reticulum (ER) phospholipids, primarily phosphatidylcholine (PC), then removed from the phospholipid pool and channeled into triacylglycerols through any of several coenzyme A (CoA)-dependent or -independent acyltransferase reactions.

efficient channeling of novel fatty acids into storage TAG. We have focused primarily on the diacylglycerol acyltransferases (DGATs), the enzyme family that catalyzes the final acylation reaction of the Kennedy pathway, converting acyl-CoA and diacylglycerol (DAG) to TAG. At least three unrelated classes of DGAT enzymes exist in higher plants; two are targeted to the membrane of the endoplasmic reticulum (ER) (DGAT1 and DGAT2) [20,21], and the other is the soluble DGAT3 class [22]. Our work has compared the properties of the DGAT1 and DGAT2 enzymes from castor and tung and to the orthologous enzymes from yeast and *Arabidopsis*, both of which do not synthesize hydroxylated or conjugated fatty acids. Tung DGAT1 and castor DGAT1 showed either no preference or only slight preference for substrates containing the appropriate novel fatty acid, and the expression patterns of both genes compared rather poorly to the time course of seed TAG synthesis [23–26], although the protein expression profile for castor DGAT1 does match the oil synthetic timeline much better [27].

However, both tung and castor DGAT2 enzymes showed strong preferences for their respective novel fatty acids. When expressed in yeast fed tung oil fatty acids, tung DGAT2 synthesized substantially more trieleostearin than did tung DGAT1, yeast DGAT, or either *Arabidopsis* DGAT1 or DGAT2 [24]. These data suggested that tung DGAT2 may play a more important role than tung DGAT1 in trieleostearin synthesis in developing tung seeds. Besides the apparent preference of tung DGAT2 for substrates containing conjugated fatty acids, cell biological analysis of the tung DGATs also suggested that each enzyme might play a different role: each is localized to discreet subdomains within the ER membrane [24]. These experiments were among the first to definitively demonstrate compartmentalization of enzymes within discreet, nonoverlapping regions of the ER membrane and support long-standing hypotheses that suggested colocalization of sets of enzymes dedicated to the same function [28]. We are currently analyzing other tung acyltransferases with regard to their targeting within plant cells to extend the depth of our understanding of this fascinating process.

Castor DGAT2 has also proven very useful in manipulating fatty acid content. It was initially characterized with respect to its ability to increase the levels of HFAs present in the seed oils of transgenic *Arabidopsis* expressing the castor hydroxylase FAH12. Singular expression of FAH12 in *Arabidopsis* results in HFA levels of 17%–18%, and repeated efforts to find better FAH12 plant lines have consistently failed [11,12,29]. Coexpression of castor DGAT2, however, resulted in multiple independent lines that produced up to 30% HFA, by far the highest level of HFA yet reported for a transgenic system [23]. As shown in Figure 2.2, detailed analysis of the TAG species produced by the *FAH12/RcDGAT2* (*Ricinus communis* DGAT2) double-transgenic lines showed that castor DGAT2 coexpression dramatically shifted the TAG fatty acid profile toward lipid species containing two or more HFAs; the *FAH12/RcDGAT2* double-transgenic plants contain higher levels of all eight major TAG species present in native castor oil (including triricindein) than do the *FAH12* single transgenics.

The preference of castor DGAT2 for ricinoleic acid-containing substrates was demonstrated directly by enzyme assay after overexpression in yeast. The castor enzyme exhibited a 10-fold preference for diricinolein as acyl acceptor over two

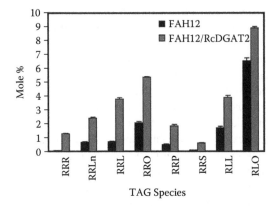

TAG Species

FIGURE 2.2 Quantitative analysis of individual triacylglycerol (TAG) species from *FAH12*- and *FAH12/RcDGAT2*-transgenic *Arabidopsis* seeds. Relative amounts of eight individual TAGs, corresponding to the eight TAG species that make up more than 97% of castor oil, were identified and quantified in the seed oils of CL7 and line 544 by liquid chromatography tandem mass spectrometry (LC/MS/MS; see ref. 19 for details). Data are means plus or minus the standard deviation of three independent measurements. P, palmitate (16:0); S, stearate (18:0); O, oleate (18:1); L, linoleate (18:2); Ln, linolenate (18:3); and R, ricinoleate (18:1, 12-OH).

other traditional DAG substrates [23]. These findings were extended even further by Kroon et al. [25], who demonstrated that RcDGAT2 could synthesize triricinolein, the primary TAG species in castor oil. Collectively, these data strongly suggest that endogenous *Arabidopsis* DGAT activities are poorly suited to accommodate HFA-containing substrates, and that RcDGAT2 is a major determinant of castor oil fatty acid composition.

2.4.2 GPAT, LPAT, LPCAT, PDAT, AND CPT: AN ALPHABET SOUP OF ENZYMES THAT MAY HOLD THE KEYS TO ULTIMATE SUCCESS

2.4.2.1 The Other Kennedy Pathway Acyltransferases

The analysis of plant DGATs, especially the DGAT2 class, has certainly provided powerful tools for the manipulation of seed oil fatty acid content. However, the results from the *FAH12/RcDGAT2* coexpression experiments suggest that even more enzymes of the Land's cycle and Kennedy pathway must be included to achieve industrially meaningful levels of novel fatty acids. The task of identifying which relevant genes to pursue is complicated by several factors:

1. Several enzymatic reactions could directly or indirectly influence the quantity and fatty acid composition of the TAG pool.
2. Only a few of the genes responsible for the enzymatic steps represented in Figure 2.1 have been cloned from any plant species; most have not been studied at the molecular level.
3. Many of the enzymes in question are present in plant genomes as large families.

The conjugases and other diverged desaturases (like tung FADX and castor hydroxy-lase) are typically single-copy genes, and each of the three DGAT classes seems to be represented by one or two genes. However, using the sequenced genome of *Arabidopsis* as a guide, many of the other lipid biosynthetic enzymes are present as large gene families. This is especially true for the glycerol-3-phosphate acyltrans-ferases (GPATs) and lysophosphatidic acid acyltransferases (LPATs). These enzymes successively catalyze the first and second glycerol acylations of the Kennedy path-way and as such are likely very important components of the channeling machin-ery responsible for packaging novel fatty acids into TAG. *Arabidopsis* contains at least eight *GPAT* genes [30,31] and at least five *LPAT* genes [32], plus several other uncharacterized genes that display homology to conserved sequence motifs com-mon to acyltransferases of this type [33]. The protein sequences of this collection of enzymes are compared phylogenetically in Figure 2.3. The eight *GPAT* genes all cluster together rather tightly, while the five characterized *LPATs* are more highly diverged. Many of the other genes align somewhere in between these two groups,

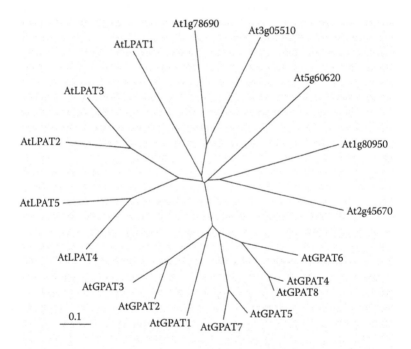

FIGURE 2.3 Phylogenetic comparison of known and putative *GPAT* (glycerol-3-phos-phate acyltransferase) and *LPAT* (lysophosphatidic acid acyltransferase) gene families from *Arabidopsis*. Characterized *LPAT* genes are designated based on the nomenclature suggested by Kim et al. [30], while *GPAT* genes are named as described by Zheng et al. [28]. All other uncharacterized genes that contain acyltransferase motifs [31] are included, using their MIPS locus names. The branch lengths are proportional to the degree of divergence, with the scale of 0.1 representing 10% change.

even when the tree is constructed with a true outgroup, making predictions of their possible enzymatic function almost impossible. Some of these genes (and their orthologs in other species) may indeed be useful for oilseed engineering, but much initial work must be done to define the enzymatic role, tissue-specific and temporal expression pattern, and other relevant properties of each gene before making that determination. Despite the anticipated difficulties, this task is necessary; previous studies have shown that other Kennedy pathway acyltransferases can positively affect the yield of novel fatty acids [34]. Efforts are currently under way to identify important *GPAT* and *LPAT* genes from tung tree.

2.4.2.2 The Land's Cycle Enzymes and Other Acyltransferases

Much of our focus, and that of other groups, has been dedicated to studying the Kennedy pathway acyltransferases. However, any enzyme that could affect the flux of fatty acids between the acyl-CoA, phospholipid, and neutral lipid pools must be considered for a potential role in directing the flow of novel fatty acids into TAG. Most types of unusual fatty acids (including oxidized or otherwise damaged acyl groups, which occur naturally in all organisms) must be rapidly removed from their site of synthesis in the phospholipid pool. Unusual fatty acids generally have deleterious effects on membrane bilayer integrity and function and therefore cannot be allowed to accumulate in membranes [35]. Oxidized fatty acids are removed from phospholipids by phospholipase activities and degraded [36]. Plants that accumulate unusual fatty acids seem to have built on this common housekeeping biochemistry to create multiple pathways to direct unusual fatty acids from membrane lipids into neutral lipids. Some of the proposed pathways involve generation of free fatty acids, and some maintain the ester linkage between the novel fatty acid and glycerol (summarized in Cahoon et al. [4]).

Two such intriguing possibilities are CPT (choline phosphotransferase) and PDAT (phospholipid:diacylglycerol acyltransferase). CPT catalyzes the reversible removal of the head group from phosphatidylcholine (PC), producing DAG and phosphocholine. The possible utility of this activity is obvious; molecules of PC containing unusual fatty acids could immediately be converted to DAG, which can be used either by DGAT or PDAT. Unfortunately, very little is known about the roles of CPT since relatively few reports of cloning of CPT genes from plants exist [37]. PDAT transfers fatty acids directly from phospholipids to the *sn*-3 position of DAG, forming TAG and lysophospholipid. This enzyme activity has been studied in several plant species, and genes for PDAT have been cloned from yeast [38] and *Arabidopsis* [39]. PDAT activities in castor and *Vernonia galamensis* prefer PC substrates containing their respective novel fatty acids and therefore may play major roles in channeling fatty acids from phospholipid to TAG [40]. In many transgenic systems, high-level production of novel fatty acids is stymied due to much of the synthesized product becoming trapped in the phospholipid fraction [14]. Manipulation of CPT and PDAT activities could prove valuable in overcoming this bottleneck.

Production of lysophospholipids may cause convergence between PDAT activity and the Land's cycle, whose hallmark is the activity of lysophosphatidylcholine acyltransferase (LPCAT). LPCAT in rat lung and at least some oilseeds appears to be reversible [18]. In plants, the Land's cycle may participate in repairing damage to oxidized lipids as discussed, but more importantly may drive the channeling of polyunsaturated fatty acids from phospholipids and galactolipids into neutral lipids [18]. Phospholipase activity, likely in the form of phospholipase A_2, is responsible for direct removal of unusual fatty acids from the phospholipid pool in some novel oilseeds, including castor [40]. Free fatty acids generated via this mechanism would require activation to acyl-CoA esters by long-chain acyl-CoA synthetase [41], thus entering the acyl-CoA pool where they can be used by any of the numerous acyl-transferase activities. These acyl-CoA molecules might need to be sequestered into dedicated oil-synthesizing domains to avoid the creation of futile cycling back into membrane lipid synthetic pathways.

2.5 CONCLUSIONS

It is obvious how the collective interplay between this large set of enzyme activities could result in rapid and extensive enrichment of unusual fatty acids in the TAG pool of developing oilseeds. The challenge of finding the important gene family members from this population has only just begun. Representative genes for many of the enzyme activities described here have not been isolated. These goals are the focus of our current efforts, with the hope that additional insights into the mechanisms of TAG synthesis and packaging will soon follow. The collective knowledge gained from chemical, biochemical, cell biological, and physiological approaches will ultimately determine the necessary gene sets required to achieve successful pathway engineering.

REFERENCES

1. Kinney, A.J. Development of genetically engineered soybean oils for food applications. *Journal of Food Lipids* 3, 273–292, 1996.
2. Damude, H.G., and Kinney, A.J. Engineering oilseeds to produce nutritional fatty acids. *Physiologia Plantarum* 132, 1399, 1992.
3. Graham, I.A., Larson, T., and Napier, J.A. Rational metabolic engineering of transgenic plants for biosynthesis of omega-3 polyunsaturates. *Current Opinions in Biotechnology* 18, 142–147, 2007.
4. Cahoon, E.B., Shockey, J.M., Dietrich, C.R., Gidda, S.K., Mullen, R.T., and Dyer, J.M. Engineering oilseeds for sustainable production of industrial and nutritional feedstocks: solving bottlenecks in fatty acid flux. *Current Opinions in Plant Biology* 10, 236–244, 2007.
5. Badami, R.C., and Patil, K.B. Structure and occurrence of unusual fatty acids in minor seed oils. *Progress in Lipid Research* 19, 119–153, 1981.
6. Caupin, H.-J. Products from castor oil: past, present, and future. In *Lipid Technologies and Applications*, Gunstone, F.D., and Padley, F.B., Eds., Dekker, New York, pp. 787–795, 1997.
7. Sonntag, N.O.V. Composition and characteristics of individual fats and oils. In *Bailey's Industrial Oil and Fat Products*, 4th Edition, Vol. 1, Swern, D., Ed. Wiley, New York, pp. 289–477, 1979.

8. Voelker, T.A., Worrell, A.C., Anderson, L., Bleibaum, J., Fan, C., Hawkins, D.J., Radke, S.E., and Davies, H.M. Fatty acid biosynthesis redirected to medium chains in transgenic oilseed plants. *Science* 257, 72–74, 1992.

9. van de Loo, F.J., Broun, P., Turner, S., and Somerville, C. An oleate 12-hydroxylase from *Ricinus communis* L. is a fatty acyl desaturase homolog. *Proceedings of the National Academy of Sciences of the United States of America* 92, 6743–6747, 1995.

10. Dyer, J.M., Chapital, D.C., Kuan, J.C., Mullen, R.T., Turner, C., McKeon, T.A., and Pepperman, A.B. Molecular analysis of a bifunctional fatty acid conjugase/desaturase from tung. Implications for the evolution of plant fatty acid diversity. *Plant Physiology* 130, 2027–2038, 2002.

11. Smith, M.A., Moon, H., Chowrira, G., and Kunst, L. Heterologous expression of a fatty acid hydroxylase gene in developing seeds of *Arabidopsis thaliana*. *Planta* 217, 506–517, 2003.

12. Broun, P., and Somerville, C. Accumulation of ricinoleic, lesquerolic, and densipolic acids in seeds of transgenic *Arabidopsis thaliana* plants that express a fatty acyl hydroxylase cDNA from castor bean. *Plant Physiology* 113, 933–942, 1997.

13. Cahoon, E.B., Carlson, T.J., Ripp, K.G., Schweiger, B.J., Cook, G.A., Hall, S.E., and Kinney, A.J. Biosynthetic origin of conjugated double bonds: production of fatty acid components of high-value drying oils in transgenic soybean embryos. *Proceedings of the National Academy of Sciences of the United States of America* 96, 12935–12940, 1999.

14. Cahoon, E.B., Dietrich, C.R., Meyer, K., Damude, H.G., Dyer, J.M., and Kinney, A.J. Conjugated fatty acids accumulate to high levels in phospholipids of metabolically engineered soybean and *Arabidopsis* seeds. *Phytochemistry* 67, 1166–1176, 2006.

15. Lee, M., Lenman, M., Banas, A., Bafor, M., Singh, S., Schweizer, M., Nilsson, R., Liljenberg, C., Dahlqvist, A., Gummeson, P.O., Sjodahl, S., Green, A., and Stymne, S. Identification of non-heme diiron proteins that catalyze triple bond and epoxy group formation. *Science* 280, 915–918, 1998.

16. Larson, T.R., Edgell, T., Byrne, J., Dehesh, K., and Graham, I.A. Acyl-coA profiles of transgenic plants that accumulate medium-chain fatty acids indicate inefficient storage lipid synthesis in developing oilseeds. *Plant Journal* 32, 519–527, 2002.

17. Lands, W.E.M. Metabolism of glycerolipids. II. The enzymatic acylation of lysolecithin. *Journal of Biological Chemistry* 235, 2233–2237, 1960.

18. Stymne, S., and Stobart, A.K. Involvement of acyl exchange between acyl-CoA and phosphatidylcholine in the remodeling of phosphatidylcholine in microsomal preparations of rat lung. *Biochemica et Biophysica Acta* 837, 239–250, 1985.

19. Kennedy, E.P. Biosynthesis of complex lipids. *Federation Proceedings* 20, 934–940, 1961.

20. Hobbs, D.H., Lu, C., and Hills, M.J. Cloning of a cDNA encoding diacylglycerol acyltransferase from *Arabidopsis thaliana* and its functional expression. *FEBS Letters* 452, 145–149, 1999.

21. Lardizabal, K.D., Mai, T.M., Wagner, N., Wyrick, A., Voelker, T., and Hawkins, D.J. DGAT2 is a new diacylglycerol acyltransferase gene family. *Journal of Biological Chemistry* 276, 38862–38869, 2001.

22. Saha, S., Enugutti, B., Rajakumari, S., and Rajasekharan, R. Cytosolic triacylglycerol biosynthetic pathway in oilseeds. Molecular cloning and expression of peanut cytosolic diacylglycerol acyltransferase. *Plant Physiology* 141, 1533–1543, 2006.

23. Burgal, J.J., Shockey, J., Chaofu, L., Dyer, J., Larson, T., Graham, I., and Browse, J. Metabolic engineering of hydroxy fatty acid production in plants: RcDGAT2 drives dramatic increases in ricinoleate levels in seed oil. *Plant Biotechnology Journal* 6, 819–831, 2008.

24. Shockey, J.M., Gidda, S.K., Chapital, D.C., Kuan, J.C., Dhanoa, P.K., Bland, J.M., Rothstein, S.J., Mullen, R.T., and Dyer, J.M. Tung tree DGAT1 and DGAT2 have nonredundant functions in triacylglycerol biosynthesis and are localized to different subdomains of the endoplasmic reticulum. *Plant Cell* 18, 2294–2313, 2006.

25. Kroon, J.T., Wei, W., Simon, W.J., and Slabas, A.R. Identification and functional expression of a type 2 acyl-CoA:diacylglycerol acyltransferase (DGAT2) in developing castor bean seeds which has high homology to the major triglyceride biosynthetic enzyme of fungi and animals. *Phytochemistry* 67, 2541–2549, 2006.

26. He, X., Turner, C., Chen, G.Q., Lin, J.T., and McKeon, T.A. Cloning and characterization of a cDNA encoding diacylglycerol acyltransferase from castor bean. *Lipids* 39, 311–318, 2004.

27. He, X., Chen, G.Q., Lin, J.-T., and McKeon, T.A. Regulation of diacylglycerol acyltransferase in developing seeds of castor. *Lipids* 39, 865–871, 2004.

28. Vogel, G., and Browse, J. Cholinephosphotransferase and diacylglycerol acyltransferase (substrate specificities at a key branch point in seed lipid metabolism). *Plant Physiology* 110, 923–931, 1996.

29. Lu, C., Fulda, M., Wallis, J.G., and Browse, J. A high-throughput screen for genes from castor that boost hydroxy fatty acid accumulation in seed oils of transgenic *Arabidopsis thaliana*. *Plant Journal* 45, 847–856, 2006.

30. Zheng, Z., Xia, Q., Dauk, M., Shen, W., Selvaraj, G., and Zou, J. *Arabidopsis* AtGPAT1, a member of the membrane-bound glycerol-3-phosphate acyltransferase gene family, is essential for tapetum differentiation and male fertility. *Plant Cell* 15, 1872–1887, 2003.

31. Li, Y., Beisson, F., Koo, A.J.K., Molina, I., Pollard, M., and Ohlrogge, J. Identification of acyltransferases required for cutin biosynthesis and production of cutin with suberin-like monomers. *Proceedings of the National Academy of the United States of America* 104, 18339–18344, 2007.

32. Kim, H.U., Li, Y., and Huang, A.H. Ubiquitous and endoplasmic reticulum-located lyso-phosphatidyl acyltransferase, LPAT2, is essential for female but not male gametophyte development in Arabidopsis. *Plant Cell* 17, 1073–1089, 2005.

33. Lewin, T.M., Wang, P., and Coleman, R.A. Analysis of amino acid motifs diagnostic for the *sn*-glycerol-3-phosphate acyltransferase reaction. *Biochemistry*, 5764–5771, 1999.

34. Knutzon, D.S., Hayes, T.R., Wyrick, A., Xiong, H., Davies, H.M., and Voelker, T.A. Lysophosphatidic acid acyltransferase from coconut endosperm mediates the insertion of laurate at the *sn*-2 position of triacylglycerols in lauric rapeseed oil and can increase total laurate levels. *Plant Physiology* 120, 739–746, 1999.

35. Millar, A.A., Wrischer, M., and Kunst, L. Accumulation of very-long-chain fatty acids in membrane glycerolipids is associated with dramatic alterations in plant morphology. *Plant Cell* 10, 1889–1902, 1998.

36. van den Berg, J.J., Op den Kamp, J.A., Lubin, B.H., and Kuypers, F.A. Conformational changes in oxidized phospholipids and their preferential hydrolysis by phospholipase A2: a monolayer study. *Biochemistry* 32, 4962–4967, 1993.

37. Dewey, R.E., Wilson, R.F., Novitzky, W.P., and Goodea, J.H. The AAPT1 gene of soybean complements a cholinephosphotransferase-deficient mutant of yeast. *Plant Cell* 6, 1495–1507, 1994.

38. Stähl, U., Carlsson, A.S., Lenman, M., Dahlqvist, A., Huang, B., Banas, W., Banas, A., and Stymne, S. Cloning and functional characterization of a phospholipid:diacylglycerol acyltransferase from *Arabidopsis*. *Plant Physiology* 135, 1324–1335, 2004.

39. Dahlqvist, A., Stähl, U., Lenman, M., Banas, A., Lee, M., Sandager, L., Ronne, H., and Stymne, S. Phospholipid:diacylglycerol acyltransferase: an enzyme that catalyzes the acyl-CoA-independent formation of triacylglycerol in yeast and plants. *Proceedings of the National Academy of Sciences of the United States of America* 97, 6487–6492, 2000.

40. Bafor, M., Smith, M.A., Jonsson, L., Stobart, K., and Stymne, S. Ricinoleic acid biosynthesis and triacylglycerol assembly in microsomal preparations from developing castorbean (*Ricinus communis*) endosperm. *Biochemical Journal* 280, 507–514, 1991.
41. Shockey, J.M., Fulda, M.S., and Browse, J.A. *Arabidopsis* contains nine long-chain acylcoenzyme A synthetase genes that participate in fatty acid and glycerolipid metabolism. *Plant Physiology* 129, 1710–1722, 2002.

3 Accumulation of Epoxy Fatty Acids in Plant Oils

David Hildebrand, Runzhi Li, Keshun Yu, and Tomoko Hatanaka

CONTENTS

Key Words: Diacylglycerol acyltransferase; epoxygenase; industrial oil; resins; triacylglycerol; vernoleate.

3.1 INTRODUCTION

The use of plants as chemical factories such as for the production of epoxy compounds covered in this chapter promises to enhance the variety and quality of products available in the future. This requires no fossil fuel for the actual biosynthesis reactions, instead using sunlight, H_2O, and atmospheric CO_2 for the hydrocarbon syntheses. Engineering oilseeds for high-epoxy fatty acid accumulation in triacylglycerol (TAG) is a way to achieve this goal. This would greatly increase the value of such improved seeds and reduce epoxide production costs. Safety concerns associated with industrial epoxidation of oils would also be eliminated. A considerable market currently exists for epoxy fatty acids, particularly for epoxy coatings and plasticizers. Presently, most of these are derived from petroleum. Soybean and linseed oil are currently utilized to some extent to produce epoxidized oil by introducing an epoxy group across the double bonds of polyunsaturated fatty acids. This is a costly process, and it would be more economical if the biosynthetic reactions in major oilseed crops were altered such that the seeds themselves converted the polyunsaturated fatty acids into epoxy fatty acids. There is no known way to produce a commercial oilseed that accumulates epoxy fatty acids by conventional breeding and genetics. Certain genotypes of several plant species, however, accumulate high levels of epoxy fatty acids in the seed oil. The best examples of this are *Vernonia galamensis, Stokesia laevis,* and *Euphorbia lagascae*.

The *V. galamensis, S. laevis,* or *E. lagascae* cannot readily be grown to produce seed on an industrial scale, limiting their current potential to replace much fossil fuel use as sources of epoxy compounds. Epoxygenase genes have been cloned from several accumulators of vernolic acid, and high expression in developing *Arabidopsis* or soybean embryos results in up to about 10% vernoleate in the seed oil, but this is insufficient for commercial production of epoxides in oilseeds. The effective channeling or selective accumulation of epoxy fatty acids into TAG is one important step in the effective commercialization of this process. Our prior experiments indicated diacylglycerol acyltransferase (DGAT), especially DGAT2, to be of particular importance to the accumulation of epoxy fatty acids in seed oil TAGs.

The biochemical reaction catalyzed by epoxygenase in plants combines the common oilseed fatty acids, linoleic or linolenic acids, with O_2, forming only H_2O and epoxy fatty acids as products (CO_2 and H_2O are utilized to make linoleic or linolenic acids). A considerable market currently exists for epoxy fatty acids, particularly for resins, epoxy coatings, and plasticizers. The U.S. plasticizer market is estimated to be about 2 billion pounds per year (Hammond 1992). Presently, most of this is derived from petroleum. In addition, there is industrial interest in use of epoxy fatty acids in durable paints, resins, adhesives, insecticides and insect repellants, crop oil concentrates, and the formulation of carriers for slow-release pesticides and herbicides (Perdue 1989, Ayorinde et al. 1993). Also, epoxy fatty acids can readily and economically be converted to hydroxy and dihydroxy fatty acids and their derivatives, which are useful starting materials for the production of plastics as well as for detergents, lubricants, and lubricant additives. Such renewable derived lubricant and lubricant additives should facilitate use of plant/biomass-derived fuels. Examples of plastics that can be produced from hydroxy fatty acids are polyurethanes and polyesters (Weber et al. 1994). As commercial oilseeds are developed that accumulate epoxy fatty acids in the seed oil, it is likely that other valuable products would be developed to use this as an industrial chemical feedstock in the future.

Currently, epoxidized soybean and linseed oils are produced by introducing an epoxy group across the double bonds of polyunsaturated fatty acids. This is a costly process, and it would be more economical if the biosynthetic reactions in oilseed crops such as soybean are altered so that the seeds themselves converted the polyunsaturated fatty acids into epoxy fatty acids. There is no known way to produce a commercial oilseed that accumulates epoxy fatty acids by conventional breeding and genetics. However, certain genotypes of several plant species accumulate high levels of epoxy fatty acids in the seed oil. Epoxide fatty acids, like vernolic acid and coronaric acid, have been found as a component of the seed oil of species represented by a number of plant families, such as the Compositae, Euphorbiaceae, Onagraceae, Dipsacaceae, and Valerianaceae (Sen Gupta 1974). Certain plants are known to have epoxygenases that transform unsaturated fatty acids into epoxy fatty acids that accumulate in seed oil (Gardner 1991). The best examples of this are *Vernonia galamensis* and *Euphorbia lagascae,* which accumulate an epoxy fatty acid known as vernolic acid (*cis*-12-epoxyoctadeca-*cis*-9-enoic acid) (Perdue 1989, Pascual and Correal 1992, Bafor et al. 1993). Some accessions of *V. galamensis* have as much as 40% oil comprised of 80% vernolic acid (Thompson et al. 1994b). There are ongoing efforts to develop *Vernonia* and *Euphorbia* as agronomic crops as sources of vernolic acid for industry, but many problems remain for their large-scale commercial production

(Thompson et al. 1994a). Although vernoleate is produced in phosphatidylcholine (PC) via epoxygenase, it selectively accumulates in TAG, not PC (Seither 1997).

The process by which the seeds of certain species of *Vernonia* and *Euphorbia* produce epoxy fatty acids is due to an epoxygenase not present in most plants, including major commercial oilseeds. Early biochemical studies by Stymne's group indicated the epoxygenase in developing *Euphorbia* seeds was a P450 monooxygenase that converts linoleic acid esterified to the second position of PC into vernolic acid at the second position of PC in a one-step reaction (Bafor et al. 1993). Our further studies with developing *Euphorbia* seeds were consistent with this enzyme being a P450 monooxygenase (Seither et al. 1997), and this was further confirmed by Cahoon et al. (2002). In our studies with *V. galamensis*, we have established that the *Vernonia* epoxygenase is also a microsomal enzyme that utilizes linoleoyl-PC as the substrate. However, the *Vernonia* epoxygenase was found to be distinctly different from the *Euphorbia* epoxygenase and typical P450 monooxygenase enzymes and was determined to be a desaturase-like enzyme (Seither et al. 1997, Hitz 1998). Similar results were obtained by Lee et al. (1998) with the epoxygenase of *Crepis palestina* and *Stokesia* epoxygenase (Hatanaka and Hildebrand 2001, Hatanaka et al. 2004).

There are multiple possible limiting steps from the formation of the epoxygenase product vernoyl-PC and the final accumulation of the epoxy moiety in TAG. Among the most important enzymes in TAG biosynthesis, including in developing soybean seeds, is DGAT (Settlage et al. 1998), and there is strong evidence that DGATs can be selective of the acyl groups or DAGs incorporated into TAG. It has been discovered that at least some plants, such as *Arabidopsis*, contain two DGAT gene classes with little or no homology to each other. The second DGAT group is now known as DGAT2 and the original one as DGAT1 (Cases et al. 2001). DGAT1 is expressed in several tissues in *Arabidopsis* in addition to developing seeds but most abundantly in developing seeds (Lu et al. 2003). In *Arabidopsis*, DGAT1 accounts for 25%–45% of seed oil, as indicated by analysis of a DGAT1-deficient mutant (Katavic et al. 1995, Routaboul et al. 1999).

DGATs from unusual fatty accumulators in TAG can show high specificity toward the coenzyme A (CoA) of the unusual fatty acids they accumulate and little or no activity toward unusual fatty acid-CoAs they do not accumulate (Cao and Huang 1987, Taylor et al. 1991, Wiberg et al. 1994); DGATs from unusual fatty acid accumulators can also show preference for diacylglycerols (DAGs) containing the corresponding unusual fatty acids (Wiberg et al. 1994, Vogel and Browse 1996, He et al. 2004). On the other hand, cytidine diphosphate (CDP)-choline:DAG cholinephosphotransferase, which can supply DAGs for TAG synthesis, showed little or no selectivity across a range of different DAG substrates (Vogel and Browse 1996).

3.2   RESULTS AND DISCUSSION

Our previous studies showed that *Vernonia galamensis* and *Stokesia laevis* developing seeds have DGAT activity with a strong preference for vernoloyl-CoA, *sn*-1,2-divernoloylglycerol, and *sn*-2-monovernoloylglycerol for incorporation into TAG, whereas soybean DGAT has much greater activity with oleoyl-CoA and

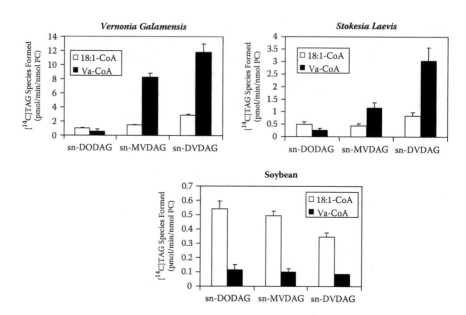

FIGURE 3.1 TAGs formed from *Vernonia galamensis*, *Stokesia laevis*, and soybean microsomal DGAT assays. The microsomes (eq. of 0.5 nmol microsomal PC) from developing seeds were fed with either [^{14}C]18:1-CoA or [^{14}C]vernoloyl-CoA (Va-CoA) together with 100 μM of 1,2-dioleoyl-*sn*-glycerol (*sn*-DODAG), 1-palmitoyl-2-vernoloyl-*sn*-glycerol (*sn*-MVDAG), or 1,2-divernoloyl-*sn*-glycerol (*sn*-DVDAG). Values are mean ±S.D. (n = 6) of three experiments each with two replications.

sn-1,2-dioleoylglycerol (Figure 3.1). These experiments indicated that DGATs are likely important in the high accumulation of vernolic acid in developing *Vernonia* and *Stokesia* seeds (Yu et al. 2006).

We isolated six novel complementary DNA (cDNA) clones: soybean DGAT1a and DGAT1b, *Euphorbia* DGAT1a and DGAT1b, and *Vernonia* DGAT1a and DGAT1b (Figure 3.2). Cluster analysis was performed to investigate the relationship among the known plant DGAT1s (total 14). Soybean DGAT1a and DGAT1b were grouped with that of *L. corniculatus*, a legume plant. *Euphorbia* DGAT1a and DGAT1b were grouped with castor DGAT, a member of the same family as *Euphorbia*. *Vernonia* DGAT1a and DGAT1b also were grouped together. Their accession numbers are as follows: *Arabidopsis thaliana* DGAT1 accession number AAF19262; *Brassica napus* DGAT accession number AAF64065; *Nicotiana tabacum* DGAT accession number AAF19345; *Olea europaea* DGAT accession number AAS01606; *Ricinus communis* (castor) DGAT accession number AAR11497 ; *Oryza sativa* (rice) DGAT accession number AAW47581; *Lotus corniculatus* DGAT accession number AAW51456; *Perilla frutescens* DGAT accession number AF298815; soybean DGAT1a accession number AB257589; soybean DGAT1a accession number AF257090; *Vernonia galamensis* DGAT1a accession number EF653276; and DGAT1b accession number EF653277. The sequences of *Euphorbia lagascae* DGAT1a and DGAT1b have not been submitted to the National Center for Biotechnology Information (NCBI) database yet.

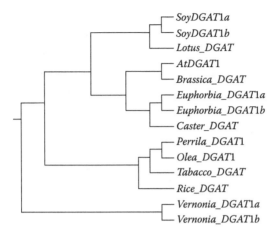

FIGURE 3.2 Phylogenetic tree of 14 different plant DGAT1s. The tree was established based on the DNA sequences of 4 DGAT1 cDNA clones by the CLUSTAL W program (version 1.82).

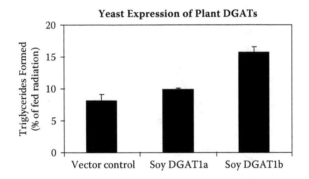

FIGURE 3.3 TAG biosynthesis activity of yeast microsomes expressing soybean DGAT1a and DGAT1b relative to the vector control. The activity of yeast microsomes expressing *Arabidopsis* DGAT1 was not different from the vector control.

Insect cells expressing *Vernonia* DGAT1a have 30 times as much TAG as vector control cells (Hatanaka et al. 2002). *Vernonia* DGAT1a and DGAT1b and *Euphorbia* DGAT1a and DGAT1b were also expressed in yeast along with soybean DGAT1a and DGAT1b and *Arabidopsis* DGAT1 as controls. The results showed that *Arabidopsis* DGAT1 has no significant activity above the vector control, and soybean DGAT1b has twice the activity of the vector controls (Figure 3.3).

Vernonia DGAT1a and DGAT1b exhibited manyfold higher activity than soybean DGAT1b. However, none of the DGAT1s showed substrate preference toward vernoloyl-CoA or 1,2-divernoloyl-*sn*-glycerol (Yu et al. 2008). Instead, oleoyl-CoA and 1,2-dioleoyl-*sn*-glycerol were preferred substrates (Figure 3.4).

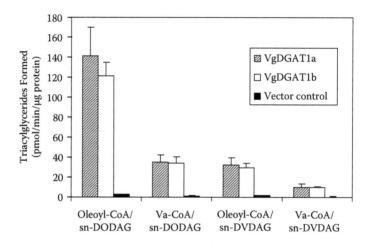

FIGURE 3.4   DGAT activities with different substrates expressing *Vernonia* DGAT1a and DGAT1b (VgDGAT1a and VgDGAT1b). Va-CoA, vernoyl-CoA; DODAG, dioleoyl-diacyl-glycerol (DAG); DVDAG, divernoyl-DAG.

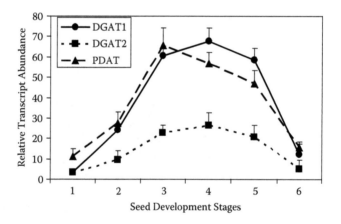

FIGURE 3.5   Relative abundance of DGAT1, DGAT2, and PDAT transcripts during castor (*Ricinus communis*) seed development.

The expression patterns of DGAT1s and DGAT2s in high-vernolic acid accumulators indicate that DGAT2 is of key importance in ricinoleic acid and vernolic acid accumulation (Figure 3.5).

Shockey et al. (2005, 2006) similarly reported that DGAT2 rather than DGAT1 is the main enzyme responsible for selective accumulation of the unusual conjugated fatty acid, eleostearic acid, in tung oil. A 370-bp DGAT2 fragment from *Vernonia* was cloned, and RACE (rapid amplification of cDNA ends) procedures were performed to clone the full-length cDNA. Southern analyses indicated that there likely are more than one DGAT2 in *Vernonia*. *Vernonia* DGAT2 groups with castor and tung DGAT2 based on sequence homology (Figure 3.6).

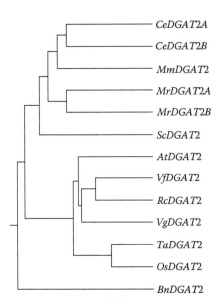

FIGURE 3.6 Phylogenetic grouping of DGAT2s. Ce, *Caenorhabditis elegans*; Mm, *Mus musculus* (mouse); Mr, *Mortierella ramanniana*; Sc, *Saccharomyces cerevisiae* (yeast); At, *Arabidopsis thaliana*; Vf, *Vernicia fordii* (tung tree); Rc, *Ricinus communis* (castor); Ta, *Triticum aestivum* (wheat); Os, *Oryza sativa* (rice); BnDGAT2:AF155224, *Brassica napus* (canola or rapeseed); VgDGAT2, *Vernonia galamensis*.

Other genes may be necessary for high vernolic acid accumulation in oilseeds such as soybeans in addition to epoxygenase and DGAT. A candidate for this is phospholipid:diacylglycerol acyltransferase (PDAT) (Stobart et al. 1997, Dahlqvist et al. 1998, 2000), although our biochemical experiments to date showed no PDAT activity with developing *Vernonia* seed microsomes that have abundant DGAT activity. Lipolytic acyl hydrolase (LAH) or a specific hydrolase such as phospholipase A_2 (PLA_2) active in releasing epoxy fatty acids made in *Vernonia, Stokesia,* or *Euphorbia* PC is another possibility, although such enzymes that selectively remove oxygenated fatty acids from membrane lipids are common in cells (Dhondt et al. 2002, Holk et al. 2002, Noiriel et al. 2004, Stahl et al. 1999). In addition, it is anticipated that acyl-CoA synthetases (ACSs) do not effectively act on epoxy fatty acid substrates in soybeans or *Arabidopsis,* but *Vernonia* and *Euphorbia* have ACSs that efficiently utilize epoxy fatty acids. Plants have multiple ACSs that participate in lipid metabolism (Schnurr et al. 2002, Shockey et al. 2002). McKeon et al. (2006) reported on a castor ACS (one among many ACSs found in castor) with high selectivity for the unusual fatty acid ricinoleic acid, which accumulates to as much as 90% of castor seed oil.

3.3 CONCLUSIONS

DGAT1 is the only enzyme proven to contribute to seed oil synthesis to date. Soybeans, *Euphorbia,* and *Vernonia* have at least two DGAT1s. DGAT1 but not

DGAT2 expression is consistent with playing a role in soybean oil biosynthesis. DGAT1, DGAT2, and overall PDAT expression is consistent with all three enzymes playing a role in *Vernonia, Stokesia,* and castor oil biosynthesis. The activity of *Vernonia* DGAT1s in yeast microsomes indicates that *Vernonia* DGAT1a or *Vernonia* DGAT1b does not play a role in the selective accumulation of Va in TAG. Biochemical and molecular data implicate DGAT2 as most important among TAG biosynthetic enzymes in high accumulation of epoxy and hydroxy fatty acid in oil of seeds that naturally accumulate high levels of these unusual fatty acids.

REFERENCES

Ayorinde, F.O., C.P. Nwaonicha, V.N. Parchment, K.A. Bryant, M. Hassan, and M.T. Clayton. 1993. Enzymatic synthesis and spectroscopic characterization of 1,3-divernoloyl glycerol from *Vernonia galamensis* seed oil. *J. Am. Oil Chem. Soc.* 70: 129–132.

Bafor, M., M.A. Smith, L. Jonsson, K. Stobart, and S. Stymne. 1993. Biosynthesis of vernoleate (*cis*-12-epoxyoctadeca-*cis*-9-enoate) in microsomal preparations from developing endosperm of *Euphoribia lagascae*. *Arch. Biochem. Biophys.* 303: 145–151.

Cahoon, E.B., K.G. Ripp, S.E. Hall, and B. McGonigle. 2002. Transgenic production of epoxy fatty acids by expression of a cytochrome P450 enzyme from *Euphorbia lagascae* seed. *Plant Physiol.* 128: 615–624.

Cao, Y., and A.H.C. Huang. 1987. Acyl coenzyme A preference of diacylglycerol acyltransferase from the maturing seeds of Cuphea, maize, rapeseed, and canola. *Plant Physiol.* 84: 762–765.

Cases, S., S.J. Stone, P. Zhou, E. Yen, B. Tow, K.D. Lardizabal, T. Voelker, and R.V. Farese, Jr. 2001. Cloning of DGAT2, a second mammalian diacylglycerol acyltransferase, and related family members. *J. Biol. Chem.* 276: 38870–38876.

Dahlqvist, A., A. Banas, and S. Stymne. 1998. Selective channelling of unusual fatty acids into triacylglycerols. In: J. Sanchez, E. Cerda-Olmedo, and E. Martinez-Force, Eds., *Advances in Plant Lipid Research*, Universidad de Sevilla, Seville, Spain, pp. 211–214.

Dahlqvist, A., U. Stahl, M. Lenman, A. Banas, M. Lee, L. Sandager, H. Ronne, and S. Stymne. 2000. Phospholipid:diacylglycerol acyltransferase: an enzyme that catalyzes the acyl-CoA-independent formation of triacylglycerol in yeast and plants. *Proc. Natl. Acad. Sci. U.S.A.* 97: 6487–6492.

Dhondt, S., G. Gouzerh, A. Muller, M. Legrand, and T. Heitz. 2002. Spatio-temporal expression of patatin-like lipid acyl hydrolases and accumulation of jasmonates in elicitor-treated tobacco leaves are not affected by endogenous levels of salicylic acid. *Plant J.* 32: 749–762.

Gardner, H.W. 1991. Recent investigations into the lipoxygenase pathway of plants. *Biochim. Biophys. Acta* 1084: 221–239.

Hammond, E. 1992. *Oil Used for Industrial Products.* Special Report 92 of the Iowa Agriculture and Home Economics Experiment Station, Iowa State University Press, Ames, IA.

Hatanaka, T., and D. Hildebrand. 2001. Expression of epoxy fatty acid synthesis genes. ASPB meeting abstract presentation, Providence, RI. http://abstracts.aspb.org/aspp2001/public/P38/0735.html.

Hatanaka, T., R. Shimizu, and D. Hildebrand. 2004. Expression of a *Stokesia laevis* epoxygenase gene. *Phytochemistry* 65: 2189–2196.

Hatanaka, T., K. Yu, and D.F. Hildebrand. 2002. Cloning and expression of *Vernonia* and *Euphorbia* diacylglycerol acyltransferase cDNAs. Proceedings of the International Society of Plant Lipids (ISPL), Okazaki, Japan, May.

He, X., C. Turner, G.Q. Chen, J.T. Lin, and T.A. McKeon. 2004. Cloning and characterization of a cDNA encoding diacylglycerol acyltransferase from castor bean. *Lipids* 39: 311–318.

Heimli, H., K. Hollung, and C.A. Drevon. 2003. Eicosapentaenoic acid-induced apoptosis depends on acyl CoA-synthetase. *Lipids* 38: 263–268.

Hitz, W.D. 1998. Fatty acid modifying enzymes from developing seeds of *Vernonia galamenensis*. U.S. Patent 5,846,784.

Holk, A., S. Rietz, M. Zahn, H. Quader, and G.F.E. Scherer. 2002. Molecular identification of cytosolic, patatin-related phospholipases A from *Arabidopsis* with potential functions in plant signal transduction. *Plant Physiol.* 130: 90–101.

Katavic, V., D.W. Reed, D.C. Taylor, E.M. Giblin, D.L. Barton, J. Zou, S.L. Mackenzie, P.S. Covello, and L. Kunst. 1995. Alteration of seed fatty acid composition by an ethyl methanesulfonate-induced mutation in *Arabidopsis thaliana* affecting diacylglycerol acyltransferase activity. *Plant Physiol.* 108: 399–409.

Lee, M., M. Lenman, A. Banas, M. Bafor, S. Singh, M. Schweizer, R. Nilsson, C. Liljenberg, A. Dahlqvist, P.-O. Gummeson, S. Sjodahl, A. Green, and S. Stymne. 1998. Identification of non-heme diiron proteins that catalyze triple bond and epoxy group formation. *Science* 280: 915–918.

Lu, C.F.L., S.B. de-Noyer, D.H. Hobbs, J.L. Kang, W.C. Wen, D. Krachtus, and M.J. Hills. 2003. Expression pattern of diacylglycerol acyltransferase-1, an enzyme involved in triacylglycerol biosynthesis, in *Arabidopsis thaliana*. *Plant Mol. Biol.* 52: 31–41.

McKeon, T., S.-T. Kang, C. Turner, X. He, G. Chen, and J.-T. Lin. 2006. Enzymatic synthesis of intermediates in castor oil biosynthesis. 97th AOCS Annual Meeting and Expo, p. 15.

Noiriel, A., P. Benveniste, A. Banas, S. Stymne, and P. Bouvier Nave. 2004. Expression in yeast of a novel phospholipase A1 cDNA from *Arabidopsis thaliana*. *Eur. J. Biochem.* 271: 3752–3764.

Pascual, M.J., and E. Correal. 1992. Mutation studies of an oilseed spurge rich in vernolic acid. *Crop Sci.* 32: 95–98.

Perdue, R.E. 1989. *Vernonia*—bursting with potential. *Agric. Eng.* 70: 11–13.

Routaboul, J., C. Benning, N. Bechtold, M. Caboche, and L. Lepiniec. 1999. The *TAG1* locus of *Arabidopsis* encodes for a diacylglycerol acyltransferase. *Plant Physiol. Biochem.* 37: 831–840.

Schnurr, J.A., J.M. Shockey, G.-J. de Boer, and J.A. Browse. 2002. Fatty acid export from the chloroplast. Molecular characterization of a major plastidial acyl-coenzyme A synthetase from *Arabidopsis*. *Plant Physiol.* 129: 1700–1709.

Scholthof, H.B. 1999. Rapid delivery of foreign genes into plants by direct rub-inoculation with intact plasmid DNA of a tomato bushy stunt virus gene vector. *J. Virol.* 73: 7823–7829.

Seither, C. 1997. Characterization of epoxy fatty acid synthesis in *Vernonia galamensis* and the isolation of candidate cDNA clones. Master's thesis, University of Kentucky, Lexington.

Seither, C., S. Avdiushko, and D.F. Hildebrand. 1997. Isolation of cytochrome P-450 genes from *Vernonia galamensis*. In: J.P. Williams, M.U. Khan, and N.W. Lem, Eds., *Physiology, Biochemistry and Molecular Biology of Plant Lipids*, Kluwer Academic, Dordrecht, pp. 389–391.

Sen Gupta, A.K. 1974. Fatty acids of soybean phosphatides and glycerides with special reference to the oxidized fatty acids. *Fette Seifen Anstrichm.* 76: 440–442.

Settlage, S.B., P. Kwanyuen, and R.F. Wilson. 1998. Relation between diacylglycerol acyltransferase activity and oil concentration in soybean. *J. Am. Oil Chem. Soc.* 75: 775–781.

Shockey, J.M., M.S. Fulda, and J.A. Browse. 2002. *Arabidopsis* contains nine long-chain acylcoenzyme A synthetase genes that participate in fatty acid and glycerolipid metabolism. *Plant Physiol.* 129: 1710–1722.

Shockey, J.M., S.K. Gidda, D.C. Chapital, J.-C. Kuan, P.K. Dhanoa, J.M. Bland, S.J. Rothstein, R.T. Mullen, and J.M. Dyer. 2006. Tung tree DGAT1 and DGAT2 have nonredundant functions in triacylglycerol biosynthesis and are localized to different subdomains of the endoplasmic reticulum. *Plant Cell* 18: 2294–2313.

Shockey, J., S. Gidda, D. Chapital, J.-C. Kuan, S.J. Rothstein, R. Mullen, and J. Dyer. 2005. Type 1 and type 2 diacylglycerol acyltransferases from developing seeds of tung (*Vernicia fordii*) display different affinities for eleostearic acid-containing substrates. Paper presented at North American Plant Lipids Conference, Fallen Leaf Lake, CA.

Stahl, U., M. Lee, S. Sjodahl, D. Archer, F. Cellini, B. Ek, R. Iannacone, D. Mackenzie, L. Semeraro, and E. Tramontano. 1999. Plant low-molecular-weight phospholipase A2s (PLA2s) are structurally related to the animal secretory PLA2s and are present as a family of isoforms in rice (*Oryza sativa*). *Plant Mol. Biol.* 41: 481–490.

Stobart, K., M. Mancha, M. Lenman, A. Dahlqvist, and S. Stymne. 1997. Triacylglycerols are synthesized and utilized by transacylation reactions in microsomal preparations of developing safflower (*Cartamus tinctorius* L.) seeds. *Planta* 203: 58–66.

Taylor, D.C., N. Weber, D.L. Barton, E.W. Underhill, L.R. Hogge, R.J. Weselake, and M.K. Pomeroy. 1991. Triacylglycerol bioassembly in microspore-derived embryos of *Brassica napus* L. cv Reston. *Plant Physiol.* 97: 65–79.

Thompson, A.E., D.A. Dierig, and K.R. Kleiman. 1994a. Germplasm development of *Vernonia galamensis* as a new industrial oilseed crop. *Ind. Crops* 3: 185–200.

Thompson, A.E., D.A. Dierig, and K.R. Kleiman. 1994b. Variation in *Vernonia galamensis* flowering characteristics, seed oil and vernolic acid contents. *Ind. Crops* 3: 175–184.

Vogel, G., and J. Browse. 1996. Cholinephosphotransferase and diacylglycerol acyltransferase. Substrate specificities at a key branch point in seed lipid metabolism. *Plant Physiol.* 110: 923–931.

Weber, N., E. Fehling, K.D. Mukhergee, K. Vosmann, B. Dahlke, S. Hellbardt, and W.H. Zech. 1994. Hydroxylated fatty acids: oleochemicals from plant oils. *Inform* 5: 475.

Wiberg, E., E. Tillberg, and S. Stymne. 1994. Substrates of diacylglycerol acyltransferase in microsomes from developing oil seeds. *Phytochemistry* 36: 573–577.

Yu, K., R. Li, T. Hatanaka, and D. Hildebrand. 2008. Cloning and functional analysis of two type 1 diacylglycerol acyltransferases from *Vernonia galamensis*. *Phytochemistry* 69: 1119–1127.

Yu, K., C.T. McCracken, Jr., R. Li, and D.F. Hildebrand. 2006. Diacylglycerol acyltransferases from *Vernonia* and *Stokesia* prefer substrates with vernolic acid. *Lipids* 41: 557–566.

4 High-Erucic Acid, Low-Glucosinolate Rapeseed (HEAR) Cultivar Development in Canada

Peter B. E. McVetty, W. G. Dilantha Fernando, Genyi Li, Muhammad Tahir, and Carla Zelmer

CONTENTS

Key Words: *Brassica napus*; cultivar development; erucic acid; industrial oil; rapeseed.

4.1 INTRODUCTION

Rapeseed (*Brassica napus* and *B. rapa*) was first grown commercially in Canada during World War II to provide erucic acid oil for use as marine lubricating oil in steamships. The government of Canada provided price guarantees for rapeseed during the war to ensure production; however, after the war price declines resulted in near cessation of rapeseed production in Canada. Small-scale rapeseed production for edible oil use in Asian markets was revived in Canada in the late 1940s by exporters using production contracts. The research and development that resulted in large-scale domestic use of rapeseed in Canada and large-scale seed, oil, and meal exports from Canada started from small beginnings in the 1950s (Stefansson and Downey 1995). Canada changed from a major importer of edible and industrial vegetable oils in the 1950s to a net exporter of edible and industrial oils based primarily on the success of rapeseed cultivar development and production.

Rapeseed cultivar development in Canada has evolved significantly over the last 60 years to include a wide range of oil types for both edible and industrial oil markets. This chapter provides a brief overview of the development of rapeseed types in Canada with a primary focus on the development of industrial oil types.

4.2 RAPESEED CULTIVAR DEVELOPMENT

After cultivar development began after World War II, rapeseed breeding programs were gradually established in numerous public sector research institutions, including several Agriculture and Agri-Food Canada research stations, the University of Manitoba, and the University of Alberta in the 1950s and 1960s (Stefansson and Downey 1995). The initial breeding objectives were to develop locally adapted rapeseed cultivars with improved productivity and seed quality. Initial rapeseed cultivar development in Canada used selection from spring habit open-pollinated population landrace introductions of *B. napus* from Argentina and from spring habit open-pollinated population landrace introductions of *B. rapa* from Poland. Selection in both species was for maturity, seed yield, uniformity, lodging resistance, oil content, and iodine value in the oil. Canada's first registered *B. napus* rapeseed cultivar Golden was developed by Agriculture and Agri-Food Canada, Saskatoon, and released in 1954 (Stefansson and Downey 1995). It had erucic acid concentrations of 40% to 45% and glucosinolate concentrations of over 100 µmol/g air-dried, oil-free meal. Several additional *B. napus* rapeseed cultivars, each with incremental improvements in agronomic or seed quality, were developed and released in Canada, including Nugget (1961), Tanka (1963), Target (1966), and Turret (1970) (Stefansson and Downey 1995). All of these rapeseed cultivars had erucic acid concentrations of 40% to 45% and glucosinolate concentrations of over 100 µmol/g air-dried, oil-free meal. Canada's first *B. rapa* rapeseed cultivar, Arlo, was developed by Svalof AB in Sweden using selection from open-pollinated populations and released in 1958 (Stefansson and Downey 1995). It also had erucic acid concentrations of 40% to 45%

and glucosinolate concentrations of over 150 µmol/g air-dried, oil-free meal. Two additional *B. rapa* rapeseed cultivars with incremental improvements in agronomic performance or seed quality were developed and released in Canada: Echo (1964) and Polar (1969) (Stefansson and Downey 1995). These rapeseed cultivars also had erucic acid concentrations of 40% to 45% and glucosinolate concentrations of over 100 µmol/g air-dried, oil-free meal. The oil from all of these rapeseed cultivars was used for edible and industrial oil purposes. Reports in the 1950s from feeding large amounts of rapeseed oil to animals suggested that erucic acid might constitute a health hazard. In response to these reports, rapeseed breeders screened *B. napus* and *B. rapa* rapeseed germplasm for genetic sources of low erucic acid. Genetic sources of low erucic acid were identified in an unlicensed *B. napus* forge rape cultivar Liho (Stefansson et al. 1961) and in a single plant selection from crosses in *B. rapa* (Downey 1964).

4.2.1 Low-Erucic Acid, High-Glucosinolate Rapeseed Cultivar Development

The development of low-erucic acid rapeseed edible oil cultivars began after the identification of the discussed genetic sources of low erucic acid in both rapeseed species. Generic breeding objectives were still to develop locally adapted rapeseed cultivars with improved productivity and seed quality. Selection in both species was for maturity, seed yield, uniformity, lodging resistance, oil content, and iodine value in the oil. Seed quality improvement now included seed oil containing less than 5% erucic acid.

Oro was Canada's first *B. napus* low-erucic acid rapeseed cultivar, developed by Agriculture and Agri-Food Canada, Saskatoon, and released in 1968 (Stefansson and Downey 1995). Oro was developed by pedigree selection in the progeny from the cross of Nugget × Liho. It had an erucic acid concentration of less than 5% and a high glucosinolate concentration of over 150 µmol/g air-dried, oil-free meal.

Span was Canada's first *B. rapa* low-erucic acid rapeseed cultivar, developed by Agriculture and Agri-Food Canada, Saskatoon, and released in 1971. Span was developed by pedigree selection in the progeny from the cross Arlo × (low-erucic acid *B. rapa* selection) (Stefansson and Downey 1995). It had an erucic acid concentration of less than 5% and a high glucosinolate concentration of over 150 µmol/g air-dried, oil-free meal.

Zephyr was the second and final low-erucic acid, high-glucosinolate *B. napus* rapeseed developed by Agriculture and Agri-Food Canada, Saskatoon, and was released in 1971 (Stefansson and Downey 1995). Zephyr was developed by pedigree selection in the progeny from a cross of a (high-oil, high-seed yield *B. napus* selection) × ({Liho × Nugget} F_4). It had high seed yield, an erucic acid concentration of less than 5%, and a high glucosinolate concentration of over 150 µmol/g air-dried, oil-free meal. These low-erucic acid, high-glucosinolate rapeseed cultivars were developed for domestic and export edible oil markets. The division of the Canadian rapeseed breeding effort into low-erucic acid types for edible oil markets and high-erucic acid types for industrial oil markets occurred in Canada in the 1960s.

4.2.2 HIGH-ERUCIC ACID, HIGH-GLUCOSINOLATE RAPESEED CULTIVAR DEVELOPMENT

The continuing need for high-erucic acid industrial oil from rapeseed led to the development of Canada's first high-erucic acid, high-glucosinolate rapeseed cultivar, R-500, developed by Agriculture and Agri-Food Canada, Saskatoon, and released in Canada in 1975 (Stefansson and Downey 1995). R-500 was a *B. rapa* cultivar developed by single-plant selection from the Yellow Sarson (from the Indian subcontinent) subspecies of *B. rapa*. Selection was for maturity, seed yield, uniformity, lodging resistance, oil content, and erucic acid content. R-500 had a high-erucic acid concentration of over 50% and a high glucosinolate concentration of over 100 μmol/g air-dried, oil-free meal. R-500 had the highest erucic acid concentration of any rapeseed cultivar grown in Canada in the 1970s. Unfortunately, R-500 was highly susceptible to white rust, *Albugo candida,* and was not commercially successful (Stefansson and Downey 1995). No further *B. rapa* high-erucic acid, high-glucosinolate cultivars were developed in Canada. No high-erucic acid industrial oil *B. napus* rapeseed cultivars with high-erucic acid, high-glucosinolate concentration were ever developed in Canada.

4.2.3 DOUBLE-LOW (LOW ERUCIC ACID AND LOW GLUCOSINOLATE) RAPESEED (CANOLA) CULTIVAR DEVELOPMENT

The development of low-erucic acid rapeseed cultivars for edible oil purposes created a high-quality vegetable oil from low-quality rapeseed oil. However, the nutritional value of rapeseed meal was limited by plant secondary metabolites known as glucosinolates commonly found in *Brassica* species (Kimber and McGregor 1995). Glucosinolates affected the thyroid gland of animals fed rapeseed meal, and the glucosinolates reduced the palatability of rapeseed meal. Rapeseed breeders screened rapeseed germplasm for genetic sources of low-glucosinolate concentration. A genetic source of low-glucosinolate concentration was discovered in a *B. napus* cultivar from Poland (Bronowski) at the Agriculture and Agri-Food Canada, Saskatoon, research station in 1967 (Stefansson and Downey 1995).

This single genetic source of the low-glucosinolate trait was used to develop all low-glucosinolate rapeseed cultivars in both *B. napus* and *B. rapa* throughout the world. The identification of a genetic source of low glucosinolates in *B. napus* permitted the expansion of rapeseed seed-quality breeding objectives to include both low erucic acid in the oil and low glucosinolates in the meal. The development of low-erucic acid, low-glucosinolate (double-low) edible oil rapeseed (canola) cultivars began in Canada in the 1960s, resulting in the world's first double-low *B. napus* rapeseed (canola) cultivar, Tower, developed by the University of Manitoba and released in 1974 (Stefansson and Downey 1995).

Tower was developed by pedigree selection in the progeny from the cross (Bronowski × {Turret × Turret}) × (Liho × {Turret × Turret}) (Stefansson and Kondra 1975). It had an erucic acid concentration of less than 1% in the seed oil and a glucosinolate concentration of less than 30 μmol/g of air-dried, oil-free meal. The first double-low *B. rapa* cultivar, Candle, was developed by Agriculture and Agri-Food Canada in Saskatoon and released in 1978 (Stefansson and Downey 1995). Candle was developed by pedigree selection in the progeny of interspecific crosses involving *B. rapa, B. juncea,* and *B. napus.* It had an erucic acid concentration of less than 2% in the seed oil and less than 30 μmol/g glucosinolates in the air-dried, oil-free meal. These double-low rapeseed cultivars had improvements in both oil quality for edible oil purposes and meal quality for animal feed purposes. The term *canola* was adopted in Canada in 1979 to describe double-low-quality rapeseed cultivars in both *B. napus* and *B. rapa.*

Hundreds of additional *B. napus* and *B. rapa* double-low rapeseed (canola) cultivars have been developed and released in Canada since the 1970s. These new double-low rapeseed (canola) cultivars have incremental improvements in agronomic performance or seed quality.

4.2.4 SPECIALTY OIL PROFILE DOUBLE-LOW RAPESEED (CANOLA) CULTIVAR DEVELOPMENT

Double-low rapeseed (canola) contains over 10% linolenic acid in the seed oil (Ratanayake and Daun 2004). Linolenic acid contains three double bonds, which reduces the oil stability and shelf life of double-low rapeseed (canola) oil. A new breeding objective of modifying the fatty acid profile of double-low rapeseed (canola) oil to increase the stability of the oil emerged in the 1980s. The initial focus was on reduction of the linolenic acid concentration in the oil. The world's first low-linolenic acid canola cultivar, Stellar, was developed by the University of Manitoba and released in 1987 (Scarth et al. 1987). Stellar was developed by pedigree selection in progeny from the cross (M11 × Regent) × Regent. Regent was a double-low rapeseed (canola) cultivar developed at the University of Manitoba; M11 was a low-linolenic acid mutant line in the cultivar Oro (Rakow 1973). Stellar had a linolenic acid concentration of less than 3% versus double-low rapeseed (canola) cultivars, with a linolenic acid concentration of over 10%. This reduction in linolenic acid concentration improved the stability of the oil. Unfortunately, the seed yield of Stellar was not competitive with double-low rapeseed (canola) cultivars of the day, and significant commercial production did not develop (Stefansson and Downey 1995).

Two additional low-linolenic acid double-low rapeseed (canola) cultivars with incremental improvements in agronomic performance, Apollo (Scarth et al. 1995) and Allons (Scarth et al. 1997), were developed and released by the University of Manitoba. There was limited commercial production of these low-linolenic acid, double-low rapeseed (canola) cultivars in Canada. More recently, high-stability oil double-low rapeseed (canola) cultivars that have low linolenic acid concentrations (<3%) and high oleic acid concentrations (>70%) have been developed. These new high-oleic acid, low-linolenic acid, high-stability oil, double-low rapeseed (canola) cultivars have competitive seed yields and are commercially successful (Canola Council of Canada 2008).

4.2.5 Disease-Resistant Rapeseed Cultivar Development

Diseases in rapeseed fields in the early years of rapeseed production in Canada were rare (Stefansson and Downey 1995). As production of rapeseed expanded in Canada, several diseases, including white mold (*Sclerotinia sclerotiorum*), white rust (*Albugo candida*), and blackleg (*Leptosphaeria maculans*), increased in incidence and importance to rapeseed seed breeders. Breeding for disease resistance in rapeseed began in the mid-1960s at Agriculture and Agri-Food Canada, Saskatoon, and in the early 1970s at the University of Manitoba. Even though all Canadian *B. napus* rapeseed cultivars were and are resistant to white rust (Downey and Röbbelen 1989), all Canadian *B. rapa* cultivars were susceptible to white rust. Canada's first white rust-resistant *B. rapa* cultivar, Tobin, was developed and released by Agriculture and Agri-Food Canada, Saskatoon, in 1980 using pedigree selection in the progeny of interspecific crosses of *B. napus* and *B. rapa*.

All Canadian *B. rapa* rapeseed cultivars were susceptible to blackleg, while *B. napus* cultivars ranged from moderately resistant to highly susceptible. Significant progress in the development of blackleg-resistant, double-low rapeseed (canola) cultivars has occurred in Canada, with several blackleg resistance genes deployed in different cultivars (Stefansson and Downey 1995). Many double-low rapeseed (canola) cultivars and rapeseed grown in Canada are now fully blackleg resistant. However, since the identification of new pathogenicity groups (PG) (PG-3, PG-4, PGT) of the blackleg fungus in the prairie provinces and in North Dakota in 2003 and 2004 (Fernando and Chen 2003, Chen and Fernando 2005, Bradley et al. 2005, Chen and Fernando 2006), there has been a concerted effort to breed for resistance against these new blackleg strains that could otherwise devastate the crop as most existing double-low rapeseed (canola) cultivars only possess PG-2 resistance genes (Dusabenyagasani and Fernando 2008). Since the host–pathogen interaction is governed by a gene-for-gene (*R* genes in host and *Avr* genes in the pathogen) interaction, it is likely that existing double-low rapeseed (canola) cultivars are susceptible to the new pathogenicity groups (Fernando et al. 2007). Molecular markers for several blackleg resistance genes are being developed at the University of Manitoba (Wang 2007, Dusabenyagasani and Fernando 2008). Molecular markers for blackleg resistance genes will facilitate the pyramiding of several blackleg resistance genes in canola/rapeseed cultivars and lines.

All Canadian rapeseed double-low rapeseed (canola) cultivars are susceptible to white mold (Thomas 1992). Selected Chinese rapeseed cultivars have field tolerance to sclerotinia, and crosses of these cultivars to Canadian cultivars are in progress to transfer this tolerance.

4.2.6 New Traits and New Cultivar Types in Rapeseed and Double-Low Rapeseed (Canola) Cultivar Development

Novel herbicide tolerance to a number of broad-spectrum herbicides, including bromoxynil, glyphosate, glufosinate ammonium, imidolozinones, and triazines, has been added to new rapeseed and double-low rapeseed (canola) cultivars developed and released in Canada in recent years (McVetty and Zelmer 2007). These herbicide tolerances have greatly simplified the production of canola and rapeseed in Canada, and they have been readily accepted by Canadian canola growers. Over 95% of canola and rapeseed production in Canada currently is of herbicide-tolerant types (McVetty and Zelmer 2007).

Rapeseed and double-low rapeseed (canola) cultivars in Canada have traditionally been open-pollinated population cultivars in both *B. napus* and *B. rapa*. More recently, hybrid *B. napus* double-low rapeseed (canola) cultivars have been developed and released in Canada. Most of these hybrids are also herbicide tolerant. Nearly 50% of double-low rapeseed (canola) production in Canada in 2005 was of hybrid *B. napus* types (Brandt and Clayton 2005). This proportion increases annually.

4.2.7 High-Erucic Acid, Low-Glucosinolate Rapeseed Cultivar Development

High-erucic acid, low-glucosinolate *B. napus* cultivar development began at the University of Manitoba in the early 1970s as a parallel and complementary breeding program to the double-low rapeseed (canola) breeding program (Scarth et al. 1992). Generic breeding objectives were still to develop locally adapted rapeseed cultivars with improved productivity and seed quality. Selection for earlier maturity, reduced plant height, increased oil content, reduced glucosinolate concentration, and increased seed yield was used. The new high-erucic acid, low-glucosinolate cultivars were intended to provide a source of industrial oil and a meal low in glucosinolates. At the beginning, many of the same crosses combined with pedigree selection were used to develop both double-low rapeseed (canola) and high-erucic acid, low-glucosinolate rapeseed (HEAR) cultivars. The double-low rapeseed (canola) cultivar Regent (Canada's second open-pollinated population *B. napus* canola cultivar, developed and released by the University of Manitoba in 1977) and Reston (Canada's first *B. napus* high-erucic acid, low-glucosinolate, open-pollinated population rapeseed cultivar), developed and released in 1982 by the University of Manitoba, both originated from the same series of crosses (Stefansson and Downey 1995). Reston had an erucic acid concentration of 40% to 45% and a glucosinolate concentration of less than 30 µmol/g in the air-dried, oil-free meal. It had 3% higher oil concentration and 1% higher protein concentration and seed yield compared to Regent, the best canola cultivar of the day. It had moderate resistance to blackleg. Reston was a commercial success and was grown as the exclusive Canadian source of high-erucic acid rapeseed oil for several years (Stefansson and Downey 1995).

Increases in erucic acid concentration were added to the breeding objectives and selection list for HEAR cultivars. Canada's second high-erucic acid, low-glucosinolate, open-pollinated population rapeseed cultivar was Hero, developed and released by the University of Manitoba in 1989 (Scarth et al. 1991). Hero was developed by pedigree selection in the progeny from a cross of a very high-erucic acid, high-glucosinolate rapeseed line of Swedish origin and Reston. It had erucic acid concentration of over 50% and a glucosinolate concentration of less than 20 µmol/g glucosinolates in the air-dried, oil-free meal. It had similar oil concentration and protein concentration as the double-low rapeseed (canola) cultivars Regent and Westar. Hero had similar seed yield to Regent but was 13% lower yielding than Westar, the predominant open-pollinated population double-low rapeseed (canola) cultivar in Canada in 1989.

A series of HEAR rapeseed cultivars with incremental improvements in agronomic performance seed quality followed the HEAR cultivar Hero. The open-pollinated population HEAR cultivars Mercury (Scarth et al. 1994), Venus (McVetty et al. 1996a), Neptune (McVetty et al. 1996b), Castor (McVetty et al. 1998), MillenniUM 01 (McVetty et al. 1999), MillenniUM 02 (McVetty et al. 2000a), and MillenniUM 03 (McVetty et al. 2000b) were all developed and released by the University of Manitoba. Mercury, Venus, Neptune, and Castor were developed by pedigree selection in the progeny from HEAR cultivar/line × HEAR cultivar/line crosses using exclusively University of Manitoba germplasm. All of these HEAR cultivars were moderately resistant to blackleg.

A new breeding objective of full resistance to blackleg was added to the HEAR breeding objectives list in the 1990s. MillenniUM 01, MillenniUM 02, and MillenniUM 03 were developed by pedigree selection in the progeny from crosses of HEAR lines and superior-performing, blackleg-resistant, double-low rapeseed (canola) cultivars from Canadian or European breeding organizations. Screening for blackleg resistance was done on seedlings grown in the greenhouse and then confirmed in disease-screening nurseries grown in the field. The MillenniUM series of HEAR cultivars was fully resistant to blackleg.

The primary improvement for the Reston-to-MillenniUM 03 series of HEAR cultivars was increased yield. There were also simultaneous incremental improvements in oil and protein concentration, a very rare occurrence in *Brassica* cultivar development. Erucic acid concentration in the oil increased from 40% to 45% in Reston to 50% to 55% in the MillenniUM series of HEAR cultivars. Yield increases of approximately 30% coupled with oil concentration increases of approximately 2.5% and protein concentration increases of 2.5% were achieved from Reston to Millennium 03 (Figure 4.1, Figure 4.2, and Figure 4.3). In addition, blackleg resistance improved

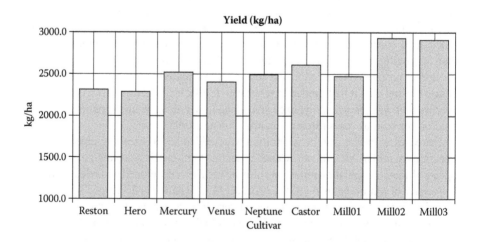

FIGURE 4.1   Seed yield of University of Manitoba HEAR cultivars compared to Reston, the first HEAR cultivar registered in Canada in 1982.

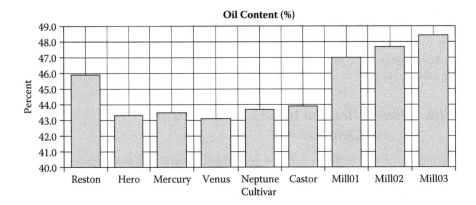

FIGURE 4.2 Oil content of University of Manitoba HEAR cultivars compared to Reston, the first HEAR cultivar registered in Canada in 1982.

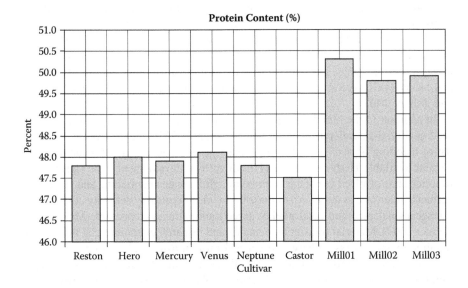

FIGURE 4.3 Protein content of University of Manitoba HEAR cultivars compared to Reston, the first HEAR cultivar registered in Canada in 1982.

from moderate resistance in Reston to full resistance in the MillenniUM series of HEAR cultivars. All of these HEAR cultivars were commercially successful, although in many cases their commercial lifetime was short since they were replaced quickly by steadily improved new HEAR cultivars.

A genomics research-and-development component has been added to the University of Manitoba HEAR breeding program. The development of molecular markers using sequence-related amplified polymorphisms (SRAP) (Li and Quiros 2001) to provide marker-assisted selection (MAS) capabilities to the breeding effort is a major part of the genomics research-and-development program. The recent development of genome-specific erucic acid gene molecular markers for the two erucic acid-controlling genes in *B. napus* (Rahman 2007) has permitted the selection of homozygous high-erucic acid genotypes in segregating generations and backcross generation progeny of HEAR × canola crosses, greatly improving breeding efficiency.

4.2.8 HERBICIDE-TOLERANT HIGH-ERUCIC ACID RAPESEED CULTIVAR DEVELOPMENT

Competition from weeds is a major production concern for double-low rapeseed (canola)/rapeseed growers in western Canada (Kimber and McGregor 1995, Orson 1995). To address this concern, *Brassica* breeders have developed double-low (canola) and rapeseed cultivars with novel, broad-spectrum herbicide tolerance. Five different types of novel herbicide-tolerant canola cultivars have been developed and released in Canada starting in the 1970s (McVetty and Zelmer 2007). Double-low rapeseed (*B. napus*) cultivars tolerant to triazine, glufosinate ammonium, glyphosate, imidazolinone, or bromoxynil herbicides have been developed and released in Canada. Herbicide tolerance to the triazines is conferred by a spontaneous mutation found in *B. rapa* that was backcrossed into *B. napus* canola (Beversdorf et al. 1980). Herbicide tolerance to the imidazolinones was created by microspore mutagenesis in *B. napus* (Swanson et al. 1989). Herbicide tolerance to glufosinate ammonium, glyphosate, and bromoxynil were all created using plant transformation technologies (McVetty and Zelmer 2007).

In addition to the breeding objectives of enhancing agronomic performance and seed quality, the development of herbicide-tolerant HEAR cultivars was added to the list of breeding objectives in the 1990s once genetic sources of herbicide tolerance became available. Herbicide-tolerant HEAR cultivar development at the University of Manitoba has involved use of the bromoxynil, glufosinate ammonium, and glyphosate herbicide tolerance systems. All of these herbicide tolerance systems are conferred by transgenes patented and controlled by large multinational companies (McVetty and Zelmer 2007). Material transfer agreements and research or commercial licenses are required to gain access to these transgenes for use in *Brassica* breeding.

Development of bromoxynil-tolerant HEAR cultivars began at the University of Manitoba in the mid-1990s. This herbicide-tolerant HEAR cultivar development program used the OXY 235 transgene developed and patented by Rhône-Poulenc Agrochimie (Freyssinet et al. 1989; Freyssinet et al. 1996). The University of Manitoba had a research agreement with Rhône-Poulenc Agrochimie authorizing its use of this transgene in *Brassica* breeding. The OXY 235 transgene was transferred from bromoxynil-tolerant double-low rapeseed (canola) lines to HEAR lines using backcrossing combined with pedigree selection. Selection for maturity, seed yield, uniformity, lodging resistance, oil content, protein content, blackleg resistance, erucic acid concentration in the oil, and bromoxynil tolerance was done. In addition, multilocation tolerance testing to demonstrate that the new bromoxynil-tolerant HEAR lines were fully tolerant to bromoxynil was conducted. The world's first bromoxynil-tolerant HEAR lines were supported for registration in Canada in

February 2002; however, Bayer CropScience, the OXY 235 transgene patent holder at that time, decided to withdraw bromoxynil-tolerant double-low rapeseed (canola) and rapeseed from the Canadian market in April 2002. As a consequence, bromoxynil-tolerant HEAR cultivars were not commercialized in Canada.

Development of glyphosate-tolerant HEAR cultivars began in the late 1990s. This herbicide-tolerant HEAR cultivar development program uses the RT73 glyphosate tolerance transgene (Barry et al. 1992) developed and patented by Monsanto. The University of Manitoba has a research agreement with Monsanto authorizing its use of this transgene in *Brassica* breeding. The RT73 transgene was transferred from glyphosate-tolerant double-low rapeseed (canola) tolerant lines to HEAR lines using backcrossing and pedigree selection. Selection for maturity, seed yield, uniformity, lodging resistance, oil content, protein content, blackleg resistance, erucic acid concentration in the oil, and glyphosate tolerance was done. In addition, multilocation tolerance testing to demonstrate that the new glyphosate-tolerant HEAR lines were fully tolerant to glyphosate was conducted. Extensive testing for the presence of the RT73 transgene and the absence of the RT200 transgene in all breeding lines is conducted to ensure construct purity for RT73. Further development of glyphosate-tolerant HEAR cultivars utilized crosses and pedigree selection of glyphosate-tolerant HEAR lines from the University of Manitoba and glyphosate-tolerant double-low rapeseed (canola) cultivars from Canadian breeding organizations. Selection for maturity, seed yield, uniformity, lodging resistance, oil content, protein content, blackleg resistance, erucic acid concentration in the oil, and glyphosate tolerance was done. In addition, multilocation tolerance testing to demonstrate that the new glyphosate-tolerant HEAR lines were fully tolerant to glyphosate was conducted. Extensive testing for the presence of the RT73 transgene and the absence of the RT200 transgene in all breeding lines is conducted to ensure construct purity for RT73. The world's first two glyphosate-tolerant HEAR cultivars, Red River 1826 (McVetty et al. 2006a) and Red River 1852 (McVetty et al. 2006b), were developed and released in 2006. They are the world's first commercially successful glyphosate-tolerant HEAR cultivars.

Development of glufosinate ammonium-tolerant HEAR cultivars began at the University of Manitoba in 2002. This herbicide-tolerant HEAR cultivar development program uses the Rf3 glufosinate ammonium tolerance transgene (Williams et al. 1997) developed and patented by Bayer CropScience. The University of Manitoba has a material transfer agreement/research agreement with Bayer CropScience authorizing its use of this transgene in *Brassica* breeding. The Rf3 transgene was transferred from a single glufosinate ammonium-tolerant double-low rapeseed (canola) line provided by Bayer CropScience to HEAR lines using backcrossing accelerated by MAS. Further development of glufosinate ammonium-tolerant HEAR cultivars utilized crosses and pedigree selection of glufosinate ammonium-tolerant HEAR cultivars/lines to conventional HEAR cultivars/lines from the University of Manitoba. Selection for maturity, seed yield, uniformity, lodging resistance, oil content, protein content, blackleg resistance, and erucic acid concentration and glufosinate ammonium tolerance in the oil was done. In addition, multilocation tolerance testing to demonstrate that the new glufosinate ammonium-tolerant HEAR lines were fully tolerant to glufosinate ammonium was conducted. There have been several glufosinate ammonium tolerance transgenes previously developed and used in *Brassica* breeding globally. To ensure Rf3 transgene construct purity, no glufosinate ammonium-tolerant canola cultivars or lines from outside the University of Manitoba are introduced into this herbicide-tolerant HEAR breeding program. The world's first two glufosinate ammonium-tolerant HEAR lines may receive support for registration in Canada in February 2008. These will be the first glufosinate ammonium-tolerant HEAR lines in the world to reach this development stage.

4.2.9 HYBRID HERBICIDE-TOLERANT HIGH-ERUCIC ACID RAPESEED CULTIVAR DEVELOPMENT

Hybrids in double-low rapeseed (canola) and rapeseed display significant heterosis for seed yield. Heterosis for seed yield in Canadian *B. napus* double-low rapeseed (canola) hybrids in the 20% to 120% range has been reported (Sernyk and Stefansson 1983, Brandle and McVetty 1989, Grant and Beversdorf 1985, Riungu and McVetty 2004). Similarly, Cuthbert (2006) reported high parent heterosis of up to 100% for seed yield in HEAR hybrids. The reports of high parent heterosis for seed yield in double-low rapeseed (canola) and rapeseed have resulted in the development of hybrids as a major breeding objective in *B. napus* breeding programs globally. Since *B. napus* is a perfect flowered species, initial hybrid double-low rapeseed (canola) and rapeseed research focused on pollination control system development. Research on a number of pollination control systems, including cytoplasmic male sterility (CMS) systems (McVetty 1997), genetic male sterility (GMS) systems (Sawhney 1997), and self-incompatibility (SI) systems (McCubbin and Dickinson 1997), has been conducted in recent decades. For CMS systems, the *nap* (Thompson 1972), *pol* (Fu 1981), *mur* (Hinnata and Konno 1979), *ogu* (Ogura 1968), and *ogu* Institut National de la Recherche Agronomique (INRA) CMS (Pelltier et al. 1983) systems have been developed for use in hybrid canola/rapeseed seed production. The *ogu* INRA CMS system is the most commonly used pollination control system for hybrid double-low rapeseed (canola) cultivar development in Canada.

Hybrid HEAR cultivar development at the University of Manitoba began in the 1990s using the *mur* CMS system (Riungu and McVetty 2003). While this CMS system was initially very promising, it was later determined that there was a biological cost to the *mur* CMS system that resulted in reduced oil content in the hybrids, making this pollination system less desirable. The University of Manitoba obtained a commercial license for use of the *ogu* INRA CMS system in 2003. The development of hybrid HEAR cultivar at the University of Manitoba now uses the *ogu* INRA CMS system. Parent lines from different heterotic gene pools are developed using selfing and pedigree selection or doubled-haploid line production (Palmer and Keller 1997). Selection in the parent lines for maturity, seed yield, uniformity, lodging resistance, oil content, protein content, glucosinolate concentration, blackleg resistance, and erucic acid concentration in the oil is done. The combining ability of the pedigree selection developing inbred lines or doubled-haploid lines is done using test crosses of the top-performing lines in different heterotic gene pools. Selection in the hybrids for maturity, seed yield, uniformity, lodging resistance, oil content, protein content, glucosinolate concentration, blackleg resistance, and erucic acid concentration in the oil is done. Non-herbicide-tolerant, glyphosate-tolerant, and glufosinate ammonium-tolerant HEAR parent lines are being developed to permit the development of non-herbicide-tolerant, glyphosate-tolerant, and glufosinate ammonium-tolerant hybrid HEAR cultivars. The *ogu* INRA CMS HEAR hybrids will be developed entirely by the University of Manitoba.

In addition, a collaborative hybrid HEAR cultivar research and development program has begun. Lembke Research Limited and the University of Manitoba are collaboratively developing hybrid HEAR cultivars using Lembke's proprietary Male Sterile Lembke (MSL) pollination control system (Frauen and Paulmann 1999). Lembke Research Limited has developed spring habit HEAR cultivars and lines that are in a heterotic pool different from the University of Manitoba HEAR cultivars and lines. In this collaboration, Lembke Research Limited is developing non-herbicide-tolerant MSL female HEAR lines using pedigree selection, while the University of Manitoba is developing non-herbicide-tolerant, glyphosate-tolerant, and glufosinate ammonium-tolerant male HEAR lines using pedigree selection and doubled-haploid development. Lembke Research Limited MSL female HEAR lines and University of Manitoba male MSL HEAR lines will be combined to produce MSL hybrid HEAR cultivars.

4.2.10   SUPER-HIGH-ERUCIC ACID RAPESEED, LOW-GLUCOSINOLATE GERMPLASM DEVELOPMENT

The University of Manitoba HEAR breeding program has an additional breeding objective of developing super-high-erucic acid, low-glucosinolate rapeseed germplasm. Super-high-erucic acid rapeseed (SHEAR) will have an erucic acid concentration of over 66% in the seed oil. Erucic acid concentration in *B. napus* is limited to a maximum of 66% since erucic acid is rarely, if ever, esterified in the *sn*-2 position of the glycerol molecule, the backbone of triglycerides in seed storage oils (Taylor et al. 1994). The discovery of *B. oleracea* genotypes with erucic acid esterified in the *sn*-2 position (Taylor et al. 1994) and the discovery of *B. rapa* genotypes with erucic acid esterified in the *sn*-2 position (D. C. Taylor personal communication) provided genes for *sn*-2-position erucic acid esterification in the parental species of *B. napus*. Interspecific crosses of *B. oleracea* and *B. rapa* followed by chromosome doubling using colchicine-produced new (resynthesized) *B. napus* genotypes that contained the *sn*-2-position erucic acid esterification genes from both parental species. Crosses among selected resynthesized *B. napus* lines combined with pedigree selection have been used in attempts to develop SHEAR germplasm. The maximum erucic acid content so far observed in the progeny of these resynthesized *B. napus* HEAR line crosses has been 64%. While this value is both the highest erucic acid content value reported and is very close to the 66% erucic acid content maximum expected for *B. napus*, it does not break the 66% erucic acid content barrier to create SHEAR genotypes. Microspore mutagenesis in the progeny of the resynthesized

B. napus line crosses is being used to create doubled-haploid lines with homozygous mutations for fatty acid profile variations, including possible super high-erucic acid types. Mutagenic SHEAR genotypes, once developed, will provide the starting genetic materials for SHEAR cultivar development. The first resynthesized *B. napus* SHEAR genotypes developed will be high in glucosinolates, so crosses to low-glucosinolate *B. napus* lines will be required to develop SHEAR cultivars.

A second and complementary approach to the development of SHEAR germplasm is being pursued in collaboration with the Plant Biotechnology Institute in Saskatoon, Saskatchewan. Taylor and colleagues at the Plant Biotechnology Institute have transformed *B. napus* HEAR genotypes with transgenes that increase *sn*-2-position erucic acid esterification (Taylor et al. 1995), increase long chain fatty acid synthesis, and block the oleic-to-linoleic synthesis step. These transgenes have been combined in HEAR genotypes in single-, double-, and triple-gene combinations at the University of Manitoba using crosses of selected transgenic lines and pedigree selection. The resulting progeny have been grown in confined field trials. The maximum erucic acid content so far observed in these transgenic *B. napus* HEAR genotypes has been 62%. The transgenic SHEAR lines developed from HEAR cultivars or lines will be SHEAR types, potentially suitable for use as new SHEAR cultivars. Development of transgenic super high-erucic acid rapeseed genotypes at the Plant Biotechnology Institute also continues using the 64% erucic acid content progeny from crosses of resynthesized *B. napus* lines developed at the University of Manitoba. A transgenic SHEAR genotype derived from resynthesized *B. napus* lines will be high in glucosinolates, so crosses to low-glucosinolate *B. napus* lines will be required to develop SHEAR cultivars.

4.3   MOLECULAR MARKER DEVELOPMENT AND MARKER-ASSISTED SELECTION FOR ENHANCING BREEDING EFFICACY

Marker-assisted selection is becoming more attractive to plant breeders as the genes underlying important traits in major crops have been cloned and functionally characterized. With the known genes, completely linked high-throughput molecular markers could be easily developed for MAS. The recent development of genome-specific erucic acid gene molecular markers targeting the mutation positions in E1 and E2 genes through Bacterial Artificial Chromosome (BAC) library construction and BAC sequencing has permitted the selection of homozygous high-erucic acid genotypes in segregating generations and backcross generation progeny of HEAR × canola crosses, greatly improving breeding efficiency. Thousands of plants from segregating populations have been screened for selecting high-erucic acid genotype in the HEAR breeding program at the University of Manitoba.

Moreover, an ultradense genetic recombination map with over 13,300 SRAP markers has been constructed at the University of Manitoba (Sun et al. 2007), which dramatically accelerates the speed of gene cloning and molecular marker development. Currently, six genes for glucosinolate biosynthesis, three genes for blackleg disease resistance, three genes for yellow seed color (Rahman et al. 2007), and two genes for erucic acid have been mapped on the ultradense map. These SRAP markers flanking the aforementioned genes can be used for MAS of low-glucosinolate content, gene pyramiding of blackleg disease resistance genes, and developing yellow-seeded rapeseed and canola. In addition, the mapping of sclerotinia tolerance QTL and high oil content QTL on the ultradense map is under way.

4.4 THE FUTURE

The emerging emphasis on renewal energy, chemical feedstocks, industrial oils, and novel uses of vegetable oils and the steadily growing bioeconomy will provide significant growth opportunities for high-erucic acid oil and for super-high-erucic acid oil. Double-digit annual growth in demand for high-erucic acid rapeseed oil and super-high-erucic acid rapeseed oil is anticipated. Since HEAR production in Canada must compete with canola production, the development of HEAR cultivars must parallel the development of canola cultivars. Future HEAR cultivars will be high-yielding, herbicide-tolerant, high-oil-content, hybrid cultivars. These cultivars will have to incorporate the best biotic and abiotic stress genes or transgenes available in canola to remain competitive. Future SHEAR cultivars will also have to provide competitive returns to growers to ensure their production. The first SHEAR cultivars will be open-pollinated population or doubled-haploid line cultivars without herbicide tolerance or any of the biotic or abiotic stress genes or transgenes available in the HEAR cultivars. Therefore, premiums for SHEAR production will be required to establish this specialty crop in Canada.

The approaches to HEAR and SHEAR breeding will change in the future. The development of molecular markers for major genes and quantitative trait loci (QTL) will greatly improve the efficiency of HEAR and SHEAR breeding and facilitate development of novel breeding objectives. Molecular markers for QTL for oil content are being developed currently. These molecular markers will speed the development of increased oil concentration double-low rapeseed (canola) and rapeseed cultivars. Molecular markers for several blackleg resistance genes have been developed (Wang 2007) that will permit the pyramiding of two or more blackleg resistance genes in HEAR and SHEAR cultivars. HEAR and SHEAR cultivars with pyramided blackleg resistance genes are expected to have long-lasting, possibly durable blackleg resistance in the face of ever-changing blackleg races in western Canada (Chen and Fernando 2006).

The genomics techniques are also being used to characterize the genetic variability in HEAR cultivars and lines to determine the genetic distance among HEAR cultivars and lines. Vincent (2008) reported very high correlations between genetic distance among HEAR parents and yield of the derived hybrid. This will greatly facilitate the selection of HEAR parents for use in hybrid HEAR cultivar development and speed the development of hybrid HEAR cultivars. Pedigree selection in the progeny of crosses, backcrosses, and doubled-haploid lines combined with MAS for major genes and for genetic distance among HEAR cultivars and lines will improve the efficiency of HEAR and SHEAR cultivar development in the future. The challenges are numerous but the opportunities unlimited for HEAR and SHEAR cultivar development and production in Canada.

REFERENCES

Barry, G., G. Kishore, M. Taylor, et al. 1992. Inhibitors of amino acid biosynthesis: strategies for imparting glysphosate tolerance to crop plants. In Biosynthesis and Molecular Regulation of Amino Acids in Plants. B.K. Singh, H.E. Flores, and J.C. Shannon, Eds. *American Society of Plant Physiologists*, Rockville, MD, pp. 139–145.

Beversdorf, W.D., J. Weiss-Lermann, L.R. Erikson, and V. Souza Machado. 1980. Transfer of cytoplasmically-inherited triazine resistance from birds rape to cultivated oilseed rape (*Brassica campestris* and *Brassica napus*). *Canadian Journal of Genetics and Cytology* 22:167–172.

Bradley, C., P.S. Parks, Y. Chen, and W.G.D. Fernando. 2005. First report of pathogenicity groups 3 and 4 of *Leptosphaeria maculans* on canola in North Dakota. *Plant Disease* 89:776.

Brandle, J.E., and P.B.E. McVetty. 1989. Heterosis and combining ability in hybrids derived from oilseed rape cultivars and inbred lines. *Crop Science* 29:1191–1195.

Brandt, S., and G. Clayton. 2005. Good value from certified hybrid canola seed. Manitoba canola Producers Commission. http://www.mcgacanola.org/CertifiedHybridCanolaSeedValue. html.

Canola Council of Canada. 2008. Canadian canola industry, overview of Canada's Canola industry. http://www.canola-council.org/ind_overview.aspx.

Chen, Y., and W.G.D. Fernando. 2005. First report of canola blackleg caused by pathogenicity group 4 of *Leptosphaeria maculans* in Manitoba. *Plant Disease* 89:339.

Chen, Y., and W.G.D. Fernando. 2006. Prevalence and pathogenicity groups of *Leptosphaeria maculans* in western Canada and North Dakota. *Canadian Journal of Plant Patholology* 28:533–539.

Cuthbert, R.D. 2006. Assessment of selected traits in hybrid HEAR. M.Sc. thesis, University of Manitoba, Winnipeg, Manitoba, Canada.

Downey, R.K. 1964. A selection of *Brassica campestris* L. containing no erucic acid in its seed oil. *Canadian Journal of Plant Science* 58:977–981.

Downey, R.K., and G. Röbbelen. 1989. *Brassica* species. In *Oil crops of the world*. G. Röbbelen, R.K. Downey, and A. Ashri, Eds. New York: McGraw-Hill, pp. 339–362.

Dusabenyagasani, M., and W.G.D. Fernando. 2008. Development of a SCAR marker to track canola resistance against blackleg caused by *Leptosphaeria maculans* pathogenicity group 3. *Plant Disease* 92:903–908.

Fernando, W.G.D., and Y. Chen. 2003. First report on the presence of *Leptosphaeria maculans* pathogenicity group-3, causal agent of blackleg of canola in Manitoba. *Plant Disease* 87:1268.

Fernando, W.G.D., Y. Chen, and K. Ghanbarnia. 2007. Blackleg disease in canola/rapeseed caused by *Leptosphaeria maculans*. In The Biology, Epidemiology and Management of Blackleg. Rapeseed Breeding. *Advances in Botanical Research*, Vol. 45. S.K. Gupta, Ed. San Diego, CA: Elsevier, pp. 272–300.

Frauen, M., and W. Paulmann. 1999. Breeding of hybrids varieties of winter oilseed rape based on the MSL-system. Tenth International Rapeseed Congress, Canberra, Australia, pp. 258–263.

Freyssinet, M., P. Creange, M. Renard, P.B.E. McVetty, P. Derose, and G. Freyssinet. 1996. Development of transgenic oilseed rape resistant to oxynil herbicides. In *Crop Protection: Forecasting and Chemical Control. International Rapeseed Congress*, pp. 974–976.

Freyssinet, G., B. Leroux, M. Lebrun, B. Pelissier, A. Sailland, and K. E. Pallet. 1989. Transfer of bromoxynil resistance into two crops. Brighton Crop Protection Conference—*Weeds*, pp. 1225–1234. British Crop Protection Council, Farnham, U.K.

Fu, T.D. 1981. Production and research of rapeseed in the People's Republic of China. *Eucarpia Cruciferae Newsletter* 6:6–7.

Grant, I., and W.D. Beversdorf. 1985. Heterosis and combining ability in spring planted oilseed rape. *Genome* 32:1044–1047.

Hinata, K., and N. Konno. 1979. Studies on a male sterile system having *B. campestris* nucleus and *D. muralis* cytoplasm. Breeding and some characteristics of this strain. *Japanese Journal of Breeding* 29:305–311.

Kimber, D.S., and D.I. McGregor. 1995. *Brassica* oilseeds. The species and their origin, cultivation and world production. In Brassica *Oilseeds Production and Utilization*. D.S. Kimber and D.I. McGregor, Eds. Wallingford, U.K.: CAB International.

Li, G., and C.F. Quiros. 2001. Sequence-related amplified polymorphism (SRAP), a new marker system based on a simple PCR reaction: its application to mapping and gene tagging in *Brassica*. *Theoretical and Applied Genetics* 103:455–461.

McCubbin, A., and H.G. Dickinson. 1997. Self-incompatibility. In *Pollen Biotechnology for Crop Production and Improvement*. K.R. Shivanna and V.K. Sawhney, Eds. Cambridge, U.K.: Cambridge University Press, pp. 199–217.

McVetty, P.B.E. 1997. Cytoplasmic male sterility. In *Pollen Biotechnology for Crop Production and Improvement*. K.R. Shivanna and V.K. Sawhney, Eds. Cambridge, U.K.: Cambridge University Press, pp. 155–182.

McVetty, P.B.E., W.G.D. Fernando, R. Scarth, and G. Li. 2006a. Red River 1826 Roundup Ready high erucic acid low glucosinolate summer rape. *Canadian Journal of Plant Science* 86:1179–1180.

McVetty, P.B.E., W.G.D. Fernando, R. Scarth, and G. Li. 2006b. Red River 1852 Roundup Ready high erucic acid low glucosinolate summer rape. *Canadian Journal of Plant Science* 86:1181–1182.

McVetty, P.B.E., S.R. Rimmer, and R. Scarth. 1998. Castor high erucic low glucosinolate summer rape. *Canadian Journal of Plant Science* 78:305–306.

McVetty, P.B.E., R. Scarth, S.R. Rimmer, and C.G.J. Van Den Berg. 1996a. Venus high erucic low glucosinolate summer rape. *Canadian Journal of Plant Science* 76:341–342.

McVetty, P.B.E., R. Scarth, S.R. Rimmer, and C.G.J. Van Den Berg. 1996b. Neptune high erucic low glucosinolate summer rape. *Canadian Journal of Plant Science* 76:343–344.

McVetty, P.B.E., R. Scarth, and S.R. Rimmer. 1999. MillenniUM 01 high erucic low glucosinolate summer rape. *Canadian Journal of Plant Science* 79:251–252.

McVetty, P.B.E., R. Scarth, and S.R. Rimmer. 2000a. MillenniUM 02 high erucic low glucosinolate summer rape. *Canadian Journal of Plant Science* 80:609–610.

McVetty, P.B.E., R. Scarth, and S.R. Rimmer. 2000b. MillenniUM 03 high erucic low glucosinolate summer rape. *Canadian Journal of Plant Science* 80:611–612.

McVetty, P.B.E., and C.D. Zelmer. 2007. Breeding herbicide tolerant oilseed rape cultivars. In Rapeseed Breeding. *Advances in Botanical Research*, Vol. 45. S.K. Gupta, Ed. San Diego, CA: Elsevier, pp. 234–270.

Ogura, H. 1968. Studies on a new male sterility system in Japanese radish with special reference to utilization of the sterility toward the practical raising of hybrid seed. *Memoirs of the Faculty of Agriculture*, Kagoshima University 6:39–78.

Orson, J. 1995. Weeds and their control. In Brassica *Oilseeds Production and Utilization*. D.S. Kimber and D.I. McGregor, Eds. Wallingford, U.K.: CAB International, pp. 93–109.

Palmer, C.E., and W.F. Keller. 1997. Pollen embryos. In *Pollen Biotechnology for Crop Production and Improvement*. K.R. Shivanna and V.K. Sawhney, Eds. Cambridge, U.K.: Cambridge University Press, pp. 392–422.

Pelletier, G., C. Primard, F. Vendel, P. Chetrit, P. Rouselle, and M. Renard. 1983. Intergeneric cytoplasmic hybridization in Cruciferae by protoplast fusion. *Molecular Genetics* 191:244–250.

Rahman, M. 2007. Development of molecular markers for marker assisted selection for seed quality traits in oilseed rape. Ph.D. thesis, University of Manitoba, Winnipeg, Manitoba, Canada.

Rahman, M., P.B.E. McVetty, and G. Li. 2007. Development of SRAP, SNP and multiplexed SCAR molecular markers for the major seed coat color gene in *Brassica rapa*. *Theoretical and Applied Genetics* 115:1101–1107.

Rakow, G. 1973. Selektion auf Linol- und Linolensauregehalt in Rapssamen nach mutagener Behandlung. *Zeitschrift für Pflanzenzuchtg.* 69:62–82.

Ratanayake, W.M.N., and J.K. Daun. 2004. Chemical composition of canola and rapeseed oils. In *Rapeseed and Canola Oil Production, Processing, Properties and Uses*. F.D. Gunstone, Ed. Oxford, U.K.: Blackwell, chap. 3.

Riungu, T.C., and P.B.E. McVetty. 2003. Inheritance of maintenance and restoration of the *Diplotaxis muralis* (*mur*) cytoplasmic male sterility system in summer rape. *Canadian Journal of Plant Science* 83:515–518.

Riungu, T.C., and P.B.E. McVetty. 2004. Comparison of the effect of *mur* and *nap* cytoplasms on the performance of intercultivar summer rape hybrids. *Canadian Journal of Plant Science* 84:731–738.

Sawhney, V.K. 1997. Genic male sterility. In *Pollen Biotechnology for Crop Production and Improvement*. K.R. Shivanna and V.K. Sawhney, Eds. Cambridge, U.K.: Cambridge University Press, pp. 183–188.

Scarth, R., P.B.E. McVetty, and S.R. Rimmer. 1994. Mercury high erucic low glucosinolate summer rape. *Canadian Journal of Plant Science* 74:205–207.

Scarth, R., P.B.E. McVetty, S.R. Rimmer, and J.K. Daun. 1992. Breeding for specialty oil quality in canola rapeseed: the University of Manitoba program. In *Seed Oils for the Future*. S.L. MacKenzie and D.C. Taylor, Eds. Champaign, IL: AOCS, pp. 171–176.

Scarth, R., P.B.E. McVetty, S.R. Rimmer, and B.R. Stefansson. 1987. Stellar low linolenic-high linoleic acid summer rape. *Canadian Journal of Plant Science* 68:509–511.

Scarth, R., P.B.E. McVetty, S.R. Rimmer, and B.R. Stefansson. 1991. Hero summer rape. *Canadian Journal of Plant Science* 71:865–866.

Scarth, R., P.B.E. McVetty, and S.R. Rimmer. 1997. Allons low linolenic summer rape. *Canadian Journal of Plant Science* 77:125–126.

Scarth, R., S.R. Rimmer, and P.B.E. McVetty. 1995. Apollo low linolenic summer rape. *Canadian Journal of Plant Science* 75:203–205.

Sernyk, J.L., and B.R. Stefansson. 1983. Heterosis in summer rape. *Canadian Journal of Plant Science* 63:407–413.

Stefansson, B.R., and R.K. Downey. 1995. Rapeseed. In *Harvest of Gold*. A.E. Slinkard and D.R. Knott, Eds. Saskatoon, Canada: University Extension Press, University of Saskatchewan, pp. 140–152.

Stefansson, B.R., F.W. Hougen, and R.K. Downey. 1961. Note on the isolation of rapeseed plants with seed oil essentially free from erucic acid. *Canadian Journal of Plant Science* 41:218–219.

Stefansson, B.R., and Z.P. Kondra. 1975. Tower summer rape. *Canadian Journal of Plant Science* 55:343–344.

Sun, Z., Z. Wang, J. Tu, J. Zhang, F. Yu, P.B.E. McVetty, and G. Li. 2007. An ultra dense genetic recombination map for *Brassica napus*, consisting of 13551 SRAP markers. *Theoretical and Applied Genetics* 114:1305–1317.

Swanson, E.B., M.J. Hergesell, M. Arnaldo, D.W. Sipell, and R.S.C. Wong. 1989. Microspore mutagenesis and selection—canola plants with field tolerance to the imidazolinones. *Theoretical and Applied Genomics* 78, 525–530.

Taylor, D.C., D.L. Barton, E.M. Giblin, S.L. MacKenzie, C.G.J. Van Den Berg, and P.B.E. McVetty. 1995. Microsomal lyso-phosphatidic acid acyltransferase from a *Brassica oleracea* cultivar incorporates erucic acid into the *sn*-2 position of seed triacylglycerols. *Plant Physiology* 109:409–420.

Taylor, D.C., S.L. MacKenzie, A.R. McCurdy, P.B.E. McVetty, E.M. Giblin, E.W. Pass, S.J. Stone, R. Scarth, S.R. Rimmer, and M.D. Pickard. 1994. Stereospecific analyses of seed triglycerols from high erucic acid Brassicaceae: detection of erucic acid at the *sn*-2 position in *Brassica oleracea* L. genotypes. *Journal of the American Oilseed Chemists Society* 71:163–167.

Thompson, K.F. 1972. Cytoplasmic male sterility in oilseed rape. *Heredity* 29:253–257.

Thomas, P. 1992. *Canola Growers' Manual*. Winnipeg, Canada: Canola Council of Canada.

Vincent, M.R.L. 2008. Genetic diversity and its relationship to hybrid performance in high eructic acid rapeseed. M. Sc. Theseis. University of Manitoba, Winnipeg, Manitoba, Canada.

Wang, Z. 2007. Marker development and gene identification for blackleg (*Leptosphaeria maculans*) disease resistance in canola (*Brassica napus*). Ph.D thesis, University of Manitoba, Winnipeg, Manitoba, Canada.

Williams, M.E., J. Leemans, and F. Michaels. 1997. Male sterility through recombinant DNA technology. In *Pollen Biotechnology for Crop Production and Improvement*. K.R. Shivanna and V.K. Sawhney, Eds. Cambridge, U.K.: Cambridge University Press, pp. 237–258.

5 Approaches for Acquired Tolerance to Abiotic Stress of Economically Important Crops

Wei-Chin Lin and Pi-Fang Linda Chang

CONTENTS

Key Words: Abiotic stress; acquired tolerance; antioxidant; breeding; crop plants; cross protection; functional genomics; heat shock; heat shock factors; heat shock proteins; hydrogen peroxide; oxidative stress; reactive oxygen species; signaling pathways; stress tolerance; transgenic plants.

5.1 INTRODUCTION

Crop management has advanced to a state at which production and quality can normally be ensured in a satisfactory manner. However, new challenges for food production still exist. The goal is not only to provide food for the essential needs of humans but also to meet the new challenge of ensuring food safety and security along the supply chain. Modern consumers are interested in obtaining health benefits through eating readily available foods. Recent progress in agricultural production has been made in reducing the use of pesticides and fungicides, while attempting to improve food quality and safety to human health. Phytochemicals and vitamins are just some examples of desirable food attributes.

In view of the continuous warnings about climate change worldwide, current crops that have been optimized for modern agriculture may not be able to meet the future challenge of potential drastic climate changes in agricultural production areas. To crop scientists, the ability of crops to tolerate abiotic stresses (e.g., extreme temperature, rainfall) has been a continuous challenge. To policy makers, extreme weather conditions may cause problems such as interruptions in the food supply. Such an unpredictable food supply could increase the severity of hunger in some parts of the world while creating higher food prices in other parts.

Tolerance to abiotic stress has drawn the attention of many scientific disciplines continuously attempting to improve productivity and quality in crop production. This chapter is intended to review current knowledge of abiotic stresses in the agricultural context and to provide some examples of enhancing crop tolerance to such environmental stress. This review does not cover all aspects of scientific literature but emphasizes acquired tolerance in the context of crop production and food quality.

5.2 CURRENT KNOWLEDGE

5.2.1 Common Mechanisms

5.2.1.1 Reactive Oxygen Species

All crop plants are subjected to some kind of stress during their life span, whether aerial parts or root systems. It is recognized that all stresses involve the generation of reactive oxygen species (ROS), also named active oxygen species or reactive oxygen intermediates (Mittler et al. 2004). Abiotic stresses such as drought stress and desiccation, salt stress, chilling, heat shock (HS), heavy metals, ultraviolet radiation, air pollutants, mechanical stress, nutrient deprivation, and high light stress enhance the production of superoxide radicals ($O_2^{\bullet-}$), hydrogen peroxide (H_2O_2), or hydroxyl radicals (HO•) (Mittler 2002). The removal of ROS depends on enzymatic and nonenzymatic antioxidant mechanisms. The major scavenging mechanisms in plants are superoxide dismutase (SOD), ascorbate peroxidase (APX), and catalase (CAT). Nonenzymatic antioxidants such as ascorbic acid and glutathione are crucial for plant defense against oxidative stress (Apel and Hirt 2004). In *Arabidopsis thaliana*, various environmental stresses can be separated into two groups according to induced enzyme activities: stresses (high temperature, high light, and water deficit) that induce dehydroascorbate reductase (DHAR) activity and stresses (ozone, sulfur dioxide, chilling temperature, and ultraviolet B [UV-B]) that induce APX activity (Kubo et al. 1999).

In viewing the complexity and similarity among various environmental stresses, a general adaptation syndrome response has been proposed (Leshem and Kuiper 1996). The application of this hypothesis implies that different types of stress (e.g., heat, cold, drought, salinity, anoxia, etc.) would evoke a similar or even identical stress-coping mechanism. The hypothesis therefore implies that tolerance to one type of stress can be induced by sublethal exposure of the organism to a different type of stress. Free-radical scavenging and antioxidant measures have been recognized as important aspects of the stress-coping mechanism (Smirnoff 1998).

Chilling tolerance in the aerial part of maize and cucumber can be induced by exposing seedling radicles to an aqueous solution of salicylic acid (SA) (Kang and Saltveit 2002b). Such induced tolerance to chilling was associated with increased activity of glutathione reductase (GR) and guaiacol peroxidase (GPX). Chilling stress in cucumber leaves caused increased levels of H_2O_2 due to deactivation of CAT. Chilling-enhanced activities of APX and GR suggest that these may be key antioxidant enzymes in removal of H_2O_2 (Lee and Lee 2000).

ROS can be elevated above normal concentrations due to either biotic or abiotic stress, resulting in oxidative damage at a cellular level (Apel and Hirt 2004). In long-term NaCl stress, plants developed a complex antioxidant system involving increased activity of SOD and APX to mitigate and repair damage by oxidative stress (Gomez et al. 2004). Short-term salt stress caused oxidative stress, which was repeated during recovery in pea leaves. Oxidative stress was indicated by increased activities of SOD and APX and by an increase in lipid peroxidation (Hernandez and Almansa 2002).

Drought stress also caused oxidative stress, as indicated by increase of H_2O_2 and increased antioxidant enzyme activities of SOD, GPX, and APX in *Populus* cuttings (Lei et al. 2007). APX activity was increased by drought, with or without abscisic acid (ABA), which is known to induce ROS, in young leaves of the ABA-deficient mutant *notabilis*. These data demonstrated that ROS are created in drought stress and affect antioxidant activities (Unyayar and Cekic 2005).

ROS are also generated by heat stress (Dat et al. 1998, 2000, Larkindale and Knight 2002). It has been suggested that ROS act as signaling molecules during heat stress (Suzuki and Mittler 2006).

5.2.1.2 H_2O_2 as Signaling Molecule

Both biotic and abiotic stresses result in H_2O_2 generation. H_2O_2 is removed from cells through a number of antioxidant mechanisms, both enzymatic and nonenzymatic. Both biotic and abiotic stresses can also induce nitric oxide (NO) synthesis. Both H_2O_2 and NO function as signaling molecules in plants, yet the origin of NO in plants is not fully resolved (Neill et al. 2002a); therefore, our discussion focuses on H_2O_2. The functions of ROS as signaling molecules controlling adaptation in plants have been reviewed recently (Gevec et al. 2005).

H_2O_2 has a dual role: It acts as a messenger molecule (at low concentrations), and it orchestrates programmed cell death (at high concentrations). A preexposure of

1–2 h to high light is sufficient to initiate an acclamatory process that builds a long-lasting protective mechanism against H_2O_2-induced cell death (Vandenabeele et al. 2003). A comprehensive analysis of H_2O_2-induced gene expression in transgenic CAT-deficient tobacco plants revealed the induction of protein kinases and several transcription factors, such as two WRKYs, two SCARECROW proteins, one EREBP, and a NAM-like protein during the first hour of high-light treatment. Abiotic stress-induced increase of H_2O_2 also initiates signaling responses that include enzyme activation, gene expression, programmed cell death, and cellular damage (Neill et al. 2002b). Xiang and Oliver (1999) reviewed glutathione and its central role in mitigating plant stress. Glutathione synthesis is induced by H_2O_2 (CAT-deficient plant) and exposure to cadmium. The glutathione synthesis is regulated at the transcriptional level. In addition to H_2O_2 as a signaling molecule, certain soluble sugars appear to be produced and involved in ROS balance and in response to oxidative stress in plants (Couee et al. 2006).

The determination of H_2O_2 function by application of exogenous H_2O_2 is not as straightforward. Effects of exogenous H_2O_2 depend on the rate of degradation in the cell location where the effect is expected to take place (Neill et al. 2002a). When rice seedlings were pretreated with low levels of H_2O_2, they acquired tolerance to salt and heat stress. Antioxidant enzyme activities and expressions of transcripts for stress-related genes were both increased (Uchida et al. 2002). In short, plant cells produce ROS, particularly superoxide and H_2O_2, as second messengers in signal transduction cascades in processes of mitosis, tropism, and cell death (Foyer and Noctor 2005). ROS are oxidative signaling agents by which plant cells can sense the environment and make appropriate adjustments to gene expression, metabolism, and physiology.

5.2.2 TOLERANCE TO ABIOTIC STRESS

Crops are vulnerable to various kinds of biotic and abiotic stress at given times during their life span. There are two broad aspects of acquired stress tolerance: stress-associated genes and proteins on one hand and stress-associated metabolites on the other (Vinocur and Altman 2005). The former include genes involved in signaling cascades and in transcriptional control, heat shock proteins (HSPs), late embryogenesis abundant (LEA) protein families, ion and water transport, and ROS scavenging and detoxification. The latter includes osmolytes and osmoprotectants, polyamines, carbon metabolism, and other metabolites involved in response mechanisms (e.g., apoptosis). Different environmental stresses induce the synthesis of new proteins, which possibly provide evolutionary value to the plants for enhanced survival under adverse environmental conditions (Dubey 1999). Such stress-induced proteins have been well documented under salinity stress, osmotic stress, HS, low-temperature treatment, anaerobiosis, infection with pathogens, wounding, gaseous pollutants, and UV radiation. Why study protein synthesis? The different sources of stress, their duration, and severity lead to differential expression of genetic information, including messenger RNA (mRNA) and proteins. Although these proteins are synthesized when plants are subjected to stress and can be revealed in tissues of plants adapted to stress, specific metabolic functions for most of these proteins have not been established in terms of how they confer adaptability toward stress. Joshi

(1999) reviewed genetic factors affecting abiotic stress tolerance in crop plants and suggested that molecular tools be used to revolutionize the genetic analysis of crop plants.

Recent progress has been made at the biochemical level. The changes in carbohydrate metabolism during extreme temperatures play a central role in acquired thermotolerance, chilling tolerance, or cold acclimation (Guy et al. 2008). Many of the transcripts altered by exogenous trehalose treatment were associated with ethylene (ET) and methyl jasmonate (MeJA)-signaling pathways. These pathways are associated with abiotic and biotic stresses (Bae et al. 2005). It would be highly desirable that some simple mechanism can be found and be applicable to most or all abiotic stresses in all plant species. The evidence seems to suggest otherwise. The signal transduction pathways controlling abiotic stress responses are very complex (Yamaguchi-Shinozaki and Shinozaki 2006). For example, paraquat-tolerant (*Rehmannia glutinosa*) plants displayed changes in antioxidant enzymes in response to oxidative stresses and hormones (Choi et al. 2004). The authors pointed out that the differential responses to paraquat, H_2O_2, SA, and ethephon are plant specific, implying each plant species has its own concerted mechanism to detoxify ROS. In coping with cross talk in environmental stress, a computational approach has been suggested to establish networks of transcription factors (Chen and Zhu 2004).

5.2.2.1 Chilling Tolerance

Chilling, temperatures below 10°C yet above freezing, imposes limitations on production and storage of most tropical and subtropical crops. Chilling injury is associated with oxidative stress. A chilling temperature of 4°C for 4 days in darkness caused an accumulation of H_2O_2 in aerial parts of maize seedlings (Anderson et al. 1995). Acclimation at 14°C for 3 days in darkness prevented this H_2O_2 accumulation. Catalase3 (CAT3) was elevated in acclimated seedlings and may represent the first line of defense from mitochondria-generated H_2O_2. Nine of the most prominent peroxidase isoenzymes were induced by acclimation. Three isoenzymes of GR were greatly affected by acclimation. Total SOD and APX activities were not affected by acclimation or chilling. A similar benefit from acclimation was observed in zucchini squash. Preconditioning at 15°C for 2 days prior to storage at 5°C reduced development of chilling injury in zucchini squash, and the reduction in chilling injury was associated with enhanced activities of ascorbate-related enzymes, such as ascorbate free radical reductase, APX, and DHAR. The enhanced chilling resistance in squash by temperature preconditioning may involve changes in the enzyme activities in the ascorbate antioxidant system (Wang 1996).

The roles of H_2O_2 in chilling tolerance have been demonstrated with exogenous applications. H_2O_2 spray induced chilling tolerance of mung bean, and the tolerance was associated with glutathione accumulation (Yu et al. 2002). The H_2O_2-elevated glutathione content is independent of the ABA mechanism of chilling protection (Yu et al. 2003).

Cultivars differ in chilling tolerance. Chilling tolerance in rice cultivars was closely linked to the cold stability of CAT and APX when chilling sensitive cultivar K-sen 4 was compared to tolerant cultivar Dunghan Ahali during a 7-day exposure to 5°C (Saruyama and Tanida 1995). Often, the less-tolerant cultivars benefit more from H_2O_2 application. The most chilling-sensitive tomato cultivar benefited the most from H_2O_2 and ethephon sprays (Al-Haddad and Al-Jamali 2003). In potatoes, H_2O_2 induced freezing tolerance in the less-tolerant potato cultivar Atlantic but not in the freezing-tolerant cultivar Alpha (Mora-Herrera et al. 2005).

Chilling tolerance of cucumber seedlings can be induced by an osmotic stress and a heat treatment (Kang et al. 2005), implying that a moderate stress can be used to induce subsequent tolerance to chilling. The results confirmed previous results that chilling injury was reduced in cucumber and rice seedlings by HS applied after chilling (Saltveit 2001). A hot water dip at 53°C for 3 min or a 3-day heat-conditioning treatment at 37°C with air at 90%–95% relative humidity reduced chilling injury in mandarin fruit. This may be due to the induction of CAT activity and its persistence during storage at 2°C (Sala and Lafuente 2000), suggesting CAT may be the major antioxidant involved in defense of mandarin fruits against chilling.

5.2.2.2 Salt Tolerance

Soil salinity is a major agricultural problem that significantly reduces productivity of a broad range of crops (Ehret and Plant 1999). Salt stress affects crop physiology in gas exchange and carbon allocation, cellular water relations, ionic conditions, effects on roots and rhizosphere factors, and effects on reproductive structures. Salt stress induces changes at the molecular level, as evidenced by analyses of protein synthesis and gene expression. Selection for salt tolerance has been extremely complicated because it is unlikely that any one gene is responsible for salt tolerance. Although biotechnology holds some promise for generation of salt-tolerant plants, the progress has been slow.

Similar to chilling, salt injury is associated with oxidative stress. Harinasut et al. (2003) reported that a 150-mM NaCl treatment caused increased H_2O_2, SOD activity, APX activity, GR activity, and expression of peroxidase activity, although CAT activity did not respond to salinity. The results indicated increased activities of peroxidase and the SOD/ascorbate-glutathione cycle. They suggested that the activities of these antioxidant enzymes in this cycle may be a good indicator for selecting salt-tolerant genotypes of plants. The H_2O_2-enhanced benefit to chilling tolerance is also similar to salt injury. Soaking with up to 120 μM of H_2O_2 alleviated oxidative damage caused by salt, possibly partly through expression of HSPs of 32 and 52 kDa (Wahid et al. 2007).

5.2.2.3 Drought Tolerance

Drought is the most important environmental stress in agriculture (Cattivelli et al. 2008). Similar to salt damage, drought is associated with oxidative stress. Selote and Khanna-Chopra (2006) observed that drought-acclimated wheat seedlings modulated growth by maintaining turgor potential and relative water content and were able to limit H_2O_2 accumulation and membrane damage by systematic upregulation of H_2O_2-metabolizing enzymes, especially APX, and by maintaining the ascorbate-glutathione redox pool. It is indicated that drought acclimation conferred "enhanced oxidative stress tolerance" by well-coordinated induction of antioxidant defenses at both the chloroplast and mitochondrial levels.

5.2.2.4   Heat Tolerance

Heat stress may cause devastating effects on plant growth and metabolism. Heat stress causes membrane leakage, accelerated cellular metabolism, and protein denaturation, which results in yield loss in crops (Boyer 1982, Chang and Lin 2000). Heat tolerance induced in plants, called *acquired thermotolerance*, is the ability of plants to induce tolerance to higher temperatures after they are subjected to short-term sublethal temperatures (Lin et al. 1984, Kotak et al. 2007). For example, pretreatment with heat stress to soybean (Jinn et al. 1997), pea (Srikanthbabu et al. 2002), and sunflower (Senthil-Kumar et al. 2003) seedlings can increase thermotolerance. When plants (actually all living organisms) are exposed to sublethal high temperatures, the expression of HSPs is induced, and normal protein synthesis is at least partly inhibited. Thus, HSPs play an important role in thermotolerance (Vierling 1991). There are five major groups of HSPs in all organisms: HSP100, HSP90, HSP70, HSP60, and small HSPs (sHSPs). Compared to other organisms, plants possess a large number of sHSPs, which may be related to the inability of movement of plants under stresses. Based on the sequence similarity, immunological cross-reactivity, and cellular localization, there are at least six classes of sHSPs in plants, including those in cytosol (classes I and II), plastid/chloroplasts, mitochondria, endoplasmic reticulum, and endomembrane (LaFayette et al. 1996, Waters et al. 1996). Recently, when analyzing the sHSPs in *Arabidopsis*, Scharf et al. (2001) identified other sHSP genes that indeed may define a new class (cytosolic class III). The rapid activation of HSP transcription and protein accumulation under HS suggest the strong involvement of HSPs with thermotolerance (Vierling 1991, Chang and Lin 2000). Accumulating data have shown that HSPs function as molecular chaperones to assist protein refolding, prevent protein aggregation, or resolubilize protein aggregates under heat stress (Wang et al. 2004).

In addition to heat stress, it has been shown that HSP genes were also induced in plants by various stresses (Chang and Lin 2000, Krishna and Gloor 2001, Sun et al. 2002), such as chilling (van Berkel et al. 1994, Krishna et al. 1995, Sabehat et al. 1998, Ukaji et al. 1999, Soto et al. 1999, Sung et al. 2001, Cheong et al. 2002), salt (Yabe et al. 1994, Sun et al. 2001); drought (Almoguera et al. 1993, Alamillo et al. 1995, Pla et al. 1998, Cho and Hong 2006); osmotic stress (Sun et al. 2001, Cheong et al. 2002); wounding (Cheong et al. 2002, Chang et al. 2007); alcohol (Kuo et al. 2000, Guan et al. 2004); amino acid analogs (Lee et al. 1996, Guan et al. 2004); and heavy metals such as arsenic, cadmium, and nickel (Lin et al. 1984, Edelman et al. 1988, Tseng et al. 1993, Yabe et al. 1994, Milioni and Hatzopoulos 1997, Guan et al. 2004, Gullì et al. 2005). Some HSPs are also shown to be induced by ABA (Alamillo et al. 1995, Pla et al. 1998, Cho and Hong 2006); brassinosteroid (Wilen et al. 1995); indoleacetic acid (IAA; Yabe et al. 1994); SA (Cheong et al. 2002, Chang et al. 2007); jasmonic acid (JA; Cheong et al. 2002; light, photoperiod, and light and dark transitions (Cheong et al. 2002, Krishna et al. 1992, Felsheim and Das 1992); and even by pathogen infection (Cheong et al. 2002). In addition, some HSPs accumulate under non-HS conditions during specific developmental stages of plants (Marrs et al. 1993, Krishna et al. 1995, Sung et al. 2001, Guan et al. 2004, Lin 2005, Chang et al.

in preparation). For example, the accumulation of sHSPs during seed maturation is similar to that of the LEA proteins (Wehmeyer et al. 1996).

HSPs could also be induced by oxidative stress. The rice class I sHSPs could accumulate under H_2O_2 even though the induction level was less than 20% of that induced by HS (Lin 2005, Chang et al. in preparation). In addition, UV irradiation could induce the level of HSP60 but not those of HSP90, HSC70 (HSP73), and sHSP-CI (Lin 2005, Chang et al. in preparation). Similar reports have suggested that some HSPs could be induced by ozone (Eckey-Kaltenbach et al. 1997), H_2O_2 (Banzet et al. 1998, Pla et al. 1998, Desikan et al. 2001, Guan et al. 2004), and γ-irradiation (Banzet et al. 1998). ROS may mediate adaptive responses to heat and various stresses through the induction of HSPs and other defense mechanisms (Dat et al. 2000, Larkindale et al. 2005a, 2005b, Suzuki and Mittler 2006).

5.2.3 TOLERANCE TO PATHOGENS AND INSECTS

The cross talk between resistance to insects and pathogens and abiotic stress has been reviewed (Bostock 2005) and is beyond the scope of our discussion. However, some examples are presented.

A protein phosphatase 2C gene, *OsBIPP2C1*, appears to play a role in enhanced disease resistance and abiotic tolerance (Hu et al. 2006). In peppers, the over-expressed CALTPI and CALTPIII genes may be involved in genes inducible to inhibit pathogen growth as well as to adapt to environmental stress. It implies that environmental stress (drought, high salt, low temperature, and wounding) triggered gene expression that is similar to defense-related responses (Jung et al. 2003). An interesting discovery is that both biotic and abiotic stresses induced reserveratrol synthesis in peanut leaves (Chung et al. 2003). Reserveratrol has antifungal activity in plants and is beneficial to human health in fighting atherosclerosis and carcinogenesis. It is believed that plant stress hormones are involved. Also, another case exists in which a pathogen-induced gene, encoding an RAV transcription factor, functions as a transcription activator triggering resistance to bacterial infection and tolerance to osmotic stress (Sohn et al. 2006). In systemic acquired resistance (SAR) to disease, expression of the gene *OsPI-PLC1* can be induced by various chemical and biological inducers in the plant defense pathway, including benzothiadiazole (BTH), SA, JA, and its methyl esters (MeJA). All these treatments were shown to induce resistance in rice against blast disease. It was illustrated that *OsPI-PLC1* was activated by BTH, which resulted in SAR (Song and Goodman 2002).

In postharvest, use of pesticides to reduce decay is considered highly undesirable. Hot water treatment (53°C) for 4 min prior to a storage of 28 days at 8°C alleviated chilling injury and decay in pepper fruit (Gonzalez-Aguilar et al. 2000). Polyamine levels were observed to increase immediately after hot water treatment and are believed to be involved in the protection mechanism. Similarly, heat treatment (i.e., hot water) and UV irradiation have been part of commercial practice to reduce decay, although these nonfungicidal methods are not as effective as use of fungicides (Ben-Yehoshua 2004).

The examples of abiotic stress-related resistance to insects are difficult to find, but nevertheless do exist. Proteinase inhibitors in plant tissues are associated with their resistance to insect attack. The trypsin inhibitor level of common bean seeds was influenced by year-to-year variation in abiotic stress: The increase of trypsin inhibitor was associated with drought stress during the growing season (Piergiovanni and Pignone 2003). *Amaranthus hypochondriacus* is a C4 dicotyledonous pseudocereal and contains high levels of proteinase inhibitors, trypsin inhibitor, and α-amylase inhibitor (Sanchez-Hernandez et al. 2004). Trypsin inhibitor activity was increased by exposure of plants to diverse treatments, particularly water stress.

5.3 POTENTIAL APPROACHES

5.3.1 BREEDING AND MOLECULAR TECHNIQUES

Current progress on breeding for tolerance to abiotic stress, including direct breeding, marker-assisted selection, and genomic technologies, has been reviewed (Cattivelli et al. 2008, Whitcombe et al. 2008). Intensive molecular-assisted breeding and genetic engineering have been considered to be at the forefront (Vinocur and Altman 2005).

5.3.1.1 Breeding

Traditional breeding approaches for abiotic stress tolerance have met with limited success (Tester and Bacic 2005). There are a number of contributing factors, including (1) the focus has been on yield rather than specific traits; (2) there is difficulty in breeding for tolerance traits, which include complexities introduced by genotype interacting with environment; and (3) desired traits can only be introduced from closely related species. In breeding for tolerance to abiotic stress in rice, three obvious reasons for slow progress include identification of few accessions, poorly understood genetic basis, and multiple coexisting abiotic stresses (Ali et al. 2006). In addition, breeders often evaluate new cultivars under ideal conditions, which may also contribute to the slow progress in successfully developing commercial tolerant cultivars (Yamaguchi and Blumwald 2005). Intensive breeding has resulted in a narrowed genetic base in many crops, which renders modern crop varieties more vulnerable to biotic and abiotic stresses. Similarly, most cereal crops are sensitive to a wide range of abiotic stresses, and variability of the gene pool appears to be relatively small and may provide only a few opportunities for major step changes in tolerance (Tester and Bacic 2005). Backcrossing and phenotyping were demonstrated to be powerful tools in developing promising lines of rice varieties with improved tolerances to many abiotic stresses.

5.3.1.2 Marker-Assisted Selection

Of potentially larger impact is the use of genetic manipulation technologies (Cushman and Bohnert 2000). More achievable increases in tolerance may be introgressed into commercial lines from tolerant landraces using marker-assisted breeding approaches. The use of quantitative trait loci (QTL) in breeding for drought resistance has been reviewed (Cattivelli et al. 2008). The application of QTL mapping is one approach to dissecting the complex issues in plant stress tolerance (Vinocur and Altman 2005). The advent of molecular markers enables variation of a complex trait to be dissected into the effects of QTL and assists the transfer of these QTL into desired cultivars or lines. QTL associated with abiotic stress have been identified in rice (salt stress), cotton (drought stress), and *Salix* (cold stress). Similarly, an alternative oxidase gene was used as a marker of genetic variation in cell reprogramming and yield stability (Arnholdt-Schmitt et al. 2006). The direct introduction of a small number of genes by genetic engineering, however, seems to be a more attractive and rapid approach to improve stress tolerance.

5.3.1.3 Transgenic Plants

In 1997, tissue culture and gene regulation were used to try to enhance salt tolerance in crop plants (Winicov and Bastola 1997, Winicov 1998). Improved salinity tolerance has been considered a quantitative trait and has been resistant to improvements by plant breeding. The strategy was to combine single-step selection of salt-tolerant cells in culture, regeneration of salt-tolerant plants, and identification of genes important in conferring salt tolerance. The hypothesis was that enhanced expression of genes for physiological systems that become limiting under salt stress or a boost in physiological systems that have evolved to protect plants from acute stress is likely to provide increased salt tolerance to the cells.

Improving plant resistance to environmental stress has also been attempted by making transgenic plants (Cattivelli et al. 2008). This approach faces two issues (Smirnoff 1998). First, lack of proper physiological analysis of transgenic plants: The transgenic plants will not yield their full potential until they are analyzed in collaboration with physiologists. Second, one has to consider whether the crop performance due to genetic manipulation and testing in the laboratory (e.g., we let plants wilt and beyond in the growth chambers) will match the results under field conditions (e.g., the crop hardly wilts at all in the drought-prone fields). The progress in actual production of transgenic plants with improved abiotic stress tolerance has been slow (Tester and Bacic 2005).

Transgenic plants have been developed either to upregulate the general stress response or to reproduce specific metabolic or physiological processes previously reported to be related to drought tolerance by classical physiological studies (Cattivelli et al. 2008). Transcription factors as well as components of the signal transduction pathways that coordinate expression of downstream regulons are thought to be ideal targets for engineering complex traits of stress tolerance. The rice *Osmyb4* gene, encoding for an MYB transcription factor, is expressed at low levels in rice under normal conditions and high levels when at 4°C (Vannini et al. 2007). However, when it was overexpressed in tomato, the acquired tolerance was limited to drought and not seen in chilling. The data indicated that the specificity and the degree of activity depend on the host plant (tomato) genomic background. Metabolic engineering has been used to increase proline, trehalose, and mannitol for increased drought tolerance (Cattivelli et al. 2008). Genetic manipulation of biochemical processes has also been demonstrated by increased tolerance to the abiotic stresses of drought, salt, and oxidative stress in tomato plants through trehalose biosynthesis (Cortina and Culianez-Macia 2005).

With the advance of new technologies in proteomics, there is a good potential to apply proteomics data to practical crop breeding (Salekdeh and Komatsu 2007). A functional genomic approach can lead to identification of stress-responsive genes, understanding of stress tolerance mechanisms (regulator genes), and genomic-assisted breeding (molecular markers; Screenivasulu et al. 2007). Designing and developing transgenic plants with tolerance to combined stresses under field conditions is one of the future directions. To effectively screen genetically modified lines, several methods have been routinely used to identify and better understand phenotypes associated with abiotic stress resistance (Verslues et al. 2006). Among them are electrolyte leakage and assays for accumulation of ROS. A comprehensive and careful evaluation of mapping populations and transgenic plants is needed to provide reliable information on the effectiveness of QTL candidate genes and transgenes.

At first, the general approach of employing genomics aimed to dissect single genes in many pathways to determine the gene's place in the stress response cascade and to gauge its contribution to tolerance acquisition (Bohnert et al. 2006). Much of this gene-by-gene approach has been guided by physiological and biochemical knowledge. A second approach has emerged that emphasizes integrated analysis of stress-dependent behavior by the entire plant. This functional genomics strategy is beginning to become a bridge to whole-plant physiology, agronomy, and crop breeding. In a review by Tester and Bacic (2005), the possible routes of breeding abiotic stress tolerance in grasses have been discussed, including use of genetic manipulation technologies from model plants (e.g., *Arabidopsis*) to crop plants. The importance of evaluation of transgenic plants under stress conditions and understanding the physiological effect of the inserted genes at the whole-plant level remains a major challenge to overcome (Zhang et al. 2000, Bhatnagar-Mathur et al. 2008).

In illustrating the applications of genomics in breeding for abiotic stress tolerance, two examples are given: one with a specific crop, and the other with a specific group of genes. First, in dealing with abiotic stress tolerance of tomato, breeding and molecular approaches were suggested (Foolad 2005). Traditional breeding for stress tolerance (i.e., salt, cold, heat) has progressed, with available resistance genes identified within the same species or related species in the same genus. In tomato, transgenic approaches have been successful in improving salt tolerance, but actual efforts to further develop new varieties for enhanced salt tolerance have been rather limited. Breeding for abiotic stress resistance may involve combining different, apparently unrelated, approaches for including several tolerance mechanisms into a specific crop (Blumwald et al. 2004).

Second, we focus on some examples of transgenic plants showing stress tolerance involving HSPs and their regulating heat shock factors (HSFs). Of course, stress tolerance can also be achieved by overexpression of other gene candidates, which has been reviewed in a book by Jenks and Hasegawa (2005). Overexpression of HSPs or HSFs could increase tolerance of plants to different abiotic stresses. Overexpression of the HSP70 in *Arabidopsis thaliana* is correlated to acquired thermotolerance (Lee and Schöffl 1996). It was demonstrated that the overexpression of plant sHSP in *Escherichia coli* (Yeh et al. 1995, 1997, Soto et al. 1999), *Daucus carota* (Malik et al. 1999), and tobacco (Sanmiya et al. 2004) can enhance tolerance against heat.

Overproduction of the HSF1 (Lee et al. 1995) or HSF3 (Prändl et al. 1998) increased basal thermotolerance in transgenic plants. Enhanced thermotolerance was observed in transgenic soybean plants overexpressing *GmHsfA1*, which activated *GmHsp70*

(Zhu et al. 2006). Overexpression of Athsp101 enhanced thermotolerance in basmati rice (*Oryza sativa* L.) (Katiyar-Agarwal et al. 2003). Overexpression of a plant sHSP in *E. coli* also showed cold tolerance to 4°C for 10 days (Soto et al. 1999).

Overexpression of *Oshsp17.7* (CI) had significantly greater resistance to heat (50°C for 2 h) and UV-B (302 nm, 3,000 mJ cm^{-1}) stresses than untransformed control plants. The level of increased thermotolerance and resistance to UV-B correlated with the level of increased Oshsp17.7 protein in the transgenic rice plants (Murakami et al. 2004). The recombinant Oshsp18.0-II protein, a class II sHSP of rice, conferred both thermotolerance and UV tolerance in *E. coli* (Chang et al. in preparation).

In AtHsfA2-overexpressing transgenic plants, AtHsfA2 enhanced tolerance to heat and oxidative stress in *Arabidopsis* (Li et al. 2005). Overexpression of a chloroplast-localized HSP21 enhanced thermotolerance in transgenic *Arabidopsis* under high-light conditions (Härndahl et al. 1999). At-HSP17.6A (a class II sHSP) could enhance osmotolerance, including withholding water for 17 days and 75 mM NaCl treatment for 3 weeks, upon overexpression (Sun et al. 2001). Overexpression of the *Arabidopsis* HsfA2 gene to a high level not only conferred thermotolerance and salt/osmotic stress tolerance but also enhanced callus growth. The authors suggested that HsfA2 was involved in multiple-stress tolerance. In addition to stress tolerance, HsfA2 also plays an important role in cell proliferation (Ogawa et al. 2007).

In transgenic tobacco, overexpression of a chloroplast-localized sHSP of sweet pepper (CaHSP26) alleviated photoinhibition of PSII and PSI during chilling stress under low irradiance (Guo et al. 2007). Transgenic tobacco plants with elevated levels of tobacco NtHSP70-1 contributed to drought tolerance (Cho and Hong 2006). Enhanced tolerance to drought stress (withholding water for 6 days or 30% polyethylene glycol [PEG] treatment for 3 days) was reported by Sato and Yokoya (2008) in transgenic rice plants overexpressing an sHSP17.7 (CI) of rice. Overexpression of a halotolerant cyanobacterium (*Aphanothece halophytice*) DnaK (a HSP70) in transgenic tobacco increased tolerance to salt (Sugino et al. 1999) and heat stress (Ono et al. 2001). *E. coli* transformant expressing a TCP-1α (CCTα) homologue of a mangrove (*Bruguiera sexangula*) plant displayed enhanced salt- and osmo-tolerance (Yamada et al. 2002).

Increased accumulation of HSP70 in transgenic *Arabidopsis* could alleviate heat stress (Sung and Guy 2003) and endoplasmic reticulum (ER) stress caused by tunicamycin treatment (Leborgne-Castel et al. 1999). Similarly, overexpressing LeHSP21.5 (an ER sHSP) in transgenic tomato plants greatly alleviated the tunicamycin-induced ER stress (Zhao et al. 2007). Enhanced accumulation of BiP (a HSP70) in transgenic tobacco protoplast increased the tolerance to drought stress caused by tunicamycin and may prevent endogenous oxidative stress (Alvim et al. 2001). Net photosynthesis was protected from heavy metal (nickel and lead) stress by increases of chloroplast sHSPs in a heat-tolerant *Agrostis stolonifera* line expressing additional chloroplast sHSPs in comparison with a near-isogenic heat-sensitive line (Heckathorn et al. 2004). A correlation of AtHSP17.4 reduction and intolerance to desiccation in mutant *Arabidopsis* seeds suggested a protective role of the sHSP in seed development (Wehmeyer and Vierling 2000).

5.3.2 SHORT-TERM TREATMENTS

Many agricultural applications involve acquired tolerance. It has been shown that exposing plants to one stress can induce tolerance to other stresses, so-called cross tolerance (Nover 1991, Sanchez et al. 1992). In agricultural practice, a previous condition often offers some relief of subsequent stress. It is possible to expose pepper plants to high day temperatures to prevent adverse low night temperature injury (Pressman et al. 2006). This has implications for reducing fuel consumption. In greenhouse-grown cucumber fruit, elevated growing temperatures during the day improved chilling tolerance of fruit and enhanced the activity of antioxidant enzymes SOD and CAT (Kang et al. 2002). It is likely that high day/night temperature (30°C/22°C) results in high phenolics and high antioxidant capacities, which are associated with chilling tolerance (Wang and Zheng 2001). Also, a precondition can offer tolerance to subsequent chilling. During cold storage (6 weeks at 2°C), hybrid Fortune mandarins developed chilling injury unless the fruits were preconditioned at 37°C for 3 days, illustrating heat-induced chilling tolerance (Sala and Lafuente 1999).

Similarly, heating tomato fruit (*Lycoperiscon esculentum*) at 38°C for 48 h prevented the development of chilling injury after 21 days at 2°C compared to the highly injured unheated fruit (Sabehat et al. 1996). Heat-induced chilling tolerance in tomato fruits was later confirmed by Adnan et al. (1998). Preconditioning of rice (*Oryza sativa* L.) seedlings at 42°C for 24 h could prevent the development of chilling injury at 5°C for 7 days. The longer the period of heat treatment, the higher the level of chilling tolerance induced (Sato et al. 2001). A 3-min HS at 45°C reduced growth inhibition of rice seedling radicles caused by chilling at 5°C for 2 days; thus, chilling tolerance was induced by HS (Kang and Saltveit 2002a). The longer the HS treatment was given in this case, however, the lower was the level of maximum chilling tolerance induced (Saltveit 2002). In cucumber seedlings, 45°C HS for 10 min induced protection to chilling at 2.5°C for 4 days. The growth inhibition of radicles caused by chilling was reduced from 60% to 42% (Kang and Saltveit 2001).

There are other preconditioning techniques in addition to heat that could lead to chilling tolerance. Irrigation with saline water (0.25% NaCl w/v) improves carotenoid content and antioxidant activity of tomato (De Pascale et al. 2001) and may lead to enhanced chilling tolerance. In sweet potato cultivars, enhanced coloration as a result of soil condition and water availability also led to high antioxidant potential (Philpott et al. 2003).

Heat can also lead to other types of tolerance besides chilling. A significant increase in drought tolerance was found if rice seedlings (*Oryza sativa* L.) were preconditioned at 42°C for 24 h (Sato and Yokoya 2008). HS at 40°C for 2 h prior to cadmium treatment provided significant cross protection of *Vigna mungo* (black gram, mung bean) seedlings to Cd^{2+} damage for up to 48 h (Kochhar and Kochhar 2005). A significant increase in UV-B tolerance was found in rice (*Oryza sativa* L.) seedlings preexposed to a high temperature (42°C) for 24 h. The level of UV-B resistance correlated well with the duration of heat treatment (Murakami et al. 2004). After HS (45°C for 30 min) or mild H_2O_2 (2 mM) pretreatments, tomato suspension cells could be protected from lethal oxidative stress caused by 10 mM of H_2O_2 (Banzet et al.

1998). In rice seedlings, heat pretreatment also increased tolerance to subsequent H_2O_2 (oxidative stress) (Lin 2005, Chang et al. in preparation). In postharvest of perishable horticultural crops, ozone technology has been used to reduce ET in the cold room and for shelf life extension of fruits (apples, pears) and vegetables (broccoli) (Skog and Chu 2001). It was reported that ozone enhances the effectiveness of negative air ions in reducing storage decay (Forney et al. 2001).

It is likely that H_2O_2 plays a pivotal role in cross tolerance (Bowler and Fluhr 2000). In addition to ROS, calcium, ABA, SA, and ET may be involved in the signaling network associated with stress tolerance (Desikan et al. 2001, Larkindale et al. 2005a, 2005b, Suzuki and Mittler 2006). These signaling molecules may have practical roles to play in agriculture. Imbibition of bean and tomato seeds with aqueous solutions of SA or acetyl SA induces multiple stress tolerance to heat, chilling, and drought (Senaratna et al. 2000). Other data suggested that SA induces abiotic stress tolerance through signaling (Horvath et al. 2007). A synergistic effect of low selenium and UV irradiation in the enhanced growth of rye grass and lettuce was at least partially associated with increased activities of the antioxidant enzymes glutathione peroxidase and CAT (Xue and Hartikainen 2000).

The concept of abiotic stress and the associated antioxidant mechanism has been applied to agricultural practices. A novel chemical cocktail called *Alethea*, used to resist fungal infection of frosty pod rot causing low yield of cocoa bean plants, has been developed for commercial application (O'Driscoll 2007). The product is based on the concept of abiotic stress-associated antioxidants. In considering employment of a stress as preconditioning, the duration of such preconditioning has to be determined. It has been recognized that there are differences in physiological and molecular responses during short and long exposures to stress. Also, the evaluation of results must involve practical factors such as biomass, yield data, and survival (Vinocur and Altman 2005).

5.3.3 New Perspectives

In a review by Grierson (1999), the beneficial effects of plant stress were presented, and the dual nature of plant stress as both good and bad was voiced. Most research on stress focuses on ill effects, and potential beneficial aspects are often ignored. Most funding agencies have not often supported this development of nonchemical methods for modern agriculture, for example, mitigation of chilling injury syndrome achieved by heat treatment. A judiciously applied stress could be a very valuable tool in mitigating chilling injury. Even within a given species or cultivar, chilling susceptibility can vary considerably with the growing district. The protective effect of Karoo Desert treatment (samples inside a metal shed for 2 days; temperatures over 32°C at noon and over 40°C late afternoon) extended successful storage at 4.5°C from 10 to 90 days when applied to summer squash and *Opuntia* "cactus pears." Stress is not only something to be combated but also one of nature's essential tools.

Water stress may upregulate the phenyl propanoid synthesis pathway and hence enhance the level of anthocyanin and hydroxycinnamic acids contributing to elevated levels of antioxidants in sweet potato roots (Philpott et al. 2003). Drought and cold stresses elevated (–)-epicatechin and some flavonoid levels in hawthorn leaves and thereby increased their antioxidant capacity (Kirakosyan et al. 2003). Cold stress caused increases in levels of vitexin-2″-O-rhamnoside, acetylvitexin-2″-O-rhamnoside, hyperoside, and quercetin in two *Crataegus* species (Kirakosyan et al. 2004). A serine proteinase inhibitor with trypsin-inhibitory activity, designated SPLTI (sweet potato leaf trypsin inhibitor), was partially purified from sweet potato leaves under water deficit (Wang et al. 2003). SPLTI genes were upregulated by water deficiency and chilling as well as osmoticant treatment in the PI-I (proteinase inhibitor I) family in plants. Data suggested that SPLTI could participate in defense systems against invasions of insects or bacteria as do other PI-Is.

The challenges and opportunities of breeding for stress tolerance to abiotic stress have been reviewed by Blumwald et al. (2004). They include new promoters for gene expression, expanding basic plant systems (*Arabidopsis*, tobacco, and rice) into crop plants, the expression of a transgene in a specific genetic environment, and combining genes (similar to gene stacking). Su and Wu (2004) have reported using a "stress-inducible promoter" that was activated under stress conditions and not only showed increased tolerance to salt or water stress but also showed higher biomass production in third-generation transgenic rice seedlings as compared to plants with a constitutive promoter. A combination of stresses may be the focus in the future (Mittler 2006). The recent advancement in genomics gained through use of a model plant system is ready to be transferred into other important crop plants (Food and Agriculture Organization of the United Nations 2003).

In addition to investigation into stress independently, the interaction between biotic and abiotic stress has begun to draw some attention. Recent evidence suggests several molecules, including transcriptional factors and kinases, as promising candidates for common players that are involved in cross talk between stress-signaling pathways (Fujita et al. 2006). In fact, two subclasses of transcription factors, DREB1/CBF and DREB2, have been identified that are induced by cold and dehydration, respectively. Not only do they have the benefit of controlling more than one gene, which can lead to sustained abiotic stress tolerance, they are also involved in biotic stress-signaling pathways (Agarwal et al. 2006). The transcription factors probably regulate various stress-inducible genes either cooperatively or separately (Seki et al. 2003). Hormone (ABA, SA, JA, ET) and ROS signaling pathways seem to play key roles in such cross talk between biotic and abiotic signaling.

It appears that breeding for stress-tolerant crops is on the rise. A few examples are given here for the issuing of patents in this area. First, the use of overexpression of a molybdenum cofactor sulfurase in plants was patented by the University of Arizona (Zhu and Xiong 2006, U.S. Patent 7038111, May 2, 2006) claiming increased tolerance to drought, freezing, and salt. Another example is the patent granted to Pioneer Hi-Bred International (Johnston, IA) specifically for wheat variety 26R87 (Edge et al. 2008, U.S. Patent 20080072346, March 20, 2008) claiming resistance to herbicide, insects, disease, and water stress. The methods included transforming wheat plants with a transgene encoding mannitol-1-phosphate dehydrogenase or LEA proteins (Edge et al. 2008). Evogene of Israel has identified key genes for improving plant yield and tolerance to abiotic stress; these genes are currently being introduced to field crops such as rice, soybean, corn, canola, and cotton through collaboration with other companies (Evogene 2007).

5.4 CONCLUSIONS

Abiotic stress exists in conjunction with the growth and development of crop plants and is unavoidable. When such stresses are perceived by plants, signaling transduction follows. Plants generate metabolic changes that can lead to stress tolerance. Abiotic stress has been associated with oxidative stress, which is conventionally viewed as causing ill effects. Recently, oxidative stress has been viewed in the holistic context of signaling pathways and in turn may lead to stress tolerance. Modern biotechnologies enhance the understanding of stress response mechanisms at the molecular level and show promise to lead to the identification of tolerance genes that can be applied to breeding. In practical agricultural practice, various stresses have been applied to improve crop quality, to enhance defense systems to insects and pathogens, and to improve tolerance to subsequent stresses. The most encouraging application is that some aspects of knowledge gained through molecular biology have been utilized in breeding stress-tolerant lines of major crops while maintaining or even increasing yields, an idea that was previously considered impossible.

ACKNOWLEDGMENTS

We thank Glenn S. Block, Yi-Hung Lin, and Ying-Ru Chen for assistance in preparation of the manuscript. The research of Pi-Fang Linda Chang is supported by grants from the National Science Council, Council of Agriculture, and National Chung Hsing University of Taiwan.

REFERENCES

Adnan, S., L. Susan, and D. Weiss. 1998. Isolation and characterization of a heat-induced gene, hcit2, encoding a novel 16.5 kDa protein: expression coincides with heat-induced tolerance to chilling stress. *Plant Mol. Biol.* 36: 935–939.

Agarwal, P.K., P. Agarwal, M.K. Reddy, and S.K. Sopory. 2006. Role of DREB transcription factors in abiotic and biotic stress tolerance in plants. *Plant Cell Rep.* 25: 1263–1274.

Alamillo, J., C. Almoguera, D. Bartels, and J. Jordano. 1995. Constitutive expression of small heat shock proteins in vegetative tissues of the resurrection plant *Craterostigma plantagineum. Plant Mol. Biol.* 29: 1093–1099.

Al-Haddad, J.M., and A.F. Al-Jamali. 2003. Effect of H_2O_2 and ethephon spray on seedling chilling tolerance in three tomato cultivars. *Food Agric. Environ.* 1: 219–221.

Ali, A.J., J.L. Xu, A.M. Ismail, et al. 2006. Hidden diversity for abiotic and biotic stress tolerances in the primary gene pool of rice revealed by a large backcross breeding program. *Field Crops Res.* 97: 66–76.

Almoguera, C., M.A. Coca, and J. Jordano. 1993. Tissue-specific expression of sunflower heat shock proteins in response to water stress. *Plant J.* 4: 947–958.

Alvim, F.C., S.M.B. Carolino, J.C.M. Cascardo, et al. 2001. Enhanced accumulation of BiP in transgenic plants confers tolerance to water stress. *Plant Physiol.* 126: 1042–1054.

Anderson, M.D., T.K. Prasad, and C.R. Stewart. 1995. Changes in isozyme profiles of catalase, peroxidase, and glutathione reductase during acclimation to chilling in mesocotyles of maize seedlings. *Plant Physiol.* 109: 1247–1257.

Apel, K., and H. Hirt. 2004. Reactive oxygen species: metabolism, oxidative stress, and signal transduction. *Annu. Rev. Plant Biol.* 55: 373–399.

Arnholdt-Schmitt, B., J.H. Costa, and R.F. de Melo. 2006. AOX—A functional marker for efficient cell reprogramming under stress? *Trends Plant Sci.* 11: 281–287.

Bae, H., E. Herman, B. Bailey, H.J. Bae, and R. Sicher. 2005. Exogenous trehalose alters *Arabidopsis* transcripts involved in cell wall modification, abiotic stress, nitrogen metabolism, and plant defense. *Physiol. Plant.* 125: 114–126.

Banzet, N., C. Richaud, Y. Deveaux, M. Kazmaier, J. Gagnon, and C. Traintaphylides. 1998. Accumulation of small heat shock proteins, including mitochondrial HSP22, induced by oxidative stress and adaptive response in tomato cells. *Plant J.* 13: 519–527.

Ben-Yehoshua, S. 2004. Effects of postharvest heat and UV applications on decay, chilling injury and resistance against pathogens of citrus and other fruits and vegetables. *Acta Hort.* 599: 159–173.

Bhatnagar-Mathur, P., V. Vadez, and K.K. Sharma. 2008. Transgenic approaches for abiotic stress tolerance in plants: retrospect and prospects. *Plant Cell Rep.* 27: 411–424.

Blumwald, E., A. Grover, and A.G. Good. 2004. Breeding for abiotic stress resistance: challenges and opportunities. *Proceedings of the 4th International Crop Science Congress,* 26 Sep–1 Oct 2004, Brisbane, Australia. CD-ROM. http://www.cropscience.org.au.

Bohnert, H.J., Q. Gong, P. Li, and S. Ma. 2006. Unraveling abiotic stress tolerance mechanisms—getting genomics going. *Curr. Opin. Plant Biol.* 9: 180–188.

Bostock, R.M. 2005. Signal crosstalk and induced resistance: straddling the line between cost and benefit. *Annu. Rev. Phytopathol.* 43: 545–580.

Bowler, C., and R. Fluhr. 2000. The role of calcium and activated oxygens as signals for controlling cross-tolerance. *Trends Plant Sci.* 5: 241–245.

Boyer, J.S. 1982. Plant productivity and environment. *Science* 218: 443–448.

Cattivelli, L., F. Rizza, F.W. Badeck, et al. 2008. Drought tolerance improvement in crop plants: an integrated view from breeding to genomics. *Field Crops Res.* 105: 1–14.

Chang, P.F.L., and C.Y. Lin. 2000. The discovery of the heat shock response in plants. In *Discoveries in Plant Biology,* Vol. 3, ed. S.D. Kung and S.F. Yang, pp. 347–370. Singapore: World Scientific.

Chang, P.F.L., T.L. Jinn, W.K. Huang, Y. Chen, H.M. Chang, and C.W. Wang. 2007. Induction of a cDNA clone from rice encoding a class II small heat shock protein by heat stress, mechanical injury, and salicylic acid. *Plant Sci.* 172: 64–75.

Chen, W.J., and T. Zhu. 2004. Networks of transcription factors with roles in environmental stress response. *Trends Plant Sci.* 9: 592–596.

Cheong, Y.H., H.S. Chang, R. Gupta, X. Wang, T. Zhu, and S. Luan. 2002. Transcriptional profiling reveals novel interactions between wounding, pathogen, abiotic stress, and hormonal responses in *Arabidopsis*. *Plant Physiol.* 129: 661–677.

Cho, E.K., and C.B. Hong. 2006. Over-expression of tobacco NtHSP70-1 contributes to drought-stress tolerance in plants. *Plant Cell Rep.* 25: 349–358.

Choi, D.G., N.H. Yoo, C.Y. Yu, B. de los Reyes, and S.J. Yun. 2004. The activities of antioxidant enzymes in response to oxidative stresses and hormones in paraquat-tolerant *Rehmannia glutinosa* plants. *J. Biochem. Mol. Biol.* 37: 618–624.

Chung, I.M., M.R. Park, J.C. Chun, and S.J. Yun. 2003. Reserveratrol accumulation and reserveratrol synthase gene expression in response to abiotic stress and hormones in peanut plants. *Plant Sci.* 164: 103–109.

Cortina, C., and F.A. Culianez-Macia. 2005. Tomato abiotic stress enhanced tolerance by trehalose biosynthesis. *Plant Sci.* 169: 75–82.

Couee, I., C. Sulmon, G. Gouesbet, and A. El Amrani. 2006. Involvement of soluble sugars in reactive oxygen species balance and responses to oxidative stress in plants. *J. Exp. Bot.* 57: 449–459.

Cushman, J.C., and H.J. Bohnert. 2000. Genomic approaches to plant stress tolerance. *Curr. Opin. Plant Biol.* 3: 117–124.

Dat, J.F., H. Lopez-Delgado, C.H. Foyer, and I.M. Scott. 1998. Parallel changes in H_2O_2 and catalase during thermotolerance induced by salicylic acid or heat acclimation in mustard seedlings. *Plant Physiol.* 116: 1351–1357.

Dat, J., S. Vandenbeele, E. Vranova, M. van Montagu, D. Inze, and F. van Breusegm. 2000. Dual action of the active oxygen species during plant stress responses. *Cell Mol. Life Sci.* 57: 779–795.

De Pascale, S., A. Maggio, V. Fogliano, P. Ambrosino, and A. Ritieni. 2001. Irrigation with saline water improves carotenoids content and antioxidant activity of tomato. *J. Hort. Sci. Biotechnol.* 76: 447–453.

Desikan, R., S.A.H. Mackerness, J.T. Hancock, and S.J. Neill. 2001. Regulation of the *Arabidopsis* transcriptome by oxidative stress. *Plant Physiol.* 127: 159–172.

Dubey, R.S. 1999. Protein synthesis by plants under stressful conditions. In *Handbook of Plant and Crop Stress*, ed. M. Pessarakli, pp. 365–397. New York: Dekker.

Eckey-Kaltenbach, H., E. Kiefer, E. Grosskopf, D. Ernst, and H. Sandermann, Jr. 1997. Differential transcript induction of parsley pathogenesis-related proteins and of a small heat shock protein by ozone and heat shock. *Plant Mol. Biol.* 33: 343–350.

Edelman, L., E. Czarnecka, and J.L. Key. 1988. Induction and accumulation of heat shock-specific poly (A+) RNAs and proteins in soybean seedlings during arsenite and cadmium treatments. *Plant Physiol.* 86: 1048–1056.

Edge, B.E., G.C. Marshall, W.J. Laskar, and K.J. Lively. 2008. Wheat variety 26R87. U.S. Patent 20070072346. March 20. http://www.freepatentsonline.com/20080072346.html ?highlight=wheat,26r87,r26&stemming=on (accessed April 4, 2008).

Ehret, D.L., and A.L. Plant. 1999. Salt tolerance in crop plants. In *Environmental Stress in Crop Plants*, ed. G.S. Dhaliwal and R. Arora, pp. 69–120. New Delhi: Commonwealth.

Evogene. 2007. Improved abiotic stress tolerance and yield. http://www.evogene.com/project. asp?project_id=4.

Felsheim, R.F., and A. Das. 1992. Structure and expression of a heat-shock protein 83 gene of *Pharbitis nil*. *Plant Physiol.* 100: 1764–1771.

Food and Agriculture Organization of the United Nations. 2003. Applications of molecular biology and genomics to genetic enhancement of crop tolerance to abiotic stress. A discussion document. http://www.ScienceCouncil.cgiar.org/publications/Pdf/GENOMICSGC.pdf.

Foolad, M.R. 2005. Breeding for abiotic stress tolerance in tomato. In *Abiotic Stress: Plant Resistance Through Breeding and Molecular Approaches*, ed. M. Ashraf and P.J.C. Harris, pp. 619–667. New York: Food Products Press.

Forney, C.F., L. Fan, P.D. Hildebrand, and J. Song. 2001. Do negative air ions reduce decay of fresh fruits and vegetables? *Acta Hort.* 553: 421–424.

Foyer, C.H., and G. Noctor. 2005. Oxidant and antioxidant signaling in plants: a re-evaluation of the concept of oxidative stress in a physiological context. *Plant Cell Environ.* 28: 1056–1071.

Fujita, M., Y. Fujita, Y. Noutoshi, et al. 2006. Crosstalk between abiotic and biotic stress responses: a current view from the points of convergence in the stress signaling networks. *Curr. Opin. Plant Biol.* 9: 436–442.

Gevec, T., I. Gadjev, S. Dukiandjiev, and I. Minkov. 2005. Reactive oxygen species as signalling molecules controlling stress adaptation in plants. In *Handbook of Photosynthesis*, ed. M. Pessarakli, pp. 209–220. Boca Raton, FL: CRC Press.

Gomez, J.M., A. Jimenez, E. Olmos, and F. Sevilla. 2004. Location and effects of long-term NaCl stress on superoxide dismutase and ascorbate peroxidase isoenzymes of pea (*Pisum sativum* cv. Puget) chloroplasts. *J. Exp. Bot.* 55: 119–130.

Gonzalez-Aguilar, G.A., L. Gayosso, R. Cruz, J. Fortiz, R. Baez, and C.Y. Wang. 2000. Polyamines induced by hot water treatments reduce chilling injury and decay in pepper fruit. *Postharvest Biol. Technol.* 18: 19–26.

Grierson, W. 1999. Beneficial aspects of stress on plants. In *Handbook of Plant and Crop Stress*, ed. M. Pessarakli, pp. 1185–1198. New York: Dekker.

Guan, J.C., T.L. Jinn, C.H. Yeh, S.P. Feng, Y.M. Chen, and C.Y. Lin. 2004. Characterization of the genomic structures and selective expression profiles of nine class I small heat shock protein genes clustered on two chromosomes in rice (*Oryza sativa* L.). *Plant Mol. Biol.* 56: 795–809.

Gullì, M., P. Rampino, E. Lupotto, N. Marmiroli, and C. Perrotta. 2005. The effect of heat stress and cadmium ions on the expression of a small *hsp* gene in barley and maize. *J. Cereal Sci.* 42: 25–31.

Guo, S.J., H.Y. Zhou, X.S. Zhang, X.G. Li, and Q.W. Meng. 2007. Overexpression of *CaHSP26* in transgenic tobacco alleviates photoinhibition of PSII and PSI during chilling stress under low irradiance. *J. Plant Physiol.* 164: 126–136.

Guy, C., F. Kaplan, J. Kopka, J. Selbig, and D. K. Hincha. 2008. Metabolomics of temperature stress. *Physiol. Plant.* 132: 220–235.

Harinasut, P., D. Poonsopa, K. Roengmongkol, and R. Charoensataporn. 2003. Salinity effects on antioxidant enzymes in mulberry cultivar. *Sci. Asia* 29: 109–113.

Härndahl, U., R.B. Hall, K.W. Osteryoung, E. Vierling, J.F. Bornman, and C. Sundby. 1999. The chloroplast small heat shock protein undergoes oxidation-dependent conformational changes and may protect plants from oxidative stress. *Cell Stress Chaperones* 4: 129–138.

Heckathorn, S.A., J.K. Mueller, S. LaGuidice, et al. 2004. Chloroplast small heat-shock proteins protect photosynthesis during heavy metal stress. *Am. J. Bot.* 91: 1312–1318.

Hernandez, J.A., and M.S. Almansa. 2002. Short-term effects of salt stress on antioxidant systems and leaf water relations of pea leaves. *Physiol. Plant.* 115: 251–257.

Horvath, E., G. Szalai, and T. Janda. 2007. Induction of abiotic stress tolerance by salicylic acid signaling. *J. Plant Growth Regul.* 26: 290–300.

Hu, X., F. Song, and Z. Zheng. 2006. Molecular characterization and expression analysis of a rice protein phosphatase 2C gene, *OsBIPP2C1*, and overexpression in transgenic tobacco conferred enhanced disease resistance and abiotic tolerance. *Physiol. Plant.* 127: 225–236.

Jenks, M.A., and P.M. Hasegawa. 2005. *Plant Abiotic Stress.* Oxford, U.K.: Blackwell.

Jinn, T.L., P.F.L. Chang, Y.M. Chen, J.L. Key, and C.Y. Lin. 1997. Tissue type-specific heat shock response and immunolocalization of class I low molecular weight heat shock proteins in soybean. *Plant Physiol.* 114: 429–438.

Joshi, A.K. 1999. Genetic factors affecting abiotic stress tolerance in crop plants. In *Handbook of Plant and Crop Stress*, ed. M. Pessarakli, pp. 795–826. New York: Dekker.

Jung, H.W., W. Kim, and B.K. Hwang. 2003. Three pathogen-inducible genes encoding lipid transfer protein from pepper are differentially activated by pathogens, abiotic, and environmental stresses. *Plant Cell Environ.* 26: 915–928.

Kang, H.M., and M.E. Saltveit. 2001. Activity of enzymatic antioxidant defense systems in chilled and heat shocked cucumber seedling radicles. *Physiol. Plant.* 113: 548–556.

Kang, H.M., and M.E. Saltveit. 2002a. Antioxidant enzymes and DPPH-radical scavenging activity in chilled and heat-shocked rice (*Oryza sativa* L.) seedlings radicles. *J. Agri. Food Chem.* 50: 513–518.

Kang, H.M., and M.E. Saltviet. 2002b. Chilling tolerance of maize, cucumber and rice seedling leaves and roots are differentially affected by salicylic acid. *Physiol. Plant.* 115: 571–576.

Kang, H.M., K.W. Park, and M.E. Saltveit. 2002. Elevated growing temperatures during the day improve chilling tolerance of greenhouse-grown cucumber (*Cucumis sativus*) fruit. *Postharvest Biol. Technol.* 24: 49–57.

Kang, H.M., K.W. Park, and M.E. Saltveit. 2005. Chilling tolerance of cucumber (*Cucumis sativus*) seedling radicles is affected by radicle length, seedling vigor, and induced osmotic- and heat-shock proteins. *Physiol. Plant.* 124: 485–492.

Katiyar-Agarwal, S., M. Agarwal, and A. Grover. 2003. Heat-tolerant basmati rice engineered by over-expression of hsp101. *Plant Mol. Biol.* 51: 677–686.

Kirakosyan, A., P. Kaufman, S. Warber, et al. 2004. Applied environmental stresses to enhance the level of polyphenolics in leaves of hawthorn plants. *Physiol. Plant.* 121: 182–186.

Kirakosyan, A., E. Seymour, P.B. Kaufman, S. Warber, S. Bolling, and S.C. Chang. 2003. Antioxidant capacity of polyphenolic extracts from leaves of *Crataegus laevigata* and *Crataegus monogyna* (Hawthorn) subjected to drought and cold stress. *J. Agric. Food Chem.* 51: 3973–3976.

Kochhar, S., and V.K. Kochhar. 2005. Expression of antioxidant enzymes and heat shock proteins in relation to combined stress of cadmium and heat in *Vigna mungo* seedling. *Plant Sci.* 168: 921–929.

Kotak, S., J. Larkindale, U. Lee, P. von Koskull-Doring, E. Vierling, and K.D. Scharf. 2007. Complexity of the heat stress response in plants. *Curr. Opin. Plant Biol.* 10: 310–316.

Krishna, P., and G. Gloor. 2001. The Hsp90 family of proteins in *Arabidopsis thaliana. Cell Stress Chaperones* 6: 238–246.

Krishna, P., R.F. Felsheim, J.C. Larkin, and A. Das. 1992. Structure and light-induced expression of a small heat-shock protein gene of *Pharbitis nil. Plant Physiol.* 100: 1772–1779.

Krishna, P., M. Sacco, J.F. Cherutti, and S. Hill. 1995. Cold-induced accumulation of Hsp90 transcripts in *Brassica napus. Plant Physiol.* 107: 915–923.

Kubo, A., M. Aono, N. Nakajima, H. Saji, K. Tanaka, and N. Kondo. 1999. Differential responses in activity of antioxidant enzymes to different environmental stresses in *Arabidopsis thaliana. J. Plant Res.* 112: 279–290.

Kuo, H.F., Y.F. Tsai, L.S. Young, and C.Y. Lin. 2000. Ethanol treatment triggers a heat shock-like response but no thermotolerance in soybean (*Glycine max* cv. Kaohsiung No. 8) seedlings. *Plant Cell Environ.* 23: 1099–1108.

LaFayette, P.R., R.T. Nagao, K. O'Grady, E. Vierling, and J.L. Key. 1996. Molecular characterization of cDNAs encoding low-molecular-weight heat shock proteins of soybean. *Plant Mol. Biol.* 30: 159–169.

Larkindale, J., and M.R. Knight. 2002. Protection against heat stress induced oxidative damage in *Arabidopsis* involves calcium, abscisic acid, ethylene, and salicylic acid. *Plant Physiol.* 128: 682–695.

Larkindale, J., J.D. Hall, M.R. Knight, and E. Vierling. 2005a. Heat stress phenotypes of *Arabidopsis* mutants implicate multiple signaling pathways in the acquisition of thermotolerance. *Plant Physiol.* 138: 882–897.

Larkindale, J., M. Mishkind, and E. Vierling. 2005b. Plant responses to high temperature. In *Plant Abiotic Stress*, ed. M.A. Jenks and P.M. Hasegawa, pp. 100–144. Oxford, U.K.: Blackwell.

Leborgne-Castel, N., E.P.W.M. Jelitto-Van Dooren, A.J. Crofts, and J. Denecke. 1999. Overexpression of BiP in tobacco alleviates endoplasmic reticulum stress. *Plant Cell* 11: 459–470.

Lee, D.H., and C.B. Lee. 2000. Chilling stress-induced changes of antioxidant enzymes in the leaves of cucumber: in gel enzyme activity assays. *Plant Sci.* 159: 75–85.

Lee, J.H., and F. Schöffl. 1996. A Hsp70 antisense gene affects the expression of HSP70/HSC70, the regulation of HSF and the acquisition of thermotolerance in transgenic *Arabidopsis thaliana. Mol. Gen. Genet.* 252: 11–19.

Lee, J.H., A. Hübel, and F. Schöffl. 1995. Derepression of the activity of genetically engineered heat shock factor causes constitutive synthesis of heat shock proteins and increased thermotolerance in transgenic *Arabidopsis. Plant J.* 8: 603–612.

Lee, Y.R.J., R.T. Nagao, C.Y. Lin, and J.L. Key. 1996. Induction and regulation of heat-shock gene expression by an amino acid analog in soybean seedlings. *Plant Physiol.* 110: 241–248.

Lei, Y., C. Yin, and C. Li. 2007. Adaptive responses of *Populus przewalskii* to drought stress and SNP application. *Acta Physiol. Plant.* 29: 519–526.

Leshem, Y.Y., and P.J.C. Kuiper. 1996. Is there a GAS (general adaptation syndrome) response to various types of environmental stress? *Biol. Plant.* 38: 1–18.

Li, C., Q. Chen, X. Gao, et al. 2005. AtHsfA2 modulates expression of stress responsive genes and enhances tolerance to heat and oxidative stress in *Arabidopsis. Sci. China C. Life Sci.* 48: 540–550.

Lin, C.Y., J.K. Roberts, and J.L. Key. 1984. Acquisition of thermotolerance in soybean seedlings: synthesis and accumulation of heat shock proteins and their cellular localization. *Plant Physiol.* 74: 152–160.

Lin, P.L. 2005. The accumulation of heat shock proteins in rice plants and their roles in protecting the rice seedlings from heat and hydrogen peroxide stresses. Master's thesis, National Chung Hsing University, Taichung City, Taiwan.

Malik, M.K., J.P. Slove, C.H. Hwang, and J.L. Zimmerman. 1999. Modified expression of a carrot small heat shock protein gene, HSP17.7, results in increased or decreased thermotolerance. *Plant J.* 20: 89–99.

Marrs, K.A., E.S. Casey, S.A. Capitant, et al. 1993. Characterization of two maize HSP90 heat shock protein genes: expression during heat shock, embryogenesis and pollen development. *Dev. Genet.* 14: 27–41.

Milioni, D., and P. Hatzopoulos. 1997. Genomic organization of Hsp90 gene family in *Arabidopsis. Plant Mol. Biol.* 35: 955–961.

Mittler, R. 2002. Oxidative stress, antioxidants and stress tolerance. *Trends Plant Sci.* 7: 405–410.

Mittler, R. 2006. Abiotic stress, the field environment and stress combination. *Trends Plant Sci.* 11: 15–19.

Mittler, R., S. Vanderauwera, M. Gollery, and F. Van Breusegem. 2004. Reactive oxygen gene network of plants. *Trends Plant Sci.* 9: 490–498.

Mora-Herrera, M.E., H. Lopez-Delgado, A. Castillo-Morales, and C.H. Foyer. 2005. Salicylic acid and H_2O_2 function by independent pathways in the induction of freezing tolerance in potato. *Physiol. Plant.* 125: 430–440.

Murakami, T., S. Matsuba, H. Funatsuki, et al. 2004. Over-expression of a small heat shock protein, sHSP17.7, confers both heat tolerance and UV-B resistance to rice plants. *Mol. Breed.* 13: 165–175.

Neill, S.J., R. Desikan, A. Clarke, R.D. Hurst, and J.T. Hancock. 2002a. Hydrogen peroxide and nitric oxide as signaling molecules in plants. *J. Exp. Bot.* 53: 1237–1247.

Neill, S., R. Desikan, and J. Hancock. 2002b. Hydrogen peroxide signaling. Curr. Opin. Plant Biol. 5: 388–395.

Nover, L. 1991. *Heat Shock Response.* Boca Raton, FL: CRC Press.

O'Driscoll, C. 2007. Plant therapy. *Chem. Ind.* 4: 21–23.

Ogawa, D., K. Yamaguchi, and T. Nishiuchi. 2007. High-level overexpression of the *Arabidopsis HsfA2* gene confers not only increased thermotolerance but also salt/osmotic stress tolerance and enhanced callus growth. *J. Exp. Bot.* 58: 3373–3383.

Ono, K., T. Hibino, T. Kohinata, et al. 2001. Overexpression of DnaK from a halotolerant cyanobacterium *Aphanothece halophytica* enhances the high-temperature tolerance of tobacco during germination and early growth. *Plant Sci.* 160: 455–461.

Philpott, M., K.S. Gould, K.R. Markham, S.L. Lewthwaite, and L.R. Ferguson. 2003. Enhanced coloration reveals high antioxidant potential in new sweet potato cultivars. *J. Sci. Food Agric.* 83: 1076–1082.

Piergiovanni, A.R., and D. Pignone. 2003. Effect of year-to-year variation and genotype on trypsin inhibitor level in common bean (*Phaseolus vulgaris* L.) seeds. *J. Sci. Food Agric.* 83: 473–476.

Pla, M., G. Huguet, D. Verdaguer, et al. 1998. Stress proteins co-expressed in suberized and lignified cells and in apical meristems. *Plant Sci.* 139: 49–57.

Prändl, R., K. Hinderhofer, G. Eggers-Schumacher, and F. Schöffl. 1998. HSF3, a new heat shock factor from *Arabidopsis thaliana*, derepresses the heat shock response and confers thermotolerance when overexpressed in transgenic plants. *Mol. Gen. Genet.* 258: 269–278.

Pressman, T., R. Shaked, and N. Firon. 2006. Exposing pepper plants to high day temperatures prevents the adverse low night temperature symptoms. *Physiol. Plant.* 126: 618–626.

Sabehat, A., S. Lurie, and D. Weiss. 1998. Expression of small heat-shock proteins at low temperatures: a possible role in protecting against chilling injuries. *Plant Physiol.* 117: 651–658.

Sabehat, A., D. Weiss, and S. Lurie. 1996. The correlation between heat shock protein accumulation and persistence and chilling tolerance in tomato fruit. *Plant Physiol.* 110: 531–547.

Sala, J.M., and M.T. Lafuente. 1999. Catalase in the heat-induced chilling tolerance of cold-stored hybrid fortune mandarin fruits. *J. Agric. Food Chem.* 47: 2410–2414.

Sala, J.M., and M.T. Lafuente. 2000. Catalase enzyme activity is related to tolerance of mandarin fruits to chilling. *Postharvest Biol. Technol.* 20: 81–89.

Salekdeh, A.G., and S. Komatsu. 2007. Crop proteomics: aim at sustainable agriculture of tomorrow. *Proteomics* 7: 2976–2996.

Saltveit, M.E. 2001. Chilling injury is reduced in cucumber and rice seedlings and in tomato pericarp discs by heat-shocks applied after chilling. *Postharvest Biol. Technol.* 21: 169–177.

Saltveit, M.E. 2002. Heat shocks increase the chilling tolerance of rice (*Oryza sativa*) seedling radicles. *J. Agric. Food Chem.* 50: 3232–3235.

Sanchez, Y., J. Taulien, K.A. Borkovich, and S. Lindquist. 1992. Hsp104 is required for tolerance to many forms of stress. *EMBO J.* 11: 2357–2364.

Sanchez-Hernandez, C., N. Martinez-Gallardo, A. Guerrero-Rangel, S. Valdes-Rodriguez, and J. Delano-Frier. 2004. Trypsin and alpha-amylase inhibitors are differentially induced in leaves of amaranth (*Amaranthus hypochondriacus*) in response to biotic and abiotic stress. *Physiol. Plant.* 122: 254–264.

Sanmiya, K., K. Suzuki, Y. Egawa, and M. Shono. 2004. Mitochondrial small heat-shock protein enhances thermotolerance in tobacco plants. *FEBS Lett.* 557: 265–268.

Saruyama, H., and M. Tanida. 1995. Effect of chilling on activated oxygen-scavenging enzymes in low temperature-sensitive and tolerant cultivars of rice (*Oryza sativa* L.). *Plant Sci.* 109: 105–113.

Sato, Y., and S. Yokoya. 2008. Enhanced tolerance to drought stress in transgenic rice plants overexpressing a small heat-shock protein, sHSP17.7. *Plant Cell Rep.* 27: 329–334.

Sato, Y., T. Murakami, H. Funatsuki, S. Matsuba, H. Saruyama, and M. Tanida. 2001. Heat shock-mediated APX gene expression and protection against chilling injury in rice seedlings. *J. Exp. Bot.* 52: 145–151.

Scharf, K.D., M. Siddique, and E. Vierling. 2001. The expanding family of *Arabidopsis thaliana* small heat stress proteins and a new family of proteins containing α-crystallin domains (Acd proteins). *Cell Stress Chaperones* 6: 225–237.

Screenivasulu, N., S.K. Sopory, and P.B. Kavi Kishor. 2007. Deciphering the regulatory mechanisms of abiotic stress tolerance in plants by genomic approaches. *Gene* 388: 1–13.

Seki, M., A. Kamei, K. Yamaguchi-Shinozaki, and K. Shinozaki. 2003. Molecular responses to drought, salinity and frost: common and different paths for plant protection. *Curr. Opin. Biotechnol.* 14: 194–199.

Selote, D.S., and R. Khanna-Chopra. 2006. Drought acclimation confers oxidative stress tolerance by inducing co-ordinated antioxidant defense at cellular and subcellular level in leaves of wheat seedlings. *Physiol. Plant.* 127: 494–506.

Senaratna, T., D. Touchell, E. Bunn, and K. Dixon. 2000. Acetyl salicylic acid (Aspirin) and salicylic acid induce multiple stress tolerance in bean and tomato plants. *Plant Growth Regul.* 30: 157–161.

Senthil-Kumar, M., V. Srikanthbabu, B. Mohan Raju, Ganeshkumar, N. Shivaprakash, and M. Udayakumar. 2003. Screening of inbred lines to develop a thermotolerant sunflower hybrid using the temperature induction response (TIR) technique: a novel approach by exploiting residual variability. *J. Exp. Bot.* 54: 2569–2578.

Skog, L.J., and C.L. Chu. 2001. Ozone technology for shelf life extension on fruits and vegetables. *Acta Hort.* 553: 431–432.

Smirnoff, N. 1998. Plant resistance to environmental stress. *Curr. Opin. Biotechnol.* 9: 214–219.

Sohn, K.H., S.C. Lee, H.W. Jung, J.K. Hong, and B.K. Hwang. 2006. Expression and functional roles of the pepper pathogen-induced transcription factor RAV1 in bacterial disease resistance and drought and salt tolerance. *Plant Mol. Biol.* 61: 897–915.

Song, F., and R.M. Goodman. 2002. Molecular cloning and characterization of a rice phosphoinositide-specific phospholipase C gene, *OsPI-PLC1*, that is activated in systemic acquired resistance. *Physiol. Mol. Plant Pathol.* 61: 31–40.

Soto, A., I. Allona, C. Collada, et al. 1999. Heterologous expression of a plant small heat-shock protein enhances *Escherichia coli* viability under heat and cold stress. *Plant Physiol.* 120: 521–528.

Srikanthbabu, V., G. Kumar, B.T. Krishnaprasad, R. Gopalakrishna, M. Savitha, and M. Udayakumar. 2002. Identification of pea genotypes with enhanced thermotolerance using temperature induction response (TIR) technique. *J. Plant Physiol.* 159: 535–545.

Su, J., and R. Wu. 2004. Stress-inducible synthesis of proline in transgenic rice confers faster growth under stress conditions than that with constitutive synthesis. *Plant Sci.* 166: 941–948.

Sugino, M., T. Hibino, Y. Tanaka, N. Nii, T. Takabe, and T. Takabe. 1999. Overexpression of DnaK from a halotolerant cyanobacterium *Aphanothece halophytice* acquires resistance to salt stress in transgenic tobacco plants. *Plant Sci.* 146: 81–88.

Sun, W., C. Bernard, B. van de Cotte, M. van Montagu, and N. Verbruggen. 2001. At-HSP17.6A, encoding a small heat-shock protein in *Arabidopsis*, can enhance osmotolerance upon overexpression. *Plant J.* 27: 407–415.

Sun, W., M. van Montagu, and N. Verbruggen. 2002. Small heat shock proteins and stress tolerance in plants. *Biochim. Biophys. Acta* 1577: 1–9.

Sung, D.Y., and C.L. Guy. 2003. Physiological and molecular assessment of altered expression of Hsc70-1 in *Arabidopsis*. Evidence for pleiotropic consequences. *Plant Physiol.* 132: 979–987.

Sung, D.Y., E. Vierling, and C.L. Guy. 2001. Comprehensive expression profile analysis of the *Arabidopsis* Hsp70 gene family. *Plant Physiol.* 126: 789–800.

Suzuki, N., and R. Mittler. 2006. Reactive oxygen species and temperature stresses: a delicate balance between signaling and destruction. *Physiol. Plant.* 126: 45–51.

Tester, M., and A. Bacic. 2005. Abiotic stress tolerance in grasses. From model plants to crop plants. *Plant Physiol.* 137: 791–793.

Tseng, T.S., S.S. Tzeng, K.W. Yeh, et al. 1993. The heat-shock response in rice seedlings: isolation and expression of cDNAs that encode class I low-molecular-weight heat-shock proteins. *Plant Cell Physiol.* 34: 165–168.

Uchida, A., A.T. Jagendorf, T. Hibino, T. Takabe, and T. Takabe. 2002. Effects of hydrogen peroxide and nitric oxide on both salt and heat stress tolerance in rice. *Plant Sci.* 163: 515–523.

Ukaji, N., C. Kuwabara, D. Takezawa, K. Arakawa, S. Yoshida, and S. Fujikawa. 1999. Accumulation of small heat-shock protein homologs in the endoplasmic reticulum of cortical parenchyma cells in mulberry in association with seasonal cold acclimation. *Plant Physiol.* 120: 481–489.

Unyayar, S., and F.O. Cekic. 2005. Changes in antioxidant enzymes of young and mature leaves of tomato seedlings under drought stress. *Turk. J. Biol.* 29: 211–216.

Van Berkel, J., F. Salamini, and C. Gebhardt. 1994. Transcripts accumulating during cold storage of potato (*Solanum tuberosum* L.) tubers are sequence related to stress-responsive genes. *Plant Physiol.* 104: 445–452.

Vandenabeele, S., K. van der Kelen, J. Dat, et al. 2003. A comprehensive analysis of hydrogen peroxide-induced gene expression in tobacco. *Proc. Natl. Acad. Sci. U.S.A.* 100: 16113–16118.

Vannini, C., M. Campa, M. Iriti,, et al. 2007. Evaluation of transgenic tomato plants ectopically expressing the rice *Osmyb4* gene. *Plant Sci.* 173: 231–239.

Verslues, P.E., M. Agarwal, S. Katiyar-Agarwal, J. Zhu, and J.K. Zhu. 2006. Methods and concept in quantifying resistance to drought, salt and freezing, abiotic stress that affect plant water status. *Plant J.* 45: 523–539.

Vierling, E. 1991. The roles of heat shock proteins in plants. *Annu. Rev. Plant Physiol. Plant Mol. Biol.* 42: 579–620.

Vinocur, B., and A. Altman. 2005. Recent advances in engineering plant tolerance to abiotic stress: achievements and limitations. *Curr. Opin. Biotechnol.* 16: 123–132.

Wahid, A., M. Perveen, S. Gelani, and S.M.A. Basra. 2007. Pretreatment of seed with H_2O_2 improves salt tolerance of wheat seedlings by alleviation of oxidative damage and expression of stress proteins. *J. Plant Physiol.* 164: 283–294.

Wang, C.Y. 1996. Temperature preconditioning affects ascorbate antioxidant systems in chilled zucchini squash. *Postharvest Biol. Technol.* 8: 29–36.

Wang, H.Y., Y.C. Huang, S.F. Chen, and K.Y. Yeh. 2003. Molecular cloning, characterization and gene expression of a water deficiency and chilling induced proteinase inhibitor I gene family from sweet potato (*Ipomoea batatas* Lam.) leaves. *Plant Sci.* 165: 191–203.

Wang, S.Y., and W. Zheng. 2001. Effect of plant growth temperature on antioxidant capacity in strawberry. *J. Agric. Food Chem.* 49: 4977–4982.

Wang, W., B. Vinocur, O. Shoseyov, and A. Altman. 2004. Role of plant heat-shock proteins and molecular chaperones in the abiotic stress response. *Trends Plant Sci.* 9: 244–252.

Waters, E.R., G.J. Lee, and E. Vierling. 1996. Evolution, structure and function of the small heat shock proteins in plants. *J. Exp. Bot.* 47: 325–338.

Wehmeyer, N., and E. Vierling. 2000. The expression of small heat shock proteins in seeds responds to discrete developmental signals and suggests a general protective role in desiccation tolerance. *Plant Physiol.* 122: 1099–1108.

Wehmeyer, N., L.D. Hernandez, R.R. Finkekstein, and E. Vierling. 1996. Synthesis of small heat-shock proteins is part of the developmental program of late seed maturation. *Plant Physiol.* 112: 747–757.

Whitcombe, J.R., P.A. Hollington, C.J. Howarth, S. Reader, and K.A. Steele. 2008. Breeding for abiotic stresses for sustainable agriculture. *Philos. Trans. R. Soc. B* 363: 703–716.

Wilen, R.W., M. Sacco, L.V. Gusta, and P. Krishna. 1995. Effects of 24-epibrassinolide on freezing and thermotolerance of bromegrass (*Bromus inermis*) cell cultures. *Physiol. Plant.* 95: 195–202.

Winicov, I. 1998. New molecular approaches to improving salt tolerance in crop plants. *Ann. Bot.* 82: 703–710.

Winicov, I., and D.R. Bastola. 1997. Salt tolerance in crop plants: new approaches through tissue culture and gene regulation. *Acta Physiol. Plant.* 19: 435–449.

Xiang, C., and D.J. Oliver. 1999. Glutathione and its central role in mitigating plant stress. In *Handbook of Plant and Crop Stress*, ed. M. Pessarakli, pp. 697–707. New York: Dekker.

Xue, T., and H. Hartikainen. 2000. The association of antioxidant enzymes with the synergistic effect of selenium and UV irradiation in enhancing plant growth. *Agric. Food Sci. Finland* 9: 177–186.

Yabe, N., T. Takahashi, and Y. Komeda. 1994. Analysis of tissue-specific expression of *Arabidopsis thaliana* Hsp90-family gene HSP81. *Plant Cell Physiol.* 35: 1207–1219.

Yamada, A., M. Sekiguchi, T. Mimura, and Y. Ozeki. 2002. The role of plant CCTα in salt- and osmotic-stress tolerance. *Plant Cell Physiol.* 43: 1043–1048.

Yamaguchi, T., and E. Blumwald. 2005. Developing salt-tolerant crop plants: challenges and opportunities. *Trends Plant Sci.* 10: 615–620.

Yamaguchi-Shinozaki, K., and K. Shinozaki. 2006. Transcriptional regulatory networks in cellular responses and tolerance to dehydration and cold stresses. *Annu. Rev. Plant Biol.* 57: 781–803.

Yeh, C.H., P.F.L. Chang, K.W. Yeh, W.C. Lin, Y.M. Chen, and C.Y. Lin. 1997. Expression of a gene encoding a 16.9-kDa heat-shock protein, Oshsp16.9, in *Escherichia coli* enhances thermotolerance. *Proc. Natl. Acad. Sci. U.S.A.* 94: 10967–10972.

Yeh, C.H., K.W. Yeh, S.H. Wu, P.F.L. Chang, Y.M. Chen, and C.Y. Lin. 1995. A recombinant rice 16.9 kDa heat shock protein can provide thermoprotection in vitro. *Plant Cell Physiol.* 36: 1341–1348.

Yu, C.W., T.M. Murphy, and C.H. Lin. 2003. Hydrogen peroxide-induced chilling tolerance in mung beans mediated through ABA-independent glutathione accumulation. *Funct. Plant Biol.* 30: 955–963.

Yu, C.W., T.M. Murphy, W.W. Sung, and C.H. Lin. 2002. H_2O_2 treatment induces glutathione accumulation and chilling tolerance in mung bean. *Funct. Plant Biol.* 29: 1081–1087.

Zhang, J., N.Y. Klueva, Z. Wang, R. Wu, T.D. Ho, and H.T. Nguyen. 2000. Genetic engineering for abiotic stress resistance in crop plants. *In Vitro Cell. Dev. Biol. Plant* 36: 108–114.

Zhao, C., M. Shono, A. Sun, S. Yi, M. Li, and J. Liu. 2007. Constitutive expression of an endoplasmic reticulum small heat shock protein alleviates endoplasmic reticulum stress in transgenic tomato. *J. Plant Physiol.* 164: 835–841.

Zhu, B., C. Ye, H. Lü, et al. 2006. Identification and characterization of a novel heat shock transcription factor gene, *GmHsfA1*, in soybeans (*Glycine max*). *J. Plant Res.* 119: 247–256.

Zhu, J.K., and L. Xiong. 2006. Method for increasing stress tolerance in plants. U.S. Patent 7038111. May 2. http://www.freepatentsonline.com/7038111.html (accessed April 4, 2008).

6 Engineering Plants for Industrial Uses

Thomas A. McKeon

CONTENTS

6.1 INTRODUCTION

Agriculture has served as a source of food and fiber and for other uses from ancient times. In addition to food, plants provide a broad array of products that serve nonfood uses. Plants are renewable resources from which numerous bio-based products can be derived. The ability to genetically engineer plants allows several different types of alteration to be introduced. These traits can expand the uses for plant-derived products. One broad type of change includes effects on agronomic performance. Such changes have already been commercialized and include herbicide and insect resistance. Among the first genetically engineered commodity crops were corn, soybean, and cotton. These were introduced in 1996 and were minor crops, with cotton at 13% of the crop, soybean 2%, and corn less than 1%. In 2008 in the United States, 80% of the corn crop, 92% of the soybean crop, and 86% of the cotton crop planted were genetically engineered with one or more traits (http://www.ers.usda.gov/data/biotechcrops). While initially there was limited acceptance by growers internationally, in the United States the advantages of reduced investment in chemical controls and improved yields led to rapid acceptance of the technology by growers. Since the introduction of these crops, they had been approved for planting in 23 countries [1], from two major (China, United States) and four minor genetically modified (GM) crop-producing countries in 1996 (Canada, Argentina, Australia, and Mexico), when virus-resistant tobacco was the major GM crop [2].

New introductions, including herbicide-resistant canola, virus-resistant papaya, and other food crops, have been accepted. Soybean producing an oil containing the ω-3 fatty acid stearidonate has been developed for the nutraceutical and fish farm feed market [3]. Interestingly, the earliest genetically engineered oilseed crop was canola with a high content of laurate in the oil [4], with the soaps and detergent industry the initial target market. As the need for biofuels continues to grow, it can be seen that the major genetically engineered crops corn, soybean, cotton, and canola can meet an expanding need for industrial applications, with cotton for fiber and the others for biodiesel or ethanol.

The major traits incorporated in the genetically engineered crops currently grown are considered input traits. That is, they reduce the need for plant treatments, primarily by enhancing herbicide or insect resistance. The result is a reduction in the level of chemical treatment with concomitant improvement in worker and environmental safety and enhanced yield as a result of weed- and pest-free growth. Other input trait modifications would include engineering with genes for nitrogen fixation or genes that more effectively extract nutrients from the soil.

A second class of engineered crops carries agronomic traits that could improve yield by enhancing growth characteristics. Since there are no antiviral treatments for plants, crops that have been engineered for viral resistance would be considered to be in this class. Drought tolerance, dwarfing, and other agronomic traits that affect yield would be included in this class of engineered crops.

From the time that genetically engineering crops became feasible, designed output traits have been a major interest. The early approach was to incorporate a gene for an enzyme that would alter the composition of the crop's product. One of the first such commercialized crops was laurate canola, with the 12-carbon fatty acid replacing much of the 18-carbon fatty acid present in commodity canola [5]. Another early commercialized crop with an altered output trait was the Flavr Savr tomato [6]. This food crop was altered to suppress the polygalacturonase involved in softening, allowing harvest and shipping at a later development stage than other fresh market tomatoes. Although neither crop achieved great commercial success, they remain a clear demonstration of the potential to be realized from engineering output traits.

Regardless of the difficult path to successful commercialization, the concept of altering output traits remains attractive, and engineered pathways can provide products that include novel oleochemicals, carbohydrates, polymers, and polyisoprenoids. Another class of changes can affect processing of the crop, including improved residue safety by removal of toxins or allergens, enhanced degradability for biofuel production, or reduced lignin content for paper production. Biopharmaceuticals and nutraceuticals represent another class of genetic enhancements that are currently being developed in plants. The availability of such genetic changes can expand the utilization of crops for nonfood purposes and help to expand the development of bio-based products [7]. Table 6.1 provides a list of general changes that are of current interest as introduced changes to crop plants.

6.2   OIL COMPOSITION AS A TARGET OF GENETIC ENGINEERING

Modification of oil composition remains a major interest of plant breeders, both traditional and molecular. The oil represents a major component present in the seed and is easily isolated. As a result of research on biochemistry and genetics of oil biosynthesis, there is considerable background information that elucidates the process of triacylglycerol (TAG) (oil) biosynthesis. Oils have both food and industrial uses, and

TABLE 6.1

Engineered Output Traits Currently in Development

Seed oils	Fatty acids
Starch	Novel branching
Fiber	Cotton, flax, blends, spidroin
Lumber	Lignin content, paper
Proteins	Enzymes
Chemicals	Sugars
Polymers	Polyhydroxyalkanoates, fructans
Nutraceuticals	Vitamins, essential fatty acids
Pharmaceuticals	Vaccines, therapeutic proteins
Fuel	Ethanol, biodiesel
Cut flowers	New colors, fragrances
Phytoremediation	Metal ions, aromatic compounds

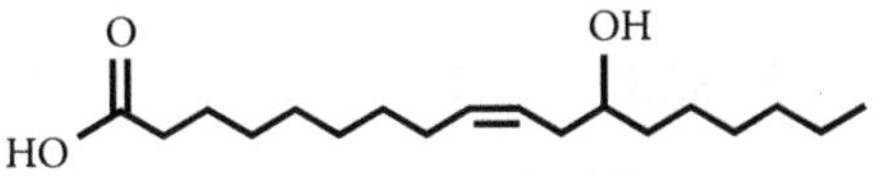

RICINOLEIC ACID

FIGURE 6.1 Ricinoleic acid.

crops that produce oils in surplus could be reconfigured to produce novel oils that would expand the market and eliminate the surplus of oil.

Analysis of the basic biochemistry of fatty acid biosynthesis in oilseeds pointed to the key role of single enzymes in altering the fatty acid composition of oils. This optimistic approach suggested that insertion of a single gene introducing a new enzyme would result in production of an oil containing a desired fatty acid. As mentioned, laurate canola was the first commercialized version of an oilseed demonstrating this concept. The choice of laurate was a result of fluctuations in the availability and price of tropical oils containing this fatty acid that is a major feedstock for the soap and detergent industry.

In developing an approach to engineer oilseeds, it is thus important to choose a "target" fatty acid that will readily fill a commercial need. In addition to laurate, another such fatty acid is ricinoleate (Figure 6.1), produced by the castor seed and containing 90% ricinoleate (12-hydroxy oleate), a unique oil in its high content of a single fatty acid and a producer of a hydroxy fatty acid. Castor oil is a chemical feedstock for a wide array of industrial products (Table 6.2). It was generally thought that, by analogy to the success in producing laurate canola by expressing a single gene, a castor oil substitute could be produced by expressing the gene for the oleate hydroxylase, the enzyme that produces ricinoleate. This chapter discusses the industrial applications of castor oil and our research efforts to elucidate the molecular process that occurs in the castor seed to produce castor oil.

TABLE 6.2
Products Derived from Castor Oil

Lubricants	Lithium grease
	Heptanoate esters for engines
Coatings	Nonyellowing drying oil
	Low-VOC oil-based paints
Surfactants	Turkey red oil
Plasticizers	Blown oil for polyamides, rubber
	Heptanoates, low-temperature uses
Cosmetics	Lipstick
Pharmaceuticals	Laxative
Polymers	Polyesters, from sebacic acid
	Polyamides, Nylon 11, Nylon 6,10
	Polyurethanes
Perfumes	Odorants include 2-octanol, heptanal, and undecenal
Fungicides	Undecenoic acid and derivatives

6.3   CASTOR AND OTHER HYDROXY FATTY ACID OILS

Due to their physical and chemical properties, hydroxy fatty acids are of considerable commercial interest. Castor oil is currently the only major source of hydroxy fatty acids. The desirable features of castor oil include the high content of oil (up to 60%) and the high content of ricinoleate in the oil (90%). The crop itself is highly adapted to broad climate regions, can tolerate drought, and can provide a higher yield than any annual oilseed crop given the same agronomic inputs. However, the presence in the seed of a hyperallergenic protein and the potent toxin ricin make processing of the castor seed problematic [8]. Therefore, alternate sources of ricinoleate or other hydroxy fatty acids are of considerable commercial interest.

6.4   SOURCES OF HYDROXY FATTY ACIDS

The only commodity source of castor oil is the castor plant (*Ricinus communis* L.); however, there are other plant sources of hydroxy fatty acids. The "new crop" *Lesquerella fenderlii* has been developed by the U.S. Department of Agriculture Agricultural Research Service (USDA-ARS) as an alternate source of hydroxy fatty acid for U.S. production. Although this plant grows in desert climates and other marginal agricultural land, it has not yet reached the volume needed to be considered a commercial success or to meet industry needs. Species of *Lesquerella* contain up to 55% lesquerolic acid, the 20-carbon analogue of ricinoleate, and some species contain up to 80% hydroxy fatty acid content. Since lesquerolic acid is biosynthetically derived from an elongation of ricinoleate, gene suppression of the elongase represents a potential pathway to an alternate source of high ricinoleate oil. To the extent that lesquerolic acid can substitute for castor oil, unmodified *Lesquerella* can fill an important niche. The lubricant and cosmetic properties of lesquerella oil and hydrogenated lesquerolate are similar to those of ricinoleate [9], and it has no apparent physiological (laxative) effect. However, lesquerolic acid cannot directly substitute for the chemical derivatives produced by processes that break down ricinoleate, such as pyrolysis for synthesis of 10-undecenoate, the precursor for 11-amino undecanoate, the precursor for Nylon 11.

At one time, castor was grown in the United States and supplied about half of the U.S. demand for castor oil. The rest of the need was filled by imported castor beans for processing. As a result of the loss of crop parity and the rising cost of energy needed to detoxify and deallergenize castor meal, castor oil production in the United States ceased. The issuance of Presidential Executive Order 13134 in August 1999 supporting a drive to generate more products from biological sources provided considerable impetus to reintroduce castor. As a result, efforts to detoxify and deallergenize castor seed using genetic engineering have arisen to meet expanding needs for a safe source of castor oil [8].

6.5   USES OF CASTOR OIL AND RICINOLEATE

The presence of the hydroxyl group in a midchain position of the fatty acid ricinoleate has a dramatic effect on the physical and chemical properties of castor oil. Interchain hydrogen bonding of the hydroxyl groups greatly increases the viscosity of castor oil in comparison to a common vegetable oil. The hydroxyl group also enhances interaction with metals, while the fatty acid chain still retains compatibility with nonpolar components, thus providing excellent properties for use in greases and lubricants. Chemically, castor oil has multiple sites available for chemical modifications, including formation of interpenetrating polymer networks that depend on the presence of additional polymerizing groups added to the fatty acyl chain [10].

Ricinoleate itself is a rich source of monomers for novel bio-based plastics (Figure 6.2). The chemical bonds next to the hydroxyl group are susceptible to quantitative cleavage, producing an array of products depending on the chemistry of the reaction. The monomers produced include 11-amino undecanoic acid and sebacic acid, which are used to make Nylon 11 and Nylon 6, 10, respectively [11].

Current uses for castor oil, ricinoleate, and derivatives obtained from them include surfactants, drying oils, greases, emollients, engineering plastics, plasticizers, anticorrosion treatments, deodorizers, thermoplastics, and insulators [11], with uses for castor oil limited by its availability. Methyl esters of castor oil added at 0.5% (v/v) to petroleum diesel enhance lubricity significantly, eliminating the need for the sulfur-based lubricity additives currently used [12]. Given the midchain hydroxy adding oxygen content reducing soot output and the low cloud point of castor oil methyl esters, castor oil methyl esters could be ideal diesel additives that even reduce air pollution. However, at 0.5% of the total diesel use, the volume of castor oil needed to supply just U.S. needs would be almost twice the total world production of castor oil.

The presence of the hydroxyl group allows a host of modifications, including sulfonation, leading to what is known as turkey red oil, the first synthetic surfactant [11]. Acid-catalyzed dehydration of castor oil produces a nonyellowing drying oil containing a high proportion of conjugated linoleic acid. The epoxidized oil is useful as a plasticizer and replaces the petroleum-derived epoxide compounds used in

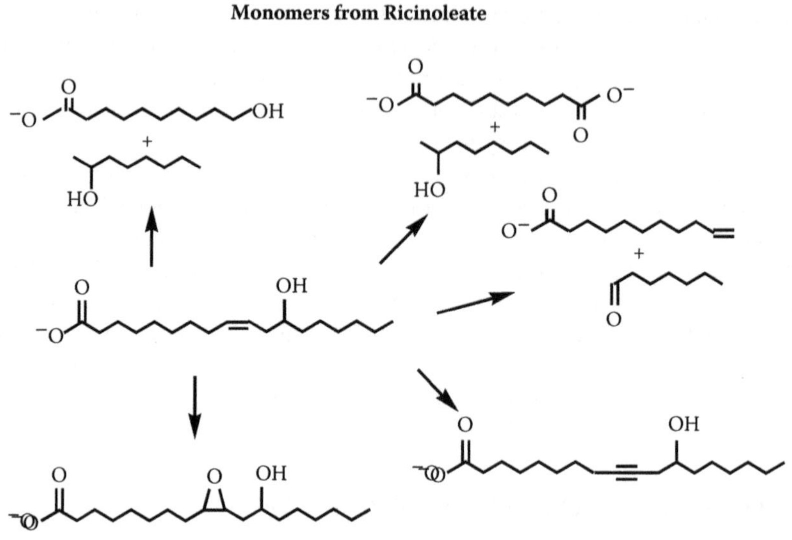

Monomers from Ricinoleate

FIGURE 6.2 Chemicals produced from ricinoleate presently or potentially used in polymers or for other purposes.

formulating paints and coatings, resulting in a low VOC (volatile organic carbon) formulation for surface treatments. Additional uses include estolides as thickening agents in lubricants, alkylated ether derivatives as nonionic detergents, and other potential uses that have been described but not been implemented due to the limited supply of castor oil [11].

6.6   BIOSYNTHESIS OF RICINOLEATE

The biosynthesis of castor oil has long been a matter of biochemical interest due to the unusual chemical nature of the reaction—a stereospecific, position-specific hydroxylation of a hydrocarbon. Research that led to the cloning of the gene for oleoyl-12-hydroxylase provided interesting insights into lipid biosynthesis and control of the fatty acid composition of oil. However, while the castor seed contains up to 60% oil with 90% ricinoleate content, transgenic plants that express the hydroxylase gene produce oils containing less than 20% hydroxy fatty acid [13].

Since the hydroxylase gene alone was not sufficient to elicit high levels of hydroxy fatty acid production, it seemed that there must be other enzymes required to achieve high ricinoleate levels in oil [14]. Based on intermediates that accumulated during in vitro castor oil biosynthesis carried out by castor seed microsomes, several enzymatic steps that appear to be important for high ricinoleate levels have been identified [14]. The pathway derived from this research is shown in Figure 6.3.

Further studies [15] indicated that ricinoleate is preferentially incorporated into TAG by a factor of 6 over oleate, leading us to identify the final step in oil biosynthesis (Figure 6.4) as a key step in maintaining high ricinoleate content while

Castor Oil Pathway
Oleoyl-PC
1
2-Ricinoleoyl-PC
3
Ricinoleate
2
5
Lyso-PC
+
Oleoyl-CoA
2-Linoleoyl-PC
4
Oleate
1,2-Diricinoleoyl-sn-glycerol
6
7
Triricinolein
(1) Oleoyl-12-Hydroxylase
(2) Oleoyl-12-Desaturase
(3) Phospholipase A2
(4) Acyl-CoA Synthetase
(5) Lyso-PC acyltransferase
(6) Diacylglycerol Acyltransferase
(7) PC: Diacylglycerol Acyltransferase

FIGURE 6.3 Biosynthetic pathway for castor oil.

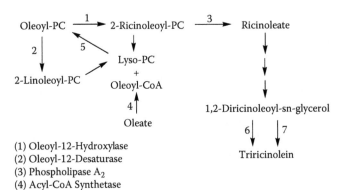

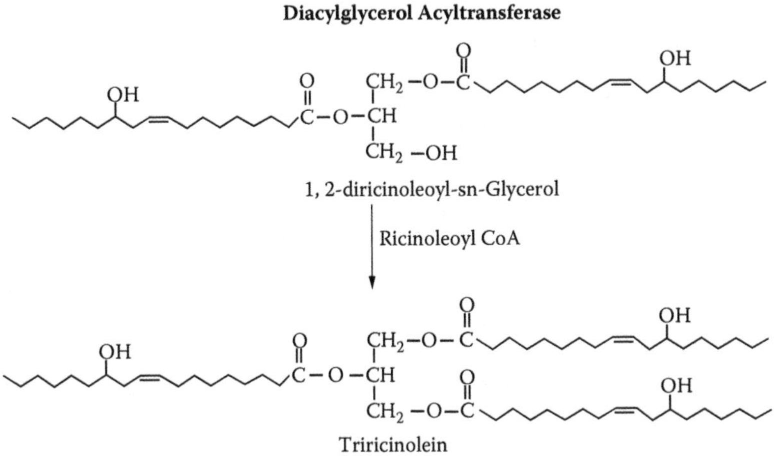

Diacylglycerol Acyltransferase
1, 2-diricinoleoyl-sn-Glycerol
Ricinoleoyl CoA
Triricinolein

FIGURE 6.4 Diacylglycerol acyltransferase reaction.

minimizing oleate incorporation into the TAG fraction. The diacylglycerol acyltransferase (DGAT) is a transmembrane enzyme that catalyzes the acylation of diacylglycerol (DAG) to TAG, using acyl-CoA (coenzyme A) as the source for the final acyl group. This step has long been considered to be rate limiting in oil biosynthesis, and considerable evidence has accumulated to indicate that altered DGAT activity levels dramatically affect the yield of oil [16].

Indeed, as a result of our cloning of the DGAT from developing castor seed, we were able to demonstrate that the activity and protein level of the cloned DGAT is closely correlated with the onset of oil biosynthesis in the seed [17]. Moreover, the

DGAT enzyme displayed a twofold preference for using diricinolein versus other nonhydroxylated fatty acyl DAG [18]. The acyl-CoA synthetase (ACS) produces the acyl donor for the DGAT reaction. We cloned several ACSs from castor and found one, designated ACS2, that displays a threefold preference for using condensing ricinoleate versus oleate with CoA-SH [19]. We are currently evaluating the role of this enzyme in seed development and oil production.

The elucidation of how the castor seed "decides" to make so much oil with such a high proportion of ricinoleate continues to provide a challenge [20]. The ultimate benefits to be gained by understanding this process include the ability to engineer ricinoleate production and insight in how to engineer the production of other uncommon and industrially useful fatty acids.

REFERENCES

1. James, C. (2008) Global Status of Commercialized Transgenic Crops: 2007 Executive Summary International Service for the Acquisition of Agri-Biotech Applications. Briefing 37. http://www.isaaa.org.
2. James, C. (1997) Global Status of Transgenic Crops in 1997. International Service for the Acquisition of Agri-Biotech Applications. ISAAA Briefing 5. http://www.isaaa.org.
3. Eckert, H., LaVallee, B., Schweiger, B.J., Kinney, A.J., Cahoon, E.B., and Clemente, T. (2006) Co-expression of the borage Δ^6 desaturase and the *Arabidopsis* Δ^{15} desaturase results in high accumulation of stearidonic acid in the seeds of transgenic soybean. *Planta* 224, 1050–1057.
4. McKeon, T.A. (2005) Genetic modification of seed oils for industrial applications. In: *Industrial Uses of Vegetable Oils*, ed. S.Z. Erhan, AOCS Press, Champaign, IL, pp. 1–13.
5. Del Vecchio, A.J. (1996) High-laurate canola. *INFORM* 7, 230–243.
6. Kramer, M.G., and Redenbaugh, K. (1994) Commercialization of a tomato with an antisense polygalacturonase gene: the FLAVR SAVR™ tomato story. *Euphytica* 79, 293–297.
7. McKeon, T.A. (2003) Genetically modified crops for industrial products and processes and their effects on human health. *Trends in Food Science and Technology* 14, 229–241.
8. McKeon, T.A., Lin, J.T., and Chen, G. (2002) Developing a safe source of castor oil. *INFORM* 13, 381–385.
9. Arquette, J.G., and Brown, J.H. (1993) Development of a cosmetic grade oil from *Lesquerella fendleri* seed. In: *New Crops*, ed. J. Janick and J.E. Simon, Wiley, New York, pp. 367–371.
10. Barrett, L.W., Sperling, L.H., Murphy, C.J. (1993) Naturally functionalized triglyceride oils in interpenetrating polymer networks. *JAOCS* 70, 523–534.
11. Caupin, H.J. (1997) Products from castor oil—past, present, and future. In: *Lipid Technologies and Applications,* ed. F.D. Gunstone and F.B. Padley, Dekker, New York, pp. 787–795.
12. Goodrum, J.W., and Geller, D.P. (2005) Influence of fatty acid methyl esters from hydroxylated vegetable oils on diesel fuel lubricity. *Bioresource Technology* 96, 851–855.
13. Broun, P., and Somerville, C. (1997) Accumulation of ricinoleic, lesquerolic, and densipolic acids in seeds of transgenic *Arabidopsis* plants that express a fatty acyl hydroxylase cDNA from castor bean. *Plant Physiology* 113, 933–942.

14. McKeon, T.A., and Lin, J.T. (2002) Biosynthesis of ricinoleic acid for castor oil production. In: *Lipid Biotechnology,* ed. T.M. Kuo and H.W. Gardner, Dekker, New York, pp. 129–139.

15. Lin, J.T., Chen, J.M., Liao, L.P., and McKeon, T.A. (2002) Molecular species of acylglycerols incorporating radiolabeled fatty acids from castor (*Ricinus communis* L.) microsomal incubations. *Journal of Agricultural and Food Chemistry* 50, 5077–5081.

16. He, X., Chen, G.Q., Lin, J.T., and McKeon, T.A. (2005) Molecular characterization of the acyl-CoA-dependent diacylglycerol acyltransferase in plants. *Recent Research Developments in Applied Microbiology and Biotechnology* 2, 69–86.

17. He, X., Chen, G.Q., Lin, J.T., and McKeon, T.A. (2004) Regulation of diacylglycerol acyltransferase in developing seeds of castor. *Lipids* 39, 865–871.

18. He, X., Turner, C., Chen, G.Q., Lin, J.T., and McKeon, T.A. (2004) Cloning and characterization of a cDNA encoding diacylglycerol acyltransferase from castor bean. *Lipids* 39, 311–318.

19. He, X., Chen, G.Q., Kang, S.T., McKeon, T.A. (2007) *Ricinus communis* contains an acyl-CoA synthetase that preferentially activates ricinoleate to its CoA ester. *Lipids* 42, 931–938.

20. McKeon, T., He, X., Chen, G.Q., Ahn, Y.J., Kang, S.T., and Lin, J.T. (2007). The castor plant as a dedicated industrial crop: elucidation of the enzymology of castor oil biosynthesis. In: *Biocatalysis and Biotechnology for Functional Foods and Industrial Products*, AOCS Press and CRC Press, Boca Raton, FL, pp. 545–552.

7 Improvement of Plant Transformation

Alex Boyko and Igor Kovalchuk

CONTENTS

Key Words: *Agrobacterium tumefaciens*; double-strand breaks; homologous recombination; MS medium; plant transformation; tissue regeneration.

7.1 INTRODUCTION

During the past decade, *Agrobacterium*-mediated genetic transformation became one of the dominant technologies used to produce a variety of genetically modified transgenic plants. The successful introduction of this technology relies on the fact that *Agrobacterium tumefaciens* is a typical plant pathogen in soil. However, in contrast to many other plant pathogens, it has a unique ability of trans-kingdom DNA transfer (reviewed in Gelvin 2003). The wide popularity of this biotechnology tool is reflected by the growing number of *Agrobacterium*-related patents published yearly (Roa-Rodriguez and Nottenburg 2003). A broad range of host species from plant to yeast and even to human cells (Tzfira and Citovsky 2003, reviewed in Lacroix et al. 2006) has been shown to be transformed with *Agrobacterium*.

Successful infection requires the activity of a number of host factors (reviewed in Citovsky et al. 2007). In particular, the last step of transformation, the actual integration of T-DNA into the plant genome, is almost completely host dependent. The activity of host factors determines the precision of transgene integration (Tzfira et al. 2004). Using yeast, it was demonstrated that mutations inactivating DNA repair proteins result in an increased resistance to *Agrobacterium*-mediated transformation or in complete inhibition of it (van Attikum et al. 2001, van Attikum and Hooykaas 2003). Unfortunately, there are still some blank spots and data controversy regarding mechanisms and factors employed during this stage. Nevertheless, the majority of studies agreed that the T-DNA preferentially integrates into the sites of DNA double-strand breaks (DSBs) using the activity of host DNA repair proteins (Chilton and Que 2003, Tzfira et al. 2003, reviewed in Citovsky et al. 2007).

DNA DSBs present in the host genome can be repaired using one of two alternative DNA repair pathways: nonhomologous end joining (NHEJ) or homologous recombination (HR). Both pathways differ in the fidelity of DNA repair (Gorbunova and Levy 1999, Puchta 2005). This may affect the intactness of T-DNA during integration. HR-dependent transformation is one of the techniques for generation of site-specific insertions and gene targeting (Vergunst and Hooykaas 1999, Puchta 2002, Hanin and Paszkowski 2003, Reiss 2003, Lida and Terada 2004). Unfortunately, the majority of DSBs in a plant cell are repaired via NHEJ, and the contribution of HR is very low (Gorbunova and Levy 1999, Puchta 2005). This is a major drawback in the development of technology for site-specific transgene integration in plant cells. Consequently, a number of studies conducted in the past several years have focused on identifying host DNA repair factors involved in T-DNA integration, revealing mechanisms that control the choice of DSB repair pathway, and manipulating these mechanisms during T-DNA integration in the host genome (reviewed in Tzfira et al. 2004, Tzfira and Citovsky 2006, Shrivastav et al. 2008).

In our study, we analyzed the effect of various chemicals capable of enhancing the HR frequency (HRF) prior to transformation. We found that these chemicals can also improve transformation efficiency. We have developed a system that allowed identification of growth media compositions that influence HRF. We have demonstrated the ability of such modified growth media to improve the transformation frequency.

7.2 EXPERIMENTAL SYSTEM ALLOWING RAPID IDENTIFICATION OF CHEMICALS THAT INFLUENCE HRF

The identification of chemical compounds that could induce HRF in plants without having a negative impact on plant physiology is an elaborate and time-consuming process. In our lab, we were able to establish a reliable system allowing rapid identification of chemical compounds inducing HRF.

Our system is based on growing transgenic *A. thaliana* plants on the control and modified Murashige and Skoog (MS) medium containing various chemicals of interest or their combinations. The presence of the β-*glucuronidase gene* (GUS)-based

HR substrate in the plant genome allows sensitive detection of HR events. The influence of chemicals present in the media on the HRF can be effectively monitored (Ilnytskyy et al. 2004). In addition, the effect of a tested chemical on other physiological parameters, including genotoxicity, plant biomass, and the like, can be also evaluated. Moreover, this system allows testing the influence of various chemical compounds on the HRF at different developmental stages (Boyko et al. 2006b) and in different plant organs (Boyko et al. 2006a). This makes the system a useful tool for adjusting pretransformation growth conditions for various plant species. It is important since transformation of these species relies on distinctly different approaches (i.e., cotyledons vs. leaf disks vs. root transformation methods).

High sensitivity of this system to genotoxic damage allows selection of chemicals that have no or a negligible impact on plant genome integrity (Kovalchuk et al. 2001). Therefore, this test system makes it possible to develop "clean" transformation techniques in contrast to those that usually result in extensive DNA damage (Kohler et al. 1989, Leskov et al. 2001).

Chemicals influencing HR can be further tested for their effect on the transformation frequency. Successful chemicals can be introduced as an active part of a novel growth medium composition. Subsequently, the combination of such active medium components can be tested for the potentiating effect on plant transformation. Thus, a growth medium composition enhancing plant transformation can be developed (Boyko and Kovalchuk 2006, U.S. Patent 11/466,184).

7.3 STRATEGIES FOR THE DEVELOPMENT OF A GROWTH MEDIUM ENHANCING HRF

The conventional MS medium composition (Murashige and Skoog 1962) represents a reliable system widely used by plant physiologists for growing various plant species. No extensive modifications of this medium have been done so far. Numerous studies conducted in our lab indicated that the chemical composition of growth media can influence DNA repair and induce HRF (unpublished data). Several different approaches can be suggested for improving the growth medium composition regarding the following transformation efficiency. This includes enrichment of growth media with various chemicals or depletion of these chemicals. The combination of both is also feasible. We discuss examples illustrating these approaches and show how modification of plant growth medium composition prior to transformation can improve the transformation frequency. Considering that the chemicals discussed in this report are currently being patented for their application in growth media, we use coded names and do not state the final concentrations used in these experiments.

7.3.1 ANALYSIS OF HRF IN THE MS MEDIUM DEPLETED OF CERTAIN CHEMICALS

Based on our previous studies, we hypothesized that lowering the concentration of chemicals in the standard MS medium could induce HRF. To test our hypothesis, we grew *Arabidopsis* line 11 plants on different modified media. The content of various

chemicals in them was lower than in the standard MS medium. The modified media
were named A, B, C, D, E, and F. The standard MS medium served as the control.
The HRF was analyzed 3 weeks postgermination. To analyze the recombination
rate (RR), we divided the HRF by the total number of genomes present in plants.
This was done deliberately since chemicals could influence DNA replication and
cell division.

Our results indicated that four of six modified media compositions led to the statistically significant induction of HR rate (RR) as compared to the conventional MS
medium (Student's test, $\alpha = .05$) (Figure 7.1). It is noteworthy that composition E led
to a 40% decrease in RR as compared to the conventional medium (Figure 7.1). It is
possible that compound E had a positive influence on HRF. It can be suggested that
enriching the MS medium with E compound could increase HRF.

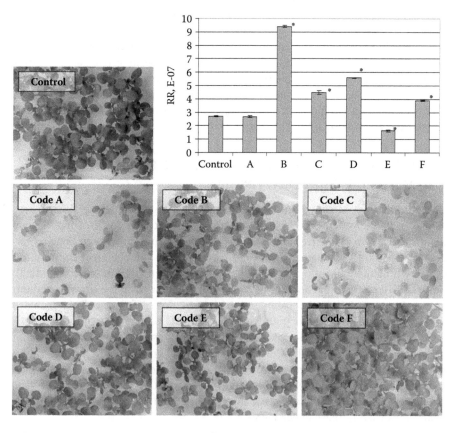

FIGURE 7.1   Effects of the modified growth medium depleted of various components on RR
in *Arabidopsis*. The graph shows RR in plants grown on the media (codes A–F) lacking various components. Values represent the mean ± standard deviation. Asterisks show a statisti-
cally significant difference from plants grown on the control MS medium (Student's test, $\alpha = .05$). Pictures show a phenotype of plants grown on the control and different modified media.
Pictures were taken at 22 days postgermination.

In short, three of six media tested in this experiment (B, D, and F) have poten-tial for application in future studies. In contrast, the C medium led to a significant 1.7-fold induction of RR and inhibited plant growth (Figure 7.1). The B medium showed the most promising results by yielding a significant 3.5-fold induction of RR. Importantly, plants grown on this medium looked healthy and were similar to those grown on the control MS medium.

This experiment allowed the fast identification of several media compositions that enhanced HR without having a negative effect on plant physiology. The B medium was chosen for further experiments.

7.3.2 ANALYSIS OF HRF IN THE MEDIUM ENRICHED WITH VARIOUS COMPOUNDS

Our previous experiments with the modified media depleted of various chemicals demonstrated that some of them might inhibit HRF. Next, we hypothesized that enrichment of the growth medium with these chemicals can enhance HR. To test this hypothesis, we grew *Arabidopsis* plants on the modified medium supplemented with various concentrations of chemicals chosen for the experiment. These chemi-cals were named G and H. The standard MS medium served as the control. HRF was analyzed 3 weeks postgermination.

In accordance with our hypothesis, plants grown at the 2.5-fold concentration of G medium (G2.5×) and 5-fold concentration of G medium (G5×) medium had RR induced by 7.0- and 7.2-fold, respectively, compared to plants grown on the MS medium (Figure 7.2A). In contrast, a very high concentration of G compound (i.e., 10×) decreased HRF drastically; it consisted of 40% of that observed in plants grown on the normal MS medium. Moreover, high concentrations of G chemical were toxic to plants and resulted in severe growth inhibition (Figure 7.2). Besides the fact that the G5× medium resulted in the highest HRF, it also triggered visible changes in a plant phenotype. Plants grown on the G5× medium were generally smaller and had a lower chlorophyll content compared to plants grown on the control MS or G2.5× media, respectively (Figure 7.2). In contrast, plants grown on the G2.5× medium grew better than control plants, indicating a positive effect of G compound not only on HR but also on plant metabolism in general (Figure 7.2). These findings indicate a need to analyze an active concentration range for each chemical tested. Selected concentrations should allow for the maximum induction of HRF without compromising plant physiology.

To ensure that the G compound did not result in genotoxic damage, we measured the number of DNA DSBs in plants grown on the control MS medium and on media containing 2× and 3× of G compound. We found no significant difference in the number of DNA DSBs in plants grown on the control and G2× media (Figure 7.2B). In fact, a statistically significant decrease in the number of DSBs was observed in plants grown on the G3× medium (Student's test, $\alpha = .05$). This suggests that the G chemical changes the balance between NHEJ and HR (Figure 7.2B).

In parallel to the results obtained from testing the G chemical, elevated levels of H chemical in the growth medium resulted in the induction of HRF in a dose-depen-dent manner ($r = .98$, $p < .05$). A gradual increase in the H compound concentration

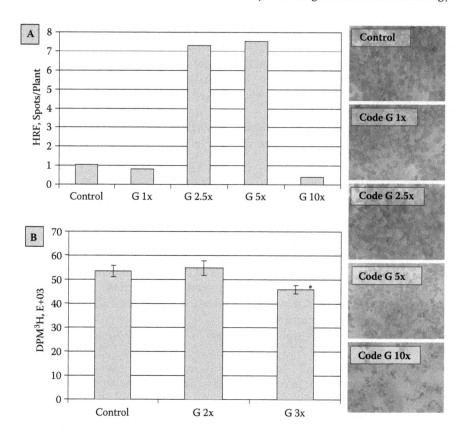

FIGURE 7.2 Effects of the modified growth medium supplemented with various quantities of G chemical on the HRF and DSB levels in *Arabidopsis*. (A) HRF in plants grown on the control MS medium and on the medium supplemented with 1×, 2.5×, 5×, and 10× of G chemical. (B) The number of DNA DSBs (*DPM 3H*) in plants grown on the control MS medium and in the presence of 2× and 3× of G chemical. Values represent the mean ± standard deviation. Asterisks show a statistically significant difference from plants grown on the control MS medium (Student's test, $\alpha = .05$). Pictures show a phenotype of plants grown on the control and different modified media. Pictures were taken 22 days postgermination.

from 1× to 2× led to a 3.9-fold induction of HRF (Figure 7.3A). However, in contrast to the G compound, the increase in concentrations of H chemical resulted in significant DNA damage, as reflected by the higher number of DNA DSBs in plants grown on the 1.5× and 2× H-containing media (Student's test, $\alpha = .05$) (Figure 7.3B). The negative effect of a high concentration of H compound on the plant genome sets restrictions for the application of this chemical for plant transformation. In this respect, the H1× medium resulted in the 2.0-fold induction of HRF and had no significant negative influence on the plant genome (Figure 7.3A, 7.3B). Thus, it can be used in combination with other factors to potentiate their effects on HRF.

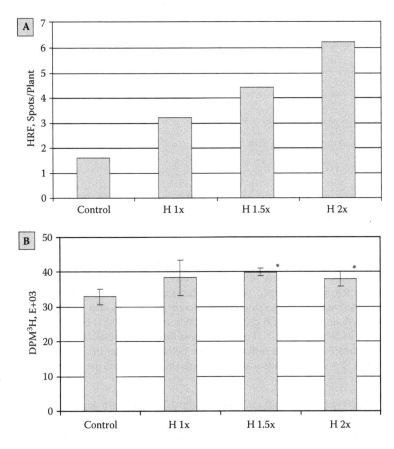

FIGURE 7.3 Effects of the modified growth medium supplemented with different amounts of H chemical on the HRF and DSB levels in *Arabidopsis*. (A) HRF in plants grown on the control MS medium and in the presence of 1×, 1.5×, and 2× of H chemical. (B) The number of DNA DSBs (*DPM 3H*) in plants grown on the control MS medium and in the presence of 1×, 1.5×, and 2× of H chemical. Values represent the mean ± standard deviation. Asterisks show a statistically significant difference from plants grown on the control MS medium (Student's test, $\alpha = .05$).

7.3.3   THE INFLUENCE OF COMBINATIONS OF VARIOUS CHEMICALS ON HRF

Using the described strategies, two media compositions, B and J, were identified that influence HR. The effect of the B medium was described. The effect of the J compound is similar to the previously described H compound, and it has a minor positive influence on HRF. We hypothesized that the simultaneous application of these two modifications in one growth medium could potentiate their effects on HR. To test this hypothesis, we grew *Arabidopsis* plants on the B medium and on the J medium of three various types (1×, 2×, and 3×) and on the combination of the B medium with

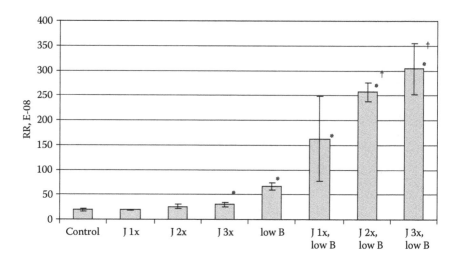

FIGURE 7.4 The potentiation effect of the combined application of J chemical and B medium on RR in *Arabidopsis*. The graph shows changes in RR in plants grown on the control MS medium; on the medium supplemented with 1×, 2×, and 3× of J chemical; on the growth medium depleted of B chemical (B medium); and on the B medium supplemented with 1×, 2×, and 3× of J chemical. Values represent the mean ± standard deviation. Asterisks show a statistically significant difference from plants grown on the control MS medium (Student's test, α = .05). †A statistically significant difference from the B medium (Student's test, α = .05).

each of three types of J medium. The standard MS medium served as the control. HRF was tested 3 weeks postgermination.

We found that 1×, 2×, and 3× J media resulted in a 1- to 1.6-fold increase in HRF, whereas the B medium induced RR by 3.5-fold (Student's test, α = .05) (Figure 7.4). In contrast, we observed a drastic induction of RR by 8.6-, 13.5-, and 16.0-fold in plants grown on the BJ1×, BJ2×, and BJ3× media, respectively, as compared to the control MS medium (Student's test, α = .05) (Figure 7.4). Importantly, RR in plants grown on the BJ2× and BJ3× media was 3.9- and 4.6-fold higher, respectively, than in plants grown on the B medium only (Student's test, α = .05) (Figure 7.4). This supports the potentiation effect that the combined application of two modifications has on HR. These experiments confirmed that the modification of several chemicals can effectively increase HRF.

7.3.4 EFFECTS OF GROWTH MEDIUM ENRICHED WITH G CHEMICAL ON STABLE TRANSFORMATION FREQUENCY

To check our hypothesis that an increase in HRF should be paralleled with an increase in transformation rate, we tested the influence of a G compound on *Agrobacterium*-mediated transformation efficiency. *Nicotiana tabacum* plants were germinated and grown on the G2× and G3× media. Plants grown on the MS medium were used as the control. Leaves of 6-week-old plants were harvested and used for *Agrobacterium*-mediated transformation, as previously published (Horsch et al.

1985). *Agrobacteria* used for transformation carried the T-DNA with the luciferase transgene and the herbicide-resistant gene. Transformed plants were regenerated under active selection, and the number of calli regenerated per single leaf disk used for transformation was calculated. It represented the calli regeneration efficiency (CRE). Stable integration events were confirmed by measuring luciferase expression in plants transplanted to soil. The number of luciferase-expressing plants regenerated per single leaf disk used for transformation represented the stable transformation frequency (STF).

We found that enrichment of a plant growth medium with compound G significantly induced the CRE under active selection conditions (Figure 7.5A). Leaf disks obtained from plants grown on the G2× and G3× media prior to transformation showed higher and faster regeneration compared to leaf disks derived from plants grown on the conventional MS medium. Overall, the CRE was 1.7- and 2.5-fold higher in the G2×- and G3×-derived plants, respectively, as compared to the CRE from the MS medium-derived plants (Student's test, $\alpha = .05$) (Figure 7.5A). Increasing the concentration of G compound from 2× to 3× resulted in a significant 1.5-fold induction of the CRE (Student's test, $\alpha = .05$) (Figure 7.5A). This is consistent with the previously observed dose-dependent induction of HRF.

A similar pattern was observed while analyzing the STF. Plants grown on the G3× medium showed a 271% increase in the STF as compared to plants grown on the control medium (Figure 7.5B). The STF in plants derived from the G2× medium was significantly 1.9-fold higher than in plants derived from the control medium (Student's test, $\alpha = .05$). Finally, a significant 1.4-fold difference was found when the STF in the G2×- and G3×-derived plants was compared (Student's test, $\alpha = .05$) (Figure 7.5B). In general, a strong positive correlation was revealed between STF and HRF ($r = .86$, $p < .05$), supporting an important role of the increased HRF for *Agrobacterium*-mediated transformation.

7.4 MATERIALS AND METHODS

7.4.1 DESCRIPTION OF A TRANSGENIC LINE USED FOR EXPERIMENTS

Identification and analysis of growth medium compositions that influence HRF were performed using the previously described transgenic *Arabidopsis* line 11 plants (Swoboda et al. 1994, Ilnytskyy et al. 2004). *Arabidopsis* plants carried the HR substrate that consisted of two overlapping truncated nonfunctional copies of the GUS gene cloned in direct orientation under the 35S cauliflower mosaic virus (CaMV) promoter. Repair of DSBs in a region of homology via HR results in a recombination event that restores the reporter gene, thereby activating the GUS gene. The GUS gene activity was visualized using a histochemical staining procedure described previously (Ilnytskyy et al. 2004).

7.4.2 CALCULATION OF HRF AND RR

The HRF was calculated by counting the number of HR events (sectors) in each plant separately, summing, and then relating it to the number of plants in tested

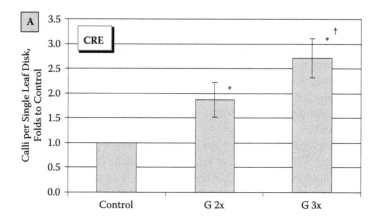

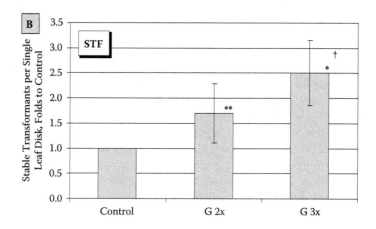

FIGURE 7.5 The effect of G chemical on calli regeneration efficiency (CRE) and stable transformation frequency (STF) in *N. tabacum* plants grown on media supplemented with 2× and 3× of G chemical. Leaf tissues for *Agrobacterium*-mediated transformation were harvested from *N. tabacum* plants grown on the control or modified MS medium supplemented with 2× and 3× of G chemical. (A) CRE represents the total number of calli regenerated (transgenic and nontransgenic) per single leaf disk used for transformation. CRE from plants grown on the control MS medium was standardized to 1.0. CRE on the growth medium containing G chemical showed changes in folds as related to the control MS medium. (B) STF represents the total number of transgenic *luciferase*-expressing plants regenerated per single leaf disk used for transformation. STF from plants grown on the control MS medium was standardized to 1.0. STF on the growth medium containing G chemical showed changes in folds as related to the control MS medium. Values represent the mean ± standard deviation. Asterisks show a statistically significant difference from the control MS medium (Student's test, α = .05). †Statistically significant difference from G 2× medium (Student's test, α = .05).

populations. The RR was calculated by relating the HRF to the number of haploid genomes per single plant. It represented the number of HR events per single-haploid genome. Calculation of RR is important especially for evaluating chemicals that may significantly affect plant growth and thereby the total number of genomes (i.e., reporter gene copies) in a single plant. The number of haploid genomes per single plant was calculated by relating the total DNA yield (in micrograms per plant) to the mean DNA content (0.16 pg) of an *A. thaliana* haploid cell (Swoboda et al. 1993) and the number of plants used for DNA preparation. To avoid bias during DNA preparation, DNA was extracted by two different methods as described previously (Boyko et al. 2005).

7.4.3 GROWTH MEDIA GENOTOXICITY ASSESSMENT

To ensure that chemicals used in the media modification experiments do not result in extensive DNA damage, the number of DNA DSBs was measured. Quantification of 3′OH DNA breaks was performed using the random oligonucleotide primed synthesis (ROPS) assay (Basnakian and James 1996). The assay is based on the ability of the *Klenow fragment* polymerase to initiate random oligonucleotide-primed synthesis from reannealed 3′OH ends of single-stranded DNA (ssDNA). After a denaturation-reassociation step, ssDNA serves as its own primer by randomly reassociating itself or associating with other ssDNA molecules. Under strictly defined reaction conditions, the incorporation of [³H]-dCPT into newly synthesized DNA will be proportional to the initial number of 3′OH ends (breaks). Radiation levels, ³H decays per minute (DPM), reflecting the number of incorporation events can be detected by a scintillation counter.

7.4.4 PLANT TRANSFORMATION

Nicotiana tabacum wild-type plants were transformed with the *Agrobacterium* strain GV3101 that carried the T-DNA containing the active *luciferase* (LUC) gene driven by the *N-gene* promoter and the *hph* gene that confers resistance for antibiotic hygromycin as a selection marker. We harvested 6-week-old plants grown on various modified and control media for *Agrobacterium*-mediated transformation using the leaf-disk transformation procedure described previously (Horsch et al. 1985). Transformed plants were regenerated under active selection conditions (hygromycin, 25 mg/L) and transplanted to soil. The LUC expression was detected using a charge-coupled device (CCD) camera as described previously (Ilnytskyy et al. 2004). Luciferase-expressing plants were counted as stable transformants, and their number was related to the total number of plants regenerated to obtain an STF. In addition, shoots produced on the calli-inducing medium were scored, and the resulting number represented the CRE.

7.5 CONCLUSIONS

Plant transformation is a complex process that depends on a number of various factors. The activity of the host DNA repair machinery represents one of the key factors that influence transgene integration efficiency. In our study, we demonstrated that high activity of HR at the time of transformation can significantly improve the STF. We suggest that the modification of a plant growth medium, using factors that affect

HRF, represents a highly efficient and low-cost strategy for enhancing plant transformation. The type of chemicals influencing HRF can be successfully identified using transgenic plants that carry a recombination substrate in their genome. This system made it possible to identify and describe several chemicals that increase HRF. Furthermore, the application of a selected chemical resulted in a significant increase in the number of transgenic plants produced in a single round of transformation. Active concentrations of these compounds for each individual plant species and, perhaps, for each transformation method should be identified experimentally.

ACKNOWLEDGMENTS

We thank Valentina Titova for proofreading the manuscript and acknowledge the financial support of AARI, NSERC, and AVAC.

REFERENCES

Basnakian, A.G., James, S.J. 1996. Quantification of 3'OH DNA breaks by random oligonucleotide primed synthesis (ROPS) assay. *DNA Cell Biol* 15(3):255–262.

Boyko, A., Filkowski, J., Hudson, D., Kovalchuk, I. 2006a. Homologous recombination in plants is organ specific. *Mutat Res* 595(1–2):145–155.

Boyko, A., Filkowski, J., Kovalchuk, I. 2005. Homologous recombination in plants is temperature and day-length dependent. *Mutat Res* 572(1–2):73–83.

Boyko, A., Kovalchuk, I. Composition and method for enhancing plant transformation and homologous recombination. U.S. Patent 11/466,184. August 22, 2006.

Boyko, A., Zemp, F., Filkowski, J., Kovalchuk, I. 2006b. Double-strand break repair in plants is developmentally regulated. *Plant Physiol* 141(2):488–497.

Chilton, M.-D.M., Que, Q. 2003. Targeted integration of T-DNA into the tobacco genome at double-strand breaks: new insights on the mechanism of T-DNA integration. *Plant Physiol* 133:956–965.

Citovsky, V., Kozlovsky, S.V., Lacroix, B., Zaltsman, A., Dafny-Yelin, M., Vyas, S., Tovkach, A., Tzfira, T. 2007. Biological systems of the host cell involved in *Agrobacterium* infection. *Cell Microbiol* 9(1):9–20.

Gelvin, S.B. 2003. *Agrobacterium*-mediated plant transformation: the biology behind the "gene-jockeying" tool. *Microbiol Mol Biol Rev* 67:16–37.

Gorbunova, V., Levy, A.A. 1999. How plants make ends meet; DNA double strand break repair. *Trends Plant Sci* 4:263–269.

Hanin, M., Paszkowski, J. 2003. Plant genome modification by homologous recombination. *Curr Opin Plant Biol* 6(2):157–162.

Horsch, R.B., Fry, J.E., Hoffman, N.L., Eichholtz, D., Rogers, S.G., Fraley, R.T. 1985. A simple and general method for transferring genes into plants. *Science* 227(4691):1229–1231.

Ilnytskyy, Y., Boyko, A., Kovalchuk, I. 2004. Luciferase-based transgenic recombination assay is more sensitive than beta-glucoronidase-based. *Mutat Res* 559(1–2):189–197.

Kohler, F., Cardon, G., Pohlman, M., Gill, R., Schider, O. 1989. Enhancement of transformation rates in higher plants by low-dose irradiation: are DNA repair systems involved in incorporation of exogenous DNA into the plant genome? *Plant Mol Biol* 12:189–199.

Kovalchuk, O., Titov, V., Hohn, B., Kovalchuk, I. 2001. A sensitive transgenic plant system to detect toxic inorganic compounds in the environment. *Nature Biotechnol* 19:568–572.

Lacroix, B., Tzfira, T., Vainstein, A., Citovsky, V. 2006. A case of promiscuity: *Agrobacterium's* endless hunt for new partners. *Trends Genet* 22:29–37.

Leskov, K.S., Criswell, T., Antonio, S., Yang, C.R., Kinsella, T.J., Boothman, D.A. 2001. When X-ray-inducible proteins meet DNA double strand break repair. *Semin Radiat Oncol* 11:352–372.

Lida, S., Terada, R. 2004. A tale of two integrations, transgene and T-DNA: gene targeting by homologous recombination in rice. *Curr Opin Biotechnol* 15:132–138.

Murashige, T., Skoog, F. 1962. A revised medium for rapid growth and bioassays with tobacco tissue cultures. *Physiol Plant* 15:473–497.

Puchta, H. 2002. Gene replacement by homologous recombination in plants. *Trends Plant Sci* 1:340–348.

Puchta, H. 2005. The repair of double-strand breaks in plants: mechanisms and consequences for genome evolution. *J Exp Bot* 56:1–14.

Reiss, B. 2003. Homologous recombination and gene targeting in plant cells. *Int Rev Cytol* 228:85–139.

Roa-Rodriguez, C., Nottenburg, C. 2003. *Agrobacterium*-mediated transformation in plants. *CAMBIA*. http://www.bios.net/Agrobacterium.

Shrivastav, M., De Haro, L.P., Nickoloff, J.A. 2008. Regulation of DNA double-strand break repair pathway choice. *Cell Res* 18(1):134–147.

Swoboda, P., Gal, S., Hohn, B., Puchta, H. 1994. Intrachromosomal homologous recombination in whole plants. *Eur Mol Biol Org J* 13:484–489.

Swoboda, P., Hohn, B., Gal, S. 1993. Somatic homologous recombination *in planta*: the recombination frequency is dependent on the allelic state of recombining sequences and may be influenced by genomic position effects. *Mol Gen Genet* 237(1–2):33–40.

Tzfira, T., Citovsky, V. 2003. The *Agrobacterium*-plant cell interaction. Taking biology lessons from a bug. *Plant Physiol* 133(3):943–947.

Tzfira, T., Citovsky, V. 2006. *Agrobacterium*-mediated genetic transformation of plants: biology and biotechnology. *Curr Opin Biotechnol* 17(2):147–154.

Tzfira, T., Frankman, R., Vaidya, M., Citovsky, V. 2003. Site-specific integration of *Agrobacterium* T-DNA via double-stranded intermediates. *Plant Physiol* 133:1011–1023.

Tzfira, T., Li, J., Lacroix, B., Citovsky, V. 2004. *Agrobacterium* T-DNA integration: molecules and models. *Trends Genet* 20:375–383.

van Attikum, H., Bundock, P., Hooykaas, P.J.J. 2001. Non-homologous end-joining proteins are required for *Agrobacterium* T-DNA integration. *Eur Mol Biol Org J* 20:6550–6558.

van Attikum, H., Hooykaas, P.J.J. 2003. Genetic requirements for the targeted integration of *Agrobacterium* T-DNA in *Saccharomyces cerevisiae*. *Nucleic Acids Res* 31:826–832.

Vergunst, A.C., Hooykaas, P.J.J. 1999. Recombination in the plant genome and its application in biotechnology. *Crit Rev Plant Sci* 18:1–31.

8 Bifidobacterial Lacto-*N*-biose/ Galacto-*N*-biose Pathway Involved in Intestinal Growth

Motomitsu Kitaoka and Mamoru Nishimoto

CONTENTS

Key Words: *Bifidobacterium longum*; bifidus factor; *Clostridium perfringens*; lacto-*N*-biose I; lacto-*N*-biose phosphorylase; Leloir pathway; LNB/GMB pathway; milk oligosaccharides.

8.1 INTRODUCTION

The terms *probiotics* and *prebiotics* have become key words in the food technology industry (Chow 2002, Collins and Gibson 1999, Schrezenmeir and de Vrese 2001). Probiotic describes a live microbial food ingredient with potentially beneficial effects on health beyond basic nutrition. Prebiotics connotes food ingredients that potentially increase probiotic intestinal bacteria in vivo. Bifidobacteria, as well as lactobacilli, are the main targets of probiotics and prebiotics.

Intestinal colonization by bifidobacteria has been an important topic, especially in pediatrics, because the colonization seems to prevent infection by some pathogenic bacteria and diarrhea (Bezkorovainy 1989). It has been known for some time that within a week after birth, breast-fed infants form intestinal flora consisting predominantly of bifidobacteria, which account for 95%–99.9% of the intestinal flora (Benno and Mitsuoka 1986, Rotimi and Duerden 1981). Conversely, bifidobacteria

113

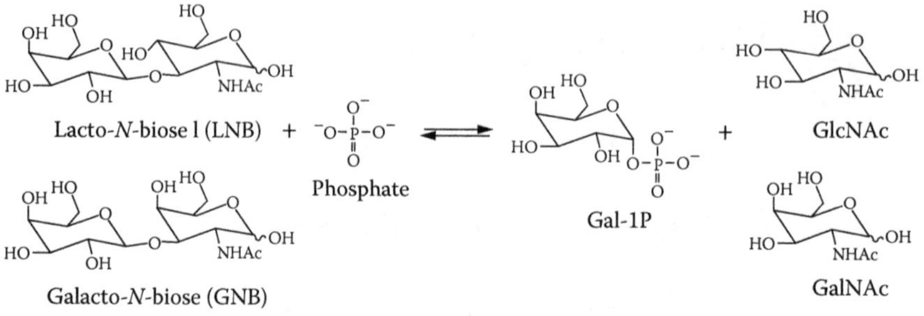

Biocatalysis and Agricultural Biotechnology

FIGURE 8.1   Reaction of lacto-*N*-biose phosphorylase.

do not readily colonize in bottle-fed infants, who often suffered infections of pathogenic bacteria early in the 20th century. To improve the growth of intestinal bifidobacteria, infant formula is now supplemented with prebiotic oligosaccharides such as lactulose, resulting in healthier bottle-fed infants. However, the intestinal flora of breast-fed infants are still different from those of bottle-fed infants, with a composition of 90% bifidobacteria and 10% enterobacteriaceae (Benno et al. 1984).

The growth factors of bifidobacteria in human milk have been investigated previously. Initially, a nitrogen-containing sugar was thought to be the required nutrient, but its success was found to be due to a special kind of *Bifidobacterium bifidum* strain that required GlcNAc for its growth (Gyorgy et al. 1954, Veerkamp 1969). Further studies revealed that oligosaccharides in human milk (human milk oligosaccharides, HMOs) are good candidates as real bifidus factors (Bezkorovainy 1989). Cow's milk contains only lactose as the sole saccharide, whereas human milk contains various oligosaccharides as well as lactose (Kunz et al. 2000, Newburg and Neubauer 1995). However, it has not been clear what oligosaccharide or residue in HMOs causes the increase in the bifidobacteria.

In 1999, β-1,3-galactosyl-*N*-acetylhexosamine phosphorylase (EC 2.4.1.211) was found in a cell-free extract of *B. bifidum*, which reversibly phosphorolyzed lacto-*N*-biose I (LNB, Galβ1 → 3GlcNAc) and galacto-*N*-biose (GNB, Galβ1 → 3GalNAc) (Figure 8.1) (Derensy-Dron et al. 1999). We named this enzyme lacto-*N*-biose phosphorylase (LNBP; Kitaoka et al. 2005). The galactose-metabolizing pathway that includes this enzyme may explain the real bifidus factor in HMOs. In this chapter, we describe our hypothesis for how bifidobacteria grow in the presence of HMOs.

8.2   PURIFICATION AND CLONING OF LNBP

Lacto-*N*-biose phosphorylase was purified from a cell-free extract of *B. bifidum* JCM1254 using hydrophobic and anion-exchange chromatography (Kitaoka et al. 2005). The purified LNBP appeared as a single band on sodium dodecyl sulfate polyacrylamide gel electrophoresis (SDS-PAGE) at 86 kDa. Its specific activity was 10 U/mg protein. The N-terminal and two internal amino acid sequences were determined to be STSGR FTIPS ESNFA EKTAE LARLW GADAV, YGYRL RPEDF VNEGS

TABLE 8.1

Homologue of the Lacto-*N*-biose Phosphorylase Gene from
***B. longum* JCM1217**

Strain	Locus	Identity (%)	Remark
Bifidobacterium longum NCC2705	BL1641	97	Intestinal
Bifidobacterium longum DJ010A	(Shotgun)	97	Intestinal
Clostridium perfringens str. 13	CPE0573	47	Intestinal/pathogen
Propionibacterium acnes KPA171202	PPA0083	46	Skin resident
Vibrio vulnificus CMCP6	VV21091	38	Pathogen
Vibrio vulnificus YJ016	VVA1614	38	Pathogen

YNSAW RVPRK, and LGGID FGEPI ADTFP VNEDV TLLRA DGGQV, respectively. On BLAST searches, all the sequences showed a high degree of similarity with a hypothetical protein encoded by the BL1641 gene of *B. longum* NCC2705, for which the complete genome sequence was available (Schell et al. 2002).

Then we cloned a gene corresponding to BL1641 from the type strain *B. longum* JCM1217. An LNBP gene, *lnpA*, was successfully amplified using the genomic DNA of *B. longum* JCM1217 as the template, with a set of primers designed based on the DNA sequence of *B. longum* NCC2705 around BL1641. The gene (AB181926) encoded a protein of 752 amino acid residues with a predicted molecular weight of 84,327. The amino acid sequence showed 97% identity to that of the BL1641 gene with the same length and no gaps. On the other hand, the LNBP gene of *B. bifidum* JCM1254 could not be amplified using the same primer set. The *lnpA* gene from *B. longum* JCM1217 was inserted into an expression vector to form pET28a-lnbp and transformed into *Escherichia coli* BL21. The recombinant enzyme was purified, and *B. longum* showed specific activity (19 U/mg) similar to that of native LNBP from *B. bifidum* (10 U/mg). It phosphorolyzed both LNB and GNB, indicating that the recombinant protein was LNBP.

BLAST searches with the amino acid sequence of LNBP gave hits for only six proteins, including BL1641 (Table 8.1). No significant identity to any proteins with known function was found, indicating that LNBP should be classified into a new family. It should be noted that no genes homologous to *lnpA* were found in other major intestinal bacteria, such as *Lactobacillus*, *Bacteroides*, and *Escherichia*.

8.3 GENE CLUSTER INCLUDING THE LNBP GENE

The LNBP gene, BL1641, seems to be located in a putative operon in the *B. longum* genome, as shown in Figure 8.2. In the operon, BL1638–1640 genes are annotated as component proteins of the adenosine triphosphate (ATP)-binding cassette (ABC)-type sugar transporter. BL1642, BL1643, and BL1644 are annotated as mucin desulfatase, galactose-1-phosphate uridylyltransferase (EC 2.7.7.10), and uridine diphosphate (UDP)-glucose 4-epimerase (EC 5.1.3.2), respectively. We cloned the gene corresponding to BL1642, BL1643, and BL1644 (*lnpB*, *lnpC*, and *lnpD*,

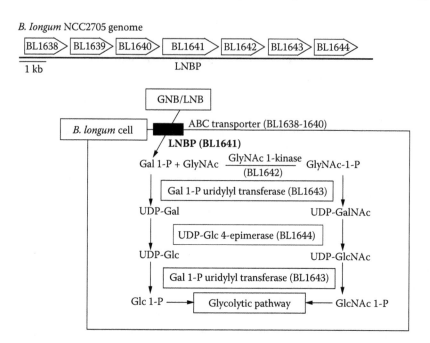

FIGURE 8.2 The operon including LNBP and the LNB/GNB pathway.

respectively) from the type strain *B. longum* JCM1217 (Nishimoto and Kitaoka 2007a).

Although BL1642 was annotated as mucin desulfatase, no homology with other desulfatases was observed. It was highly homologous to a hypothetical protein neighboring a mucin desulfatase gene from *Prevotella* sp. RS2 and moderately homologous to some kinases. LnpB protein did not hydrolyze GlcNAc 6-sulfate, suggesting that it was not a desulfatase. Then, we examined the kinase activity of LnpB on GlcNAc. Thin-layer chromatographic (TLC) analysis revealed a new spot in the position of a sugar phosphate. The compound was later confirmed to be α-GlcNAc1*P*, not GlcNAc6*P*, by both high-performance ion chromatographic (HPIC) and nuclear magnetic resonance (NMR) analysis, indicating that the protein encoded by the *lnpB* gene has a kinase activity that phosphorylates the α-anomeric hydroxyl group. This protein had similar activity on GalNAc, also yielding α-GalNAc1*P*, and weak activity on several monosaccharides. We therefore propose the name *N*-acetylhexosamine 1-kinase (NahK) for this enzyme. The optimum temperature for this reaction was 40°C and the optimum pH 8.5. The enzyme was stable up to 30°C and at pH 5.0 to 9.5. Its K_m values against GlcNAc and GalNAc were 0.118 and 0.065 m*M*, respectively. Its k_{cat} values were 1.21 and 0.752 s^{-1}, respectively, and its K_m value against ATP was 0.172 m*M*.

BL1643 was annotated to be a galactose 1-phosphate uridylyltransferase (EC 2.7.7.10) (GalT2), meaning the enzyme transfers the UMP (uridine monophosphate) unit of UTP (uridine triphosphate) to Gal1*P*. Although the LnpC protein did not have this activity, it showed UDP-glucose-hexose 1-phosphate uridylyltransferase

(EC 2.7.7.12) (GalT) activity, transferring the UMP unit from UDP-Glc to Gal1P with the specific activity of 1.08 U/mg protein. Unusually, the enzyme also transferred the UMP unit to GlcNAc1P (2.24 U/mg protein) and GalNAc1P (2.30 U/mg protein). The LnpD had UDP-glucose 4-epimerase (GalE) activity (158 U/mg), as well as epimerized UDP-GalNAc into UDP-GlcNAc (147 U/mg protein). We therefore describe the catalytic activity of the LnpC and LnpD proteins as BLGalT and BLGalE, respectively.

Taken together, the enzyme activities of the BL1641–1644 homologues suggest that this gene cluster may function as an operon for LNB/GNB. LNB and GNB generated outside the cell are thought to be transported into the cell by a protein encoded by a putative ABC transporter gene located upstream of this cluster (BL1638–1640). Subsequently, LNB/GNB can be phosphorolyzed by LNBP to form Gal1P and GlcNAc/GalNAc. Gal1P is converted to Glc1P by the interaction of BLGalT and BLGalE and then joins the glycolytic pathway. In contrast, GlcNAc/GalNAc can be phosphorylated by NahK to form GlcNAc1P/GalNAc1P, which are converted by BLGalT and BLGalE, with GlcNAc1P entering an amino sugar metabolic pathway. Thus, the operon encodes a sufficient number of enzymes to metabolize entire LNB and GNB molecules to substrates of a subsequent metabolic pathway. We call the metabolic pathway encoded by this operon the LNB/GNB pathway (Figure 8.2).

8.4   ROLE OF THE LNB/GNB PATHWAY

It is valuable to find the origins of LNB and GNB in human tissues to understand the role of the LNB/GNB pathway in intestinal growth of bifidobacteria. GNB residues exist in abundance in mucous membranes as O-linked glycoproteins. Since bifidobacteria possess endo-α-N-acetylgalactosaminidase that releases O-α-linked GNB (Fujita et al. 2005), the operon is considered to play a key role in intestinal colonization by utilizing GNB as a core 1 sugar chain in mucin.

We have also noticed that this operon clearly explains the long-unresolved question of why bifidobacteria are the predominant flora of breast-fed infants. When mature bovine milk is compared with human milk, the major difference is in the oligosaccharide content. Human milk contains various oligosaccharides having LNB residues at the nonreducing end, whereas mature bovine milk does not contain oligosaccharides other than lactose. Furthermore, the bovine colostrum contains a small amount of oligosaccharides, but LNB residues have not been found in these oligosaccharides. Because lacto-N-tetraose (Galβ1 $\rightarrow$ 3GlcNAcβ1 $\rightarrow$ 3Galβ1 $\rightarrow$ 4Glc) and lacto-N-fucopentaose I (Fucα1 $\rightarrow$ 2Galβ1 $\rightarrow$ 3GlcNAcβ1 $\rightarrow$ 3Galβ1 $\rightarrow$ 4Glc), which contain LNB residues, are predominant components in HMOs as well as 2′-fucosyllactose, it is reasonable to postulate that the LNB residues in HMOs are metabolized in the LNB/GNB pathway if bifidobacteria possess lacto-N-biosidase and α-fucosidase. Both enzymes have been found in *B. bifidum* (Wada et al. 2008, Katayama et al. 2004). Thus, it is reasonable to hypothesize that the LNB residues in HMOs act as the bifidus factor in breast-fed infants (Figure 8.3). It should be noted that human colostrums contain more oligosaccharides than mature human milk, aiding in the quick colonization of newborn infants (Asakuma et al. 2007, 2008).

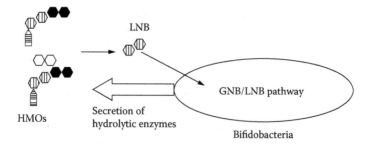

FIGURE 8.3 The LNB hypothesis. Bifidobacteria secrete hydrolytic enzymes to liberate LNB from HMOs, and the resultant LNB is transported into the bifidobacterial cells and metabolized through the LNB/GNB pathway.

8.5 LNBP HOMOLOGUE OF *CLOSTRIDIUM PERFRINGENS*

Bacterial genomic projects have revealed that *Clostridium perfringens*, *Propionibacterium acnes*, and *Vibrio vulnificus* have LNBP gene homologs. Among them, *C. perfringens* is an intestinal bacterium that is considered to be undesirable for human health. Thus, it is important to compare the characteristics of the LNBP homologue from *C. perfringens* with LNBP from a bifidobacterium to design a bifidus factor for avoiding the growth of *C. perfringens*.

Clostridium perfringens is a Gram-positive anaerobic bacterium often isolated from the gastrointestinal tract of animals and humans, soil, and sewage. It is well known that *C. perfringens* strains cause food poisoning, gas gangrene, and septicemia (McDonel 1980, Petit et al. 1999). Genomic sequences of three strains of *C. perfringens* (strain 13, ATCC13124, and SM101) are available (Shimizu et al. 2002, Myers et al. 2006). We cloned the cpf0553 gene from the type strain *C. perfringens* ATCC13124 and examined the substrate specificity of this enzyme (Nakajima et al. 2008).

The CPF0553 protein phosphorolyzed GNB much faster than LNB, whereas the bifidobacterial LNBP phosphorolyzed them with similar rates (Table 8.2). Thus, the protein should be named galacto-*N*-biose phosphorylase (GNBP). GNBP was stable up to 45°C, and its optimum pH was 6.5–7.0.

The difference in the preference of the substrates indicates that LNBP seems to be important in the metabolism of both GNB and LNB and contributes to the colonization of *B. longum* in the large intestine and usage of HMOs, whereas GNBP is

TABLE 8.2
Comparison of LNBP from *B. longum* and GNBP from *C. perfringens*

Enzyme	Activity on LNB (s^{-1})	Activity on GNB (s^{-1})
LNBP from *B. longum*	20.8	36.8
GNBP from *C. perfringens*	0.4	5.0

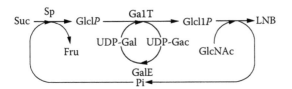

FIGURE 8.4 Enzymatic preparation of LNB from sucrose. Sucrose is phosphorolyzed by sucrose phosphorylase to form Glc1P. The resultant Glc1P is converted into Gal1P by the concerted actions of GalT and GalE in the presence of UDP-Glc. Finally, LNB is generated from the resultant Gal1P and GlcNAc. Overall, sucrose and GlcNAc are converted into LNB and fructose by the action of the four enzymes in the presence of catalytic amounts of phosphate and UDP-Glc.

important only for GNB metabolism. The substrate specificity of GNBP was reasonable for usage of a carbon source in the large intestine. GNB was abundant as the disaccharide core structure of mucin, and *C. perfringens* ATCC13124 has an endo-α-*N*-acetylgalactosaminidase (Fujita et al. 2005) gene homologue (*cpf0685*), which liberates the core 1 structure from mucin.

8.6 CONCLUDING REMARKS

The existence of the bifidobacterial LNB/GNB pathway explains how HMOs act as the bifidus factor in human milk. We have already developed an enzymatic method for large-scale LNB production (Figure 8.4) (Nishimoto and Kitaoka 2007b). Since LNB has been hypothesized to be the real bifidus factor in HMOs, it may be applicable as an additive to infant formula to produce bifidus flora in bottle-fed as well as breast-fed infants.

REFERENCES

Asakuma, S., Akahori, M., Kimura, K., Watanabe, Y., Nakamura, T., Tsunemi, M., Arai, I., Sanai, Y., and Urashima, T. 2007. Sialyl oligosaccharides of human colostrum: changes in concentration during the first three days of lactation. *Biosci. Biotechnol. Biochem.* 71: 1447–1451.

Asakuma, S., Urashima, T., Akahori, M., Obayashi, H., Nakamura, T., Kimura, K., Watanabe, Y., Arai, I., and Sanai, Y. 2008. Variation of major neutral oligosaccharides levels in human colostrum. *Eur. J. Clin. Nutr.* 62: 488–494.

Benno, Y., and Mitsuoka, T. 1986. The development of gastrointestinal micro-flora in humans and animals. *Bifidobacteria Microflora* 5: 13–25.

Benno, Y., Sawada, K., and Mitsuoka, T. 1984. The intestinal microflora of infants: composition of fecal flora in breast-fed and bottle-fed infants. *Microbiol. Immunol.* 28: 975–986.

Bezkorovainy, A. 1989. Ecology of bifidobacteria. In *Biochemistry and Physiology of Bifidobacteria*, ed. A. Bezkorovainy and R. Miller-Catchpole, pp. 29–72. Cleveland, OH: CRC Press.

Chow, J. 2002. Probiotics and prebiotics: a brief overview. *J. Ren. Nutr.* 12: 76–86.

Collins, M.D., and Gibson, G.R. 1999. Probiotics, prebiotics, and synbiotics: approaches for modulating the microbial ecology of the gut. *Am. J. Clin. Nutr.* 69: 1052S–1057S.

Derensy-Dron, D., Krzewinski, F., Brassart, C., and Bouquelet, S. 1999. Beta-1,3-galactosyl-N-acetylhexosamine phosphorylase from *Bifidobacterium bifidum* DSM 20082: characterization, partial purification and relation to mucin degradation. *Biotechnol. Appl. Biochem.* 29: 3–10.

Fujita, K., Oura, F., Nagamine, N., Katayama, T., Hiratake, J., Sakata, K., Kumagai, H., and Yamamoto, K. 2005. Identification and molecular cloning of a novel glycoside hydrolase family of core 1 type O-glycan-specific endo-alpha-N-acetylgalactosaminidase from *Bifidobacterium longum*. *J. Biol. Chem.* 280: 37415–37422.

Gyorgy, P., Rose, C.S., and Springer, G.F. 1954. Enzymatic inactivation of bifidus factor and blood group substances. *J. Lab. Clin. Med.* 43: 543–552.

Katayama, T., Sakuma, A., Kimura, T., Makimura, Y., Hiratake, J., Sakata, K., Yamanoi, T., Kumagai, H., and Yamamoto, K. 2004. Molecular cloning and characterization of *Bifidobacterium bifidum* 1,2-alpha-L-fucosidase (AfcA), a novel inverting glycosidase (glycoside hydrolase family 95). *J. Bacteriol.* 186: 4885–4893.

Kitaoka, M., Tian, J., and Nishimoto, M. 2005. Novel putative galactose operon involving lacto-N-biose phosphorylase in *Bifidobacterium longum*. *Appl. Environ. Microbiol.* 71: 3158–3162.

Kunz, C., Rudloff, S., Baier, W., Klein, N., and Strobel, S. 2000. Oligosaccharides in human milk: structural, functional, and metabolic aspects. In *Annual Review of Nutrition*, Vol. 20, ed. D. B. McCormick, D. M. Bier, and R. J. Cousins, pp. 699–722. Palo Alto, CA: Annual Reviews.

McDonel, J.L. 1980. *Clostridium perfringens* toxins (type A, B, C, D, E). *Pharmacol. Ther.* 10: 617–655.

Myers, G.S.A., Rasko, D.A., Cheung, J.K., Ravel, J., Seshadri, R., DeBoy, R.T., Ren, Q.H., Varga, J., Awad, M.M., Brinkac, L.M., Daugherty, S.C., Haft, D.H., Dodson, R.J., Madupu, R., Nelson, W.C., Rosovitz, M.J., Sullivan, S.A., Khouri, H., Dimitrov, G.I., Watkins, K.L., Mulligan, S., Benton, J., Radune, D., Fisher, D.J., Atkins, H.S., Hiscox, T., Jost, B.H., Billington, S.J., Songer, J.G., McClane, B.A., Titball, R.W., Rood, J.I., Melville, S.B., and Paulsen, I.T. 2006. Skewed genomic variability in strains of the toxigenic bacterial pathogen, *Clostridium perfringens*. *Genome Res.* 16: 1031–1040.

Nakajima, M., Nihira, T., Nishimoto, M., and Kitaoka, M. 2008. Identification of galacto-N-biose phosphorylase from *Clostridium perfringens* ATCC13124. *Appl. Microbiol. Biotechnol.* 78: 465–471.

Newburg, D.S., and Neubauer, S.H. 1995. Carbohydrate in milks: analysis, quantities, and significance. In *Handbook of Milk Composition*, ed. R.G. Jensen, pp. 273–349. San Diego, CA: Academic Press.

Nishimoto, M., and Kitaoka, M. 2007a. Identification of N-acetylhexosamine 1-kinase in the complete lacto-N-biose I/galacto-N-biose metabolic pathway in *Bifidobacterium longum*. *Appl. Environ. Microbiol.* 73: 6444–6449.

Nishimoto, M., and Kitaoka, M. 2007b. Practical preparation of lacto-N-biose I, a candidate for the bifidus factor in human milk. *Biosci. Biotechnol. Biochem.* 71: 2101–2104.

Petit, L., Gibert, M., and Popoff, M.R. 1999. *Clostridium perfringens*: toxinotype and genotype. *Trends Microbiol.* 7: 104–110.

Rotimi, V.O., and Duerden, B.I. 1981. The development of the bacterial flora in normal neonates. *J. Med. Microbiol.* 14: 51–58.

Schell, M.A., Karmirantzou, M., Snel, B., Vilanova, D., Berger, B., Pessi, G., Zwahlen, M.C., Desiere, F., Bork, P., Delley, M., Pridmore, R.D., and Arigoni, F. 2002. The genome sequence of *Bifidobacterium longum* reflects its adaptation to the human gastrointestinal tract. *Proc. Natl. Acad. Sci. U.S.A.* 99: 14422–14427.

Schrezenmeir, J., and de Vrese, M. 2001. Probiotics, prebiotics, and synbiotics: approaching a definition. *Am. J. Clin. Nutr.* 73: 361s–364s.

Shimizu, T., Ohtani, K., Hirakawa, H., Ohshima, K., Yamashita, A., Shiba, T., Ogasawara, N., Hattori, M., Kuhara, S., and Hayashi, H. 2002. Complete genome sequence of *Clostridium perfringens*, an anaerobic flesh-eater. *Proc. Natl. Acad. Sci. U.S.A.* 99: 996–1001.

Veerkamp, J.H. 1969. Uptake and metabolism of determinatives of 2-deoxy-2-amino-D-glucose in *Bifidobacterium bifidum* var. *pennsylvanicus*. *Arch. Biochem. Biophys.* 129: 248–256.

Wada, J., Ando, T., Kiyohara, M., Ashida, H., Kitaoka, M., Yamaguchi, M., Kumagai, H., Katayama, T., and Yamamoto, K. 2008. Lacto-*N*-biosidase (LnbBF) from *Bifidobacterium bifidum*, a critical enzyme for degradation of human milk oligosaccharides with type-1 structure. *Appl. Environ. Microbiol.* 74: 3996–4004.

9 Improvement of Agronomic Traits Using Different Isoforms of Ferredoxin for Plant Development and Disease Resistance

Yi-Hsien Lin, Hsiang-En Huang, and Teng-Yung Feng

CONTENTS

Key Words: Ferredoxin; redox potential; transgenic technology.

9.1 INTRODUCTION

Quality and productivity of crops are always the highest concerns of modern agriculture, and it is now possible to generate novel plant lines with functional properties via transgenic plant technology. Energy resources are becoming exhausted as the world develops, and the use of plant material as biofuel is an important substitute scheme to solve this problem. To increase plant material, strategies were developed based on enhancement of disease resistance and plant growth promotion. The improvement of agronomic traits using different isoforms of ferredoxin (Fd) for plant development and disease resistance of economical crops would be promising in agriculture.

9.2 TYPES OF FERREDOXINS IN HIGHER PLANTS

Plant-type Fd is an electron transfer protein containing a [2Fe-2S] cluster with highly negative redox potential that contains the $CX_4CX_2CX_nC$ ($n = 29$ or $n \neq 29$) pattern. It was subdivided into two different groups, photosynthetic-type Fd that generally exists in the green tissue of plants and the nonphotosynthetic-type Fd that generally exists in the stored tissue (Wada et al. 1986; Takubo et al. 2003; Gou et al. 2006). Eighty-eight different plastid-type Fds were identified from complementary DNA (cDNA) databases of [2Fe-2S]-type Fds prior to 2002 (Bertini et al. 2002). Currently, more information about Fd isoprotein is recorded in the Universal Protein Resource (UniProt) database (http://www.uniprot.org/). There are 121 different Fd isoproteins belonging to the [2Fe-2S] plant-type Fd family in this databank.

9.2.1 PHOTOSYNTHETIC FERREDOXINS

The photosynthetic-type Fd, also called leaf-type Fd (LeFd), delivers reducing equivalents from photosystem I (PSI) to Fd-NADP⁺ reductase for NADP⁺ photoreduction to generate the NADPH (nicotinamide adenine dinucleotide phosphate) needed for CO_2 fixation. Electrons are transferred to Fd from PSI, then they could be transferred through Fd:NADP⁺ reductase (FNR) and Fd:thioredoxin reductase (FTR) for the Calvin cycle (CO_2 assimilation) or proton gradient regulation (PGR5) protein for cyclic electron transfer (Figure 9.1). Fd also plays a substantial role in electron transfer to alternative plastid enzymes, such as Fd-dependent nitrate reductase (NiR), Fd-dependent sulfite reductase (SiR), Fd-dependent glutamate synthase (Fd-Gogat), FNR, and FTR. These enzymes are involved in fundamental metabolic processes, including photosynthesis, nitrogen metabolism, fatty acid synthesis, active oxygen species production, and cyclic electron flow that eliminates the excess of reducing power and prevents uncontrolled overreduced states in the stroma (Arnon 1989; Curdt et al. 2000; Buchanan and Luan 2005; Meyer 2001; Geigenberger et al. 2005; Munekage et al. 2002).

9.2.2 NONPHOTOSYNTHETIC FERREDOXINS

Nonphotosynthetic Fds include root-type Fd. Root-type Fds can receive electrons from NADPH by FNR in plastids in roots and nongreen tissues (Balmer et al. 2006). In *Arabidopsis*, root-type Fds (AtFd3) were revealed to have a less-negative charge than leaf-type Fds (AtFd1 and AtFd2) as shown in Figure 9.1 (Hanke et al. 2004).

9.3 APPLICATION OF THE *pflp* GENE IN TRANSGENIC PLANTS

We were the first laboratory to discover that Fds not only are involved in plant development but also contribute to enhance disease resistance. PFLP is the Fd-like protein

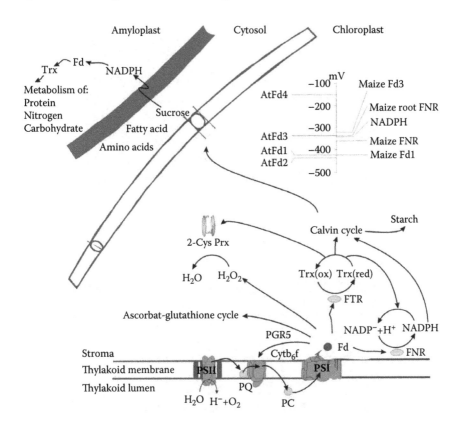

FIGURE 9.1 Plant-type ferredoxin (Fd) isoforms with diverse redox potential could donate or receive electrons involved in many metabolic processes (modified from the model of chloroplast described in Tognetti et al. 2006 and Hanke et al. 2004). PSI and PSII, photosystem I and II, respectively; FNR, Fd:NADP reductase; FTR, Fd:thioredoxin reductase; Trx, thioredoxin; PGR5, proton gradient regulation protein; 2-Cys Prx, 2-Cys peroxiredoxin; PQ, plastoquinone; PC, plastocyanin.

(11 kDa) isolated from sweet pepper in our laboratory (Dayakar et al. 2003). It shares 87% similarity with the *fer_tobac* gene encoding LeFd of tobacco (P83526) and 73.42% and 74.82% with LeFd of *Arabidopsis* AtFd1 and AtFd2, respectively. Its cDNA (*pflp* gene) was cloned, and its transgenic plants had disease resistance against many bacterial diseases. In addition, one of the Fd isoforms could promote fruit ripening and enhance the growth rate of transgenic plants.

9.3.1 MECHANISM OF PFLP FOR PLANT DISEASE RESISTANCE

In our study, the hypersensitive response (HR), the effective mechanism for disease resistance, was only observed in *pflp*-transgenic plants when a low concentration of harpin$_{Pss}$ was infiltrated into transgenic and wild-type leaves (Figure 9.2). In wild-type tobacco plants, total leaf-type Fd levels were increased by inoculation with leaf spot pathogen of bean *Pseudomonas syringae* pv. *syringae* (Pss) but were reduced by

T-SPFLP10-1   Wt

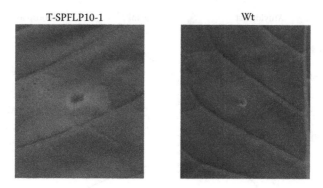

FIGURE 9.2 The hypersensitive response (HR) was induced after treatment of harpin$_{Pss}$ in *pflp*-transgenic tobacco leaves (T-SPFLP10-1), but HR was not induced in wild-type (Wt) leaves.

Wt   T-SPFLP18-1   T-SPFLP10-1

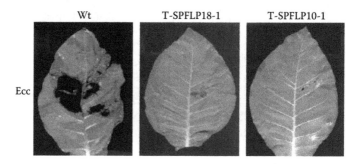

FIGURE 9.3 Resistance of *pflp*-transgenic tobacco plants (T-SPFLP18-1 and T-SPFLP10-1) against bacterial soft rot disease caused by *E. carotovora* subsp. *carotovora* (Ecc) was observed.

soft-rpt pathogen Ecc. In *pflp*-transgenic tobacco plants, HR-like necrosis and H$_2$O$_2$ accumulation were induced immediately after inoculation of the soft rot pathogen *Erwinia carotovora* subsp. *carotovora* (Ecc) in leaves. And, the proliferation of the soft rot pathogen was highly retarded after inoculation in transgenic plants (Huang et al. 2004). The *pflp*-transgenic tobacco plants also showed resistance against soft rot disease (Figure 9.3). Transgenic plants with the antisense *pflp* gene could reduce total leaf-type Fd, and transgenic plants also showed susceptibility to *Pseudomonas fluorescens* (Huang et al. 2007a). These results indicate that the immunity of transgenic plants was highly related to the HRs that were induced during pathogen attack.

9.3.2 Fds for Biotic Stress Tolerance

According to the findings of plant immunity in the tobacco plant triggered by HR, we thought that broad-spectrum disease resistance against bacterial pathogens could be induced in *pflp*-expressing transgenic plants. Many *pflp*-transgenic plants of different crops were obtained and revealed disease resistance against bacterial

FIGURE 9.4   The bacterial leaf blight symptom caused by *Xanthomonas oryzae* pv. *oryzae* on wild-type (A) and *pflp*-transgenic rice plants (B).

FIGURE 9.5   Symptoms of bacterial soft rot caused by Ecc (A) and wilt disease caused by *Ralstonia solanacearum* (B) on *pflp*-transgenic tomato plants (24-18-7) and wild-type plant (Cln1558a).

pathogens, including transgenic rice against bacterial leaf blight (Figure 9.4), transgenic broccoli against bacterial soft rot, transgenic tomato against bacterial wilt and soft rot (Figure 9.5), transgenic calla lily against bacterial soft rot (Figure 9.6), and transgenic orchid against bacterial soft rot (Tang et al. 2001, Liau et al. 2003, Huang et al. 2007b, Yip et al. 2007). These findings will provide a new strategy for selecting disease resistance cultivars of different crops efficiently.

9.3.3   Fds for Abiotic Stress Tolerance

The cross talk between abiotic and biotic stress responses in plants depends on generation of reactive oxygen species (Fujita et al. 2006, Mittler, 2002). It is reasonable to suppose that the tolerance of both biotic and abiotic stress will be influenced by the overexpressed Fd in the transgenic plants. It has been demonstrated that the level of photosynthetic-type Fd in a plant is reduced when it is under abiotic stress, such as high light intensity, low temperature, or high concentrations of H_2O_2, herbicides or heavy metals (Tognetti et al.2006, Zimmermann et al. 2004, Erdner et al.

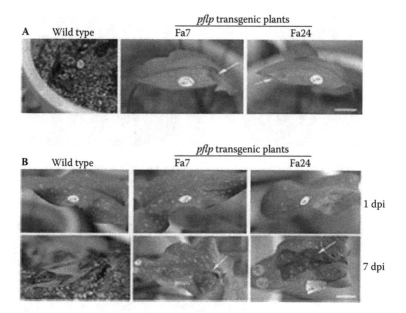

FIGURE 9.6 The bacterial soft rot symptom in the leaves of calla lily caused by Ecc on (A) 8-week old or (B) 4-week-old *pflp*-transgenic lines (Fa7 and Fa24) and wild-type plants.

1999, Mazouni et al. 2003, Debouba et al. 2006, 2007, van Thor et al. 2000, Elliott et al. 1989, John et al. 1997, Vorst et al. 1993). The overexpression of flavodoxin, a non-iron-containing redox protein acting as a substitute for Fd, would enhance the tolerance of abiotic stress, including low temperature, high light intensity, herbicide presence, and iron starvation, in the transgenic tobacco (Tognetti et al. 2006, 2007). In addition, the transgenic transplastomic tobacco-expressing photosynthetic-type Fd from *Arabidopsis* (AtFd2) in its chloroplasts was postulated to enhance tolerance of abiotic stress, particularly under conditions of low light (Yamamoto et al. 2006).

Fd is also able to affect the plant tolerance to abiotic stress through cooperation with Fd-dependent enzymes that catalyze the last step of photosynthetic electron transport in chloroplasts and drive electrons from reduced Fd to $NADP^+$. The transgenic tobacco overexpressing FNR from the pea is more tolerant to high photooxidative damage caused by both high light intensity stress and redox-cycling herbicide paraquat than the nontransgenic one (Rodriguez et al. 2007). Conversely, the FNR-deficient transgenic tobacco plant expressing the antisense FNR transcripts is abnormally prone to photooxidative injury. Exposure of the transgenic plants to moderately high irradiation resulted in rapid loss of photosynthetic capacity and accumulation of singlet oxygen in leaves (Palatnik et al. 2003). Thus, it is clear that the Fd isoproteins play some roles in abiotic stress tolerance.

9.3.4 Fd Designed as a Selective Marker for Transgenic Technology

A novel method for selection of transgenic plants utilizing the *pflp* gene as a selection marker and Ecc as the selection agent has been developed. The use of the *pflp*

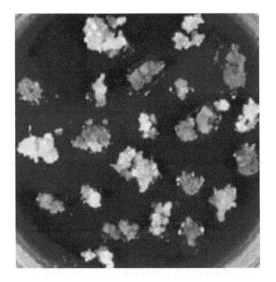

FIGURE 9.7 PFLP protein could be a selection marker for transgenic plant screening of orchid (*Oncidium*) plants. The survivability of *pflp*-transgenic lines (AG2, AG7, AG8, and AG10) and wild-type (WT) orchid plants could be observed after inoculation of soft rot pathogen.

gene as a selection marker (Figure 9.7) may facilitate the use of smaller gene constructs due to removal of bulky antibiotic selection and reporter genes (You et al. 2003).

9.3.5 Fd Isoforms as Source and Sink Switches for Crop Productivity

Fds play an essential role in a number of electron transfer processes in different metabolic pathways involving various redox partners. The redox potentials of plant Fds are typically in the range of −350 to −450 mV. Since the redox potentials of leaf-type Fds have a higher negative charge than −400 mV, the reduced leaf-type Fds tend to donate electrons as sources. In our study, we used the AtFd3 gene from *Arabidopsis*, and results indicated its transgenic *Arabidopsis* plants could enhance the growth rate. Besides, transient expression of AtFd3 was able to promote the early ripening of tomato fruit (unpublished data). These results emphasize that root-type Fd could be a sink switch for increasing crop productivity.

9.4   PERSPECTIVES

According to these new findings, the advantageous features of Fd transgenic plants are as follows:

1. Crop protection: Fd transgenic plants were resistant to many bacterial pathogen genuses, which could provide a strategy to obtain new disease resistance cultivars of different crops.
2. Environment protection: Application of disease resistance Fd transgenic plants can reduce pesticide usage.
3. Increased productivity: Since root-type Fd was able to perform as a sink switch, its application for enhancing productivity of economical crops would be promising agriculturally.
4. Biosafety: Fds are ubiquitous, and their overexpression in transgenic plants might provide lower risk to our environment and human health.

REFERENCES

Arnon, D.I. (1989). The discovery of ferredoxin: the photosynthetic path. *Trends Biochem. Sci.* 13, 30–33.

Bertini, I., Luchinat,C., Provenzani, A., Rosato, A., and Vasos, P.R. (2002). Browsing gene banks for FesS2 ferrodoxins and structural modeling of 88 plant-type sequences: An analysis of fold and function. *Proteins* 46, 110–127.

Balmer, Y., Vensel, W.H., Cai, N., Manieri, W., Schürmann, P., Hurkman, W.J., and Buchanan, B.B. (2006). A complete ferredoxin/thioredoxin system regulates fundamental processes in amyloplasts. *Proc. Natl. Acad. Sci. U.S.A.* 103, 2988–2993.

Buchanan, B.B., and Luan, S. (2005). Redox regulation in the chloroplast thylakoid lumen: a new frontier in photosynthesis research. *J. Exp. Botany* 56, 1439–1447.

Curdt, I., Singh, B.B., Jakoby, M., Hachtel, W., and Bohme, H. (2000). Identification of amino acid residues of nitrite reductase from *Anabaena* sp. PCC 7120 involved in ferredoxin binding. *Biochim. Biophys. Acta* 1543, 60–68.

Dayakar, B.V., Lin, H.-J., Chen, C.-H., Ger, M.-J., Lee, B.-H., Pai, C.-H., Chow, D., Huang, H.-E., Hwang, S.-Y., Chung, M.-C., and Feng, T.-Y. (2003). Ferredoxin from sweet pepper (*Capsicum annuum* L.) intensifying harpin$_{pss}$-mediated hypersensitive response shows an enhanced production of active oxygen species (AOS). *Plant Mol. Biol.* 51, 913–924.

Debouba, M., Gouia, H., Valadier, M.H., Ghorbel, M.H., and Suzuki, A. (2006). Salinity-induced tissue-specific diurnal changes in nitrogen assimilatory enzymes in tomato seedlings grown under high or low nitrate medium. *Plant Physiol. Biochem.* 44, 409–419.

Debouba, M., Maaroufi-Dghimi, H., Suzuki, A., Ghorbel, M.H., and Gouia, H. (2007). Changes in growth and activity of enzymes involved in nitrate reduction and ammonium assimilation in tomato seedlings in response to NaCl stress. *Ann. Bot.* (London) 99, 1143–1151.

Elliott, R.C., Dickey, L.F., White, M.J., and Thompson, W.F. (1989). Cis-acting elements for light regulation of pea ferredoxin 1 gene expression are located within transcribed sequences. *Plant Cell* 1, 691–698.

Erdner, D.L., Price, N.M., Doucette, G.J., Peleato, M.L., and Anderson, D.M. (1999). Characterization of ferredoxin and flavodoxin as markers of iron limitation in marine phytoplankton. *Mar. Ecol. Prog. Ser.* 18443–18453.

Fujita, M., Fujita, Y., Noutoshi, Y., Takahashi, F., Narusaka, Y., Yamaguchi-Shinozaki, K., and Shinozaki, K. (2006). Crosstalk between abiotic and biotic stress responses: a current view from the points of convergence in the stress signaling networks. *Curr. Opin. Plant Biol.* 9, 436–442.

Geigenberger, P., Kolbe, A., and Tiessen, A. (2005). Redox regulation of carbon storage and partitioning in response to light and sugars. *J. Exp. Botany* 56, 1469–1479.

Gou, P., Hanke, G.T., Kimata-Ariga, Y., Standley, D.M., Kubo, A., Taniguchi, I., Nakamura, H., and Hase, T. (2006). Higher order structure contributes to specific differences in redox potential and electron transfer efficiency of root and leaf ferredoxins. *Biochemistry* 45, 14389–14396.

Hanke, G.T., Kimata-Ariga, Y., Taniguchi, I., and Hase, T. (2004). A post genomic characterization of *Arabidopsis* ferredoxins. *Plant Physiol.* 134, 255–264.

Huang, H.-E., Ger, M.-J., Yip, M.-K., Chen, C.-Y., Pandeya, A.-K., and Feng, T.-Y. (2004). A hypersensitive response was induced by virulent bacteria in transgenic tobacco plants overexpressing a plant ferredoxin-like protein (PFLP). *Physiol. Mol. Plant Pathol.* 64, 103–110.

Huang, H.-E., Ger, M.-J., Chen, C.-Y., Pandey, A.-K., Yip, M.-K., Chou, H.-W., and Feng, T.-Y. (2007a). Disease resistance to bacterial pathogens affected by the amount of ferredoxin-I protein in plants. *Mol. Plant Pathol.* 8, 129–137.

Huang, H.-E., Liu, C.-A., Lee, M.-J., Kuo, C.-G., Chen, H.-M., Ger, M.-J., Tsai, Y.-C., Chen, Y.-R., Lin, M.-K., and Feng, T.-Y. (2007b). Resistance enhancement of transgenic tomato to bacterial pathogens by the heterologous expression of sweet pepper ferredoxin-I protein. *Phytopathology* 97, 900–906.

John, I., Hackett, R., Cooper, W., Drake, R., Farrell, A., and Grierson, D. (1997). Cloning and characterization of tomato leaf senescence-related cDNAs. *Plant Mol. Biol.* 33, 641–651.

Liau, C.-H., Lu, J.-C., Prasad, V., Hsiao, H.-H., You, S.-J., Lee, J.-T., Yang, N.-S., Huang, H.-E., Feng, T.-Y., Chen, W.-H., and Chan, M.-T. (2003). The sweet pepper ferredoxin-like protein (*pflp*) conferred resistance against soft rot disease in *Oncidium* orchid. *Transgenic Res.* 12, 329–336.

Mazouni, K., Domain, F., Chauvat, F., and Cassier-Chauvat, C. (2003). Expression and regulation of the crucial plant-like ferredoxin of cyanobacteria. *Mol. Microbiol.* 49, 1019–1029.

Meyer, J. (2001). Ferredoxins of the third kind. *FEBS Lett.* 509, 1–5.

Mittler, R. (2002). Oxidative stress, antioxidants and stress tolerance. *Trends Plant Sci.* 7, 405–410.

Munekage, Y., Hojo, M., Meurer, J., Endo, T., Tasaka, M., and Shikanai, T. (2002). *PGR5* is involved in cyclic electron flow around photosystem I and is essential for photoprotection in *Arabidopsis*. *Cell* 110, 361–371.

Palatnik, J.F., Tognetti, V.B., Poli, H.O., Rodriguez, R.E., Blanco, N., Gattuso, M., Hajirezaei, M.R., Sonnewald, U., Valle, E.M., and Carrillo, N. (2003). Transgenic tobacco plants expressing antisense ferredoxin-NADP(H) reductase transcripts display increased susceptibility to photo-oxidative damage. *Plant J.* 35, 332–341.

Rodriguez, R.E., Lodeyro, A., Poli, H.O., Zurbriggen, M., Peisker, M., Palatnik, J.F., Tognetti, V.B., Tschiersch, H., Hajirezaei, M.R., Valle, E.M., and Carrillo, N. (2007). Transgenic tobacco plants overexpressing chloroplastic ferredoxin-NADP(H) reductase display normal rates of photosynthesis and increased tolerance to oxidative stress. *Plant Physiol.* 143, 639–649.

Takubo, K., Morikawa, T., Nonaka, Y., Mizutani, M., Takenaka, S., Takabe, K., Takahashi, M.-A., and Ohta, D. (2003). Identification and molecular characterization of mitochondrial ferredoxins and ferredoxin reductase from *Arabidopsis*. *Plant Mol. Biol.* 52, 817–830.

Tang, K.X., Sun, X.F., Hu, Q.N., Wu, A.Z., Lin, C.H., Lin, H.-J., Twyman, R.M., Christou, P., and Feng, T.-Y. (2001). Trangenic rice plants expressing the ferredoxin-like protein (AP1) from sweet pepper show enhanced resistance to *Xanthomonas oryzae* pv. *oryzae*. *Plant Sci.* 160, 1035–1042.

Tognetti, V.B., Palatnik, J.F., Fillat, M., Melzer, M., Hajirezaei, M.-R., Valle, E.M., and Carrillo, N. (2006). Functional replacement of ferredoxin by a cyanobacterial flavodoxin in tobacco confers broad-range stress tolerance. *Plant Cell* 18, 2035–2050.

Tognetti, V.B., Zurbriggen, M.D., Morandi, E.N., Fillat, M.F., Valle, E.M., Hajirezaei, M.R., and Carrillo, N. (2007). Enhanced plant tolerance to iron starvation by functional substitution of chloroplast ferredoxin with a bacterial flavodoxin. *Proc. Natl. Acad. Sci. U.S.A.* 104, 11495–11500.

van Thor, J.J., Jeanjean, R., Havaux, M., Sjollema, K.A., Joset, F., Hellingwerf, K.J., and Matthijs, H.C. (2000). Salt shock-inducible photosystem I cyclic electron transfer in Synechocystis PCC6803 relies on binding of ferredoxin:NADP(+) reductase to the thylakoid membranes via its CpcD phycobilisome-linker homologous N-terminal domain. *Biochim. Biophys. Acta* 1457, 129–144.

Vorst, O., van Dam, F., Weisbeek, P., and Smeekens, S. (1993). Light-regulated expression of the *Arabidopsis thaliana* ferredoxin a gene involves both transcriptional and post-transcriptional processes. *Plant J.* 3, 793–803.

Wada, K., Onda, M., and Matsubara, H. (1986). Ferredoxin isolated from plant non-photosynthetic tissues: purification and characterization. *Plant Cell Physiol.* 27, 407–415.

Yamamoto, H., Kato, H., Shinzaki, Y., Horiguchi, S., Shikanai, T., Hase, T., Endo, T., Nishioka, M., Makino, A., Tomizawa, K., and Miyake, C. (2006). Ferredoxin limits cyclic electron flow around PSI (CEF-PSI) in higher plants-stimulation of CEF-PSI enhances non-photochemical quenching of Chl fluorescence in transplastomic tobacco. *Plant Cell Physiol.* 47, 1355–1371.

Yip, M.-K., Huang, H.-E., Ger, M.-J., Chiu, S.-H., Tsai, Y.-C., Lin, C.-I., and Feng, T.-Y. (2007). Production of soft rot resistant calla lily by expressing a ferredoxin-like protein gene (*pflp*) in transgenic plants. *Plant Cell Rep.* 26, 449–457.

You, S.-J., Liau, C.-H., Huang, H.-E., Feng, T.-Y., Prasad, V., Hsiao, H.-H., Lu, J.-C., and Chan, M.-T. (2003). Sweet pepper ferredoxin-like protein (*pflp*) gene as a novel selection marker for orchid transformation. *Planta* 217, 60–65.

Zimmermann, P., Hirsch-Hoffmann, M., Hennig, L., and Gruissem, W. (2004). GENEVESTIGATOR. *Arabidopsis* microarray database and analysis toolbox. *Plant Physiol.* 136, 2621–2632.

10 Biosynthesis of Unusual Fatty Acids in Microorganisms and Their Production in Plants

Xiao Qiu and Dauenpen Meesapyodsuk

CONTENTS

Key Words: *Claviceps purpurea*; lipid biosynthesis; metabolic engineering; plant oil; unusual fatty acids.

10.1 INTRODUCTION

The world petroleum reserve is finite. The current price for crude oil stays high because world demand for this limited resource is increasing. Petroleum is mostly used for producing fuel oil. However, approximately 10% to 15% of petroleum, by volume, is utilized as raw material for manufacture of lubricants, solvents, fertilizers, pesticides, plastics, and other chemicals. Because petroleum is a nonrenewable resource, there is an urgent need to search for alternative and sustainable sources for these petrochemicals. Vegetable oils can be produced in a large scale with reasonable costs and have structures similar to petroleum. Therefore, plant seed oils represent the most promising resource (green oleochemicals) to replace petrochemicals.

The utility of oil is defined by the composition of the fatty acids acylated on the glycerol backbone (Spector 1999, Voelker and Kinney 2001). The composition of fatty acids differing in structure directly affects the physical and chemical properties of oil, which in turn will influence applications and industrial value. Edible oils for human consumption are predominantly comprised of six common fatty acids: palmitate (16:0), palmitoleate (16:1-9), stearate (18:0), oleate (18:1-9), linoleate (18:2-9,12), and linolenate (18:3-9,12,15). In comparison, most industrial oils contain so-called unusual fatty acids. These fatty acids differ in chemical structure from common fatty acids in chain length, number and position of double bonds, or possession of functional groups. For instance, seed oils rich in fatty acids with hydroxyl groups have commercial value as components of biolubricants, plasticizers, adhesives, and paints, while the high-epoxy fatty acid oils are excellent additives for the manufacture of plastics and nylons (Jaworski and Cahoon 2003).

Unusual fatty acids occur widely in native plant species and sporadically in microorganisms (Dalsgaard et al. 2003). However, generally microorganisms producing these fatty acids are not easily cultured. In addition, these unusual fatty acid-producing microorganisms have poor yields and oil contents in comparison to some oleaginous yeast and fungi or to oilseed crops with high yields and oil levels. Molecular elucidation of the biosynthetic mechanisms of potentially important unusual fatty acids from microorganisms and heterologous expression of genes involved in the biosynthesis in oilseed crops would represent an attractive alternative for renewable sources of industrial oils.

10.2 MECHANISMS UNDERLYING THE BIOSYNTHESIS OF UNUSUAL FATTY ACIDS

Molecular analysis of the biosynthesis of unusual fatty acids in microorganisms and plants started in the 1980s. The first gene involved in the biosynthesis of unusual fatty acids was cloned from *Escherichia coli* and encodes an enzyme introducing a carbocyclic ring into oleic acid, resulting in formation of cyclopropane fatty acid (Grogan and Cronan 1984). The first plant gene involved in the synthesis of unusual fatty acids was cloned from seeds of the California bay tree (*Umbellularia californica*) and encodes a medium-chain fatty acyl thioesterase (UcFATB) (Voelker et al. 1992). Soon after, the biosynthesis of other unusual fatty acids was elucidated in both microorganisms and plants. A soluble Δ-4 palmitoyl acyl carrier protein (ACP) desaturase was identified from coriander (*Coriandrum sativum*), which catalyzes the synthesis of petroselinic acid (18:1-6) (Cahoon et al. 1992), and the enzyme has sequence similarity to Δ-9 stearoyl ACP desaturases found in all plant species. The first acyl-CoA (coenzyme A) condensing enzyme (3-ketoacyl-CoA synthase) (FAE1), which elongates long-chain saturated and monounsaturated fatty acids, was cloned from *Arabidopsis* by transposon insertional mutagenesis (James et al. 1995), and the homologous enzymes involved in the synthesis of very long-chain monounsaturated fatty acids such as erucic (22:1-13) and nervonic (24:1-15) acids were identified in the *Brassica* family by the homology comparison. Other naturally occurring unusual monounsaturated fatty acids with a double bond in other than the ninth position include hexadec-6-enoic acid (16:1-6) in *Thunbergia alata* and eicosa-5-enoic acid (20:1-5) in *Limnanthes douglasii*. The former is synthesized by a soluble Δ-6 16:0 ACP desaturase (Cahoon et al. 1994), while the latter results from the activity of a membrane-bound acyl-CoA desaturase that inserts a double bond at position 5 of 20:0-CoA (Cahoon et al. 2000).

The Δ-12 oleate desaturase (FAD2) is a ubiquitous enzyme found in all living organisms except for mammals, including humans. Variant forms of this desaturase with different regiospecificity are responsible for the biosynthesis of various unusual fatty acids in plants. The first divergent form of FAD2 was identified by expressed sequence tag (EST) sequencing in castor bean (*Ricinus communis*) and encodes a Δ-12 oleate hydroxylase, which introduces a hydroxyl group at the Δ-12 position of oleic acid, producing ricinoleic acid (12-OH-18:1-9) (van de Loo et al. 1995). Soon afterward, other divergent forms were identified through the use of sequence information. *Crepis alpina* seed oil contains proximately 70% crepenynic acid (9-octadecen-12-ynoic acid). In this plant, an FAD2-like enzyme (Δ-12 acetylenase) cloned by degenerate polymerase chain reaction (PCR) catalyzes the conversion of the Δ-12 double bond of linoleic acid into a triple bond (Lee et al. 1998). *Crepis palaestina* accumulates approximately 60% of its seed fatty acids as vernolic acid (12,13-epoxy-9-octadecenoic acid). A FAD2-like enzyme isolated by the same approach catalyzes the formation of the Δ-12,13 epoxy group of vernolic acid from linoleic acid (Δ-12 epoxygenase) (Lee et al. 1998). *Euphorbia lagascae* also accumulates substantial amounts of the same epoxy fatty acid in seeds; however, it utilizes a cytochrome P450 Δ-12 linoleate epoxygenase to synthesize this fatty acid (Cahoon et al. 2002).

Aspergillus nidulans uses a family of oxygenated long-chain fatty acids called *psi factor* (precocious sexual inducer), such as psiBα (8-hydroxy-18:2-9,12) and psiCα (5,8-dihydroxy-18:2-9,12) to modulate sexual and asexual spore development. Both hydroxyl fatty acids are produced by cytochrome P450-like fatty acid oxygenases using linoleic acid as a substrate (Tsitsigiannis et al. 2005).

Conjugated fatty acids are special types of unusual fatty acids that are characterized by any two double bonds separated by one single bond in the acyl chain. The most widely occurring conjugated fatty acids in plants are conjugated linolenic acids, which are 18 carbons in length with three conjugated double bonds, some or all of which are in *trans* configuration. The first gene involved in the synthesis of conjugated linolenic acids was identified in *Momordica charantia* by EST sequencing. It has sequence homology to FAD2 and can convert the Δ-12 double bond of linoleic acid into two conjugated double bonds at positions 11 and 13, producing α-eleostearic acid (18:3-9c,11t,13t) (Cahoon et al. 1999). *Calendula officinalis* seeds contain about 40% calendic acid, which has three conjugated double bonds at positions 8, 10, and 12. A FAD2-like conjugase cloned from this species could convert the Δ-9 double bonds of linoleic and oleic acids into two conjugated double bonds at positions 8 and 10, producing conjugated linolenic (18:3-8t,10t,12c) and linoleic (18:2-8t,10t) acids (Cahoon et al. 2001, Qiu et al. 2001b). In addition, a conjugase involved in the biosynthesis of punicic acid (18:3-9c,11t,13c) in *Punica granatum* was recently cloned independently by two different groups (Hornung et al. 2002, Iwabuchi et al. 2003).

Dimorphecolic acid is one of the major fatty acids in *Dimorphotheca sinuata* seeds. Two complementary DNAs (cDNAs) for divergent FAD2 enzymes were isolated from developing seeds by EST sequencing and homology searches. One codes for an oleate desaturase, which introduces a *trans* double bond at the Δ-12 position of oleic acid, while the second encodes an enzyme that can convert the resulting 18:2-9c,12t into dimorphecolic acid (9-OH-18:2-10t,12t) (Cahoon and Kinney 2004).

Conjugated linoleic acids (18:2) are popular dietary supplements in the nutraceutical markets, particularly the 18:2-9c,11t and 18:2-10t,12c isomers. These fatty acids are normally only present in microorganisms. In bacteria, the anaerobic *Butyrivibrio fibrisolvens* in animal rumens was initially found to be able to convert linoleic acid into conjugated linoleic acid 18:2-9c,11t, an intermediate in biohydrogenation, by using an enzyme called linoleate isomerase (Kepler et al. 1966). However, the gene encoding this enzyme has not yet been cloned. In addition, some dairy starter cultures such as *Propionibacterium freudenreichii* and enteric bacteria such as *Bifidobacterium* and *Lactobacillus* species can also produce conjugated linoleic acids from linoleic acid (Ogawa et al. 2005).

Carbocyclic fatty acid such as cyclopropane and cyclopropene widely occur in microorganisms and plants. The first cyclopropane fatty acid synthase was cloned from *E. coli* more than two decades ago (Grogan and Cronan 1984). This enzyme catalyzes the synthesis of 9,10-methylene octadecanoic acid from oleic acid using S-adenosylmethionine as a methylene donor. Two other cyclopropane fatty acid synthases were identified in *Mycobacterium tuberculosis*; these are responsible for cyclopropanation of the proximal and distant double bonds of mycolic acid, a major constituent of the mycobacterial cell wall complex used to provide a mechanism against potentially adverse environmental conditions, including the human defense system (George et al. 1995, Otero et al. 2003). In plants, cyclopropene fatty acids such as sterculic acid have been found in *Sterculia foetida*. EST sequencing identified a cDNA from the developing seeds that encodes an enzyme (S-adenosylmethionine methyltransferase) homologous to bacterial cyclopropane fatty acid synthase producing dihydrosterculic acid. Dihydrosterculic acid is then desaturated by a cyclopropane desaturase to sterculic acid (Bao et al. 2002, 2003). However, the cloning of this desaturase has not yet been reported.

Many lower fungal species produce substantial amounts of long-chain polyunsaturated fatty acids (PUFAs) (Ratledge 2004). Notably, the oleaginous *Mucor ramannianus*, *Mortierella alpina,* and *Thraustochytrium* produce substantial amounts of γ-linoleic acid (GLA, 18:3-6,9,12), arachidonic acid (AA, 20:5-5,8,11,14), eicosapentaenoic acid (EPA, 20:5-5,8,11,14,17), and docosahexaenoic acid (DHA, 22:6-4,7,10,13,16,19). These fatty acids have recently drawn tremendous attention in scientific and industrial communities because of their health benefits in humans and animals (Makrides and Gibson 2000). The first Δ-6 desaturase involved in the biosynthesis of PUFAs was identified from cyanobacteria in the early 1990s (Reddy et al. 1993, Reddy and Thomas 1996), and it introduced a Δ-6 double bond into linoleic acid to form GLA. Shortly afterward, several Δ-6 desaturase genes were cloned from fungal species such as *Mortierella alpina* (Huang et al. 1999) and *Pythium irregulare* (Hong et al. 2002) and from plants (Sayanova et al. 1997). The introduction of fungal Δ-6 desaturases into canola has resulted in production of more than 40% of GLA in oilseeds, an example of transgenic production of healthy fatty acids in plants by using microbial gene resources (Hong et al. 2002). The biosynthetic pathways of very long-chain polyunsaturates such as DHA have recently been elucidated in primitive fungal species such as thraustochytriads. A polyketide synthase was found in *Schizichytrium* sp., a unicellular marine fungus responsible for the biosynthesis of DHA and docosapentaenoic acid (DPA, 22:5n-6) (Metz et al. 2001), whereas *Thraustochytrium sp.*, a closely related species, utilizes the Δ-4 desaturation pathway to synthesize DHA (Qiu et al. 2001a). The entire DHA biosynthetic pathway from microorganisms has been reconstituted in oilseed crops to produce very long-chain PUFAs (Wu et al. 2005). These research studies have opened up the possibility of large-scale production of nutraceutical fatty acids in plants for dietary supplementation (Truksa et al. 2006).

10.3 PRODUCTION OF INDUSTRIAL FATTY ACIDS IN PLANTS: A MICROBIAL EXAMPLE

Claviceps purpurea, a fungal pathogen of the plant ergot disease, parasitizes young flowers of many grasses (Mey et al. 2002). During infection, this pathogen forms a specialized structure called sclerotia in the host ovary, a hard, pigmented, and compact mass of mycelia containing high levels of ricinoleic acid (12-hydroxyl-18:1-9) (Morris et al. 1966, Morris and Hall 1966) as well as ergot alkaloids (Tudzynski et al. 2001). Hydroxyl fatty acids such as ricinoleic acid are important industrial feedstock for lubricants, functional fluids, process oils, ink, paints, coatings, foams, and other polymers. At present, castor bean (*Ricinus communis*) is the only biological source for this fatty acid. Due to poor agronomic performance and the presence of highly potent toxins (ricin) and allergens in the seed, castor bean is not an ideal source for large volumes of this fatty acid. Thus, an alternative source for this hydroxyl fatty acid is highly desirable. Genes involved in ricinoleic acid biosynthesis have been identified from plant castor bean (van de Loo et al. 1995) and *Lesquerella fendleri* (Broun et al. 1998). Both genes encode an oleate 12-hydroxylase, which introduces a hydroxyl group at position 12 of oleic acid. Introduction of the caster bean oleate hydroxylase into tobacco and *Arabidopsis thaliana* resulted in ricinoleic acid accumulation at a level below 15% of the total fatty acid in seeds (Broun and Somerville 1997, Broun et al. 1998, Smith et al. 2003).

   Claviceps purpurea has long been known to produce ricinoleic acid; however, the biosynthetic pathway for the biosynthesis in the fungus was unknown. It was originally hypothesized in the 1960s that the synthesis of ricinoleic acid in *C. purpurea* is catalyzed by a hydratase using linoleic acid as a substrate, and that it does not require any molecular oxygen, a critical cofactor for the hydroxylation process (Morris et al. 1966, Morris 1970). This hypothesis was held for almost 40 years until a recent deuterium NMR study that showed the synthesis of ricinoleic acid in *C. purpurea* follows a hydroxylation process using oleic acid as the substrate and oxygen as a coreactant (Billault et al. 2004).

However, the exact nature of the hydroxylase enzyme still remains elusive. Two possibilities exist that could be used to explain the hydroxylation process. First, the hydroxylation is catalyzed by a desaturase-like hydroxylase as occurs in castor bean. Second, the hydroxylation is catalyzed by a cytochrome P450-like hydroxylase as is seen in some fungal species. For example, the fungus *Aspergillus nidulans* uses oxygenated fatty acids such as 8-hydroxy-18:2-9,12 and 5,8-dihydroxy-18:2-9,12 to modulate spore development. Both these hydroxyl fatty acids are synthesized by cytochrome P450-like fatty acid oxygenases (Tsitsigiannis et al. 2004, 2005).

We hypothesized that the biosynthesis of ricinoleic acid in *C. purpurea* might resemble that in plants, in which the hydroxylation is catalyzed by a desaturase-like enzyme; therefore, we focused on the first possibility that the hydroxylase is an oleate desaturase-like enzyme. Following the hypothesis, we designed two degenerate primers targeting fungal oleate desaturases and used RNA isolated from *C. purpurea* sclerotia as the template for RT-PCR (reverse transcription polymerase chain reaction). By using this method, we identified several cDNAs that are highly homologous to microbial oleate desaturases from fungi. One of them encodes a Δ-12 desaturase that introduces a Δ-12 double bond into oleic and palmitoleic acids. The second cDNA encodes a novel desaturase catalyzing Δ-12, Δ-15, and ω-3 desaturation activities, with ω-3 desaturation of linoleic acid predominating (Meesapyodsuk et al. 2007). The third cDNA (*CpFAH*) encodes a fatty acid hydroxylase that preferentially introduces a hydroxyl group into position 12 of oleic acid. Compared to its plant counterparts, *Claviceps* hydroxylase appears to have wider substrates. As well as showing strong hydroxylase activity on oleate and palmitoleate, it catalyzes Δ-12 desaturation of oleate and palmitoleate and, to a limited extent, is able to desaturate linoleate and ricinoleate at the ω-3 position.

To evaluate the capacity of the fungal hydroxylase for production of hydroxyl fatty acids in plants, we expressed this gene in *A. thaliana* under the control of the seed-specific promoter. The untransformed wild-type *Arabidopsis* seeds produce several fatty acids; the most abundant one is 18:2-9,12, followed by 20:1-11, 18:3-9,12,15, and 18:1-9. The *Arabidopsis* mutant line (*fad2/fae1*), which carries mutations in both the *FAD2* and *FAE1* (coding for the condensing enzyme for saturates and mono-unsaturated elongation) genes, produces a high level of oleate that can be used for the synthesis of ricinoleic acid. Expression of CpFAH in wild-type *Arabidopsis* resulted in the production of three hydroxyl fatty acids: ricinoleic acid, followed by densipolic acid, derived from the ω3 desaturation of ricinoleic acid, and lesquerolic acid, a 20C hydroxyl fatty acid derived from the elongation of ricinoleic acid. The total three hydroxyl fatty acids in transformed wild-type *Arabidopsis* ranged from 15.9% to 22.0%. Expression of CpFAH in *Arabidopsis* double-mutant line produced two hydroxyl fatty acids, ricinoleic acid and lesquerolic acid. The level of these two fatty acids in the bulked samples accounted for 17.3% to 25% of the total fatty acids, with the highest level in a single seed reaching 29.1%. These are considerably higher levels than were seen with the expression of plant hydroxylases in either wild-type or mutant *Arabidopsis* plants (Broun and Somerville 1997, Smith et al. 2003). The reason for the better performance of the fungal hydroxylase is currently unknown but might be simply that CpFAH itself has higher hydroxylase activity. Furthermore, since the fungal oleate hydroxylase has less similarity than plant counterparts to endogenous homologous desaturase genes, CpFAH might be subject to less-stringent regulatory control by the host plant.

10.4 FUTURE PERSPECTIVES

Although great progress has been made in the molecular analysis of biosynthesis of unusual fatty acids in microorganisms, there are many novel unusual fatty acids in the microbial world whose chemical structures and potential industrial value are as yet undefined. The biosynthetic pathways of many unusual fatty acids with potentially industrial value in microorganisms, especially in fungi, remain to be determined. Identification of the chemical structure and definition of biosynthetic pathways of the fatty acids could open up novel industrial applications for these fatty acids.

Because of high costs for culturing microorganisms and extracting oils from their biomass, production of a large volume of unusual fatty acids in microbial systems would not be realistic. On the other hand, oilseed crops possess good agronomic performance and have high yield and oil content; thus, they could serve as a vehicle for production of unusual fatty acids. Therefore, heterologous expression of microbial genes involved in the biosynthesis of unusual fatty acids in oilseed crops would be very attractive for large-scale production of these fatty acids for industrial applications.

Metabolic engineering of unusual fatty acids in oilseeds has been extensively attempted using plant genes. However, to date the production of unusual fatty acids through the introduction of these genes into oilseed crops has met with only limited success. Although proof of concept has been achieved by the production of many unusual fatty acids in transgenic crops, commercially viable levels have generally yet to be obtained (Jaworski and Cahoon 2003, Napier 2007). It is not yet known why the level of hydroxyl fatty acid in transgenic plants is so much lower than that in native species. It has been hypothesized that contributing factors may include the inefficient removal of newly synthesized unusual fatty acids from membrane phospholipids (Thomæus et al. 2001, Cahoon et al. 2006) and the inefficient transfer of newly synthesized unusual fatty acids to storage lipids (Napier 2007, Cahoon et al. 2007). Many unusual fatty acids, particularly those derived from modifications occurring on phospholipids of the endoplasmic reticulum, must be removed from phospholipids and then mobilized to triacylglycerols; otherwise, they may have a detrimental effect on the entity of the membrane. It has been hypothesized that newly synthesized unusual fatty acids are removed from phospholipids by phospholipases or by various acyltransferases, such as phospholipid-diacylglycerol acyltransferases (Dahlqvist et al. 2000) and lysophospholipid acyltransferase (Jain et al. 2007). A direct transacylation between phospholipids, neutral lipids, and the acyl CoA pool has also been proposed to be involved (Jaworski and Cahoon 2003). However, none of these hypotheses have been unambiguously proven. Obtaining information on enzymes involved in the biosynthesis of unusual fatty acids from microorganisms, followed by plant transgenic studies, will offer additional information or a better understanding of mechanisms underlying the intermediate flux of the biosynthesis and the assembly of unusual fatty acids into neutral lipids in plants.

REFERENCES

Bao X, Katz S, Pollard M, Ohlrogge J. (2002) Carbocyclic fatty acids in plants: biochemical and molecular genetic characterization of cyclopropane fatty acid synthesis of *Sterculia foetida*. *Proc Natl Acad Sci USA* 99: 7172–7177.

Bao X, Thelen JJ, Bonaventure G, Ohlrogge JB. (2003) Characterization of cyclopropane fatty-acid synthase from *Sterculia foetida*. *J Biol Chem* 278: 12846–12853.

Billault I, Mantle PG, Robins RJ. (2004) Deuterium NMR used to indicate a common mechanism for the biosynthesis of ricinoleic acid by *Ricinus communis* and *Claviceps purpurea*. *J Am Chem Soc* 126: 3250–3256.

Broun P, Boddupalli S, Somerville C. (1998) A bifunctional oleate 12-hydroxylase: desaturase from *Lesquerella fendleri*. *Plant J* 13: 201–210.

Broun P, Somerville C. (1997) Accumulation of ricinoleic, lesquerolic, and densipolic acids in seeds of transgenic Arabidopsis plants that express a fatty acyl hydroxylase cDNA from castor bean. *Plant Physiol* 113: 933–942.

Cahoon EB, Carlson TJ, Ripp KG, Schweiger BJ, Cook GA, Hall SE, Kinney AJ. (1999) Biosynthetic origin of conjugated double bonds: production of fatty acid components of high-value drying oils in transgenic soybean embryos. *Proc Natl Acad Sci USA* 96: 12935–12940.

Cahoon EB, Cranmer AM, Shanklin J, Ohlrogge JB. (1994) Delta 6 hexadecenoic acid is synthesized by the activity of a soluble delta 6 palmitoyl-acyl carrier protein desaturase in *Thunbergia alata* endosperm. *J Biol Chem* 269: 27519–27526.

Cahoon EB, Dietrich CR, Meyer K, Damude HG, Dyer JM, Kinney AJ. (2006) Conjugated fatty acids accumulate to high levels in phospholipids of metabolically engineered soybean and Arabidopsis seeds. *Phytochemistry* 67: 1166–1176.

Cahoon EB, Kinney AJ. (2004) Dimorphecolic acid is synthesized by the coordinate activities of two divergent delta12-oleic acid desaturases. *J Biol Chem* 279: 12495–12502.

Cahoon EB, Marillia EF, Stecca KL, Hall SE, Taylor DC, Kinney AJ. (2000) Production of fatty acid components of meadowfoam oil in somatic soybean embryos. *Plant Physiol* 124: 243–251.

Cahoon EB, Ripp KG, Hall SE, Kinney AJ. (2001) Formation of conjugated delta8,delta10-double bonds by delta12-oleic-acid desaturase-related enzymes: biosynthetic origin of calendic acid. *J Biol Chem* 276: 2637–2643.

Cahoon EB, Ripp KG, Hall SE, McGonigle B. (2002) Transgenic production of epoxy fatty acids by expression of a cytochrome P450 enzyme from *Euphorbia lagascae* seed. *Plant Physiol* 128: 615–624.

Cahoon EB, Shanklin J, Ohlrogge JB. (1992) Expression of a coriander desaturase results in petroselinic acid production in transgenic tobacco. *Proc Natl Acad Sci USA* 89: 11184–11188.

Cahoon EB, Shockey JM, Dietrich CR, Gidda SK, Mullen RT, Dyer JM. (2007) Engineering oilseeds for sustainable production of industrial and nutritional feedstocks: solving bottlenecks in fatty acid flux. *Curr Opin Plant Biol* 10: 236–244.

Dahlqvist A, Stahl U, Lenman M, Banas A, Lee M, Sandager L, Ronne H, Stymne S. (2000) Phospholipid:diacylglycerol acyltransferase: an enzyme that catalyzes the acyl-CoA-independent formation of triacylglycerol in yeast and plants. *Proc Natl Acad Sci USA* 97: 6487–6492.

Dalsgaard J, St John M, Kattner G, Muller-Navarra D, Hagen W. (2003) Fatty acid trophic markers in the pelagic marine environment. *Adv Mar Biol* 46: 225–340.

George KM, Yuan Y, Sherman DR, Barry CE III. (1995) The biosynthesis of cyclopropanated mycolic acids in *Mycobacterium tuberculosis*. Identification and functional analysis of CMAS-2. *J Biol Chem* 270: 27292–27298.

Grogan DW, Cronan JE Jr. (1984) Cloning and manipulation of the *Escherichia coli* cyclopropane fatty acid synthase gene: physiological aspects of enzyme overproduction. *J Bacteriol* 158: 286–295.

Hong H, Datla N, Reed DW, Covello PS, MacKenzie SL, Qiu X. (2002) High-level production of gamma-linolenic acid in *Brassica juncea* using a delta6 desaturase from *Pythium irregulare*. *Plant Physiol* 129: 354–362.

Hornung E, Pernstich C, Feussner I. (2002) Formation of conjugated delta11delta13-double bonds by delta12-linoleic acid (1,4)-acyl-lipid-desaturase in pomegranate seeds. *Eur J Biochem* 269: 4852–4859.

Huang YS, Chaudhary S, Thurmond JM, Bobik EG Jr, Yuan L, Chan GM, Kirchner SJ, Mukerji P, Knutzon DS. (1999) Cloning of delta12- and delta6-desaturases from *Mortierella alpina* and recombinant production of gamma-linolenic acid in *Saccharomyces cerevisiae*. *Lipids* 34: 649–659.

Iwabuchi M, Kohno-Murase J, Imamura J. (2003) Delta 12-oleate desaturase-related enzymes associated with formation of conjugated trans-delta 11, cis-delta 13 double bonds. *J Biol Chem* 278: 4603–4610.

Jain S, Stanford N, Bhagwat N, Seiler B, Costanzo M, Boone C, Oelkers P. (2007) Identification of a novel lysophospholipid acyltransferase in *Saccharomyces cerevisiae*. *J Biol Chem* 282: 30562–30569.

James DW Jr, Lim E, Keller J, Plooy I, Ralston E, Dooner HK. (1995) Directed tagging of the Arabidopsis fatty acid elongation1 (FAE1) gene with the maize transposon activator. *Plant Cell* 7: 309–319.

Jaworski J, Cahoon EB. (2003) Industrial oils from transgenic plants. *Curr Opin Plant Biol* 6: 178–184.

Kepler CR, Hirons KP, McNeill JJ, Tove SB. (1966) Intermediates and products of the biohydrogenation of linoleic acid by *Butyrinvibrio fibrisolvens*. *J Biol Chem* 241: 1350–1354.

Lee M, Lenman M, Banas A, Bafor M, Singh S, Schweizer M, Nilsson R, Liljenberg C, Dahlqvist A, Gummeson PO, Sjodahl S, Green A, Stymne S. (1998) Identification of non-heme diiron proteins that catalyze triple bond and epoxy group formation. *Science* 280: 915–918.

Makrides M, Gibson RA. (2000) Long-chain polyunsaturated fatty acid requirements during pregnancy and lactation. *Am J Clin Nutr* 71: 307S–311S.

Meesapyodsuk D, Reed DW, Covello PS, Qiu X. (2007) Primary structure, regioselectivity, and evolution of the membrane-bound fatty acid desaturases of *Claviceps purpurea*. *J Biol Chem* 282: 20191–20199.

Metz JG, Roessler P, Facciotti D, Levering C, Dittrich F, Lassner M, Valentine R, Lardizabal K, Domergue F, Yamada A, Yazawa K, Knauf V, Browse J. (2001) Production of polyunsaturated fatty acids by polyketide synthases in both prokaryotes and eukaryotes. *Science* 293: 290–293.

Mey G, Oeser B, Lebrun MH, Tudzynski P. (2002) The biotrophic, non-appressorium-forming grass pathogen *Claviceps purpurea* needs a Fus3/Pmk1 homologous mitogen-activated protein kinase for colonization of rye ovarian tissue. *Mol Plant Microbe Interact* 15: 303–312.

Morris LJ. (1970) Mechanisms and stereochemistry in fatty acid metabolism. *Biochem J* 118: 681–693.

Morris LJ, Hall SW. (1966) The structure of the glycerides of ergot oils. *Lipids* 1: 188–196.

Morris LJ, Hall SW, James AT. (1966) The biosynthesis of ricinoleic acid by *Claviceps purpurea*. *Biochem J* 100: 29C–30C.

Napier JA. (2007) The production of unusual fatty acids in transgenic plants. *Annu Rev Plant Biol* 58: 295–319.

Ogawa J, Kishino S, Ando A, Sugimoto S, Mihara K, Shimizu S. (2005) Production of conjugated fatty acids by lactic acid bacteria. *J Biosci Bioeng* 100: 355–364.

Otero J, Jacobs WR Jr, Glickman MS. (2003) Efficient allelic exchange and transposon mutagenesis in *Mycobacterium avium* by specialized transduction. *Appl Environ Microbiol* 69: 5039–5044.

Qiu X, Hong H, MacKenzie SL. (2001a) Identification of a delta 4 fatty acid desaturase from *Thraustochytrium* sp. involved in the biosynthesis of docosahexanoic acid by heterologous expression in *Saccharomyces cerevisiae* and *Brassica juncea*. *J Biol Chem* 276: 31561–31566.

Qiu X, Reed DW, Hong H, MacKenzie SL, Covello PS. (2001b) Identification and analysis of a gene from *Calendula officinalis* encoding a fatty acid conjugase. *Plant Physiol* 125: 847–855.

Ratledge C. (2004) Fatty acid biosynthesis in microorganisms being used for single cell oil production. *Biochimie* 86: 807–815.

Reddy AS, Nuccio ML, Gross LM, Thomas TL. (1993) Isolation of a delta 6-desaturase gene from the cyanobacterium *Synechocystis* sp. strain PCC 6803 by gain-of-function expression in *Anabaena* sp. strain PCC 7120. *Plant Mol Biol* 22: 293–300.

Reddy AS, Thomas TL. (1996) Expression of a cyanobacterial delta 6-desaturase gene results in gamma-linolenic acid production in transgenic plants. *Nat Biotechnol* 14: 639–642.

Sayanova O, Smith MA, Lapinskas P, Stobart AK, Dobson G, Christie WW, Shewry PR, Napier JA. (1997) Expression of a borage desaturase cDNA containing an N-terminal cytochrome b5 domain results in the accumulation of high levels of delta6-desaturated fatty acids in transgenic tobacco. *Proc Natl Acad Sci USA* 94: 4211–4216.

Smith MA, Moon H, Chowrira G, Kunst L. (2003) Heterologous expression of a fatty acid hydroxylase gene in developing seeds of *Arabidopsis thaliana*. *Planta* 217: 507–516.

Spector AA. (1999) Essentiality of fatty acids. *Lipids* 34 Suppl: S1–S3.

Thomæus S, Carlsson AS, Stymne S. (2001) Distribution of fatty acids in polar and neutral lipids during seed development in *Arabidopsis thaliana* genetically engineered to produce acetylenic, epoxy and hydroxy fatty acids. *Plant Science* 161: 997–1003.

Truksa M, Wu G, Vrinten P, Qiu X. (2006) Metabolic engineering of plants to produce very long-chain polyunsaturated fatty acids. *Transgenic Res* 15: 131–137.

Tsitsigiannis DI, Kowieski TM, Zarnowski R, Keller NP. (2004) Endogenous lipogenic regulators of spore balance in *Aspergillus nidulans*. *Eukaryot Cell* 3: 1398–1411.

Tsitsigiannis DI, Kowieski TM, Zarnowski R, Keller NP. (2005) Three putative oxylipin biosynthetic genes integrate sexual and asexual development in *Aspergillus nidulans*. *Microbiology* 151: 1809–1821.

Tudzynski P, Correia T, Keller U. (2001) Biotechnology and genetics of ergot alkaloids. *Appl Microbiol Biotechnol* 57: 593–605.

van de Loo FJ, Broun P, Turner S, Somerville C. (1995) An oleate 12-hydroxylase from *Ricinus communis* L. is a fatty acyl desaturase homolog. *Proc Natl Acad Sci USA* 92: 6743–6747.

Voelker T, Kinney AJ. (2001) Variations in the biosynthesis of seed-storage lipids. *Annu Rev Plant Physiol Plant Mol Biol* 52: 335–361.

Voelker TA, Worrell AC, Anderson L, Bleibaum J, Fan C, Hawkins DJ, Radke SE, Davies HM. (1992) Fatty acid biosynthesis redirected to medium chains in transgenic oilseed plants. *Science* 257: 72–74.

Wu G, Truksa M, Datla N, Vrinten P, Bauer J, Zank T, Cirpus P, Heinz E, Qiu X. (2005) Stepwise engineering to produce high yields of very long-chain polyunsaturated fatty acids in plants. *Nat Biotechnol* 23: 1013–1017.

11 Evaluation of Gene Flow in a Minor Crop

Safflower for the Production of Plant-Made Pharmaceuticals in Canada

Marc A. McPherson, Randall J. Weselake, and Linda M. Hall

CONTENTS

Key Words: Bioproducts; gene flow; plant molecular farming; plant with novel trait (PNT).

11.1 INTRODUCTION

Plant molecular farming (PMF) offers the capability to synthesize high-value pharmaceutical and bioindustrial products in plants at lower cost and, in some cases, superior quality than currently employed production systems (Goodman et al. 1987). However, by adopting a plant system rather than a contained cell culture-based platform, there are significant regulatory and public perception hurdles that must be overcome. While transgenic crops have been grown extensively, occupying over 120 M ha in 23 countries (James 2007), trade barriers, labeling thresholds, and international harmonization of trait acceptance still limit the market access of genetically modified (GM) crops. Successfully commercialized transgenic crops

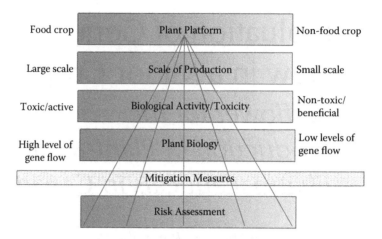

FIGURE 11.1 Overview of key factors to consider prior to choosing a plant species platform for bioproducts and aspects that may influence the risk assessment. The use of a food crop may pose a greater risk than nonfood crops. Larger-scale production (greater acreage) may increase risk relative to smaller production scale. Molecules with greater toxicity or biological activity would increase risk relative to more benign molecules. Larger amounts of pollen- and seed-mediated gene flow throughout the production cycle would increase risks. Mitigation measures (best management practices, BMP) could be incorporated into a production scheme to limit gene flow and reduce the amount of potential risk. A risk assessment will be required and will likely take into consideration all of these factors.

have been deemed substantially equivalent to commodity crops; that is, there are no significant changes to chemical composition or functionality of the products of transgenic crops. Because bioproducts will likely differ from commodity crops, it is likely that they will face additional and unpredictable regulatory hurdles. Modified seed-handling procedures may be required to ensure bioproducts are channeled to markets and segregated from commodity crops.

Because the platform crop will significantly impact the proponent ability to successfully navigate the regulatory challenges, several factors should be considered prior to the selection of a platform crop (Figure 11.1). If the bioproduct has a significant adverse effect on human or animal health or on the environment, it is very unlikely that production of the bioproduct would be permitted outside a contained facility. In addition to the bioproduct's toxicity, the public perception of the product, whether it is perceived to be a significant consumer benefit, may play a role. The scale of production required will influence the ability to segregate bioproduct crops; the larger the production, the greater the opportunity for comingling to occur with commodity food crops. Crop species have often been proposed and developed as platforms for PMF because they produce high yields, have refined cultural methods, and are well characterized for transformation and protein expression. In addition, crop plants are domesticated (Warwick and Stewart 2005), reducing concerns of escape or persistence to the natural environment. Conversely, the use of crop plants as platforms increases the opportunity for comingling or adventitious presence (AP) of plant material containing bioproducts in human or animal food because they cannot be visually distinguished. Many noncrop plant species may not be appropriate for PMF because they are weeds or undomesticated species that would be difficult to contain (Sparrow et al. 2007), and abandoned crops may need extensive agronomic and breeding development to optimize productivity. In addition, weedy species or abandoned crops may not be readily amenable to genetic engineering or protein production. Finally, the reproductive biology of the crop is a critical factor that affects the ability to segregate a crop and meet environmental biosafety regulatory requirements. Gene flow (discussed separately here) is the framework for this risk assessment. All of these factors affect the cost and ability to mitigate procedures. This chapter focuses on regulatory requirements for environmental biosafety of transgenic crops for bioproducts and the research required to quantify gene flow in potential crop platforms, using as an example the crop safflower (*Carthamus tinctorius* L.) as a platform for PMF.

11.2   REGULATION OF PLANTS WITH NOVEL TRAITS IN CANADA

Plants derived by biotechnology are regulated in Canada by two agencies, Health Canada and the Canadian Food Inspection Agency (CFIA), with the former responsible for human health and the latter safety of food, feed, and the environment. The CFIA defines a plant with novel traits (PNT) as "a plant containing a trait not present in plants of the same species already existing as stable, cultivated populations in Canada, or is present at a level significantly outside the range of that trait in stable, cultivated populations of that plant species in Canada" (CFIA 2004). This includes plants modified by mutagenesis, somaclonal variation, intra- and interspecific crosses, protoplast fusion, and recombinant DNA technology (transformation). The objective in regulating novel crops is to protect the health and safety of humans, animals, the food and feed system, and the environment.

The CFIA Plant Biosafety Office has identified five key areas, "pillars," to be assessed for environmental safety: (1) the potential of the novel plant to become a weed of the agroecosystem or invasive to natural areas; (2) gene flow to wild relatives and potential for their hybrid offspring to become more weedy or invasive; (3) potential for the novel plant to become a pest; (4) potential impact of novel plant or its gene products on nontarget organisms, including humans; and (5) potential biodiversity impacts (CFIA 2004). In addition, the CFIA is responsible for the safety of the food/feed system (CFIA Feed Section) and does not consider in its assessment the potential influence of a novel plant or its products on Canadian market access.

Although conventional agriculture is not risk free and has significant consequences to the environment, PMF may pose additional biosafety concerns; the CFIA is cautiously developing a revised framework to evaluate environmental biosafety assessment of PMF. Thresholds for adventitious presence (admixture) have not been set for deregulated transgenic plants in Canada with production traits like herbicide and insect resistance. Since the products are equally safe, they are comingled following commercial release. But, transgenic plants modified to produce bioproducts like pharmaceuticals may require thresholds and assurances of a safe environment and food/feed system.

To date, the CFIA has drafted a preliminary directive to assess the environmental safety of PNT intended for PMF (CFIA 2005) and a preliminary framework for regulating PMF (CFIA 2006a). In addition, the CFIA is developing a framework for "confined commercial production" to utilize segregation processes developed for confined research experiments, modified for large-scale production. The CFIA is continuing to formulate a biosafety assessment framework through a transparent process with multiple stakeholders (CFIA 2001). A significant hurdle to the completion of a regulatory framework for PMF has been identified by the CFIA as a socioeconomic and ethics-based policy outside its mandate. This policy needs to be drafted by the government of Canada (GoC) to gain guidance for acceptable risk of PMF in Canada. These questions include the following: Is PMF compatible with Canadian agriculture? Should PMF be allowed in confined spaces or open fields? Should food/feed crops be used for PMF? The CFIA requires the GoC socioeconomic policy and guidance on the benefits and acceptable risks of PMF to set thresholds and develop regulations (CFIA 2005, 2006a).

11.3   RISK ASSESSMENT OF NOVEL PLANTS: A POTENTIAL APPROACH

Environmental risk assessment has been refined for previous products, including pesticides and pollutants, and has been adapted for the assessment of novel plants. Scientific research informs the decision and is thus a critical part of the process of risk assessment, but the public and government must be active participants for the process to be completed successfully (Peterson and Arntzen 2004, Raybould 2006). The risk assessment process begins with input from stakeholders to identify potential hazards (harm) and is considered in conjunction with the likelihood of exposure (occurrence) (Raybould 2006). Risk to the environment and food/feed system of producing pharmaceuticals in crop plants is a function of both exposure and hazard. Hazard is related to the biological activity or toxicity of the pharmaceutical. Exposure can occur throughout the crop's production cycle or via volunteers in following years. Exposure may be high, but if the hazard is minimal, overall risk may be low and the product deemed safe. Similarly, if the exposure is minimal and hazard is moderate, risk may be low and the product considered safe.

The following experiments are presented to clarify our experimental approach for a preliminary environmental biosafety risk assessment of confined commercial production of transgenic safflower for PMF. Problem formulation (to identify harm) was developed by identifying the concerns outlined by the CFIA for the environment and food/feed system with emphasis on the potential for adventitious presence of seeds in other crop products. The vectors for exposure considered were pollen and seed because safflower is not propagated vegetatively. In addition, exposure routes were identified by considering the cultivation methods of safflower and its life cycle in an agroecosystem setting. Hypotheses were developed and tested first by surveying existing literature, and gaps in knowledge were identified. Experiments were designed to quantify aspects of transgenic and nontransgenic safflower biology and agronomy to quantify the potential for exposure to the environment and food/feed system to transgenic material. Areas for further research are identified, and mitigation measures that may reduce risk of exposure by modification of production management practices are summarized. Because biologically active plant-made pharmaceuticals may be the most carefully regulated of all crops, this research approach may inform environmental biosafety for other crops.

11.4   GENE FLOW IN TRANSGENIC SAFFLOWER

Safflower is an oilseed crop that has been grown in the Middle East and Northern Africa since 4500 B.C. (Dajue and Mundel 1996, Howard et al. 1915, McGregor and Hay 1952). Currently, safflower is grown for its seed oil throughout the Mediterranean, Europe, and the Americas, including the United States and Chile. In the United States, high oleic or high linoleic oil safflower types are grown for edible oil and the paint and varnish industry, respectively. In the Canadian prairies, early varieties developed for the short growing season have lower oil content than U.S. varieties and are thus marketed for birdseed (Mundel et al. 2004). Safflower is currently a minor crop in Alberta, with 320 to 810 ha grown annually (Mundel et al. 2004). This low acreage may aid in choosing sites appropriate for PMF safflower by permitting large isolation distances from commodity fields.

Safflower (*Carthamus tinctorius* L.; cv. "Centennial") has been transformed for PMF using constructs encoding a seed-targeted high-value protein and constitutive expressed *phosphinothricin acetyltransferase* (*pat*) to confer resistance to the broad-spectrum herbicide glufosinate (L-*phosphinothricin*). The herbicide resistance gene was used as a marker to identify and quantify pollen-mediated gene flow (PMGF) in safflower.

11.4.1   POLLEN-MEDIATED GENE FLOW

Two main routes of exposure in the agroecosystem are from seed-mediated gene flow (SMGF) and PMGF. To assess the first route, we evaluated the potential for PMGF to occur from transgenic safflower to one or more of its weedy relatives. To assess the second route, we quantified PMGF from transgenic to commodity safflower of the same variety using the glufosinate-resistant trait as a marker.

11.4.2   PMGF TO WILD/WEEDY RELATIVES

One concern of PMF in safflower is the movement of a transgene to one or more wild/weedy relatives and the introgression of that gene in those populations and the consequence to them. For hybridization to occur, two plant species must be spatially sympatric and flower synchronously. The potential for PMGF from transgenic safflower (*Cathamus tinctorius* L.) to one or more of its wild/weedy relatives in the Americas was assessed from previously published work on safflower breeding and geographic distributions. Of the 16 recognized *Carthamus* species *sensu* López-González (López-González 1989), only 5 occur in North and South America: *C. creticus* L., *C. lanatus* L., *C. leucocaulos* Sibth. et Sm., *C. oxyacanthus* Bieb., and cultivated safflower (*C. tinctorius*). Four wild safflower relatives have been introduced in some

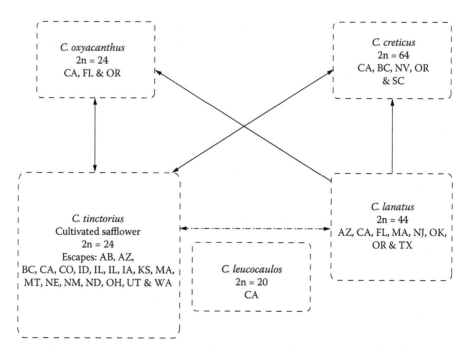

FIGURE 11.2 Summary of interspecific hybridization experiments for species found in Canada and the United States (see text for references). Solid lines indicate fertile F_1 hybrids with viable seed production. Dotted lines indicate hybridization occurred, but F_1 hybrids could not be obtained without embryo rescue or treatments with colchicine. Arrows indicate the direction of the cross (male → female).

areas of the New World where PMF with safflower has been proposed (Hickman 1993). Geographic locations (spatial) of these species have been documented in the United States and Canada (Figure 11.2). *Carthamus oxyacanthus* has been naturalized in Oregon and Florida (Hickman 1993) and is known from a single collection in California in 1978 (Keil 2006). *Carthamus lanatus* has been introduced to Oregon and Florida and reported in Arizona. It has also been reported in Massachusetts and New Jersey. *Carthamus lanatus* has been a weed in California since 1891 (Fuller 1979). Kessler (1987) documented *C. lanatus* as a serious weed growing in an isolated area of Oklahoma. *Carthamus lanatus* and *C. leucocaulos* have been documented in Texas and collected from Argentina and Chile (Correll and Johnston 1970, Hickman 1993, Marticorena and Quezada 1985). *Carthamus creticus* has been reported in California (Hickman 1993, Hoover 1970, Munz 1968). This geographic information was determined from *floras* and may overestimate the distribution of each species because it is based on occurrences from herbarium specimens collected over several decades. This literature rarely provides insights into the density or frequency of plant species in a given region. It is not known how long populations of wild or cultivated safflower persist in or outside the agroecosystem in North America. In addition, information regarding the flowering times (temporal) in these localities has not been reported, making estimates of potential floral synchrony impossible.

Of the four wild/weedy relatives that occur in the Americas, only three are cross compatible with safflower. Hybridization has been attempted with *C. leucocaulos,* but F$_1$ hybrid seed did not germinate. Cultivated safflower is cross compatible with *C. lanatus,* but the F$_1$ hybrids are unlikely to survive in nature because the seeds require embryo rescue and treatment with colchicine to produce hybrid plants. However, hybridization between safflower and both *C. oxyacanthus* and *C. creticus* are known to produce F$_1$ hybrids. Data from the literature have enabled a reduction in the number of safflower wild/weedy relatives requiring further study and identified aspects of the geography and biology of these species requiring further investigation.

Wild/weedy safflower populations would be either advantaged or disadvantaged by introgression of crop genes, including transgenes for herbicide resistance and plant-made pharmaceuticals (PMP). Domestication traits in safflower include decreased shattering, seed dormancy, pappus length, and duration of the rosette stage and could reduce fitness of weedy species. However, most alleles controlling domestication traits are recessive (Berville et al. 2005), and if a transgenic safflower were to hybridize with weedy relatives, recessive domestication traits may not be expressed. The herbicide resistance trait may also confer an enhanced ability to survive (fitness) in the agroecosystem if that herbicide is used on the population and provides a selective advantage. To reduce the risk of gene flow, transgenic safflower should be grown in areas where wild/weedy relatives known to be cross compatible and cultivated safflower escapes do not occur.

11.4.3 PMGF FROM TRANSGENIC PMP TO COMMODITY SAFFLOWER

Intraspecific PMGF, between safflower crops, could result in adventitious presence in commodity safflower crops. Although it is known that PMGF for both wind- and insect-pollinated species decreases exponentially with distance from the pollen source with the nearest neighbors receiving most of the pollen (Levin and Kerster 1974), information concerning safflower was limited in the published literature. Studies of safflower PMGF with a diversity of genotypes indicated that self-pollination rates varied by genotype from 0% to 100%. However, higher oil-yielding safflower lines had PMGF rates from 1% to 5%. Studies also showed that wind did not move pollen beyond 1.2 m, suggesting that safflower pollen movement is primarily mediated by insects (Claassen 1950). The PMGF rate of transgenic Centennial safflower, a variety developed in the United States with high seed oil content and intended for PMF in the Americas, was unknown. This cultivar readily produces seeds under greenhouse conditions when insects are excluded, suggesting it is self-pollinating.

Experiments to quantify PMGF from transgenic to nontransgenic safflower under field conditions in three different environments were performed at El Bosque (Santiago, Chile) in 2002, Westwold (British Columbia, Canada) in 2002, and Lethbridge (Alberta, Canada) in 2004 (Figure 11.3). Experimental scale was restricted by CFIA and Chilean regulations. Transgenic and nontransgenic safflowers (cv. Centennial) were used as pollen source and recipient, respectively, in these experiments. Pollination of nontransgenic safflower from the transgenic line

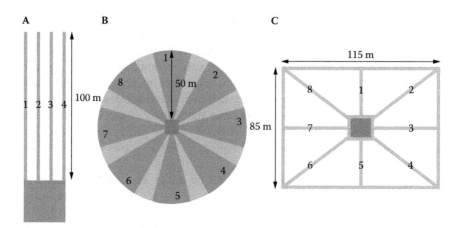

FIGURE 11.3 Experiment designs showing the spatial arrangement of the transgenic pollen source (darkest color), nontransgenic recipient plants (light and dark gray), and the area free of vegetation (white). (A) Westwold, British Columbia, Canada; (B) Lethbridge, Alberta, Canada; and (C) El Bosque, Chile. The light gray area was sampled and retained, whereas the dark gray area was not retained at harvest; numbers indicate blocks or replicates. The pollen source at Westwold was 30 × 30 m, whereas at Lethbridge and El Bosque the pollen source was 10 × 9.9 m and 10 × 11.2 m, respectively.

produced hemizygous seed resistant to the herbicide glufosinate. Seed harvested from the nontransgenic plants at various distances and directions from the transgenic pollen source were grown in subsequent years and screened for glufosinate resistance. Surviving plants were counted and the presence of the transgene confirmed at the protein and DNA levels using molecular techniques. Pollen movement as a function of distance was modeled using regression analysis and the exponential decay function $p = ae^{-bd}$, where p is the estimated frequency of PMGF, a is the intercept, b is the rate of decline, and d is the distance from the transgenic source plot.

The rate of decline of PMGF b was steepest at Lethbridge (Figure 11.4), indicating a rapid drop in PMGF with distance from the transgenic source plot. The distance at which PMGF frequency was reduced by 50% (O_{50}) ranged from 6.1 to 11.4 m; values for O_{90} ranged from 20.4 to 37.9 m. Closest to the source, the frequency (percentage) of PMGF was 0.48% at Chile, 1.67% at Westwold, and 0.62% at Lethbridge. Over all three experiments, the mean PMGF frequency ranged from 0.0% to 0.86% in the first 10 m and from 0.0% to 0.54% at 10 to 20 m from the transgenic pollen source. At distances of 50 to 100 m from the transgenic pollen source, the mean PMGF frequency ranged from 0.0024% to 0.03%. The highest rate of PMGF was 1.67% at a mean distance of 3.0 m at the Westwold site.

Conclusions from these studies are limited by the scale of the experiments and the size of the pollen source. Future studies should be conducted on a larger scale (10 to 100 ha) to determine the influence of pollen source size on PMGF in safflower in different agroecosystems.

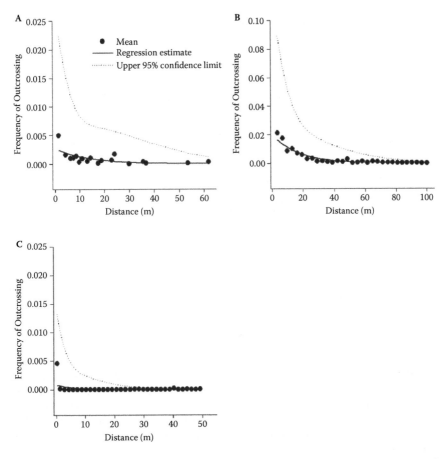

FIGURE 11.4 Frequency of safflower PMGF with increasing distance from the transgenic pollen source at (A) El Bosque, Chile, in 2002, (B) Westwold in 2002, (C) Lethbridge in 2004. Solid circles are mean PMGF frequency at each distance. The solid line was generated from a regression analysis using the exponential decay function, and the dotted line is the upper 95% confidence interval.

11.4.4 Harvest Loss and Seed Bank Persistence

To estimate the density of safflower seeds lost after harvest, five commercial fields were sampled after safflower harvest. Safflower seed harvest losses were highly variable within fields and within and among farm sites (Figure 11.5). Losses of viable safflower seeds ranged from approximately one to five times the recommended seeding rate (100 to 150 viable seeds m^{-2}). By comparison, canola harvest losses can be 9 to 56 times the recommended seeding rate (Gulden et al. 2003a) and within the range of wheat harvest losses (Anderson and Soper 2003, Clarke 1985). Harvest losses are the largest uncontrolled opportunity for movement of seed

| Harvest loss | | | | |
| | | | 95% confidence interval | |
Site	Field	Mean	Lower	Upper
Warner	1	518	287	749
	2	126	0	356
	3	134	0	365
Wrentham	4	515	344	687
	5	81	0	312

Burial study results for Ellerslie 2002

| | | | Frequency of viable seeds | | | |
Depth (cm)	a	b	Spring 1	Fall 1	Spring 2	Fall 2
0	98.33 (0.34)	0.57 (0.01)	42.77 (0.5442)	21.31 (0.004941)	5.383 (0.2377)	2.682 (0.147)
2	97.73 (0.58)	1.54 (0.03)	10.22 (0.5019)	1.546 (0.001399)	0.037 (0.0064)	0.0056 (0.0012)
15	98.10 (0.57)	1.84 (0.05)	6.546 (0.4414)	0.6796 (0.000844)	0.0077 (0.0018)	0.0008035 (0.0002359)

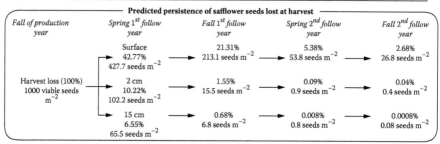

Predicted persistence of safflower seeds lost at harvest

Fall of production year	Spring 1st follow year	Fall 1st follow year	Spring 2nd follow year	Fall 2nd follow year
	Surface 42.77% 427.7 seeds m^{-2} →	21.31% 213.1 seeds m^{-2} →	5.38% 53.8 seeds m^{-2} →	2.68% 26.8 seeds m^{-2}
Harvest loss (100%) 1000 viable seeds m^{-2}	2 cm 10.22% 102.2 seeds m^{-2} →	1.55% 15.5 seeds m^{-2} →	0.09% 0.9 seeds m^{-2} →	0.04% 0.4 seeds m^{-2}
	15 cm 6.55% 65.5 seeds m^{-2} →	0.68% 6.8 seeds m^{-2} →	0.008% 0.8 seeds m^{-2} →	0.0008% 0.08 seeds m^{-2}

FIGURE 11.5 Safflower seed losses at harvest and predicted persistence in soil from field studies. The table at the top is the mean number of safflower seeds lost after harvest in commercial fields. The table in the middle is the results of the regression analysis of seed viability rate of decline over time at the Ellerslie site established in 2002 with means and standard errors for intercept (a), rate of seed viability decline (b), and frequency of seeds in spring and fall for 2 follow years. Note that no viable seeds were recovered from the burial studies after 2 years. The figure at the bottom provides estimates of the percentage and number of seeds per unit area that would remain viable if 1,000 seeds m^{-2} were lost in the fall following safflower production, based on the rate of decline of safflower seeds in the artificial seed bank (burial) studies.

by wind, water, insects and animals, and farm machinery within the agronomic system. Tillage after harvest may decrease the potential movement of seed but may increase the persistence of seed in the seed bank through the induction of secondary dormancy, as has been shown for canola (Gulden et al. 2003b) and some weed species (Batlla and Benech-Arnold 2007, Benech-Arnold et al. 2000). Artificial seed bank studies with transgenic and nontransgenic safflower seed were initiated to address this concern.

Safflower seed persistence was evaluated in artificial seed bank trials established in the fall and sampled and tested for viability throughout the following growing seasons. These data were regressed using the same exponential decay function for PMGF with distance replaced with time after burial (Figure 11.5). There were no significant differences in seed persistence in transgenic and conventional safflower.

Safflower seeds did not persist beyond 2 years after planting, and their viability declined (*b* in Figure 11.5) more rapidly when they were buried relative to those at the soil surface. Less than 50% of safflower seeds at the soil surface and about 10% of buried seeds remained viable the following spring. Thus, the Canadian winter limited safflower seed viability and potential for volunteers in the first follow year. Artificial seed banks are known to overestimate seed viability in soil. This may be due to reduced seed predation, preventing natural burial processes, and field operations (tillage, seeding, etc.), which may change seed position in the soil profile and potentially stimulate or prevent seed germination.

Using the results of the studies of safflower seed losses at harvest and seed persistence, a prediction can be made regarding the number of viable seeds per unit area that may remain over time in the Canadian agroecosystem (Figure 11.5). If a large harvest loss event occurred (about 1,000 viable seeds m^{-2}), only 42.77% of the seed on the soil surface (discounting predation) would remain viable in the next follow spring, whereas 10.22% and 6.55% of seed buried at 2 and 15 cm, respectively, would remain viable. In the spring of the second follow year, it is predicted that 5.38%, 0.09%, and 0.008% of seed at the surface and buried at 2 and 15 cm would remain viable. Therefore, seed burial by tillage not only would reduce potential seed movement but also would decrease seed persistence.

Seeds that survive to germinate the following year will form a volunteer population. To quantify safflower volunteerism on a landscape scale, surveys of fields that previously grew commercial safflower using conventional production practices were conducted. These surveys revealed that volunteer populations are limited to the first year following production. Safflower volunteer densities in fields cropped to barley, wheat (spring or winter), or chemical fallow varied throughout the growing season. There were 0.0 to 10.9 safflower plants m^{-2} in early-to-late spring, 0.0 to 9.3 safflower plants m^{-2} in summer, and 0.0 to 0.33 safflower plants m^{-2} in fall at the time of the follow crop harvest (data not shown). Some safflower volunteers set seed in two of three barley fields and one of three wheat fields, but only about 50% of these seeds were viable. In all chemical fallow fields surveyed, safflower plants did not persist beyond the flowering stage and did not set seed (data not shown). Recropping fields to cereal crops the year following safflower production limited volunteer survival and fecundity, but in some cases it is not eliminated. The few safflower volunteers in cereals can potentially recharge the seed bank and may become admixed with the cereal grain at harvest. However, these safflower volunteers have a significantly reduced fecundity relative to safflower in monocultures (data not shown). The field surveys suggest volunteer safflower populations are limited to the first follow year, but the artificial burial studies estimated viable seed persisting for two years. The discrepancy between the field surveys and the burial studies suggests that seed persistence was overestimated by artificial burial studies. Seed-to-soil contact is limited by packaging the seed in artificial seed bank studies; predation by small mammals and birds is prevented, as is mortality caused by field operation (tillage, seeding herbicides, etc.).

To quantify the potential for admixture from safflower to follow crops at harvest and to determine if herbicides and cultural practices could limit admixture, we conducted field experiments. Safflower seeds were spread on the soil surface at 50 seeds m^{-2} (half the recommended seeding rate) and either barley or canola planted. The transgenic safflower cv. Centennial intended for PMF was resistant to the broad-spectrum herbicide glufosinate but susceptible to herbicides similar to conventional safflower (data not shown). Three herbicides commonly used in-crop to control broadleaf weeds in barley were used (thifensulfuron-methyl/tribenuron-methyl at rates of 5/5 g ai ha^{-1}; bromoxynil/MCPA at rates of 280/280 g ai ha^{-1}, and of 2,4-D at a rate of 394.4 g ai ha^{-1}). All broadleaf herbicides were tank mixed with fenoxaprop-p-ethyl at a rate of 92.51 g ai ha^{-1} to control grass weeds.

Four canola systems, each with its own variety and herbicide, were examined, and one of these included a glufosinate-resistant variety, necessitating the use of glufosinate-resistant safflower in these trials. The canola systems were Liberty Link Invigor 2663® sprayed with Liberty® (glufosinate ammonium at 400 g ai ha^{-1}), Roundup Ready DKL3455® sprayed with Roundup Transorb® (glyphosate at 810 g ai ha^{-1}), Clearfield 46A76® sprayed with Odyssey® (imazamox + imazethapyr both at 14.70 g ai ha^{-1}), and a conventional nontransgenic variety Quantum II® that was sprayed with Poast Ultra®/Muster® (sethoxydim + ethametsulfuron-methyl at 211.19/ 22.23 g ai ha^{-1} and applied with 0.5% v/v Merge).

Herbicides applied to barley significantly reduced safflower volunteer density and biomass. Herbicides applied to barley significantly reduced safflower admixture with harvested barley except at Lethbridge in 2004. Glufosinate-resistant canola and conventional canola treated with glufosinate and sethoxydim plus ethametsulfuron-methyl did not significantly reduce safflower density (21 and 7.5 plants m^{-2}, respectively), biomass (85 and 37 g m^{-2}, respectively), or percentage admixture (1.02% and 1.45% w/w, respectively) of safflower seed with harvested canola seed. Relative to untreated controls, the application of imazamox plus imazethapyr on resistant canola significantly reduced safflower density (8.3 and 3.4 plants m^{-2}, respectively) and biomass (54 and 7 g m^{-2}, respectively). Although this canola system and its associated herbicides reduced safflower seed admixture with harvested canola, it was not significant (1.59% and 0.12% w/w, respectively). Relative to untreated controls, the application of glyphosate to resistant canola significantly reduced safflower density (17.4 and 1.8 plants m^{-2}, respectively), biomass (86 and 8 g m^{-2}, respectively), and the percentage admixture of safflower seeds with harvested canola seed (1.06% and 0.03% w/w, respectively). Glyphosate was the only herbicide to significantly reduce the total number of safflower seeds admixed per square meter with harvested canola. However, the viability of admixed safflower in canola was not significantly reduced by glyphosate. Safflower seeds recovered from harvested canola and barley had low viability (<50%).

11.5   DISCUSSION

Transgenic crops for pharmaceutical production may become the most carefully regulated. An understanding of both the biology of the crop and conventional management practices is required to conduct a risk assessment. Safflower is a domesticated crop, but in Canada relatively little research has been dedicated to agronomy and breeding. In western Canada and some regions of the United States, safflower has no wild or weedy relatives located within the region of adaptation, which should eliminate the potential for PMGF to other species. However, cultivated safflower escapes have been reported in Alberta, and further information is required on the distribution and persistence of these "escaped populations." Surveys of geographic areas where safflower wild/weedy relatives and safflower escapes have been identified in the North American flora should be conducted to determine if historical collections of these species indicated existing populations or random occurrences. If populations of escaped safflower and one or more safflower wild/weedy relatives do persist, additional information from these surveys could include the extent of persistence in the environment, quantification of their abundance and range, and examination of their flowering time to determine if it overlaps with domesticated safflower.

Intraspecific PMGF (from transgenic to nontransgenic safflower) was determined to be relatively low in small-scale experiments: below 1% in the first 10 m from the pollen source and below 0.05% at 50 to 100 m from the pollen source. Because the size of the pollen source influences distribution, large-scale PMGF studies from transgenic to nontransgenic safflower should be conducted to provide information to develop appropriate isolation distances of PMF production sites from commodity production.

Harvest losses of safflower can be one to five times the recommended seeding rate. Safflower seeds have some short-term sprouting resistance but no seed dormancy, which limits persistence in the soil seed bank. This oilseed is attractive to seed predators, including small mammals and birds (personal observation), which may reduce seed density but may facilitate transport of seed from the field. The distance and amount of seed transported by animals have not been quantified, and the fate of transported seed is unknown. Volunteer populations are restricted to the first follow year in conventional fields. Although they can be controlled by conventional herbicides in subsequent crops, there is the potential for survivors to flower, set viable seed, and contribute to both PMGF and SMGF.

Seed movement following harvest was outside of the scope of this research; however, in other crops, seed spillage has contributed to volunteer populations (Saji et al. 2005, Yoshimura et al. 2006). Seeding mixing during seed handing can lead to AP in other crops or commodity safflower. Seed may be a more important factor than pollen for gene flow because it is able to survive for long periods of time, and seed is traded both locally and internationally.

Based on the studies summarized here, we present a brief overview of some best management practices to mitigate gene flow from transgenic safflower intended for molecular farming of pharmaceuticals (Table 11.1). Many of these practices are similar to those employed by certified seed growers and crops currently channeled to markets requiring identity preservation. Isolation zones or growing areas free of conventional safflower will be necessary to reduce PMGF and SMGF. In Canada, safflower is grown in the southern region of the prairies, primarily in rotation with cereals, and is a low acreage minor crop. Thus, locating fields isolated from commodity safflower production should be relatively straightforward.

TABLE 11.1

Some Best Management Practices to Limit Gene Flow from Transgenic Safflower Production to the Environment and Food/Feed System During Specific Operations

Crop	Operation	Gene Flow Route	Concern	Mitigation
PMF safflower	Preplanting	Pollen	Wild/weedy relatives	Choose sites where wild/weedy relatives do not occur
	Seed treatment	Seed	Seed loss due to pathogens	Chemical treatment of seeds prior to planting
	Planting	Seed	Seed movement off site	Containment of seed and clean seeding equipment after seeding
	Crop production	Seed	Seed loss due to pathogens	In-crop fungicide application (minor use required)[a]
		Pollen	Commodity safflower	Isolation distance from commodity safflower
		Pollen	Commodity safflower	Nontransgenic safflower planted in perimeter of field removed after flowering to prevent seed production
		Pollen and seed	Off-site movement	Fallow or non-food/feed green manor grown in perimeter
	Harvesting	Seed	Seed losses	Limit seed losses by combine settings and perhaps chaff wagon[b]
	Postharvest	Seed	Seed movement with residue	Tillage to incorporate seed and limit movement by abiotic (wind, water, etc.) and biotic (human, insect, and animals) factors and to accelerater loss of viable seed in soil
	Tillage	Seed	Seed movement off site	Clean equipment prior to movement off field
Follow crop	Preplanting	Seed and pollen	Safflower seed and volunteers	Crop choices include a cereal or chemical fallow
		Seed and pollen	Safflower seed and volunteers	Herbicide(s) to control safflower volunteers
	In crop[c]	Seed and pollen	Safflower seed and volunteers	Herbicide(s) to control safflower volunteers
	Preharvesting	Seed	Admixture with harvested seed	Scouting (monitoring) fields and removing safflower volunteers

[a] The safflower capitula (inflorescence/seed head) can drop to the ground when seeds are mature when infected with certain fungal pathogens (for details, see Mundel and Huang 2003). This should be avoided to limit seed losses during production, but no fungicides have been registered for use on safflower in Canada.

[b] The efficacy of chaff wagon to reduce the number of seeds lost during harvest has not been quantified for safflower but has been shown to reduce seed bank inputs for oats (Shirtliffe and Entz 2005).

[c] Chemical fallow is preferred over a follow crop in the first year after transgenic safflower production for PMF.

As outlined, landscape-scale studies may be required to determine appropriate isolation distance. To further mitigate PMGF, the effectiveness of surrounding transgenic safflower fields with nontransgenic border plants (trap crop) could be assessed on a landscape scale. Currently in Canada, trap crops are used to reduce pollen movement from transgenic canola/oilseed rape field experiments (CFIA 2006b). The efficacy of a trap crop to reduce outcrossing has been documented for transgenic canola (*Brassica napus* L.; Morris et al. 1994, Reboud 2003, Staniland et al. 2000) and cotton (*Gossypium hirsutum* L.; Kareiva et al. 1994). Nontransgenic safflower (trap crop) should be planted in the perimeter of transgenic safflower fields and removed after flowering and prior to seed set to limit subsequent PMGF and SMGF.

During crop production, fungal pathogens should be controlled to limit SMGF because they can lead to seed head (capitulum) loss prior to harvest. However, no fungicides have been registered for safflower in Canada, and a minor use registration for chemicals licensed elsewhere would be required. Containment of seed during transportation and handling to and from the site for planting should be rigorously conducted and all seeding equipment cleaned on site prior to movement to limit SMGF.

Careful combine setting can reduce but not eliminate seed loss. A chaff wagon could be used to retain most of the seed and small residue that is lost from the combine. However, it is not clear how to best manage large amounts of residual material once it is collected. An alternative is to incorporate residual material after harvest to encourage biological degradation of seed and germination and to limit movement of seed by abiotic and biotic factors. Seed burial studies demonstrated an accelerated reduction of safflower seed persistence when the seeds are buried.

The cleaning of all tillage equipment prior to leaving the site will limit SMGF. The choice of crop to follow transgenic safflower will depend on thresholds for AP of PMP in commodity crops. In a cereal or some canola systems (glyphosate and imidazolinone resistant), field operations and herbicides limit safflower volunteer survival and fecundity, but they are not eliminated.

We recommend the use of chemical fallow in the first follow year as the most effective way to limit safflower recharge of the seed bank and prevent AP in commodity crops. Diligent scouting (monitoring) of follow fields and removal of safflower volunteers will reduce safflower volunteer survival and opportunity for PMGF and SMGF in the agroecosystem and will limit admixture (AP) in harvested follow commodity seed.

This preliminary risk assessment quantified gene flow through both seed and pollen and identified data gaps. This approach, developed for a highly sensitive crop trait, may be useful in other minor or abandoned crops to determine their suitability as a platform for PMF or other bioproducts (Figure 11.1). The PMP expressed in the transgenic safflower used in these studies has a seed-specific promoter that limits exposure to the environment from vegetative parts of the plant. In Canada and certain regions of the United States, safflower may be suitable for PMP because it is a nonfood crop grown on limited acreage and is intended for production on a small scale; from the studies presented here, low levels of gene flow are expected from safflower in the agroecosystem, and several mitigation measures can be used to limit gene flow from safflower (Figure 11.1).

REFERENCES

Anderson, R.L., and Soper, G. 2003. Review of volunteer wheat (*Triticum aestivum*) seedling emergence and seed longevity in soil. *Weed Technol.* 17(3): 620–626.

Batlla, D., and Benech-Arnold, R.L. 2007. Predicting changes in dormancy level in weed seed soil banks: Implications for weed management. *Crop Protect.* 26: 189–197.

Benech-Arnold, R.L., Sanchez, R.A., Forcella, F., Kruk, B.C., and Ghersa, C.M. 2000. Environmental control of dormancy in weed seed banks in soil. *Field Crops Res.* 67(2): 105–122.

Berville, A., Breton, C., Cunliffe, K., Darmency, H., Good, A.G., Gressel, J., Hall, L.M., McPherson, M.A., Medail, F., Pinatel, C., Vaughan, D.A., and Warwick, S.I. 2005. Issues of ferality or potential ferality in oats, olives, the pigeon-pea group, ryegrass species, safflower, and sugarcane. In: Gressel, J., ed., *Crop Ferality and Volunteerism: A Threat to Food Security in the Transgenic Era?* Taylor & Francis, Boca Raton, FL, pp. 231–255.

Canadian Food Inspection Agency. 2001. CFIA multi-stakeholder consultation on plant molecular farming: Report of the proceedings. Available: http://www.inspection.gc.ca/english/plaveg/bio/mf/reportprocede.shtml. Accessed: September 28, 2007.

Canadian Food Inspection Agency. 2004. Assessment criteria for determining environmental safety of plants with novel traits (directive Dir94–08). Available at: http://www.inspection.gc.ca/english/plaveg/bio/dir/dir9408e.shtml. Accessed May 18, 2007.

Canadian Food Inspection Agency. 2005. Assessment criteria for the evaluation of environmental safety of plants with novel traits intended for commercial plant molecular farming. Available at: http://www.inspection.gc.ca/english/plaveg/bio/mf/fracad/evaluae.shtml. Accessed: March 14, 2008.

Canadian Food Inspection Agency. 2006a. Developing a regulatory framework for the environmental release of plants with novel traits intended for commercial plant molecular farming in Canada. Available at: http://www.inspection.gc.ca/english/plaveg/bio/mf/fracad/commere.shtml. Accessed April 6, 2007.

Canadian Food Inspection Agency. 2006b. Specific terms and conditions to conduct field trials with PNT. Available at: http://www.inspection.gc.ca/english/plaveg/bio/dt/term/2006/2006e.shtml. Accessed June 4, 2007.

Claassen, C.E. 1950. Natural and controlled crossing in safflower, *Carthamus tinctorius* L. *Agron. J.* 42: 301–304.

Clarke, J.M. 1985. Harvesting losses of spring wheat in windrower combine and direct combine harvesting systems. *Agron. J.* 77(1): 13–17.

Correll, D.S., and Johnston, M.C. 1970. *Manual of the Vascular Plants of Texas.* Texas Research Foundation, Renner, TX.

Dajue, L., and Mundel, H.H. 1996. Safflower: *Carthamus tinctorius* L. International Plant Genetic Resources Institute, Rome, Italy.

Fuller, T.C. 1979. Ecology of some California weeds that also occur in Australia. *Proceedings of the 7th Asian-Pacific Weed Science Society Conference.* Sydney, Australia. Asian-Pacific Weed Science Society.

Goodman, R.M., Knauf, V.C., Houck, C.M., and Comai, L. 1987. Molecular farming. European Patent Application WO 87/00865, PCT/US86/01599.

Gulden, R.H., Shirtliffe, S.J., and Thomas, A.G. 2003a. Harvest losses of canola (*Brassica napus*) cause large seedbank inputs. *Weed Sci.* 51(1): 83–86.

Gulden, R.H., Shirtliffe, S.J., and Thomas, A.G. 2003b. Secondary seed dormancy prolongs persistence of volunteer canola in western Canada. *Weed Sci.* 51(6): 904–913.

Hickman, J.C. 1993. *The Jepson Manual: Higher Plants of California.* University of California Press, Los Angeles.

Hoover, R.F. 1970. *The Vascular Plants of San Luis Obispo County.* University of California Press, Los Angeles.

Howard, A., Howard, G.L.C., and Khan, A.R. 1915. Studies in Indian oil-seeds. In: *Memoirs of the Department of Agriculture in India*. Botanical series edition, Thacker, Spink, Calcutta, India, pp. 237–272.

James, C. 2007. ISAAA brief 37-2007—executive summary > ISAAA.org. Available at: http://www.isaaa.org/Resources/publications/briefs/37/reportsummary/default.html. Accessed April 4, 2008.

Kareiva, P., Morris, W., and Jacobi, C.M. 1994. Studying and managing the risk of cross fertilization between transgenic crops and wild relatives. *Mol. Ecol.* 3: 15–21.

Keil, D.J. 2006. Carthamus. In: *Flora of North America Editorial Committee*, ed., Flora of North America: North of Mexico. Oxford University Press, New York, pp. 178–181.

Kessler, E. 1987. *Carthamus lanatus* L. (Asteraceae: Cynareae)—a potential serious plant pest in Oklahoma. *Proc. Okla. Acad. Sci.* 67: 39–43.

Levin, D.A., and Kerster, H.W. 1974. Gene flow in seed plants. *Evol. Biol.* 7: 139–220.

López-González, G. 1989. Acerca de la clasificacion natural del genero *Carthamus* L., *s.l. Anales Del Jardin Botan*. Madrid 47: 11–34.

Marticorena, C., and Quezada, M. 1985. *Catalog de la flora Vascular de Chile*. Gayana; Botanica. Universidad de Concepcion Chile, Concepcion, Chile.

McGregor, W.G., and Hay, W.D. 1952. Safflower Canadian experiments. *Sci. Agr.* 32(4): 204–213.

Morris, W., Kareiva, P., and Raymer, P. 1994. Do Barren Zones and Pollen Traps Reduce Gene Escape from Transgenic Crops? *Ecological Applications.* (4) 157–165.

Mundel, H.H., Blackshaw, R.E., Byers, R.J., Huang, H.C., Johnson, D.L., Keon, R., Kubik, J., McKenzie, R., Otto, B., Roth, B., and Stanford, K. 2004. *Safflower Production on the Canadian Prairies: Revisited in 2004*. Agriculture and Agri-Food Canada, Lethbridge, Alberta, Canada.

Mundel, H.H., and Huang, H.C. 2003. Control of major diseases of safflower by breeding for resistance and using cultural practices. In: Huang, H.C., and Acharya, S.N., eds., *Advances in Plant Disease Manage*ment. Research Signpost, Kerala, India, pp. 293–310.

Munz, P.A. 1968. *A California Flora and Supplements*. University of California Press, Berkeley.

Peterson, R.K.D., and Arntzen, C.J. 2004. On risk and plant-based biopharmaceuticals. *Trends Biotechnol.* 22(2): 64–66.

Raybould, A. 2006. Problem formulation and hypothesis testing for environmental risk assessments of genetically modified crops. *Environ. Biosafety Res.* 5: 1–7.

Reboud, X. 2003. Effect of a gap on gene flow between otherwise adjacent transgenic *Brassica napus* crops. *Theor. Appl. Genet.* 106(6): 1048–1058.

Saji, H., Nakajima, N., Aono, M., Tamaoki, M., Kubo, A., Wakiyama, S., Hatase, Y., and Nagatsu, M. 2005. Monitoring the escape of transgenic oilseed rape around Japanese ports and roadsides. *Environ. Biosafety Res.* 4(4): 217–222.

Shirtliffe, S.J., and Entz, M.H. 2005. Chaff collection reduces seed dispersal of wild oat (*Avena fatua*) by a combine harvester. *Weed Sci.* 53: 465–470.

Sparrow, P.A.C., Irwin, J.A., Dale, P.J., Twyman, R.M., and Ma, J.K.C. 2007. Pharma-planta: Road testing the developing regulatory guidelines for plant-made pharmaceuticals. *Transgenic Res.* 16: 147–161.

Staniland, B.K., McVetty, P.B.E., Friesen, L.F., Yarrow, S., Freyssinet, G., and Freyssinet, M. 2000. Effectiveness of border areas in confining the spread of transgenic *Brassica napus* pollen. *Can. J. Plant Sci.* 80: 521–526.

Warwick, S.I., and Stewart, C.N., Jr. 2005. Crops come from wild plants—how domestication, transgenes, and linkage together shape ferality. In: Gressel, J.B., ed., *Crop Ferality and Volunteerism*. Taylor & Francis, Boca Raton, FL, pp. 9–25.

Yoshimura, Y., Beckie, H.J., and Kazuhito, M. 2006. Transgenic oilseed rape along transportation routes and port of Vancouver in western Canada. *Environ. Biosafety Res.* 5: 67–75.

Section 2

Bio-Based Industry Products

12 Production of Functional γ-Linolenic Acid (GLA) by Expression of Fungal Δ12- and Δ6-Desaturase Genes in the Oleaginous Yeast *Yarrowia lipolytica*

Lu-Te Chuang, Dzi-Chi Chen, Ying-Hsuan Chen,
Jean-Marc Nicaud, Catherine Madzak,
and Yung-Sheng Huang

CONTENTS

Key Words: Δ6-Desaturase; Δ12-desaturase; γ-linolenic acid (GLA); *Mortierella alpina*; oleaginous yeast; *Yarrowia lipolytica*.

12.1   INTRODUCTION

γ-Linolenic acid (GLA; Δ6,9,12-18:3) is one of the n-6 polyunsaturated fatty acids (n-6 PUFAs) derived from linoleic acid (LA; Δ9,12-18:2), by the action of Δ6-desaturase. GLA has been demonstrated to exert different beneficial effects in a variety of cell cultures, animal models, and human studies. These effects include the reduction of inflammation-related diseases, such as rheumatoid arthritis and cystic fibrosis, through modulating of immune response [1–4], decrease in the development of hypertension [5], and progress of other chronic diseases like diabetes and cancer [6,7]. The mechanisms by which GLA exerts these biological functions have been attributed to its modulation of n-6 PUFA metabolism and eicosanoid synthesis. In mammals, Δ6-desaturation of LA to GLA is considered as a rate-limiting step of the n-6 PUFA metabolic pathway [8]. The reaction step is readily influenced by many physiological and pathological conditions [9]. Supplementation of GLA could bypass this step and resume the subsequent n-6 PUFA metabolism.

As benefits of GLA for human health have now been well established, the general public has searched for GLA-rich oils for the purpose of supplementation or as a food ingredient. At present, the natural sources of GLA are plant seed oils from evening primrose, borage, and black currant. GLA can also be found in fungal oils from *Mucor* spp. and *Mortierella* spp. However, many of these sources are not economically well suited for use in human nutrition. The current GLA-rich oils (borage oil and evening primrose oil) in the market are still expensive and unreliable due to great variations in availability and high cost of processing. Thus, there is an increasing effort to engineer the fatty acid biosynthesis of oil seed crops to produce GLA at low cost.

Huang and his coworkers have previously introduced a vector with Δ12- and Δ6-desaturases into canola plants through *Agrobacterium*-mediated transformation and subsequently produced high-GLA canola oil (HGCO) [10]. Transgenic HGCO accumulated up to 40% of GLA, which is significantly higher than proportions found in borage oil (20%) and evening primrose oil (10%) [11]. Huang and his coworkers have also established a model of transformed *Saccharomyces cerevisiae* expressing both Δ12- and Δ6-desaturase genes from *Mortierella alpina* to synthesize *de novo* GLA without exogenous fatty acid substrates added [12]. Coexpression of Δ12- and

Δ6-desaturase genes allowed the conversion of endogenous oleic acid (OA; Δ9-18:1) to LA by the Δ12-desaturase, and the newly synthesized LA was then Δ6-desaturated to form GLA. This result indicates that single-cell oils (SCOs) from edible yeasts could be another substitute for more expensive plant oils.

Since baker's yeast is low in fat content, the overall yield of GLA was low. In this study, we chose the oleaginous yeast *Yarrowia lipolytica* as host for the production of GLA-rich yeast oil. Our approach was based on the fact that this yeast contains LA [13], accumulates large amounts of lipids, and is widely used in industrial applications with GRAS (generally regarded as safe) status. Besides, the genome of *Y. lipolytica* has been sequenced and is available on the Genolevures Web site (http://cbi.labri.fr/Genolevures/elt/YALI), and genetically modified strains and various expression vectors have been developed [14–16]. Therefore, in this study, we inserted, either separately or together, both Δ12- and Δ6-desaturase complementary DNA (cDNA) from *Mortierella alpina* in an established *Y. lipolytica* expression vector [14]. The newly constructed plasmids were then introduced into a genetically modified *Y. lipolytica* strain. By cultivating the transformed yeasts simultaneously expressing Δ12- and Δ6-desaturase genes, we obtained efficient production of GLA, which was effectively converted from endogenous OA and LA.

12.2   EXPERIMENTAL PROCEDURES

12.2.1   CHEMICALS AND CULTURE MEDIA

Gas chromatography (GC) standard (RL-461) and thin-layer chromatography (TLC) standard (18-5-A) were obtained from Nu-Chek Prep Incorporated (Elysian, MN). Yeast extract, peptone, yeast nitrogen base, and thiamine were purchased from Difco Laboratories (Detroit, MI). All the cultivation media used in this study are described in Table 12.1. Triheptadecanoin (used as an internal standard), glucose, $(NH_4)_2SO_4$, NH_4Cl, KH_2PO_4, $MgSO_4 \cdot 7H_2O$, Na_2HPO_4, and citric acid were from Sigma (St. Louis, MO). Hexane was ultraviolet (UV) grade, and other solvents were distilled-in-glass quality.

12.2.1.1   Plasmid Construction, Yeast Transformation, and Gene Expression

The cDNA sequences corresponding to Δ6-desaturase (AF110510) and Δ12-desaturase (AF110509) from *M. alpina* were obtained according to Huang et al. (1999) [12] and were inserted into pYLEX1 expression vector from a YLEX kit. The *Y. lipolytica* Expression Kit (YLEX kit for expression of recombinant protein in *Y. lipolytica*) was purchased from Yeastern Biotech Company, Limited (Taipei, Taiwan). The vectors from the kit have been previously described under alternative names, with pYLEX1 corresponding to the expression vector named pINA1269 [14].

Briefly, a pair of primers with homology to the sequences downstream of the initiation site and upstream of the stop codon of Δ6-desaturase, respectively, were designed according to the specifications of YLEX kit: forward primer Delta6-AATG, 5′Ph- AATGGCTGCTGCTCCCAGTG-3′ and reverse primer Delta6-XcmI, 5′-GCCAGAGTCGGACTGGTTACTGCGCCTTACCCATC-3′ (*Xcm*I restriction site underlined). PCR amplification was run on a DNA thermal cycler (Applied

TABLE 12.1

Complete Formula of the Culture Media Used in This Study

Phosphate-Buffered Saline (PPB) Medium (pH 6.0)*

Composition	Concentration
Glucose (g/L)	20
Yeast extract (g/L)	1.32
NH$_4$Cl (g/L)	1.32
KH$_2$PO$_4$ (g/L)	0.32
MgSO$_4$×7H$_2$O (g/L)	0.24
Thiamine (mg/L)	0.33
0.2M Na$_2$HPO$_4$ (mL/L)	63.15
0.1M Citric acid (mL/L)	36.85

Yeast Extract Peptone Dextrose (YPD)-Related Media (g/L)

	YPD1	YPD2	1/5 YP	1/10 YP	1/20 YP	1/50 YP
Yeast extract	10	10	2	1	0.5	0.2
Peptone	10	20	4	2	1	0.4
Glucose	10	20	30	30	30	30

Yeast Nitrogen Base Dextrose (YNBD) Medium

Composition	Concentration (g/L)
Yeast nitrogen base without amino acids and ammonium sulfate	1.72
(NH$_4$)$_2$SO$_4$	5
Glucose	20

^{a} pH adjusted to 6.0 by 0.2M Na$_2$HPO$_4$ and 0.1M citric acid.

Biosystem, Foster City, CA) following a program of 30 s at 94°C, 30 s at 55°C, and 1.5 min at 72°C for 35 cycles, followed by extension for 5 min at 72°C. After amplification, Δ6-desaturase PCR products were digested with *Xcm*I restriction endonuclease and ligated to pYLEX1 vector restricted with *Pml*I and *Xcm*I to construct the plasmid pYLd6 (Figure 12.1A). Similarly, the forward primer Delta12-AATG (5′Ph-AATGGCACCTCCCAACACTATC-3′) and the reverse primer Delta12-BamHI (5′-GTGGATCCTTACTTCTTGAAAAAGACCACG-3′) were used to amplify Δ12-desaturase cDNA (*Bam*HI restriction site underlined). To construct the vector expressing the Δ12-desaturase gene (pYLd12), the PCR fragment was digested with *Bam*HI restriction enzyme and ligated to pYLEX1 vector restricted with *Pml*I and *Bam*HI (Figure 12.1B). Newly constructed plasmids were screened first by restriction enzyme digestion and PCR and then by DNA sequencing.

To construct a plasmid for simultaneous expression of Δ12- and Δ6-desaturase genes, we designed a pair of primers with homology, respectively, to the 5′ end and the 3′ end of the sequence of Δ12-desaturase expression cassette (hp4d promoter/Δ12-

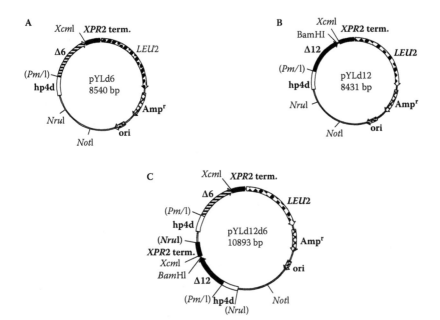

FIGURE 12.1 Plasmids used to transform *Yarrowia lipolytica* Polg strain for expression of, respectively, *Mortierella alpina* Δ6-desaturase gene (A), Δ12-desaturase gene (B), or both Δ12- and Δ6-desaturase genes (C).

desaturase gene/terminator) from pYLd12 plasmid. The forward primer Delta6-12F, 5′-GGTCCGGACCCAGTAGTAGGTTGAGGC-3′, and the reverse primer Delta6-12R, 5′-TTTCCGGACTTATCATCGATGATAAGCTCTC-3′, were used together with linear pYLd12 for a PCR amplification as described. The newly amplified blunt-ended PCR fragment and the plasmid pYLd6 digested with *Nru*I restriction enzyme were ligated to form the coexpression plasmid pYLd12d6 (Figure 12.1C). The control (empty pYLEX1 vector) and the three newly constructed plasmids were linearized using *Not*I enzyme and introduced into the genetically modified *Y. lipolytica* strain Polg (Leu⁻, ΔAEP, ΔAXP, Suc⁺, pBR322 integration platform) [14]. Transformation of *Y. lipolytica* was performed using the one-step method previously described by Chen et al. [17]. Transformants were selected by plating on YNBD (yeast nitrogen base dextrose) plates (minimal medium without leucine).

Randomly picked transformants Polg(pYLd6) were first grown overnight in YPD1 medium (Table 12.1) at 28°C. These precultures (1 × 10⁸ cells) were used to inoculate 100 mL of PPB (phosphate-buffered saline) medium for studying the activity of recombinant Δ6-desaturase. The cultivation conditions (PPB medium, 28°C) have previously been reported to give an efficient production of heterologous proteins by *Yarrowia* cells [18–20]. To examine whether culture conditions affected total lipid production by the transformed yeast Polg(pYLd6), yeast cells were incubated in various conditions of medium, incubation time, and temperature. To improve the yield of GLA by transformed yeasts, selected transformants of the newly constructed Polg(pYLd12) and Polg(pYLd12d6) strains were compared. These strains were cultivated in 1/20 YP medium at 28°C for 72 h to check for the production of recombinant desaturases. In all studies, the Polg strain transformed with the empty vector [Polg(pYLEX1)] was used as the negative control.

12.2.1.2 Lipid Extraction, TLC Separation, and Fatty Acid Analysis

After cultivation, cells were harvested by centrifugation, and the cell pellet was washed once with sterile deionized water. Total lipids from transformed yeasts were extracted following the procedure described previously [12]. Briefly, the rinsed cells were extracted with 30 mL of chloroform:methanol (2:1, v/v) at 4°C overnight. The lipids in chloroform were separated from the aqueous phase by adding 6 mL of saline solution and were collected, and solvent was evaporated using a stream of nitrogen. To visualize the lipid fractionation and GLA distribution, aliquots of total lipids in chloroform were applied onto TLC plates. The chromatoplates were developed with a solvent system of hexane-diethyl ether-acetic acid (70:30:1 v/v/v). After development, the plates were dried under nitrogen and then visualized with 2,7-dichlorofluorescein/ethanol solution under UV light. The marked lipid fractions were then scraped off individually from the plates for further analysis.

The extracted total lipids or different lipid fractions were saponified and methylated to generate fatty acid methyl esters (FAMEs). FAMEs were then analyzed by gas liquid chromatography using an Agilent 6890 gas chromatograph equipped with a flame ionization detector and a fused-silica capillary column (Omegawax; 30 m × 0.32 mm i.d., 0.25-μm film thickness, Supelco, Bellefonte, PA). Fatty acids were identified by comparing their retention times to those of known standards and quantified using the internal standard. In this study, the conversion of substrates to products (i.e., Δ6-desaturation and Δ12-desaturation) was calculated by the ratio of [Product]/([Product] + [Substrate]) × 100%.

12.2.1.3 Statistical Analysis

Data were analyzed by analysis of variance (ANOVA) and Fisher's protected least significant difference (LSD) to compare differences between means of cell mass, of total lipids, and of conversion rates. Means differences were considered significant at the $p \leq .05$ level.

12.3 RESULTS

12.3.1 Conversion of LA to GLA by Δ6-Desaturase

LA is the major PUFA synthesized by *Y. lipolytica* [13]. To convert the endogenous LA to GLA, the plasmid pYLd6 containing the Δ6-desaturase gene from *M. alpina* was introduced into the genetically modified strain of *Y. lipolytica* Polg. The transformed yeast polg(pYLd6) was then cultivated in the PPB medium at 28°C for 72 h. Results in Figure 12.2 show that a significant portion (69%) of the endogenous LA was converted to GLA. There were also trace amounts of Δ6,9-16:2 (1.6%) and Δ6,9-18:2 (12.1%), the Δ6-desaturation products of palmitoleic acid (Δ9-16:1)

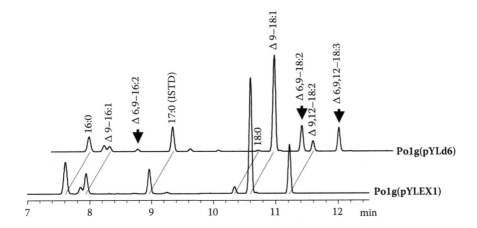

FIGURE 12.2 Gas chromatogram of fatty acids of total lipids in the yeasts transformed with empty vector [Polg(pYLEX1)] or the yeasts carrying the *Mortierella* Δ6-desaturase gene [Polg(pYLd6)]. Solid arrows show the presence of Δ6,9-16:2, Δ6,9-18:2, and γ-linolenic acid (Δ6,9,12-18:3), the Δ6-desaturation products of Δ9-16:1, Δ9-18:1, and linoleic acid (Δ9,12-18:2), respectively.

and OA (Δ9-18:1), respectively. None of these three Δ6 fatty acids was found in the control yeast transformed with empty vector [Polg(pYLEX1)] (Figure 12.2).

12.3.2 An Optimal Medium for Lipid Production by *Y. lipolytica*

To determine what type of medium would be optimal for the growth and lipid production of the transformed *Y. lipolytica*, four common yeast culture media (YNBD, PPB, YPD1, and YPD2) and four nitrogen-reduced media (1/5 YP, 1/10 YP, 1/20 YP, and 1/50 YP) were tested (Table 12.1). Results in Figure 12.3A show that different culture media significantly influenced cell mass and total cellular lipids. Among four common media, yeast cells grown in YPD2 medium exhibited the highest cell mass (9.7 g/L), followed by YPD1 (6.6 g/L), PPB (4.7 g/L), and YNBD (3.7 g/L). However, when the amounts of cellular lipids accumulated in the transformed yeast were compared, yeasts grown in YNBD exhibited the highest (25.8 mg/g yeast), followed by those grown in PPB (18.8 mg/g yeast), in YPD1 (17.2 mg/g yeast), and in YPD2 (12.0 mg/g yeast). Similar results were also observed in yeast cells cultivated in four nitrogen-reduced media. The level of cell mass of transformed *Y. lipolytica* decreased, whereas those of cellular lipids increased as the level of nitrogen nutrients (yeast extract and peptone) in the medium decreased (Figure 12.3A).

The effect of different culture media on Δ6-desaturation activity (conversion of LA to GLA) in the transformed yeast is shown in Figure 12.3B. When using four common media with the exception of YNBD, yeast cells converted at least 30% of LA into GLA. In three of the four nitrogen-reduced media (1/10 YP, 1/20 YP, and 1/50 YP) studied, yeast cells converted LA to GLA at a level of approximately 60%. However, when the level of GLA in total cellular lipids was calculated, yeast

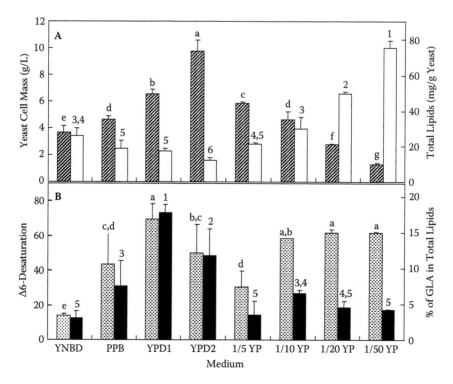

FIGURE 12.3 Effect of different media on cell mass (■), lipid content (□), rate of Δ6-desaturation of linoleic acid to γ-linolenic acid (□), and percentage of γ-linolenic acid in total lipids (▨) in the transformed *Y. lipolytica*. Δ6-Desaturation was calculated as [Product]/[Product + Substrate] × 100%. Each value point represents the mean of three independent experiments. In each category, values with different letters or numbers are significantly different from each other at $p < .05$.

cells grown in nitrogen-reduced media contained less GLA (3.5% to 6.5%) than those grown in two nitrogen-rich media (YPD1 and YPD2) (Figure 12.3B).

To produce high cellular lipid content without compromising too much the growth and GLA production, the 1/20 YP medium was selected for the cultivation of transformed yeasts in all the following studies.

12.3.3 EFFECT OF CULTIVATION TIME ON TOTAL LIPID AND GLA PRODUCTION IN THE TRANSFORMED *Y. LIPOLYTICA*

Results examining the effect of different incubation times on lipid production are shown in Figure 12.4A. Generally, cell mass of the transformed *Y. lipolytica* increased progressively with the cultivation time and reached the plateau at the fifth day. However, there were no significant changes in cellular lipid content (ranging from 37 to 49 mg/g of yeast cells) as a function of time. The length of cultivation also

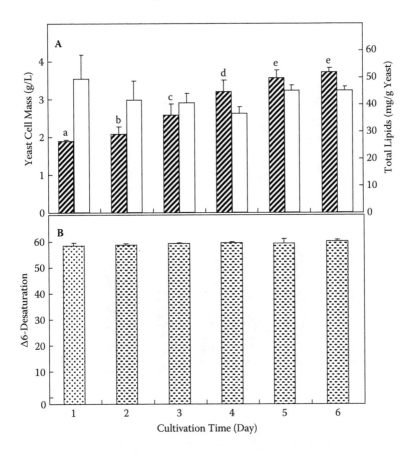

FIGURE 12.4 Effect of length of cultivation time on cell mass (■), lipid content (□), and rate of Δ6-desaturation of linoleic acid to γ-linolenic acid (□) in the transformed *Y. lipolytica*. Δ6-Desaturation was calculated as [Product]/[Product + Substrate] × 100%. Each value point represents the mean of three independent experiments. In each category, values with different letters or numbers are significantly different from each other at $p < .05$.

had no significant effect on conversion of LA to GLA (approximately 60%) in the transformed yeast Po1g(pYLd6) (Figure 12.4B).

12.3.4 EFFECT OF TEMPERATURE ON TOTAL LIPID AND GLA PRODUCTION IN THE TRANSFORMED *Y. LIPOLYTICA*

To examine the effect of different temperatures on cell growth and lipid production in Po1g(pYLd6), transformed yeast cells were cultivated in 1/20 YP medium at 15°C, 18°C, 22°C, 25°C, 28°C, 32°C, or 35°C for 72 h. Results in Figure 12.5A show that the levels of cell mass were comparable (approximately 2.8 g/L) when the transformed yeasts were cultivated at temperatures ranging from 15°C to 28°C, but the cell mass decreased rapidly to 1.5 and 0.4 g/L at 32°C and 35°C, respectively.

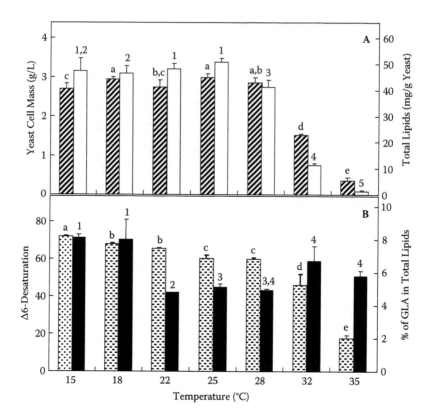

FIGURE 12.5   Effect of incubation temperature on cell mass (■), lipid content (□), rate of Δ6-desaturation of linoleic acid to γ-linolenic acid (□), and percentage of γ-linolenic acid in total lipids (▨) in the transformed *Y. lipolytica*. Δ6-Desaturation was calculated as [Product]/[Product + Substrate] × 100%. Each value point represents the mean of three independent experiments. In each category, values with different letters or numbers are significantly different from each other at $p < .05$.

A similar pattern of effect on cellular lipid contents was also found. When cultures were performed at temperatures ranging from 15°C to 25°C, the amount of total lipids was found to be around 50 mg/g in transformed yeasts. The lipid contents fell to 40 mg/g of yeast at 28°C and dropped further to 11 and 1 mg/g of yeast at 32°C and 35°C, respectively.

The incubation temperature also affected the conversion of LA to GLA and the amount of GLA formed in the transformed *Yarrowia* cells (Figure 12.5B). The rate of Δ6-desaturation decreased slowly from 70% to 60% between 15°C and 28°C, fell to 46% at 32°C, and to 18% at 35°C. Moreover, the percentage of GLA in total lipids in the transformed yeasts was around 8% at 15°C and 18°C. The level was reduced to approximately 5% at temperatures ranging from 22°C to 35°C.

TABLE 12.2
Percentages of Oleic Acid (OA), Linoleic Acid (LA), and γ-Linolenic Acid (GLA) in Total Lipids from Four Transformed Strains of *Y. lipolytica*

Transforming Plasmid	pYLEX1 (control)	pYLd12	pYLd6	pYLd12d6
OA	40.7 ± 1.5^a	19.4 ± 1.9^b	53.5 ± 0.9^c	26.3 ± 2.7^d
LA	19.4 ± 1.2^a	38.7 ± 1.1^b	3.3 ± 0.1^c	12.5 ± 0.4^d
GLA	—	—	4.9 ± 0.1^a	20.0 ± 1.9^b

Note: Values on the same row with different superior letters are significantly different ($p < .05$).

12.3.5 CONVERSION OF OA TO GLA BY Δ12-DESATURASE AND Δ6-DESATURASE IN THE TRANSFORMED *Y. LIPOLYTICA*

Results described clearly indicate that the recombinant Δ6-desaturase in Po1g(pYLd6) functioned effectively in converting LA to GLA. However, the low initial level of LA in total lipids in yeasts (Table 12.2) seemed to limit the overall production of GLA. To raise the yield of GLA, increasing the availability of LA converted from OA through inserting an additional Δ12-desaturase gene would be one possible solution. The *Y. lipolytica* Po1g strain was transformed with pYLd12 plasmid containing a Δ12-desaturase gene from *M. alpina* (Figure 12.1B). A randomly picked transformed colony was incubated in YPD1 overnight. Yeasts (10^8 cells) from this preculture were then inoculated into 1/20 YP medium, and cultures were incubated for 3 days (72 h). Total lipids were extracted and fatty acid composition analyzed. GC chromatograms from the transformed yeast (with Δ12-desaturase gene) were then compared with the control (without Δ12-desaturase gene). Results in Figure 12.6 show that a large proportion of OA (67%) in the transformed yeast was converted to LA, whereas only 32% of OA was converted to form LA in the control strain (Figure 12.6). Results in Figure 12.6 also show the appearance of a small additional peak in the transformed yeasts [Po1g(pYLd12)], but not in the control yeasts. This minor novel fatty acid was tentatively identified as Δ9,12-16:2 as discussed previously [12]. This finding indicates an efficient conversion of the endogenous OA to LA by action of the inserted recombinant Δ12-desaturase gene.

12.3.6 EFFECT OF SIMULTANEOUS EXPRESSION OF FUNGAL Δ12- AND Δ6-DESATURASE GENES IN THE TRANSFORMED *Y. LIPOLYTICA*

Results described demonstrate that recombinant Δ6- or Δ12-desaturase genes were successfully expressed in the transformed *Y. lipolytica*. We examined whether simultaneous expression of Δ12- and Δ6-desaturase genes could increase the production

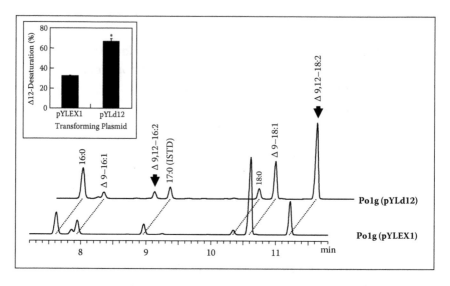

FIGURE 12.6 Gas chromatogram of fatty acids in total lipids and rate of Δ12-desaturation (inset) of control yeasts transformed with empty vector [Po1g(pYLEX1)] or yeasts carrying the *Mortierella* Δ12-desaturase gene [Po1g(pYLd12)]. Solid arrows show the presence of Δ9-16:1, the Δ12-desaturation product of Δ9-16:1, and linoleic acid (Δ9,12-18:2), the Δ12-desaturation product of oleic acid (Δ9-18:1). Δ12-Desaturation was defined as that described in Figure 12.3 legend. Each value point represents the mean of three independent experiments. Value with star symbol (*) indicates significantly different from the control at $p < .05$.

of GLA biosynthesized from the endogenous OA and LA. The plasmid pYLd12d6 was constructed by insertion of two expression cassettes containing both genes (Figure 12.1C). After transformation, the resulting yeast strain Po1g(pYLd12d6) was then cultivated under the same conditions as for Po1g(pYLd12). After harvest, total cellular lipids in yeast cells were extracted and fatty acid composition analyzed. Results in GC chromatogram (Figure 12.7) show a big peak of GLA appearing only in yeasts transformed with both desaturase genes but not in the controls. However, there were also trace amounts of four additional peaks in the chromatogram from Po1g(pYLd12d6). Three of them were identified as Δ6,9-16:2, Δ9,12-16:2, and Δ6,9-18:2 as previously described. The fourth novel fatty acid was tentatively identified as Δ6,9,12-16:3, a possible Δ6-desaturation product of Δ9,12-16:2.

12.3.7 COMPARISON OF FATTY ACID COMPOSITION IN FOUR TRANSFORMED *Y. LIPOLYTICA* STRAINS

Four different strains of transformed *Y. lipolytica* have been established in this study. Table 12.2 shows comparison of fatty acid composition in their lipids. The fungal Δ12-desaturase gene was introduced into yeast cells, generating a new yeast strain, Po1g(pYLd12). Fatty acid analysis showed that the percentage of LA in the cellular

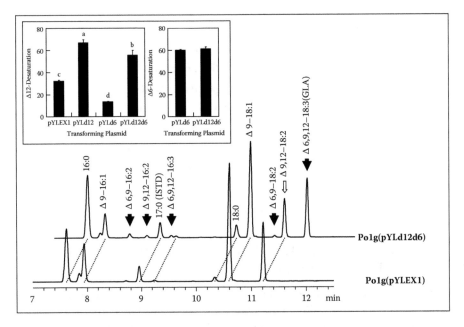

FIGURE 12.7 Gas chromatogram of fatty acids in total lipids and rates of Δ12-desaturation and Δ6-desaturation (inset) in yeasts transformed with empty vector [Po1g(pYLEX1)] or yeasts carrying both the *Mortierella* Δ12- and Δ6-desaturase genes [Po1g(pYLd12d6)]. Solid arrows show the presence of Δ6- or Δ12-desaturation products, including Δ6,9-16:2, Δ9,12-16:2, Δ6,9,12-16:3, Δ6,9-18:2, and γ-linolenic acid (Δ6,9,12-18:3). Open arrow shows the presence of linoleic acid (Δ9,12-18:2), a fatty acid found in both transformed yeasts. The Δ6- and Δ12-desaturation were defined as described in Figure 12.3 legend. Each value point represents the mean of three independent experiments. Values with different letters are significantly different from each other at $p < .05$.

lipids in this new yeast strain was twofold higher than that in the control strain (38% vs. 19%). The lipids extracted from Po1g(pYLd6), a yeast strain carrying the fungal Δ6-desaturase gene, contained a higher percentage of OA compared to the control strain (53% vs. 40%), while most of its LA was converted to GLA. When both Δ12- and Δ6-desaturase genes were coexpressed, levels of LA and GLA (12.5% and 20%, respectively) increased nearly fourfold compared to those when only the Δ6-desaturase gene was expressed (3% and 5%, respectively). Accordingly, the level of OA was decreased by around twofold [53.5% in Po1g(pYLd6) vs. 26% in Po1g(pYLd12d6)]. We have also calculated the rates of conversion from OA to LA (Δ12-desaturation) and from LA to GLA (Δ6-desaturation) in these yeast strains. Results in Figure 12.7 show that the Po1g(pYLd12) strain displayed the highest rate of converson of OA to LA (67%), followed by Po1g(pYLd12d6) (55%), Po1g(pYLEX1) (32%), and Po1g(pYLd6) (13%). However, there were no significant differences in Δ6-desaturation rate (conversion of LA to GLA) between Po1g(pYLd12d6) and Po1g(pYLd6) (Figure 12.7).

TABLE 12.3

Distribution of Total Lipids and γ-Linolenic Acid (GLA) from Po1g(pYLd12d6) Strain

	Total Lipids (%)	Total GLA (%)
Phospholipids (PLs)	27.6 ± 0.9^b	28.1 ± 1.0^2
Free fatty acids (FFAs)	15.8 ± 1.4^c	26.3 ± 3.5^2
Triacylglycerols (TAGs)	47.4 ± 2.3^a	44.2 ± 4.6^1
Others	9.2 ± 0.3^d	1.4 ± 0.3^3

Note: Values on the same column with different superior letters or numbers are significantly different ($p < .05$).

12.3.8 DISTRIBUTION OF GLA IN CELLULAR LIPID FRACTIONS IN TRANSFORMED YEASTS

Results from this study clearly demonstrated that the transformed *Y. lipolytica*, particularly Po1g(pYLd12d6), could efficiently synthesize GLA from endogenous OA and LA. To study in which cellular lipid fraction the newly formed GLA was incorporated, we examined GLA distribution in different lipid fractions by TLC. As shown in Table 12.3, the greatest percentage of GLA (47%) was found in the triacylglycerol (TAG) fraction in yeast cellular lipids, followed by phospholipids (PLs) (28%) and free fatty acids (FFAs) (16%).

12.4 DISCUSSION

GLA has been demonstrated to exhibit many beneficial effects on human health. To find alternative and reliable sources of GLA, many studies have been conducted to develop GLA-rich oil. Several studies have reported the isolation of the Δ6-desaturase gene from different species and expressing it in baker's yeast [12,21,22]. Others have utilized different fungi to produce GLA [23] or to express genes involved in GLA synthesis into higher plants [24]. However, there is no report of the use of the oleaginous yeasts as a host to synthesize high-GLA SCO. In this study, we demonstrated that the GRAS oleaginous yeast *Y. lipolytica*, transformed with both Δ12- and Δ6-desaturase cDNAs, could produce and accumulate up to 20% of GLA in total cellular lipids.

Our strategy for engineering *Y. lipolytica* to produce a high level of GLA followed the concept of coexpression of Δ6- and Δ12-desaturase genes reported by one of our previous studies [12]. However, differently from our previous approach of transforming yeasts with two plasmids simultaneously, we used a unique *Y. lipolytica* expression vector, pYLEX1, for coexpression. This vector makes use of the patented recombinant hp4d promoter for the expression of foreign genes [14,25]. In the present study, the plasmids pYLd6 and pYLd12, containing each expression cassette with the respective desaturase gene, exhibited high degrees of conversion of substrates to products (Figure 12.7), especially Δ6-desaturation of LA to GLA. Approximately 60% of LA was converted to GLA, which was significantly higher than rates of Δ6-desaturation reported previously [12,22,26]. Although the rates of Δ6-desaturation were changed significantly when the transformed yeast Polg(pYLd6) was cultivated in a variety of media and temperatures (Figures 12.3B and 12.5B), the Δ6-desaturase gene was consistently expressed without severe repression, a characteristic of hp4d promoter that makes it more versatile than other *Yarrowia* promoters requiring induction by nutrients or chemicals [15]. These findings are consistent with the previous observations that hp4d promoter exhibited a strong quasi-constitutive activity and was not repressed by culture conditions such as preferred carbon or nitrogen sources or acidic environment [14].

The expression of Δ6-desaturase gene into *Y. lipolytica* resulted in a significant decrease in the rate of Δ12-desaturation, as shown by the comparison between Polg(pYLEX1) and Polg(pYLd6) (fell from 32.5% to 17%) or, also, to a lesser extent between Polg(pYLd12) and Polg(pYLd12d6) (Figure 12.7). We cannot explain why this decrease occurred, but the reason accounting for a lesser degree of decrease between Polg(pYLd12) and Polg(pYLd12d6) was probably that both plasmids contained the expression cassette with the Δ12-desaturase gene. The efficient expression of Δ12-desaturase gene compensated in part for the possible negative influence resulting from Δ6-desaturase expression. In contrast, no effect was observed on the rate of Δ6-desaturation when the Δ12-desaturase gene was introduced into *Y. lipolytica* (Figure 12.7).

This study also revealed that the two expression cassettes inserted into pYLEX1 vector were able to function effectively and independently. This new process might be applied for expressing two (or possibly more) different genes, or several copies of one gene, in a single transformation step.

The other objective of this study was to optimize the growth conditions of transformed *Y. lipolytica* for the production of higher levels of GLA. We found higher cell mass and total lipids in transformed yeasts grown in the temperature range from 15°C to 28°C (Figure 12.4A) but not at 32°C or 35°C. Especially, when cells were cultivated at 15°C, the rate of Δ6-desaturation was the greatest (72%). These results were consistent with previous observations that *Y. lipolytica* growth was impaired when temperature exceeded 32°C to 34°C [27], and that expression of Δ6- and Δ12-desaturase genes was enhanced at 15°C [12]. In addition to growth temperature, incubation time was analyzed. Cell mass increased gradually with incubation time, but the amount of total cellular lipids (mg/g yeast) corresponding to samples fluctuated importantly (Figure 12.3A). This might be partly due to an accumulation of citric acid in the medium, which might influence lipid synthesis possibly through reserving the activity of adenosine triphosphate (ATP) citrate lyase [28].

The content and fatty acid composition of lipids produced by yeasts and fungi depend strongly on culture medium. It has been shown that yeast cells cultivated in nitrogen-limited media consumed the available nitrogen and grew slowly and then converted the remaining glucose to oil [29]. In this study, higher lipid contents were accumulated in transformed *Y. lipolytica* grown in nitrogen-reduced media (Figure 12.3A), which agreed with earlier results indicating that a high molar ratio carbon/nitrogen of the growth medium allowed oleaginous fungi to accumulate lipids [30]. In contrast, transformed yeasts grew faster in the complete media (YPD1 and YPD2) (Figure 12.3A), and less lipid accumulation was observed. The PPB medium has been optimal in some cases of heterologous protein expression [18–20] but not for lipid accumulation (Figure 12.3A). The poor performance of transformed yeasts in minimal YNBD medium (Figure 12.3A) is consistent with the previous observation that some strains of *Y. lipolytica* did not grow well in this medium [27].

In a separate study, we cultivated transformed yeast Polg(pYLd12d6) at 28°C for 72 h and estimated its growth and lipid production. Cell mass and storage lipids were approximately 4 g/L and 0.02 g/g of yeast cells, respectively, and 20% of GLA appeared in total lipids (Table 12.2). When lipid fractions were separated by TLC, only 47% of total lipids were in the triacylglycerol fraction, but they contained as high as 44% of total GLA (Table 12.3). These findings revealed that, even if yeast cells performed poorly in both biomass and lipid storage under the current culture conditions, the production of GLA and its redistribution from PLs to triacylglycerols remained efficient. Therefore, these results give us a direction for improvement of yeast growth and lipid storage.

One advantage for developing SCO is that microorganisms are easier to manipulate for optimization of growth and lipid production. The results from this study are interesting for such an application and provide optimized conditions of temperature and culture medium. To improve yields of yeast lipids or GLA, for example, a method of lower-temperature culture, such as 15°C, in a nitrogen-reduced medium will be optimal. However, *Y. lipolytica* cultivated in a flask restricted the accumulation of lipids inside cells due to the appearance and increase of citric acid during cultivation. It was shown that using a single-stage continuous culture to grow yeasts could increase biomass, minimize the effect of citric acid, and provide a controlled environment (dissolved oxygen, molar C/N ratio, pH, etc.) to accumulate a higher quantity of lipids [28]. From the information shown in Figure 12.3A, a two-stage continuous culture could be designed as follows: Yeast cells should be first inoculated in a rich medium to increase cell biomass and then transferred to a nitrogen-reduced medium to accumulate yeast lipids. However, it has been previously suggested that no practical advantage over a single-stage system was obtained when using a two-stage continuous culture to produce microbial oils [31].

From the viewpoint of genetic engineering, several strategies could be used to improve further the lipid content of transformed yeast cells. One possibility would be the introduction of an expression cassette containing the gene for diacylglycerol acyltransferase (DGAT), which is involved in diverting diacylglycerol from membrane lipid synthesis into triacylglycerol [32]. To increase the level of GLA, we could possibly insert more expression cassettes with Δ12- or Δ6-desaturase genes (carried on monocopy vectors such as pYLEX1, but using other selection markers) or change for a more adequate multiple-copy vector carrying the corresponding genes (as exemplified in Nicaud et al. [16]. Such further studies would possibly lead to improved yields of lipids and GLA.

In conclusion, by using an expression vector with the strong hp4d promoter, we demonstrated that coexpression of Δ12- and Δ6-desaturase genes in a genetically modified strain of *Y. lipolytica* resulted in biosynthesis of GLA from endogenous OA and LA. The level of GLA was up to 20% of total cellular lipids and 44% in the fraction of triacylglycerol. These yields of yeast oil and GLA could possibly be increased further through culture optimization and molecular engineering.

ACKNOWLEDGMENT

This work was supported in part by a grant from the National Science Council of Taiwan (95-2320-B-005-013-MY3).

REFERENCES

1. Calder, P.C., and Zurier, R.B., Polyunsaturated fatty acids and rheumatoid arthritis, *Curr. Opin. Clin. Nutr. Metab. Care* 2001, *4*, 115–121.
2. Rothman, D., DeLuca, P., and Zurier, R.B. Botanical lipids: effects on inflammation, immune responses, and rheumatoid arthritis, *Semin. Arthritis Rheum.* 1995, *25*, 87–96.
3. Kast, R.E., Borage oil reduction of rheumatoid arthritis activity may be mediated by increased cAMP that suppresses tumor necrosis factor-alpha, *Int. Immunopharmacol.* 2001, *12*, 2197–2199.
4. Christophe, A., Robberecht, E., Franckx, H., De Baets, F., and van de Pas, M., Effect of administration of gamma-linolenic acid on the fatty acid composition of serum phospholipids and cholesteryl esters in patients with cystic fibrosis, *Ann. Nutr. Metab.* 1994, *38*, 40–47.
5. Engler, M.M., Schambelan, M., Engler, M.B., Ball, D.L., and Goodfriend, T.L., Effects of dietary gamma-linolenic acid on blood pressure and adrenal angiotensin receptors in hypertensive rats, *Proc. Soc. Exp. Biol. Med.* 1998, *218*, 234–237.
6. Arisaka, M., Arisaka, O., and Yamashiro, Y., Fatty acid and prostaglandin metabolism in children with diabetes mellitus. II. The effect of evening primrose oil supplementation on serum fatty acid and plasma prostaglandin levels, *Prostaglandins Leukot. Essent. Fatty Acids* 1991, *43*, 197–201.
7. Das, U.N., Tumoricidal and anti-angiogenic actions of gamma-linolenic acid and its derivatives, *Curr. Pharm. Biotechnol.* 2006, *7*, 457–66.
8. Sprecher, H., Biochemistry of essential fatty acids, *Prog. Lipid Res.* 1981, *20*, 13–22.
9. Brenner, R.R., Nutritional and hormonal factors influencing desaturation of essential fatty acids, *Prog. Lipid Res.* 1981, *20*, 41–47.
10. Knutzon, D.S., Chan, G.M., Mukerji, P., Thurmond, J.M., et al., in: Altman, A., Ziv, M., and Izhar, S. (Eds.), *Proceedings of the IX International Congress of the International Association of Plant Tissue Culture and Biotechnology*, Kluwer Academic, Dordrecht, Netherlands, 1999, pp. 575–578.
11. Liu, J.-W., Huang, Y.-S., DeMichele, S.J., Bergana, M., et al., in: Huang, Y.-S., and Ziboh, V.A. (Eds.), *γ-Linolenic Acid: Recent Advances in Biotechnology and Clinical Applications*, AOCS Press, Champaign, IL, 2001, pp. 61–71.
12. Huang, Y.-S., Chaudhary, S., Thurmond, J.M., Bobik, E.G., et al., Cloning of Δ12- and Δ6-desaturase from *Mortierella alpina* and recombinant production of γ-linolenic acid in *Saccharomyces cerevisiae*, *Lipids* 1999, *34*, 649–659.
13. Ratledge, C., in: Gunstone, F.D. (Ed.), *Structured and Modified Lipids*, Dekker, New York, 2001, pp. 351–399.
14. Madzak, C., Treton, B., and Blanchin-Roland, S., Strong hybrid promoters and integrative expression/secretion vectors for quasi-constitutive expression of heterologous proteins in the yeast *Yarrowia lipolytica*, *J. Mol. Microbiol. Biotechnol.* 2000, *2*, 207–216.

15. Madzak, C., Gaillardin, C., and Beckerich, J.-M., Heterologous protein expression and secretion in the non-conventional yeast *Yarrowia lipolytica*: a review, *J. Biotechnol.* 2004, *109*, 63–81.

16. Nicaud, J.-M., Madzak, C., van den Broek, P., Gysler, C., et al., Protein expression and secretion in the yeast *Yarrowia lipolytica*, *FEMS Yeast Res.* 2002, *2*, 371–379.

17. Chen, D.-C., Beckerich, J.-M., and Gaillardin, C., One-step transformation of the dimorphic yeast *Yarrowia lipolytica*, *Appl. Microbiol. Biotechnol.* 1997, *48*, 232–235.

18. Madzak, C., Otterbein, L., Chamkha, M., Moukha, S., et al., Heterologous production of a laccase from the basidiomycete *Pycnoporus cinnabarinus* in the dimorphic yeast *Yarrowia lipolytica*, *FEMS Yeast Res.* 2005, *5*, 635–646.

19. Jolivalt, C., Madzak, C., Brault, A., Caminade, E., et al., Expression of laccase IIIb from the white-rot fungus *Trametes versicolor* in the yeast *Yarrowia lipolytica* for environmental applications, *Appl. Microbiol. Biotechnol.* 2005, *66*, 450–456.

20. Kopečný, D., Pethe, C., Šebela, M., Houba-Hérin, N., et al., High-level expression and characterization of *Zea mays* cytokinin oxidase/dehydrogenase in *Yarrowia lipolytica*, *Biochimie* 2005, *87*, 1011–1022.

21. Cho, H.P., Nakaruma, M.T., and Clarke, S.D., Cloning, expression, and nutritional regulation of the mammalian Δ-6 desaturase, *J. Biol. Chem.* 1999, *274*, 471–477.

22. Laoteng, K., Mannontarat, R., Tanticharoen, M., and Cheevadhanarak, S., Δ6-Desaturase gene of *Mucor rouxii* with high similarity to plant Δ6-desaturase and its heterologous expression in *Saccharomyces cerevisiae*, *Biochem. Biophys. Res. Commun.* 2000, *279*, 17–22.

23. Lindber, A.-M., and Hansson, L., Production of γ-linolenic acid by the fungus *Mucor rouxii* on cheap nitrogen and carbon sources, *Appl. Microbiol. Biotechnol.* 2004, *36*, 26–28.

24. Hong, H., Datla, N., Reed, D.W., Covello, P.S., et al., High-level production of γ-linolenic acid in *Brassica juncea* using a Δ6 desaturase from *Pythium irregulare*, *Plant Physiol.* 2002, *129*, 354–362.

25. Madzak, C., Blanchin-Roland, S., and Gaillardin, C., Upstream activating sequences and recombinant promoter sequences functional in *Yarrowia* and vectors containing them, European Patent Application, 1995, EP0747484A1.

26. Sakuradani, E., Kobayashi, M., and Shimizu, S., Δ6-Fatty acid desaturase from an arachidonic acid-producing *Mortierella* fungus gene cloning and its heterologous expression in a fungus, *Aspergillus*, *Gene* 1999, *238*, 445–453.

27. Barth, G., and Gaillardin, C., *Yarrowia lipolytica*, in: Worf, C. (Ed.) *Nonconventional Yeast in Biotechnology, A Handbook, Vol. 1*, Springer-Verlag, Berlin, 1996, pp. 313–388.

28. Papanikolaou, S., and Aggelis, G., Lipid production by *Yarrowia lipolytica* growing on industrial glycerol in a single-stage continuous culture, *Bioresource Technol.* 2002, *82*, 43–49.

29. Ratledge, C., in: Kamel, B.S., and Kakuda, Y. (Eds.), *Technological Advances in Improved and Alternative Sources of Lipids*, Blackie Academic and Professional, London, 1994, pp. 235–291.

30. Papanikolaou, S., Komaitis, M., and Aggelis, G., Single cell oil (SCO) production by *Mortierella isabellina* grown on high-sugar content media, *Bioresource Technol.* 2004, *95*, 287–291.

31. Hall, M.J., and Ratledge, C., Lipid accumulation in an oleaginous yeast (*Candida* 107) growing on glucose under various conditions in a one- and two-stage continuous culture, *Appl. Environ. Microbiol.* 1977, *33*, 577–584.

32. Hobbs, D.H., Lu, C., and Hills, M.J., Cloning of a cDNA encoding diacylglycerol acyltransferase from *Arabidopsis thaliana* and its functional expression, *FEBS Lett.* 1999, *452*, 145–149.

13 Production of Monoacylglycerols through Lipase-Catalyzed Reactions

Yomi Watanabe and Yuji Shimada

CONTENTS

Key Words: Bound water; esterification; ethanolysis; glycerolysis; immobilized glycerol; lipase; monoacylglycerol; regiospecific analysis.

13.1 INTRODUCTION

Monoacylglycerols (MAGs) are good emulsifiers used widely in the food, cosmetics, and drug industries. MAGs with saturated and monoenoic fatty acids (FAs) are produced industrially by chemical alcoholysis of triacylglycerols (TAGs; oils and fats) with 2 mol glycerol at high temperatures of 210°C–240°C. But, the process cannot be applied to synthesis of MAGs with unstable FAs. Meanwhile, enzymatic processes efficiently proceed under mild conditions and are effective for production of MAGs with FAs that are unstable and have nutraceutical activity.

Lipases catalyze not only hydrolysis of TAGs but also esterification of alcohols and long-chain FAs and transesterification (alcoholysis, acidolysis, and interesterification) (Figure 13.1). Synthesis of MAGs using the lipases has been actively studied since about 1990. Many of the reports were syntheses of MAGs with saturated and monoenoic FAs by hydrolysis of TAGs, esterification of FAs with glycerol, glycerolysis of TAGs, and ethanolysis of TAGs in organic solvent systems. Meanwhile, an organic solvent-free system is preferable from the viewpoint of industrial production, and several systems have been proposed: esterification of FAs with glycerol and glycerolysis of TAGs. This chapter deals with lipase-catalyzed reactions for production of MAGs as well as the reaction mechanism in organic solvent-free systems that achieve a high yield of MAGs.

1. Hydrolysis

$$R_1OCOR_2 + H_2O \longrightarrow R_1COOH + R_2OH$$

(Hydrolysis of TAG)

$$TAG + H_2O \longrightarrow MAG + FFA + Glycerol$$

2. Esterification

$$R_1COOH + R_2OH \longrightarrow R_1OCOR_2 + H_2O$$

(Esterification of FA and glycerol)

$$FA + Glycerol \longrightarrow MAG + H_2O$$

3. Transesterification

3-1 Alcoholysis

$$R_1OCOR_2 + R_3OH \longrightarrow R_1OCOR_3 + R_2OH$$

(Glycerolysis of TAG/FAEE)

$$TAG + Glycerol \longrightarrow MAG$$

$$FAEE + Glycerol \longrightarrow MAG + EtOH$$

(Ethanolysis of TAG)

$$TAG + EtOH \longrightarrow MAG + FAEE$$

3-2 Acidolysis

$$R_1OCOR_2 + R_3COOH \longrightarrow R_3OCOR_2 + R_1COOH$$

3-3 Interesterification

$$R_1OCOR_2 + R_3OCOR_4 \longrightarrow R_1OCOR_4 + R_3OCOR_2$$

FIGURE 13.1 Lipase-catalyzed reactions and production of MAG using the reactions. FA, fatty acid; MAG, monoacylglycerol; TAG, triacylglycerol; FAEE, fatty acid ethyl ester; EtOH, ethanol.

13.2   PRODUCTION OF MAG BY HYDROLYSIS

A group of lipases recognizes only ester bonds at the 1,3-positions of TAGs, termed 1,3-position-specific lipases. Hydrolysis of TAGs with this type of lipase produces 2-MAGs in principle. However, the yield of MAGs is very low (33 mol% for total FAs), even though TAGs are completely converted to MAGs. In addition, because spontaneous acyl migration occurs during the reaction, it is very difficult to control the reaction for production of a good yield of MAGs.

Several reaction systems were reported to achieve comparatively good yield. Hydrolysis of triolein with porcine pancreatic lipase immobilized on Celite produced monoolein in an 80% yield for TAG (Plou et al. 1996). Hydrolysis in microemulsion based on sodium bis(2-ethylhexyl)sulfosuccinate (AOT) using *Rhizopus oryzae* lipase was reported to be effective: Reaction in a mixture of palm oil/water/AOT/isooctane (4:8:4:83 by weight) converted the oil to MAGs in an 80% yield (Holmberg and Osterberg 1988). Hydrolysis of canola oil in supercritical carbon dioxide was also studied using immobilized *Rhizomucor miehei* lipase, but the yield of MAGs for TAGs was only 40% (Rezaei and Temelli 2000).

13.3   PRODUCTION OF MAG BY ESTERIFICATION

MAGs are synthesized by esterification of FAs and glycerol. The reaction proceeds by contact of the two substrates, but they are hydrophobic and hydrophilic. Their contact was accelerated by the use of glycerol, of which a thin layer covers the surface of silica gel; the glycerol is termed *immobilized glycerol.*

Esterification of FAs and glycerol produces not only MAGs but also diacylglycerols (DAGs) and TAGs. A reaction system in which glycerol derivatives were used as substrates suppressed by-production of DAGs and TAGs. In addition, conversion of MAGs to DAGs/TAGs was suppressed by considering physical properties of substrates and substrate specificity of lipase. These reaction systems are described in the following sections.

13.3.1   REACTION WITH IMMOBILIZED GLYCEROL

A mixture of hydrophobic FAs and hydrophilic glycerol forms a two-layer system, and expansion of the interface of the two layers results in a high degree of esterification. Based on this idea, a reaction system with glycerol immobilized on silica gel was proposed (Berger et al. 1992). In addition, esterification of FAs and glycerol was affected by organic solvents. These facts combined developed a reaction system: the reaction of oleic acid and immobilized glycerol in 2-methyl-2-butanol with immobilized *R. miehei* lipase synthesized MAG in a good yield (MAG/DAG 94:2 by weight; 45% esterification) (Bellot et al. 2001).

When immobilized glycerol was not used, immobilized lipase used as a catalyst seemed to be covered with glycerol, resulting in interference of mass transfer of FAs to the lipase catalytic site. Therefore, Castillo et al. (1997) hypothesized that efficient esterification with immobilized glycerol is due to improvement of mass transfer of FAs to the lipase catalytic site.

13.3.2 REACTION WITH GLYCEROL DERIVATIVES

Esterification of 1,2-*O*-isopropylidene glycerol and FAs with lipases from *Burkholderia cepacia* (Hess et al. 1995) and *R. miehei* (Li and Ward 1994) suppressed by-production of DAGs and produced a good yield of a regioisomer of MAGs [1(3)-MAGs], but the product changed to a mixture of 1(3)-MAGs and 2-MAGs after purification and storage because of spontaneous acyl migration. It therefore makes no sense to produce a regioisomer of MAGs except for a special purpose, such as production of structured TAGs.

MAGs also were synthesized efficiently using phenylboronic acid that protected two hydroxyl groups of glycerol. A reaction with immobilized *R. miehei* lipase in *n*-hexane containing glycerol/phenylboronic acid/stearic acid (5:5:1 by weight) proceeded simultaneous protection of glycerol and esterification of protected glycerol with FAs, resulting in production of MAGs in a good yield (Steffen et al. 1992).

Acylation of protected glycerol synthesized a good yield of MAGs. But, this process requires an additional process—cleavage of the protected group of glycerol after the reaction.

13.3.3 REACTION AT LOW TEMPERATURE

Reactions at low temperatures were first reported in glycerolysis of TAGs for the production of MAGs (see Section 13.4.2) and were then applied to esterification of FAs and glycerol. When the esterification was conducted at temperatures at which all substrates and products were in the liquid state, DAGs and TAGs were by-produced. Meanwhile, the yield of MAGs increased significantly when the reaction was conducted at lower temperatures that made the reaction mixture solidify. It was, for example, reported that MAG was synthesized efficiently by esterification of palmitic acid with glycerol immobilized on silica gel at 25°C in *n*-hexane using *R. oryzae* lipase (MAG/DAG 6:1 by weight; 80% esterification) (Kwon et al. 1995).

Even in an organic solvent-free system, MAGs were efficiently synthesized when the esterification was conducted at low temperatures. The reaction system is described next.

13.3.3.1 Production of MAG by Esterification at Low Temperature

Conjugated linoleic acid (CLA) is produced by alkali conjugation of safflower or sunflower oil in propylene or ethylene glycol. The first product is a free fatty acid (FFA) mixture that contains almost equal amounts of 9-*cis*, 11-*trans* (9*c*,11*t*) and 10*t*,12*c*-CLA (referred to as FFA-CLA). The FFA-CLA has various physiological activities and is used as a neutraceutical food. If FFA-CLA is efficiently converted to its MAG, the new product can be used as a functional emulsifier for various kinds of foods and can also be added to beverages. Hence, development of a process for producing a MAG of CLA (MAG-CLA) has strongly been desired. In light of the instability of CLA, an enzymatic process is suitable for its production.

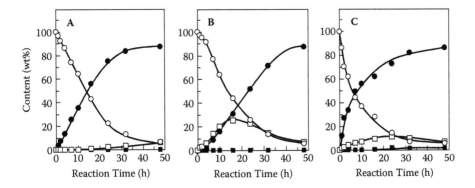

FIGURE 13.2   Production of MAG-CLA by esterification at low temperature. A mixture of 1 mol FFA-CLA, 5 mol glycerol, and lipase solution (2 wt% water) was agitated at 5°C. Reaction using (A) 200 U/g *P. camembertii* lipase; (B) 50 U/g *R. oryzae* lipase; (C) 200 U/g *C. rugosa* lipase. ○, Content of FFA-CLA; ●, MAG-CLA; □, DAG-CLA; ■, TAG-CLA.

When FFA-CLA was esterified at 30°C with 2 mol glycerol using *Penicillium camembertii* lipase, almost equal amounts of MAG- and DAG-CLAs were synthesized at the equilibrium state (Yamaguchi and Mase 1991, Watanabe et al. 2002). Meanwhile, when the reaction was conducted at 5°C for 50–70 h (>95% esterification), the yield of MAG-CLA to total acylglycerols reached more than 90% (Watanabe et al. 2004). The reaction periods were shortened to 24 h by starting dehydration at 5 mm Hg after 10 h (Watanabe et al. 2004). Lipases from *Candida rugosa* and *R. oryzae* also produced MAG-CLA efficiently at 5°C: The degree of esterification reached more than 95% at the equilibrium state, and the yield of MAG-CLA to total acylglycerols was more than 90% (Watanabe et al. 2003).

In the reactions using the above three lipases, the yield of MAG-CLA at the equilibrium state was not different significantly, but a difference was observed in the time courses for production of MAG- and DAG-CLAs (Figure 13.2). The *P. camembertii* lipase did not convert MAG to DAG in the early stage of the reaction and by-produced DAG slightly after 20 h (Figure 13.2A). On the contrary, *R. oryzae* lipase synthesized DAG faster than MAG, showing that the lipase recognized MAG more strongly than glycerol. This reaction converted DAG to MAG after 15 h (Figure 13.2B). The *C. rugosa* lipase also produced DAG gradually even in the early stage of reaction, although the amount of DAG was not as large compared with that produced by *R. oryzae*. Conversion of the DAG to MAG was observed after 24 h (Figure 13.2C).

13.3.3.2   Effect of Low Reaction Temperature: Solidification of MAG

The *P. camembertii* lipase catalyzed only esterification of FFA-CLA and glycerol when the reaction was conducted at 5°C (Figure 13.2A) (Watanabe et al. 2004), but *R. oryzae* and *C. rugosa* lipases catalyzed not only esterification of FFA-CLA and glycerol but also conversion of DAG- to MAG-CLA (glycerolysis of DAG-CLA) (Figure 13.2B and 13.2C) (Pinsirodom et al. 2004, Watanabe et al. 2003). Hence,

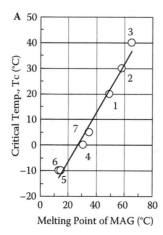 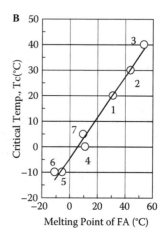

FIGURE 13.3 Correlation among critical temperatures for production of MAG and melting point of corresponding MAG and FFA. The critical temperature was defined as the highest temperature at which the reaction produced 95 wt% of MAG based on the total amount of acylglycerols (MAG and DAG) at more than 80% esterification. (A) Correlation between critical temperature and melting point of MAG. (B) Correlation between critical temperature and melting point of FFA. 1, capric acid; 2, lauric acid; 3, myristic acid; 4, oleic acid; 5, linoleic acid; 6, α-linolenic acid; 7, FFA-CLA.

P. camembertii lipase was used for studying why the low-temperature reaction is effective for production of MAG-CLA.

Efficient production of MAGs in the low-temperature reaction can be explained by assuming that synthesized MAGs are excluded from the reaction system because MAGs are solidified most easily at low temperature. Consequently, MAGs do not serve as the precursor of DAGs and are efficiently accumulated in the reaction mixture. If this hypothesis is correct, the critical temperature for MAG production, defined as the highest temperature at which DAG synthesis is repressed, should depend on the melting point of the MAGs. Hence, the effect of temperature on esterification of C_{10}–C_{18} FAs with glycerol was studied using *P. camembertii* lipase, and the critical temperatures for production of MAG were determined. The critical temperature of each MAG showed a linear correlation with melting point of the MAG, which supported the hypothesis (Figure 13.3A). In addition, because the melting point of MAG depends on that of the constituent FA, the optimal temperature for production of MAG can be predicted from the melting point of the FA used as a substrate (Figure 13.3B) (Pinsirodom et al. 2004).

High-yield synthesis of MAG at the critical temperature can be explained as follows (Figure 13.4) (Pinsirodom et al. 2004). The melting point of MAG is the highest among those of all components (FA, glycerol, MAG, and DAG) in the reaction mixture; thus, only MAG solidifies at the critical temperature. Lipase acts on liquid-state substrates strongly but on solid-state substrates very weakly. When reaction is conducted at the critical temperature that makes only MAG solidify, MAG is accumulated efficiently in the reaction mixture because MAG is not converted to DAG.

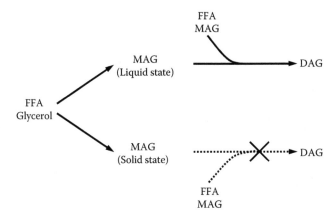

FIGURE 13.4   Reaction catalyzed by lipase at low temperature. Reaction is conducted at low temperature at which MAG solidifies. Lipase acts on liquid-state substrate strongly and on solid-state substrate weakly. Therefore, lipase scarcely converts MAG to DAG at low temperature.

13.3.4   REACTION WITH DEHYDRATION

Water is one of the products produced by esterification of FA and glycerol. Hence, removal of this by-product shifts the equilibrium to esterification, and a high degree of esterification is achieved. Actually, in the esterification with *P. camembertii* and *R. oryzae* lipases (Section 13.3.3.1), the reaction period was shortened largely by removal of water with a vacuum pump during the reaction (Watanabe et al. 2004).

Interestingly, dehydration also affected the composition of reaction products. When esterification of FFA-CLA and glycerol was conducted at 30°C with *P. camembertii* lipase, almost equal amounts of MAG- and DAG-CLAs were synthesized (Watanabe et al. 2002, Pinsirodom et al. 2004). In this reaction, the start of dehydration on the way of the reaction not only increased the degree of esterification significantly (>95%) but also suppressed conversion of MAG- to DAG-CLA (Watanabe et al. 2002). In the following sections, dehydration is shown to be very effective for a high-yield production of MAG-CLA.

13.3.4.1   Production of MAG by Esterification with Dehydration

When esterification of FFA-CLA and glycerol was conducted at 30°C without dehydration using *P. camembertii* lipase, the amounts of MAG- and DAG-CLAs were almost the same at the equilibrium state (85% esterification) (Figure 13.5Aa). The initial reaction mixture was composed of FFA-CLA, glycerol, and a water solution of the lipase. Because the reaction by-produced water, the water content increased from 2.2 to 4.1 wt% (Figure 13.5Ab). Meanwhile, when the reaction was conducted with dehydration using a vacuum pump, the contents of MAG- and DAG-CLAs at the equilibrium state were 85 and 9 wt%, respectively (Figure 13.5Ba). The water content decreased from 2.2 to 0.2 wt% (Figure 13.5Bb) (Watanabe et al. 2005).

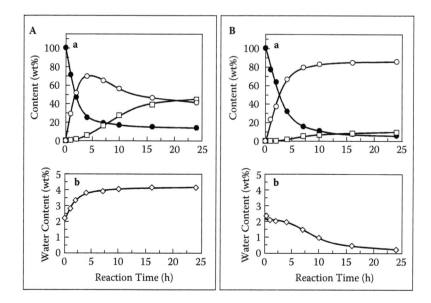

FIGURE 13.5 Time courses of esterification of FFA-CLA and glycerol with or without dehydration using *P. camembertii* lipase. A mixture of FFA-CLA/glycerol (1:2 mol/mol) and lipase solution (the final contents of water and lipase are 2 wt% and 200 U/g, respectively) was agitated at 30°C. (A) Reaction without dehydration: a, contents of FFA-CLA (●), MAG-CLA (○), and DAG-CLA (□) in the oil part; b, content of water in the reaction mixture. (B) Reaction with dehydration: a, contents of FFA-, MAG-, and DAG-CLAs in the oil part, with symbols the same as those in Figure 13.5A; b, content of water in the reaction mixture.

A regioisomer of MAG-CLA synthesized by this reaction was studied. The *P. camembertii* lipase is 1,3-position specific (Yamaguchi and Mase 1991). This fact was consistent with the report that the synthesized MAG-CLA was 1(3)-MAG (Watanabe et al. 2005). Furthermore, spontaneous acyl migration was not observed during the reaction with dehydration.

In general, large amounts of glycerol increase the ratio of MAG to total acyl-glycerols. But, the amount of glycerol in the esterification with dehydration did not affect the yield of MAG significantly, although an increase of glycerol accelerated the esterification (Watanabe et al. 2005).

These results showed that esterification with dehydration suppressed the conversion of MAG to DAG and was very effective for high-yield production of MAG. This reaction system was especially advantageous for production of MAGs of linoleic and α-linolenic acids compared with a low-temperature reaction, in which the critical temperature of their MAGs was −10°C (Figure 13.3A), because the velocity of the reaction below 0°C is very slow, and large amounts of lipase are required (Pinsirodom et al. 2004).

13.3.4.2 Effect of Dehydration on Substrate Specificity of Lipase

In acidolysis of TAG with FA using immobilized *R. oryzae* lipase, pretreatment of the lipase in a substrate mixture containing small amounts of water was necessary for expression of the activity (Shimada et al. 1996). The *C. rugosa* lipase with a limited amount of bound water did not catalyze esterification of FAs with glycerol and sterols (Watanabe et al. 2003, Nagao et al. 2005). When powdered *P. camembertii* lipase was used for esterification of FAs with glycerol in a mixture without addition of water, the activity was expressed fully after 20-h lag period (Watanabe et al. 2004). These reports indicated that bound water of lipases plays an important role in the expression of their activity.

In addition, it was shown (Section 13.3.4.1) that *P. camembertii* lipase produced MAGs and DAGs in the esterification without dehydration and only MAGs in the reaction with dehydration. This phenomenon also can be explained by the concept of bound water. When the reaction is conducted without dehydration, the lipase molecule has enough bound water [Figure 13.6, lipase without dehydration (a)]. The lipase in this state recognizes FA and MAG; thus, the reaction of FA and glycerol produces not only MAG but also DAG (Figure 13.6). Meanwhile, when the reaction is conducted with dehydration, water in the reaction mixture is removed, and the amount of bound water decreases [Figure 13.6, (b)]. This lipase recognizes FA but not MAG. Therefore, the reaction synthesizes MAG but not DAG (Figure 13.6). These discussions arrived at the hypothesis that substrate specificity of the lipase is changed by the amount of bound water (Watanabe et al. 2005).

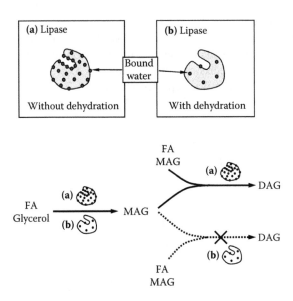

FIGURE 13.6 Effect of bound water of lipase on its substrate specificity. When esterification of FA and glycerol was conducted without dehydration, *P. camembertii* lipase (a) carried large amounts of bound water. The lipase (a) catalyzed esterification of FA with glycerol and conversion of MAG to DAG. When esterification of FA and glycerol was conducted with dehydration, the lipase (b) carried limited amounts of bound water. The lipase (b) catalyzed esterification of FA with glycerol but not conversion of MAG to DAG.

13.4   PRODUCTION OF MAG BY TRANSESTERIFICATION

Lipases catalyze not only hydrolysis and esterification but also transesterification (Figure 13.1). MAGs are therefore synthesized by glycerolysis of FA esters (e.g., TAGs, FA ethyl esters [FAEEs], and FA vinyl esters) and by ethanolysis of TAGs.

13.4.1   GLYCEROLYSIS OF TAG

MAGs are produced by glycerolysis of TAGs in a mixture with or without organic solvent. McNeill et al. (1990) reported that an organic solvent-free glycerolysis at low temperature is effective for high-yield (70%) production of MAGs from beef tallow. When the reaction was conducted at low temperature or when the reaction temperature was decreased stepwise, synthesized MAGs were solidified. Because lipases recognize solid-state substrates weakly, conversion of MAGs to DAGs scarcely occurred, resulting in high-yield production of MAGs (see Section 13.3.3.2). It was reported that MAGs are produced in high yield (*ca.* 90%) by glycerolysis of olive oil at 10°C, palm oil at 40°C, and palm stearin at 40°C using *Pseudomonas* and *Burkholderia* lipases (McNeill et al. 1991, McNeill and Yamane 1991).

Glycerolysis at low temperature was also observed in the reaction of FFA-CLA and glycerol at 5°C using *R. oryzae* and *C. rugosa* lipases (Figure 13.2B and 13.2C). These reactions synthesized MAG- and DAG-CLAs in the early stage. Meanwhile, in the late stage, the content of synthesized DAG-CLA decreased, and that of MAG-CLA increased. These results showed that glycerolysis of synthesized DAG-CLA proceeds together with esterification of FFA-CLA with glycerol in the low-temperature reaction (Watanabe et al. 2003, Pinsirodom et al. 2004).

A two-step process comprising esterification of FAs with glycerol and glycerolysis of synthesized DAGs was reported (Figure 13.7) (Watanabe et al. 2002). When FFA-CLA was esterified with 5 mol glycerol at 30°C using *P. camembertii*

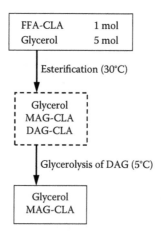

FIGURE 13.7   Production of MAG-CLA by two-step *in situ* reaction comprising esterification and glycerolysis using *P. camembertii* lipase.

lipase, almost equal amounts of MAG- and DAG-CLAs were synthesized. This reaction mixture was solidified by agitation on ice. The solidified mixture was then allowed to stand at 5°C for 2 weeks. Consequently, glycerolysis of synthesized DAG-CLA proceeded successfully, and the content of MAG-CLA increased to 90 wt%. The degree of esterification was more than 95% at the end of the reaction.

13.4.2   GLYCEROLYSIS OF FA ETHYL AND VINYL ESTERS

A by-product in esterification of FAs and glycerol is water. Meanwhile, reaction of FAEEs and glycerol by-produces ethanol (EtOH). Reaction under reduced pressure removes EtOH more easily than water; thus, glycerolysis of FAEEs proceeds more efficiently compared with esterification of FAs and glycerol. When glycerolysis of docosahexaenoic acid ethyl ester (DHAEE) was conducted under reduced pressure using *Pseudomonas* lipase immobilized on calcium carbonate, the content of partial acylglycerols reached 53 wt% (Rosu et al. 1998).

Reaction of FA vinyl esters with glycerol produces MAGs and vinyl alcohol, which is very unstable and converts easily to acetoaldehyde. Because acetoaldehyde does not participate in the reaction, the equilibrium shifts to production of MAGs. When vinyl esters of lauric, myristic, and oleic acids were allowed to react with glycerol immobilized on silica gel using lipases from *R. miehei*, *Burkholderia glumae*, *Rhizopus niveus*, and *Thermomyces lanuginosa* in *t*-butyl methyl ether, MAGs were synthesized in a high yield (Berger and Schneider 1992, Waldinger and Schneider 1992).

13.4.3   ETHANOLYSIS OF TAG

Ethanolysis of TAGs is one reaction for production of MAGs. Only 1 mol of MAGs, however, is produced from 1 mol TAGs, even though the ethanolysis proceeds completely. This system is not effective from the viewpoint of the yield for total FAs but is suitable for production of MAGs with polyunsaturated fatty acids (PUFAs) at the 2-position.

13.4.3.1   Production of 2-MAG

Ethanolysis of TAGs with a 1,3-position-specific lipase produces 2-MAGs when a high degree of conversion is achieved. Immobilized *Candida antarctica* lipase was found to be an enzyme fitting this aim.

This lipase shows intrinsically nonpositional specificity, but its positional specificity depends on the particular reactions examined. The lipase was nonspecific in alcoholysis of TAGs with small amounts of methanol (1/3 molar amounts for total FAs in TAGs) (Shimada et al. 2002) and preferentially recognized the 1,3-positions of glycerol in esterification of FAs with glycerol (Kawashima et al. 2001). However, the lipase showed 1,3-positional specificity in ethanolysis of TAGs in the presence of large amounts of EtOH (>12 molar amounts for total FAs in TAGs) (Irimescu et al. 2001a, 2001b; Shimada et al. 2003).

A single-cell oil containing arachidonic acid (AA) underwent ethanolysis with 20 mol EtOH for total FAs in the oil using immobilized *C. antarctica* lipase

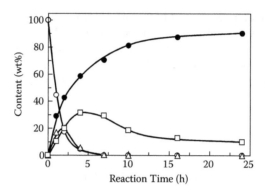

FIGURE 13.8 Time course of ethanolysis of a single-cell oil containing 40 wt% AA with immobilized *C. antarctica* lipase. A mixture of the oil and three weight parts of EtOH was shaken at 30°C with 4 wt% immobilized lipase. O, TAGs; ●, FAEEs; △, 1(3),2-DAGs; □, MAGs. Only 2-MAGs were detected before 7-h reaction, and a small amount of 1(3)-MAGs was detected thereafter.

(Figure 13.8). The ethanolysis generated 1(3),2-DAGs in the early stage but not 1,3-DAGs. The content of 2-MAGs reached a maximum value at 4 h (31 wt%; the content of FAs in 2-MAGs was 28 mol% based on total FAs in the reaction mixture), and 1(3)-MAGs were not detected during the first 4 h, although negligible amounts of 1(3)-MAGs were detected after 7 h.

The reaction in the presence of large amounts of EtOH is expected as a system for production of high-purity of 2-MAGs, but the yield of MAGs is only 33% for total FAs in the reaction mixture, even though the conversion of TAGs to 2-MAGs is 100%. This yield is extremely low compared with that in esterification or glycerolysis.

13.4.3.2 Application to Synthesis of Structured TAG-Containing PUFA

Structured TAGs with medium-chain FAs at the 1,3-positions and with long-chain FAs at the 2-position (referred to as MLM-type TAGs) are hydrolyzed to 2-MAGs and FFAs faster than natural oils and fats with long-chain FAs, resulting in efficient absorption into intestinal mucosa (Ikeda et al. 1991, Christensen et al. 1995). Furthermore, because PUFAs play a role in the prevention of a number of human diseases, MLM-type TAGs containing PUFAs are desirable as nutrition for patients with maldigestion and malabsorption of lipids and as high-value-added nutraceuticals for the elderly.

MLM-type TAGs are produced typically by acidolysis of natural oils with medium-chain FAs or by their interesterification with medium-chain FAEEs using immobilized 1,3-position-specific lipases (e.g., lipases from *R. oryzae, R. miehei,* and *T. lanuginosa*) (Figure 13.9). MLM-type TAGs containing PUFAs, such as docosahexaenoic acid (DHA), γ-linolenic acid, and AA, are produced using tuna oil, borage oil, and a single-cell oil containing AA as starting materials, respectively. However, these natural oils have PUFAs not only at the 2-position but also at the 1,3-positions.

$$\begin{bmatrix} OCOR_1 \\ OCOR_2 \\ OCOR_3 \end{bmatrix} + \begin{array}{c} C_7COOH \\ (C_7OCOEt) \end{array} \longrightarrow \begin{bmatrix} OCOC_7 \\ OCOR_2 \\ OCOC_7 \end{bmatrix} + \begin{array}{c} R_1COOH \\ R_3COOH \\ (R_1OCOEt) \\ (R_3OCOEt) \end{array}$$

FIGURE 13.9 Production of MLM-type TAGs through transesterification with medium-chain FA or its ethyl ester using immobilized 1,3-position-specific lipase.

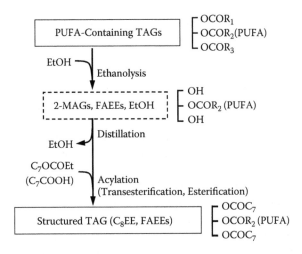

FIGURE 13.10 Production of MLM-type TAGs containing PUFAs using 2-MAGs as intermediates. Immobilized *C. antarctica* lipase is useful for the first-step ethanolysis, and immobilized *R. miehei* (*R. oryzae*) lipase is available for the second-step acylation.

In addition, because 1,3-position-specific lipases act on PUFAs weakly, all FAs at the 1,3-positions cannot be exchanged with medium-chain FAs. It is therefore difficult to produce high-purity MLM-type TAGs by transesterification of natural oils with immobilized 1,3-position-specific lipases.

To overcome this problem, Irimescu et al. (2001a, 2001b) proposed a two-step process comprising production of 2-MAGs by ethanolysis of TAGs containing PUFAs in the presence of excess amounts of EtOH using immobilized *C. antarctica* lipase (first step) and acylation of the 2-MAGs using immobilized 1,3-position-specific lipase (second step) (Figure 13.10). This process includes the risk that 2-MAGs are converted easily to 1(3)-MAGs by spontaneous acyl migration. Hence, they omitted a process of purification of 2-MAGs from the first-step reaction mixture and directly added excess amounts (20 mol for 2-MAGs) of caprylic acid ethyl ester (C_8EE) to the first-step reaction mixture after removal of EtOH by evaporation. This process produced high-purity (51%) MLM-type TAGs containing DHA from bonito oil in an 85% yield (Irimescu et al. 2001a). When trieicosapentaenoin (or tridocosahexaenoin) was used as a starting material, 1,3-capryloyl-2-eicosapentaenoyl (docosahexaenoyl) glycerol (99% purity) was synthesized in an 85% yield (Irimescu et al. 2001b).

13.4.3.3 Application to Regiospecific Analysis of TAG

Natural oils and fats are a mixture of various TAGs. In general, the TAGs contain FAs at the 1,3-positions that are saturated or have a small number of double bonds and FAs at the 2-position that have a large number of double bonds. However, complete location of a FA on the glycerol backbone is not specified. The location of FA in TAG has received a great deal of attention from the viewpoint of nutrition and TAG metabolism, and a facile method for analyzing FA composition at the 1,3- and 2-positions has been sought.

So far, the regiospecific analysis has been conducted by Grignard degradation (Becker et al. 1993) or by hydrolysis with a 1,3-position-specific lipase, such as lipases from pancreas, *R. oryzae*, and *R. miehei* (Luddy et al. 1964). After the reactions, FA compositions at the 2- and 1,3-positions can be determined based on the contents of FAs in 2-MAGs that are isolated from the reaction products.

However, the two methods have some drawbacks. Grignard degradation requires close attention because even a small amount of moisture greatly impedes the reaction. Also, the lipase method does not provide an exact analysis because the 1,3-position-specific lipases listed do not act on all FAs to a similar degree. In addition, the 2-MAGs are not accumulated efficiently in the reaction mixture because acyl migration occurs easily in the hydrolysis. Therefore, regiospecific analysis of oils containing PUFAs particularly cannot be achieved satisfactorily by this enzymatic method.

To overcome these drawbacks, a new reaction system was developed: ethanolysis of TAGs with immobilized *C. antarctica* lipase in the presence of large amounts of EtOH. In this reaction, (1) the lipase was 1,3-position specific, and TAGs were converted to 2-MAGs in a high yield (Figure 13.8); and (2) the lipase acted on saturated and unsaturated C_{14}–C_{24} FAs to a similar degree (Shimada et al. 2003). Therefore, FA composition at the 2-position in TAGs can be determined by analysis of FAs in 2-MAGs generated by the ethanolysis.

To evaluate the accuracy of the enzymatic method, regiospecific analyses of cocoa fat, borage oil, tuna oil, and AA-containing single-cell oil were conducted. Ethanolysis of the fats and oils was conducted for 4 h (see Figure 13.8; the content of 2-MAGs was maximum), and FA compositions of 2-MAGs purified from the reaction mixture were analyzed. As the Grignard method is presently believed to be the most reliable, 2-MAGs prepared by Grignard degradation were analyzed. The FA compositions determined by the two methods correlated well (Shimada et al. 2003), showing that this enzymatic method is effective for regiospecific analysis of TAGs. In addition, the reliability of this method was confirmed with the analyses of the other oils (Kawashima et al. 2004, Shen and Wijesundera 2006).

13.5 PURIFICATION OF MAG FROM REACTION MIXTURE BY SOLVENT WINTERIZATION

Among the reactions described, esterification of FAs and glycerol with dehydration can be expected as a process for production of MAGs with unstable FAs, but the

TABLE 13.1

Purification of MAG-CLA by Repeated Solvent Winterization

Procedure	Weight (g)		Composition (wt%)[a]			Recovery of MAG (%)
	Total	Glycerol	FFA	MAG	DAG	
Oil layer[b]	30.0	6.5	5.2	87.0	7.8	100
Winterization[c]	24.8	5.6	2.0	95.8	2.2	89.9
Winterization[c]	23.1	5.5	0.9	98.4	0.7	84.7

[a] The total content of MAGs, DAGs, and FFAs was expressed as 100 wt%.

[b] A mixture of FFA-CLA/glycerol (1:2 mol/mol) and *P. camembertii* lipase solution (final contents of water and lipase were 1.0 wt% and 200 U/g, respectively) was agitated at 30°C/4 mm Hg for 48 h. The reaction mixture was separated to the oil and glycerol layers by centrifugation. The oil layer included 21.6 wt% glycerol.

[c] The oil layer (1 g) was dissolved in 4 mL *n*-hexane, and the solution was kept at 0°C for 24 h. The precipitates were recovered by centrifugation, and the solvent was removed with an evaporator.

reaction mixture contains 3–7 wt% FFAs (see Section 13.3.4.1). Molecular distillation is adopted for industrial purification of MAGs synthesized by a chemical process. The distillation is useful for purification of MAGs from a mixture including small amounts of FFAs, but the distillation is not suitable for removal of considerably large amounts of FFAs because the boiling points of FFAs and MAGs are close. Solvent winterization was proposed as a process for purification of MAGs from a mixture including large amounts of FFAs (Berger and Schneider 1992, Watanabe et al. 2006).

MAG-CLA was synthesized by esterification of FFA-CLA and glycerol using *P. camembertii* lipase with dehydration (see Section 13.3.4.1). The reaction mixture contained 5 wt% FFA-CLA, which was removed by solvent winterization. The result is shown in Table 13.1. The oil layer was dissolved in *n*-hexane and kept at 0°C. The resulting precipitates were removed by centrifugation. Two-time repetition of this procedure decreased the content of FFA-CLA from 5 wt% to 0.9 wt% and increased the purity of MAG-CLA from 87 wt% to 98 wt% (85% recovery of MAG) (Watanabe et al. 2006).

13.6 CONCLUSION

Production of MAGs reported so far was reviewed based on the reactions (hydrolysis, esterification, and transesterification) catalyzed by lipases from the viewpoint of industrial production. In addition, application of the reactions to oil processing and regiospecific analysis of TAGs has been introduced. There is increased attention to lipases in the food industries and in the field of synthesis and degradation of unstable oil- and fat-related compounds. We hope this review is useful for those who intend to use lipases for oil and fat processing.

REFERENCES

Becker, C.C., Rosenquist, A., and Holmer, G. 1993. Regiospecific analysis of triacylglycerols using allyl magnesium bromide. *Lipids* 28:147–149.

Bellot, J.C., Choisnard, L., Castillo, E., and Marty, A. 2001. Combining solvent engineering and thermodynamic modeling to enhance selectivity during monoglyceride synthesis by lipase-catalyzed esterification. *Enz. Microb. Technol.* 28:362–369.

Berger, M., Laumen, K., and Schneider, M.P. 1992. Enzymatic synthesis of glycerol I. Lipase-catalyzed synthesis of regioisomerically pure 1,3-*sn*-diacylglycerols. *J. Am. Oil Chem. Soc.* 69:955–960.

Berger, M., and Schneider, M.P. 1992. Enzymatic synthesis of glycerol II. Lipase-catalyzed synthesis of regioisomerically pure 1(3)-rac-monoacylglycerols. *J. Am. Oil Chem. Soc.* 69:961–965.

Castillo, E., Dossat, V., Marty, A., Condoret, J.S., and Combes, D. 1997. The role of silica gel I. Lipase-catalyzed esterification reactions of high-polar substrates. *J. Am. Oil Chem. Soc.* 74:77–85.

Christensen, M.S., Hoy, C.E., Becker, C.C., and Redgrave, T.G. 1995. Intestinal absorption and lymphatic transport of eicosapentaenoic, docosahexaenoic, and decanoic acids: dependence on intramolecular triacylglycerols structure. *Am. Clin. Nutr.* 61:56–61.

Hess, R., Bornscheuer, U., Capewell, A., and Scheper, T. 1995. Lipase-catalyzed synthesis of monostearoylglycerol in organic solvents from 1,2-*O*-isopropylidene glycerol. *Enz. Microb. Technol.* 17:725–728.

Holmberg, K., and Osterberg, E. 1988. Enzymatic preparation to monoglycerides in microemulsion. *J. Am. Oil Chem. Soc.* 65:1544–1548.

Ikeda, I., Tomari, Y., Sugano, M., Watanabe, S., and Nagata, J. 1991. Lymphatic absorption of structured glycerolipids containing medium-chain fatty acids and linoleic acid, and their effect on cholesterol absorption in rats. *Lipids* 26:369–373.

Irimescu, R., Furihata, K., Hata, K., Iwasaki, Y., and Yamane, T. 2001a. Two-step enzymatic synthesis of docosahexaenoic acid-rich symmetrically structured triacylglycerols via 2-monoacylglycerols. *J. Am. Oil Chem. Soc.* 78:743–748.

Irimescu, R., Furihata, K., Hata, K., Iwasaki, Y., and Yamane, T. 2001b. Utilization of reaction medium-dependent regiospecificity of *Candida antarctica* lipase (Novozym 435) for synthesis of 1,3-dicapryloyl-2-docosahexaenoyl (or eicosapentaenoyl) glycerol. *J. Am. Oil Chem. Soc.* 78:285–289.

Kawashima, A., Nagao, T., Watanabe, Y., et al. 2004. Preparation of regioisomers of structured TAG consisting of one mole of CLA and two moles of caprylic acid. *J. Am. Oil Chem. Soc.* 81:1013–1020.

Kawashima, A., Shimada, Y., Yamamoto, M., et al. 2001. Enzymatic synthesis of high-purity structured lipids with caprylic acid at 1,3-positions and polyunsaturated fatty acid at 2-position. *J. Am. Oil Chem. Soc.* 78:611–616.

Kwon, S.J., Han, J.J., and Rhee, J.S. 1995. Production and *in situ* separation of mono- or diacylglycerol catalyzed by lipases in *n*-hexane. *Enz. Microb. Technol.* 17:700–704.

Li, Z.Y., and Ward, O.P. 1994. Synthesis of monoglyceride containing omega-3 fatty acids by microbial lipase in organic solvent. *J. Ind. Microbiol.* 13:49–52.

Luddy, F.E., Barford, R.A., Herb, S.F., Magidman, P., and Riemenschneider, R.W. 1964. Pancreatic lipase hydrolysis of triacylglycerides by a semimicro technique. *J. Am. Oil Chem. Soc.* 41:693–696.

McNeill, G.P., Shimizu, S., and Yamane, T. 1990. Solid phase enzymatic glycerolysis of beef tallow resulting in a high yield of monoglyceride. *J. Am. Oil Chem. Soc.* 67:779–783.

McNeill, G.P., Shimizu, S., and Yamane, T. 1991. High-yield enzymatic glycerolysis of fats and oils. *J. Am. Oil Chem. Soc.* 68:1–5.

McNeill, G.P., and Yamane, T. 1991. Further improvements in the yield of monoglycerides during enzymatic glycerolysis of fats and oils. *J. Am. Oil Chem. Soc.* 68:6–10.

Nagao, T., Kobayashi, T., Hirota, Y., et al. 2005. Improvement of a process for purification of tocopherols and sterols from soybean oil deodorizer distillate. *J. Mol. Catal. B Enzym.* 37:56–62.

Pinsirodom, P., Watanabe, Y., Nagao, T., Sugihara, A., Kobayashi, T., and Shimada, Y. 2004. Critical temperature for production of MAG by esterification of different FA with glycerol using *Penicillium camembertii* lipase. *J. Am. Oil Chem. Soc.* 81:543–547.

Plou, F.J., Barandiaran, M., Calvo, M.V., Ballesteros, A., and Pastor, E. 1996. High-yield production of mono- and di-oleylglycerol by lipase-catalyzed hydrolysis of triolein. *Enz. Microb. Technol.* 18:66–71.

Rezaei, K., and Temelli, F. 2000. Lipase-catalyzed hydrolysis of canola oil in supercritical carbon dioxide. *J. Am. Oil Chem. Soc.* 77:903–909.

Rosu, R., Iwasaki, Y., Shimizu, N., Doisaki, N., and Yamane, T. 1998. Enzymatic synthesis of glycerides from DHA-enriched PUFA ethyl ester by glycerolysis under vacuum. *J. Mol. Catal. B Enzym.* 4:191–198.

Shen, Z., and Wijesundera, C. 2006. Evaluation of ethanolysis with immobilized *Candida antarctica* lipase for regiospecific analysis of triacylglycerols containing highly unsaturated fatty acids. *J. Am. Oil Chem. Soc.* 83:923–927.

Shimada, Y., Ogawa, J., Watanabe, Y., et al. 2003. Regiospecific analysis by ethanolysis of oil with immobilized *Candida antarctica* lipase. *Lipids* 38:1281–1286.

Shimada, Y., Sugihara, A., Maruyama, K., et al. 1996. Production of structured lipid containing docosahexaenoic and caprylic acids using immobilized *Rhizopus delemar* lipase. *J. Ferment. Bioeng.* 81:299–303.

Shimada, Y., Watanabe, Y., Sugihara, A., and Tominaga, Y. 2002. Enzymatic alcoholysis for biodiesel fuel production and application of the reaction to oil processing. *J. Mol. Catal. B Enzym.* 17:133–142.

Steffen, B., Siemann, A., and Lang, S. 1992. Enzymatic monoacylation of trihydroxy compounds. *Biotechnol. Lett.* 14:773–778.

Waldinger, C., and Schneider, M. 1992. Enzymatic synthesis of glycerol III. Lipase-catalyzed synthesis of regioisomerically pure 1,3-*sn*-diacylglycerols and 1(3)-rac-monoacylglycerols derived from unsaturated fatty acids. *J. Am. Oil Chem. Soc.* 69:1513–1519.

Watanabe, Y., Shimada, Y., Yamauchi-Sato, Y., et al. 2002. Synthesis of MAG of CLA with *Penicillium camembertii* lipase. *J. Am. Oil Chem. Soc.* 79:891–896.

Watanabe, Y., Yamauchi-Sato, Y., Nagao, T., et al. 2003. Production of MAG of CLA in a solvent-free system at low temperature with *Candida rugosa* lipase. *J. Am. Oil Chem. Soc.* 80:909–914.

Watanabe, Y., Yamauchi-Sato, Y., Nagao, T., Yamamoto, T., Ogita, K., and Shimada, Y. 2004. Production of monoacylglycerol of conjugated linoleic acid by esterification followed by dehydration at low temperature using *Penicillium camembertii* lipase. *J. Mol. Catal. B Enzym.* 27:249–254.

Watanabe, Y., Yamauchi-Sato, Y., Nagao, T., et al. 2005. Production of MAG of CLA by esterification with dehydration at ordinary temperature using *Penicillium camembertii* lipase. *J. Am. Oil Chem. Soc.* 82:619–623.

Watanabe, Y., Nagao, T., Kanatani, S., Kobayashi, T., Terai, T., and Shimada, Y. 2006. Purification of monoacylglycerols with conjugated linoleic acid synthesized through a lipase-catalyzed reaction by solvent winterization. *J. Oleo Sci.* 55:537–543.

Yamaguchi, S., and Mase, T. 1991. High-yield synthesis of monoglyceride by mono- and diacylglycerol lipase from *Penicillium camembertii* U-150. *J. Ferment. Bioeng.* 72:162–167.

14 Physiological Function of a Japanese Traditional Fish Sauce, Ishiru

Hajime Taniguchi, Toshiki Enomoto, and Toshihide Michihata

CONTENTS

14.1 INTRODUCTION

Fish sauce is one of the seasonings produced and used worldwide since ancient times. These days, it is produced and consumed mainly in Southeast Asia, such as *nam pla* of Thailand, *nuoc mam* of Viet Nam, and *patis* of Philippines [1]. Although it is not a major sauce, fish sauce also is produced in Japan and is used as a seasoning for a variety of processed foods, such as instant noodles. Ishiru is one of the representative fish sauces in Japan that is produced in the peninsula area of Ishikawa Prefecture. In the area facing the Toyama Gulf, squid is the major fishery product, and its inside is used as a starting material for ishiru. In the area facing the Japan Sea, ishiru is produced from sardine.

The flow of ishiru production is shown in Figure 14.1. As a starting material, squid or sardine is used. In the case of squid (ishiru-squid), only its viscera are used, whereas in the case of sardine (ishiru-sardine) the whole body is used. Starting materials are mixed with 18% to 20% salt in tanks made of synthetic resin as shown in Figure 14.2B and left to stand for 1 or 2 years with occasional stirring. During this period, autolysis of fish proteins proceeds, and in most cases, lactic acid fermentation proceeds. As the maturation or fermentation proceeds, the contents in the tank are separated in two phases. The upper half with brown color forms residues, which contain undigested parts of fish and lipid, and is usually discarded. The lower half with a dark color constitutes ishiru. After the heat treatment and filtration, ishiru products as shown in Figure 14.2A are released in the market. Whereas soy sauce is a national seasoning in Japan, ishiru has been used

199

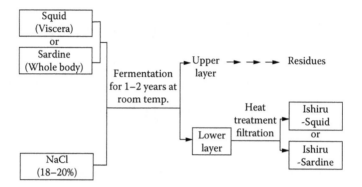

FIGURE 14.1   Flow of ishiru production.

FIGURE 14.2   Ishiru products (A) and tanks (B) producing ishiru.

only in limited areas of Japan. The important difference in these two seasonings is that whereas soy sauce is produced to meet the Japan Agricultural Standard, ishiru has no such standard; therefore, there are wide variations in its quality and its chemical composition.

14.2   CHEMICAL COMPOSITION OF ISHIRU PRODUCTS [2–4]

Chemical compositions of two types of Ishiru products are shown in Table 14.1 together with that of soy sauce. It is easily observed that both types of ishiru have a wide variety in their chemical compositions. Ishiru is produced by a number of small manufacturers, and most of them produce it according to their own traditional methods. There is no national standard on the quality of ishiru products. A remarkable feature of ishiru products is that they contain higher amounts of total nitrogen compared to soy sauce. On the other hand, their carbohydrate contents are quite low compared to soy sauce. This great difference between ishiru and soy sauce reflects differences in their starting materials, fish protein for the former and cereal carbohydrate for the latter. Ishiru also contains significant amounts of fats. One disadvantage of ishiru products is that they contain high amounts of salt. As shown in Table 14.1,

TABLE 14.1
Composition (g/100 mL) of Ishiru Products

	Ishiru-Squid (5 Products)	Ishiru-Sardine (5 Products)	Soy Sauce
Moisture (wt%)	63.79–73.35	63.75–71.50	67.1
Total nitrogen	1.73–2.49	1.42–2.16	1.23
Total carbohydrates	0.04–0.08	0.03–0.08	10.1
Crude fat	0.08–0.33	0.08–0.22	0
NaCl	14.81–26.54	25.74–27.33	14.5
Ash	15.87–26.61	25.25–28.90	15.1
pH	4.77–6.25	5.16–6.53	4.8
Specific gravity	1.14–1.22	1.20–1.23	1.2

some of them contain as high as 25% sodium chloride. High salt intake is considered to be the major cause of higher risk of cardiovascular disease among the Japanese. Reduction or removal of salt from ishiru products is an interesting challenge for further application.

Major components of total nitrogen found in ishiru products are free amino acids and peptide. In Table 14.2, free amino acid compositions of two types of ishiru and amount of peptides are shown. Roughly 80%–90% of nitrogen is free amino acids, and the remaining parts are peptides. Ishiru from squid has relatively higher amounts of free amino acids, whereas ishiru from sardine is relatively rich in peptides. Ala, Gly, Glu, Asp, and Lys form the major free amino acids in both types of ishiru products, whereas ishiru from squid contains twice as much taurine as ishiru from sardine.

Table 14.3 shows the organic acid content in ishiru products. It is seen that lactic acid forms the major organic acid in some ishiru products of both types. However, its content varies greatly depending on the products. Lactic acid contributes greatly to the umami of ishiru products.

The major components of ishiru flavor [3] are aldehydes such as 2-methyl butanal, pyrazines such as methylpyrazine, and ketones such as 2-butanone and dimethyl disulfide. Trimethylamine, a smell compound often found in fish products, was not detected. Histamine and spermidine are the two major polyamines found in ishiru products, as shown in Table 14.4.

14.3   PHYSIOLOGICAL FUNCTION OF ISHIRU

The effects of ishiru products on peroxidaion of linolenic acid were studied. To the reaction mixture containing linolenic acid, 20 μL of ishiru was added, and the extent of peroxidation was determined by the thiobarbituric acid method. Results are expressed as the relative extent of oxidation. As shown in Figure 14.3, all the ishiru products of both types inhibited linolenic acid peroxidation by around 80%. There was no significant difference in antioxidant activity between ishiru-squid and ishiru-sardine.

TABLE 14.2

Amino Acid and Peptide Compositions (mmol/100 mL) of Ishiru Products

	Ishiru-Squid (5 Products)	Ishiru-Sardine (5 Products)
Peptides	3.7–9.6	9.5–18.8
Total amino acid	62.3–81.5	44.5–76.1
Asp	0.21–8.58	0.79–6.57
Glu	6.81–8.63	5.33–9.09
Asn	Trace	Trace
Ser	0.18–5.69	0.23–5.03
Thr	0.79–4.80	1.62–4.52
Tyr	0.17–0.53	0.62–1.19
Cys	0.28–0.32	Trace to 0.37
Met	1.58–1.93	1.07–1.83
Gly	6.07–8.60	3.04–6.67
Ala	7.48–19.56	5.26–17.41
Val	4.23–5.99	2.82–5.63
Leu	2.80–4.72	3.67–4.64
Ile	2.44–3.64	1.87–3.23
Phe	1.68–2.55	1.06–2.21
Trp	Trace to 0.22	Trace to 0.52
Lys	5.95–8.16	4.56–7.91
His	Trace to 1.24	0.35–2.96
Arg	Trace to 3.00	0.07–3.32
Taurine	4.16–5.70	1.90–2.72
Citrulline	Trace to 0.29	0.19–1.11
α-Aminobutyric acid	Trace to 0.43	Trace to 1.63
γ-Aminobutyric acid	Trace to 1.54	Trace to 0.11
Ornithine	0.06–3.30	0.14–4.07

The radical scavenging activity of ishiru products was studied using DPPH (1,1-diphenyl-2-picrylhydrazyl) as a radical and gallic acid as a standard radical scavenger. Scavenging activity of variously diluted ishiru products was assayed, and results were expressed as gallic acid equivalence in milliliter samples. As shown in Figure 14.4, all of the ishiru samples showed scavenging activity of 2–4 μmol gallic acid equivalence per milliliter. Again, there was no significant difference between the two types of ishiru products. Thus, all samples of ishiru showed strong antioxidant activity in terms of inhibiting both lipid peroxidation and radical scavenging activity. These activities are supposed to be caused by peptides they contain.

Another physiological function of ishiru products is their potent inhibiting activity against ACE (angiotensin I-converting enzyme). In this experiment, variously

TABLE 14.3

Organic Acid Composition (mmol/100 mL) of Ishiru Products

	Ishiru-Squid (5 Products)	Ishiru-Sardine (5 Products)
Formic acid	0.11–1.96	0.12–1.58
Acetic acid	0.46–2.78	0.37–3.21
Lactic acid	1.15–52.72	12.11–32.22
Malic acid	0–0.03	0
Succinic acid	0.06–0.20	0.06–0.13
Pyroglutamic acid	1.90–2.94	0.72–3.07

TABLE 14.4

Polyamine Contents of Ishiru Products

	Ishiru-Squid (5 Products)	Ishiru-Sardine (5 Products)
Tyramine	0–44.9	20.1–48.8
Putrescine	6.4–40.1	4.4–12.3
Cadaverine	0–9.0	0–8.2
Histamine	13.1–37.5	33.6–120.0
Agmatine	0–31.4	2.4–5.4
Tryptamine	0–1.3	0–5.4
Spermidine	19.0–37.8	1.45–1.68

diluted ishiru samples were incubated with hippuryl-histidyl-leucine and ACE for 60 min at 37°C, and the amount of released hippuric acid was determined by high-performance liquid chromatography (HPLC). Results are shown as microliters of ishiru samples giving 50% inhibition. As shown in Figure 14.5, products of ishiru-squid showed an IC_{50} at 0.36–0.52 μL/mL, whereas those of ishiru-sardine showed an IC_{50} at 0.14–0.25 μL/mL, a stronger activity.

Then the ACE inhibitory activity of ishiru products was examined in vivo using SHR (spontaneously hypertensive rats). The 6-week-old SHR were fed with a controlled diet or diets containing 5% freeze-dried ishiru samples for up to 16 weeks, and their blood pressure was monitored weekly. As shown in Figure 14.6, the blood pressure of rats fed with the control diet steadily increased with growth and reached about 200 mm Hg around the 13th week. When fed with ishiru-squid, the blood pressure of rats increased in the same manner as the control. On the contrary, the blood pressure of rats fed with ishiru-sardine showed considerably lower values than those of control rats, and after reaching the level of about 180 mm Hg in the 13th week, the

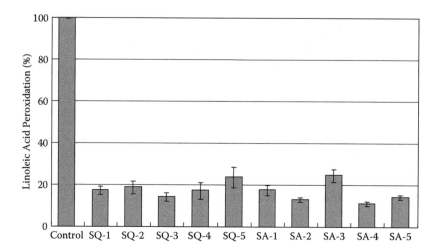

FIGURE 14.3 Inhibition of linoleic acid oxidation by ishiru products. SQ-1 to SQ-5, commercial products of ishiru-squid; SA-1 to SA-5, commercial products of ishiru-sardine.

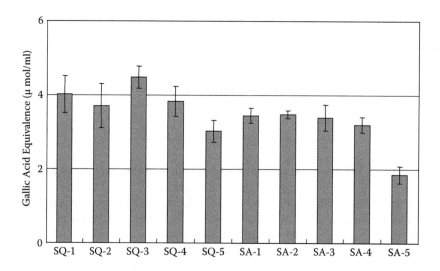

FIGURE 14.4 Radical scavenging activity of ishiru products. SQ-1 to SQ-5, commercial products of ishiru-squid; SA-1 to SA-5, commercial products of ishiru-sardine.

pressure slightly leveled off. This result is in accordance with the result of the in vitro experiment shown in Figure 14.5; that is, ishiru-sardine showed higher ACE inhibitory activity than ishiru-squid. Peptides contained in ishiru samples are presumed to exert these activities. Isolation and identification of these peptides are being undertaken.

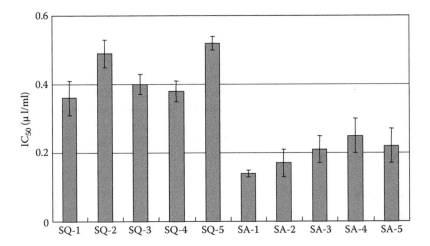

FIGURE 14.5   ACE inhibiting activity of ishiru products. SQ-1 to SQ-5, commercial products of ishiru-squid; SA-1 to SA-5, commercial products of ishiru-sardine.

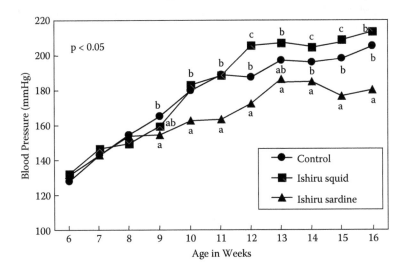

FIGURE 14.6   Effects of desalted ishiru products on spontaneously hypertensive rats.

14.4   CONCLUSIONS

Ishiru is a fish sauce that has been produced and consumed locally in Ishikawa Prefecture, Japan. Ishiru is produced from sardine or squid; the latter is a unique raw material for fish sauce. Findings regarding the physiological function of foods and increasing health consciousness among people led us to the study of this traditional food material. Ishiru contains higher amounts of peptides and amino acids, fats compared to soy sauce, which is made from soybeans and wheat. It was found that both types of ishiru exhibited high antioxidative and ACE-inhibitory activities. Of interest, ishiru made from squid exhibited higher ACE-inhibiting activity in the test tube and higher blood pressure-controlling activity in animal tests than ishiru made from sardine. Peptides contained in ishiru products are supposed to be responsible for both of these activities. It was concluded that ishiru could be an excellent food material with high physiological functions. Effective removal of sodium chloride from ishiru products is in progress.

REFERENCES

1. Fish sauce. http://en.wikipedia.org/wiki/Fish_sauce.
2. Michihata, T., Sado, Y., Yano, T., and Enomoto, T. General components, free amino acids and volatile compounds of Japanese fish sauce Ishiru [abstract]. 11th World Congress of Food Science and Technology, p. 103 (2001).
3. Michihata, T., Yano, T., and Enomoto, T. Volatile compounds of headspace gas in the Japanese fish sauce, ishiru. *Biosci. Biotechnol. Biochem. 66,* 2251 (2002).
4. Watanabe, F., Michihata, T., Takenaka, S., Kittaka-Katsura, H., Enomoto, T., Miyamoto, E., and Adachi, S. Purification and characterization of corrinoid compounds from a Japanese fish sauce. *J. Liquid Chromatogr. Related Technol. 27,* 2113 (2004).

15 Modifying Enzyme Character by Gene Manipulation
Preparation of Chimeric Genes of β-Glucosidases

Kiyoshi Hayashi and Motomitsu Kitaoka

CONTENTS

Key Words: *Cellvibrio gilvus;* gene manipulation; β-glucosidases; shuffling; *Thermotoga maritime.*

15.1 SCREENING OF ENZYMES FROM NATURE

Enzymes are proteins that catalyze chemical reactions and are essential for maintaining health. Enzymes are also used in industries that manufacture products essential for our modern life (e.g., chemical, biofuel, paper, food). The enzymes need to possess certain characteristics for application in these industries. A popular method is to identify the microorganisms that produce enzymes with the required character. The production of enzymes by isolated microorganisms is no longer essential as a result of advances in biotechnology; cloning and overexpression of enzyme genes can overcome this hurdle. Screening genes of the target enzymes directly from the soil is a promising technique since more than 99% of microorganisms are not culturable. However, all these methods are dependent on identification of naturally occurring enzymes.

15.2   IMPROVING ENZYME CHARACTER BY GENE MANIPULATION

Enzymes of required characters can also be obtained by modification of available enzymes. Since the enzyme character is dependent on the amino acid sequence, changing this sequence modifies the enzyme. Point mutation and gene shuffling are two methods for changing the amino acid sequences of enzymes. The point mutation technique is quite simple and several convenient methods have been developed for achieving this [1]. However, it is rather difficult to significantly change the enzyme character by introducing the point mutation since the structural change introduced is often negligible.

15.3   PREPARATION OF CHIMERIC GENES

Significant change in enzyme character can be achieved by gene shuffling. Random shuffling and site-specific shuffling are two methods of gene shuffling. In a random-shuffling study by Stemmer [2], error-prone polymerase chain reaction (PCR) was used to create a library of mutagenized genes, and high-throughput screening was required to select the target enzyme with improved character. Another method is gene shuffling at a designated site, which provides a limited number of chimeric genes.

15.4   SHUFFLING AT THE DESIGNATED SITES
OF β-GLUCOSIDASES

To prepare enzymes of improved character by shuffling at designated sites, two β-glucosidases were selected as parental enzymes: β-glucosidases from *Thermotoga maritima* and *Cellvibrio gilvus*. β-Glucosidase (Tm) from *T. maritima* is a highly thermostable enzyme (85°C) that displays transglycosylation activity, while β-glucosidase (Cg) from *C. gilvus* is mesophilic (35°C) and does not display trans-glycosylation activity.

Both enzymes are family 3-glycoside hydrolases, and each consists of an N-terminal domain and a C-terminal domain. A nonhomologous region comprising 72–90 amino acid residues is present between the two regions. The three-dimensional structure of these enzymes is unknown. Looking at structures from the same family, it has been reported that in 3-glycoside hydrolases the N-terminal domain forms an $(\alpha/\beta)_8$-barrel structure [3], which represents one of the typical folding patterns observed in the glycoside hydrolases. One of the catalytic residues of the nucleophile/base is located on the N-terminal domain, and the other one, a proton donor, is located on the C-terminal domain [3]. The amino acid identities between the two enzymes are 32.4% in the N-terminal domain and 36.4% in the C-terminal domain.

Four shuffling sites were selected based on two criteria. The first criterion was selection of regions where several amino acid residues were identical. These regions were considered to be the conserved regions, and often the three-dimensional structure is also conserved; it is therefore likely that there is less stress in folding the peptide

```
Cg451'   AIQAQAPNAKVVFDDGRDPARAARVAAGADVALVFANQWIGEANDAQTLALPDGQEELIT
                                              *.*.   ***.
Tm421"   GTVIKPKLPENFLSEKEIKKAAKKNDVAVVVISRISGEGYDRKPVKGDFYLSDDELELIK
                          Catalytic Residue E552 ▼
Cg511'   SVA-----GANGRTVVVLQTGGPVTM-PWLARVPAVLEAWYPGTSGGEAIANVLFGAVNPS
           *.    ....  **.*..*.*...  *  * ..*.*.*.  *  .*.** * .***
Tm481"   TVSKEFHDQGKKVVVLLNIGSPIEVASWRDLVDGILLVWQAGGQEMGRIVADVLVGKINPS
                          Catalytic Residue E524 ▲
                         Cg3Tm ──────┌──▷── Tm3Cg
Cg566'   GHLPATFPQSEQQLPRPKLDGDPKNPELQFAVDYHEGAAVGYKWFDLKGHKPLFPFGHGL
         *.**.***.   ..*. .. *.**.   *  * * *.  *▎***...*  * .*  .  **.**
Tm541"   GKLPTTFPKDYSDVPSWTFPGEPKD-NPQ-RVVYEEDIY▎GYRYYDTFGVEPAYEFGYGL
                         Tm3Cg ──▷──┘└──▶── Cg3Tm

                         Cg4Tm ──────┌──▷── Tm4Cg
Cg626'   SYTTFAYSGLSGQLKDGRLHVRFKVTNTGNVAGKDVPQVY▎AAPMSTKWEAP-KRLAAWSK
         ***.*.*..*.   .....  *.*... ****. ***.*.***   .  *  .  *  .  * .*
Tm599"   SYTKFEYKDLKIAIDGETLRVSYTITNTGDRAGKEVSQVY▎IKAPKGKIDKPFQELKAFHK
                         Tm4Cg ──▷──┘└──▶── Cg4Tm

Cg685'   VALL-PGETKEVEVAVEPRVLAMFDEKSRTWRRPKGKIRLTLAEDASAANATSVTVELPA
           .** ***..*...  * ** **.*  *   .*.....
Tm659"   TKLLNPGESEEISLEIPLRDLASFDGK--EWVVESGEYEVRVGASSRDIRLRDIFLVEGE

Cg744'   STLDARGRAR

Tm717"   KRFKP
```

FIGURE 15.1 Amino acid alignment in the C-terminal domain of β-glucosidases.

into the active catalytic form. Second, shuffling sites should be away from each other, preferably based on the available three-dimensional structure in the same family.

Two sites in the N-terminal domain (sites 1 and 2) and two sites in the C-terminal domain (sites 3 and 4) were selected as shuffling sites. The alignment of amino acids in the Cg and Tm β-glucosidase in the C-terminal domain is described in Figure 15.1.

15.5 PREPARATION OF CHIMERIC GENES BY OVERLAPPING PCR

The construction of chimeric enzyme genes was carried out using a self-priming PCR. High-fidelity DNA polymerase (KOD-Plus, Toyobo Biochemicals, Japan) was used in three-step PCR reactions to overcome incorporation of undesired errors during multiple PCR steps.

The first PCR step amplified the selected N- and C-terminal regions of the two genes (Figure 15.2). Primers 2 and 3 were specially designed for production of chimeric enzyme genes. Primer 2 consisted of 20 bases of the enzyme A gene and 10 bases of the additional DNA sequence of the enzyme B gene. Denaturation and annealing were performed at 98°C for 1 min and 55°C for 1 min, respectively. Primer extension was carried out at 68°C for 1–2 min and repeated for 20 cycles. These conditions were also employed in the second and third PCR.

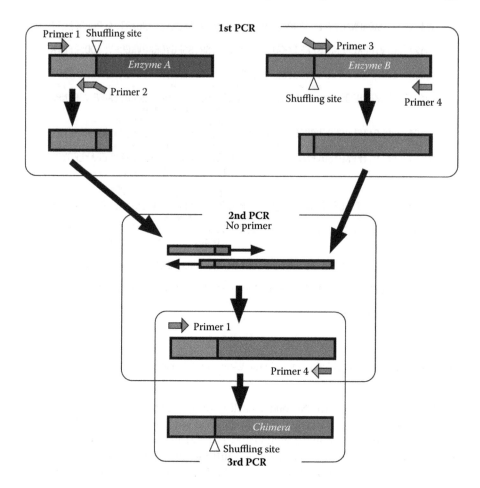

FIGURE 15.2   Chimeric gene preparation by overlapping PCR.

The PCR products purified by agarose gel electrophoresis were used as template DNA for the second stage-overlapping PCR without primers. In this process, strands having the same base pairs (20 bases) overlapped and acted as primers for each other. In the third PCR, the combined fragment (chimeric enzyme gene) was amplified using forward and reverse primers (primers 1 and 4).

15.6   EXPRESSION, PURIFICATION, AND CHARACTERIZATION OF CHIMERIC ENZYMES

To express the recombinant chimeric genes, they were ligated with *pET28a*(+) using restriction enzymes. Constructed plasmids were then transformed into *Escherichia coli* BL21(DE3) to obtain the chimeric enzyme. Transformants were grown at 30°C, and the target protein was induced by addition of isopropyl-β-D-thiogalactopyranoside (IPTG). The *E. coli* cells harvested by centrifugation were resuspended in 3-(N-morpholino)-propanesulfonic acid (MOPS) buffer (pH 6.5) and

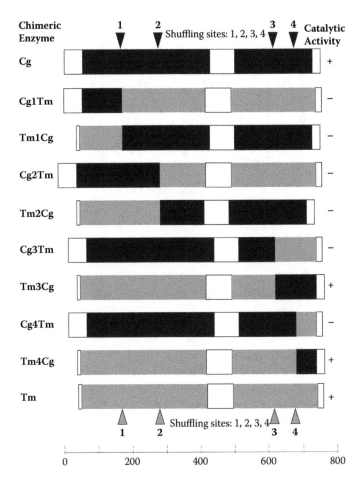

FIGURE 15.3 Eight constructed chimeric β-glucosidases.

sonicated. Expressed β-glucosidases were purified using a 6× His-tag chelate column followed by an anion exchange MonoQ HR 5/5 column (Pharmacia, Milton Keynes, U.K.).

Of the eight chimeric genes constructed, only two chimeric enzymes (Tm3Cg and Tm4Cg) were obtained as catalytically active forms, while the remaining six chimeric enzymes were insoluble (Figure 15.3). The catalytically active form is obtained only if it is shuffled at the C-terminal domain; 80% (Tm3Cg) and 88% (Tm4Cg) of their amino acid sequences originated from *T. maritima* (Figure 15.2).

The two active chimeric enzymes, Tm3Cg and Tm4Cg, displayed thermal profiles intermediate between those of the two parental enzymes; they were optimally active at 65°C and 70°C, respectively (Figure 15.4). These two chimeric enzymes were optimally active at pH 4.1 and 3.9, which is closer to that observed for the *T. maritima* enzyme (pH 3.2–3.5) than that for the *C. gilvus* enzyme (pH 6.2–6.5).

Furthermore, the kinetic parameters of the two chimeric enzymes for the substrate of *p*NP-β-D-glucopyranoside and *p*NP-β-D-xylopyranoside were closer to those of

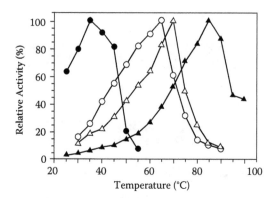

FIGURE 15.4 Temperature optimum for native and chimeric β-glucosidases. ●, Cg; ○, Tm3Cg; △, Tm4 Cg; ▲, Tm.

TABLE 15.1
Substrate Specificities of Chimeric Enzymes

	Parental and Chimeric Enzymes			
	Tm	Tm3Cg	Tm4Cg	Cg
*p*NP-B-D-glucopyranoside				
*K*m	0.0039	0.012	0.0082	0.44
*k*cat	6.4	5.6	3.8	42
*k*cat/*K*m	1,641	467	463	95
*p*NP-B-D-xylopyranoside				
*K*m	2.6	2.8	3.2	11
*k*cat	18	13	24	3.1
*k*cat/*K*m	6.9	4.6	7.5	0.3

the *T. maritima* enzyme than those of the *C. gilvus* enzyme (Table 15.1). The k*cat* value for the two chimeric enzymes was not significantly decreased, indicating that the catalytic activity of the enzyme was adequately maintained.

Transglycosylation activity was enhanced by preparing the chimeric enzyme: The activity of the Tm3Cg chimera was at a level twice that observed with the Tm enzyme; no activity was observed in Cg (Figure 15.5).

15.7 CONCLUDING REMARKS

Shuffling at designated sites enables estimation of the chimeric enzymes' characters. Many studies have suggested that the chimeric enzymes follow the character of the parental enzymes [4–6]. Therefore, the preparation of chimeric enzymes can be regarded as "molecular breeding." Three-dimensional structure is not essential for designing chimeric enzymes.

Preparation of chimeric enzymes is easy if the amino acid alignment of the two parental enzymes is highly identical (>80%). However, the character of the two

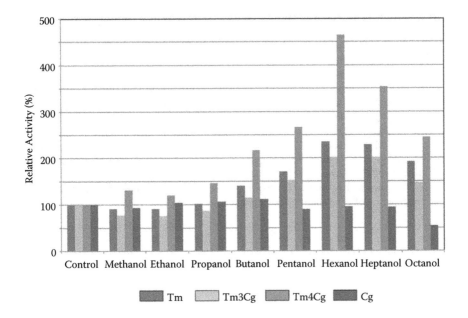

FIGURE 15.5 Transglycosylation activity of chimeric enzymes. Relative rates of p-nitrophenol production from pNP-β-D-glucopyranoside were measured in the presence of a series of straight-chain alcohols (50 mM).

parental enzymes as well as that of the prepared chimeric enzymes is very similar. If the homology is quite low (<30%), the character of the two parental enzymes is distinct, and the construction of chimeric genes is easy, but obtaining an active form is quite difficult. We tried to construct 16 chimeric enzymes using two parental enzymes whose homology was about 30%. All chimeric genes were translated into full-length peptide with no catalytic activity.

The ideal degree of homology in the alignment of amino acids is 40%–50%, where the character of the parental enzymes (e.g., enzyme stability, substrate specificity, kcat, km) are significantly different. In our experience, one-third of chimeric genes will give catalytically active forms. One method of converting the noncatalytic form into the catalytic form is introducing a point mutation in the noncatalytic chimeric genes [7].

Enzymes can be utilized not only for conversion of materials but also for energy, environmental resources, and medicines. Techniques to alter enzyme character might unlock the full capabilities of enzymes.

REFERENCES

1. R. Fujii, M. Kitaoka, and K. Hayashi. 2006. Error-prone rolling circle amplification: the simplest random mutagenesis protocol. *Nat. Protoc.* 1(5): 2493–2497.
2. W.P. Stemmer. 1994. Rapid evolution of a protein in vitro by DNA shuffling. *Nature* 370: 389–391.

3. J.N. Varghese, M. Hrmova, and G.B. Fincher. 1999. Three-dimensional structure of a barley beta-D-glucan exohydrolase, a family 3 glycosyl hydrolase. *Structure* 7: 179–190.

4. K. Goyal, Y.K. Kim, M. Kitaoka, and K. Hayashi. 2001. Construction and characterization of chimeric enzymes of the *A. tumefaciens* and *T. maritima* β-glucosidases. *J. Mol. Catal. B-Enzym.* 16: 43–51.

5. K. Goyal, P. Selvakumar, and K. Hayashi. 2001. Molecular cloning, purification and characterization of thermostable β-glucosidase from hyperthermophilic strain *Thermotoga maritima*: activity in the presence of alcohols. *J. Mol. Catal. B-Enzym.* 15: 45–53.

6. L. Ying, M. Kitaoka, and K. Hayashi. 2004. Effects of the truncation at the non-homologous region of a family 3 beta-glucosidase from *Agrobacterium tumefaciens*. *Biosci. Biotechnol. Biochem.* 68: 1113–1118.

7. S. Nirasawa and K. Hayashi. 2008. Construction of a chimeric aminopeptidase by a combination of gene shuffling and mutagenesis. *Biotechnol. Lett.* 30: 363–368.

16 Chemoenzymatic Synthesis of Chiral Pharmaceutical Intermediates

Ramesh N. Patel

CONTENTS

Key Words: Biocatalysis; chiral intermediates; enzymatic processes; synthesis of pharmaceuticals.

16.1 INTRODUCTION

The production of single enantiomers of chiral intermediates has become increasingly important in the pharmaceutical industry [1]. Single enantiomers can be produced by chemical or chemoenzymatic synthesis. The advantages of biocatalysis over chemical synthesis are that enzyme-catalyzed reactions are often highly enantioselective and regioselective. They can be carried out at ambient temperature and atmospheric pressure, thus avoiding the use of more extreme conditions, which could cause problems with isomerization, racemization, epimerization, and rearrangement. Microbial cells and enzymes derived therefrom can be immobilized and reused for many cycles. In addition, enzymes can be overexpressed to make biocatalytic processes economically efficient, and enzymes with modified activity can be tailor made. Directed evolution of biocatalysts can lead to increased enzyme activity, selectivity, and stability [2–10]. A number of review articles [11–21] have been published on the use of enzymes in organic synthesis. This review provides some examples of the use of enzymes for the synthesis of single enantiomers of key intermediates used in pharmaceutical synthesis.

16.2   ANTIANXIETY DRUG

16.2.1   Enzymatic Preparation of 6-Hydroxybuspirone

Buspirone (Buspar®; **1**, Figure 16.1) is a drug used for treatment of anxiety and depression that is thought to produce its effects by binding to the serotonin 5-HT1A receptor [22]. Mainly as a result of hydroxylation reactions, it is extensively converted to various metabolites [23], and blood concentrations return to low levels a few hours after dosing [24]. A major metabolite, 6-hydroxybuspirone **2** (Figure 16.1), produced by the action of liver cytochrome P450 CYP3A4, is present at much higher concentrations in human blood than buspirone itself. This metabolite has anxiolytic effects in an anxiety model using rat pups and binds to the human 5-HT1A receptor [25]. Although the metabolite has only about a third of the affinity for the human 5-HT1A receptor as buspirone, it is present in human blood at 30–40 times higher concentration than buspirone following a dose of buspirone and therefore may be responsible for much of the effectiveness of the drug [26].

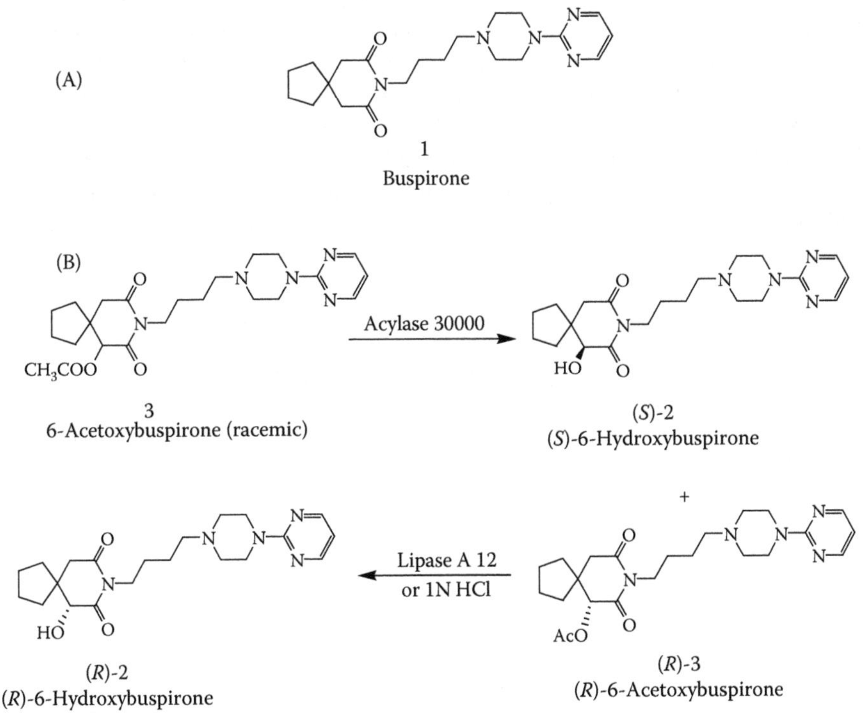

FIGURE 16.1   (A) Antianxiety drug buspirone **1**. (B) Enzymatic resolution of 6-acetoxybuspirone **3**.

For development of 6-hydroxybuspirone as a potential antianxiety drug, preparation and testing of the two enantiomers as well as the racemate were of interest. Both
the (R)- and (S)-enantiomers, isolated by chiral high-performance liquid chromatography (HPLC), were effective in tests using a rat model of anxiety [27]. Whereas
the (R)-enantiomer showed somewhat tighter binding and specificity for the 5-HT1A
receptor, the (S)-enantiomer had the advantage of being cleared more slowly from the
blood. An enzymatic process was developed for resolution of 6-acetoxybuspirone **3**
(Figure 16.1).

L-Amino acid acylase from *Aspergillus melleus* (Amano acylase 30000) was
used to hydrolyze racemic 6-acetoxybuspirone to (S)-6-hydroxybuspirone **2** in 96%
enantiomeric excess (EE) after 46% conversion. The remaining (R)-6-acetoxy-
buspirone with 84% EE was converted to (R)-6-hydroxybuspirone **2** by acid hydrolysis [28]. The EE of both enantiomers could be improved to more than 99% by
crystallization as a metastable polymorph. Direct hydroxylation of buspirone to
(S)-6-hydroxbuspirone by *Streptomyces antibioticus* ATCC 14980 has also been
described [28].

In an alternate process, enantioselective microbial reduction of 6-oxobuspirone **4**
(Figure 16.2) to either (R)- or (S)-6-hydroxybuspirone was described by Patel et al.
[29]. About 150 microorganisms were screened for the enantioselective reduction of
4. *Rhizopus stolonifer* SC 13898, *Rhizopus stolonifer* SC 16199, *Neurospora crassa*
SC 13816, *Mucor racemosus* SC 16198, and *Pseudomonas putida* SC 13817 gave
more than 50% reaction yields and more than 95% EEs of (S)-6-hydroxybuspirone.
The yeast strains *Hansenula polymorpha* SC 13845 and *Candida maltosa* SC 16112
gave (R)-6-hydroxybuspirone **2** in more than 60% reaction yield and more than 97%
EE. The NADP-dependent (R)-reductase (RHBR), which catalyzes the reduction
of 6-oxobuspirone to (R)-6-hydroxybuspirone, was purified to homogeneity from
cell extracts of *Hansenula polymorpha* SC 13845. (R)-Reductase from *Hansenula*

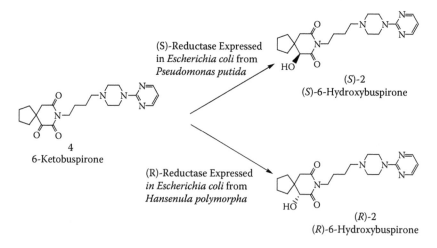

FIGURE 16.2   Enantioselective enzymatic reduction of 6-oxobuspirone **4**.

polymorpha SC 13845 was cloned and expressed in *Escherichia coli*. To regenerate the cofactor NADPH (nicotinamide adenine dinucleotide phosphate) required for reduction, we have also cloned and expressed the glucose-6-phosphate dehydrogenase gene from *Saccharomyces cerevisiae* in *Escherichia coli*.

The NAD-dependent (*S*)-reductase (SHBR), which catalyzes the reduction of 6-ketobuspirone to (*S*)-6-hydroxybuspirone **2**, was also purified to homogeneity from cell extracts of *Pseudomonas putida* SC 16269. The (*S*)-reductase from *Pseudomonas putida* SC 16269 was cloned and expressed in *Escherichia coli*. To regenerate the cofactor NADH (nicotinamide adenine dinucleotide) required for reduction, we have also cloned and expressed the formate dehydrogenase gene from *Pichia pastoris* in *Escherichia coli*. Recombinant *Escherichia coli* expressing (*S*)-reductase and (*R*)-reductase catalyzed the reduction of 6-ketobuspirone to (*S*)-6-hydroxybuspirone and (*R*)-6-hydroxybuspirone, respectively, in more than 98% yield and more than 99.9% EE [30].

16.3   ANTIDIABETIC DRUG (DIPEPTIDYL PEPTIDASE IV INHIBITOR)

16.3.1   ENZYMATIC REDUCTIVE AMINATION OF 2-(3-HYDROXY-1-ADAMANTYL)-2-OXOETHANOIC ACID

Glucagon-like peptide 1 (GLP-1) analogs and dipeptidyl peptidase IV (DPP-IV) inhibitors are two promising new approaches currently being explored for treatment of type 2 diabetes [31–35]. GLP-1, a peptide secreted by the gut in response to feeding, has the beneficial effects of increasing glucose-stimulated insulin secretion, decreasing glucagon secretion, and increasing the β-cell mass of pancreatic islets [31–33]. GLP-1 (7–36) amide is rapidly inactivated by conversion to GLP-1 (9–36) amide by DPP-IV [31–33]. To alleviate the inactivation of GLP-1, more stable analogs of GLP-1 as well as inhibitors of DPP-IV are approaches to provide improved control of blood glucose for diabetics.

Saxagliptin **5** [35] (Figure 16.3A), a DPP-IV inhibitor under development by Bristol-Myers Squibb, requires (*S*)-*N*-boc-3-hydroxyadamantylglycine **7** (Figure 16.3B) as an intermediate. Previously, we have demonstrated the conversion of keto acids to chiral amino acid by an enzymatic reductive amination process [36–38]. We have developed a process for reductive amination of the keto acids **6** to the corresponding amino acid **7** using (*S*)-amino acid dehydrogenases to prepare key chiral amino acid for the synthesis of Saxagliptin [39,40]. A modified form of a recombinant phenylalanine dehydrogenase cloned from *Thermoactinomyces intermedius* and expressed in *Pichia pastoris* or *Escherichia coli* was used for this process. NAD produced during the reaction was recycled to NADH using formate dehydrogenase. The modified phenylalanine dehydrogenase contains two amino acid changes at the C-terminus and a 12-amino acid extension of the C-terminus.

Production of multikilogram batches was originally carried out with extracts of *Pichia pastoris* expressing the modified phenylalanine dehydrogenase from

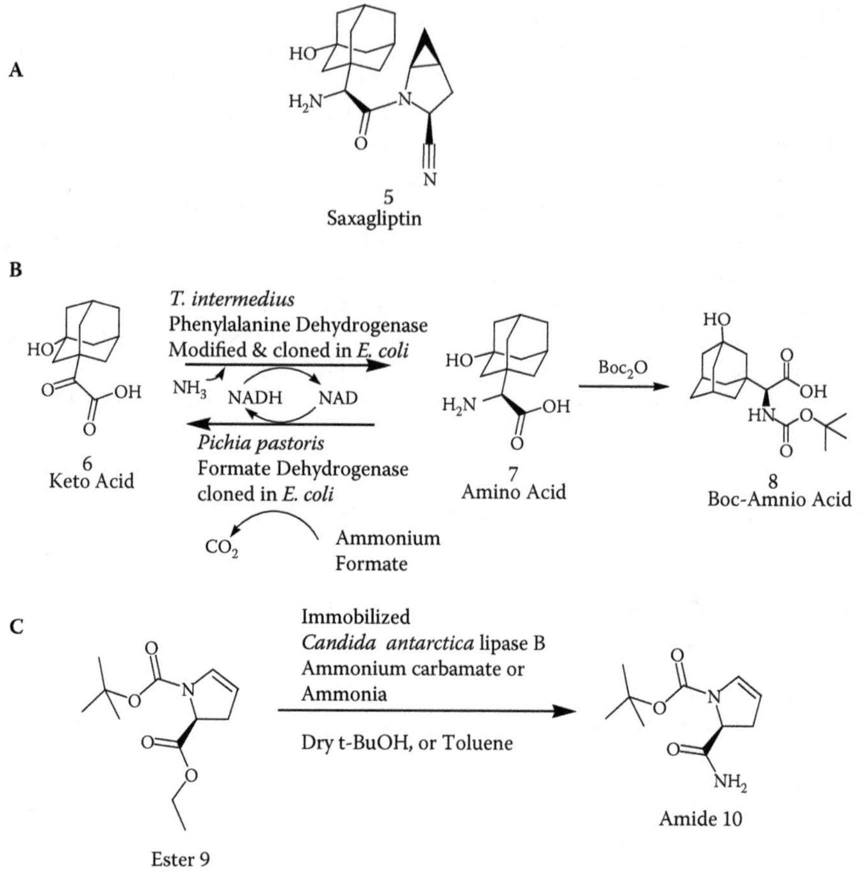

A

5
Saxagliptin

B

6
Keto Acid

7
Amino Acid

8
Boc-Amnio Acid

C

Ester 9

Amide 10

FIGURE 16.3   (A) Antidiabetic drug Saxagliptin **5**. (B) Enzymatic reductive amination of 2-(3-hydroxy-1-adamantyl)-2-oxoethanoic acid **6**. (C) Enzymatic ammonolysis of (5S)-4,5-dihydro-1H-pyrrole-1,5-dicarboxylic acid, 1-(1,1-dimethylethyl)-5-ethyl ester **9**.

Thermoactinomyces intermedius and endogenous formate dehydrogenase. The reductive amination process was further scaled up using a preparation of the two enzymes expressed in single recombinant *E. coli*. The amino acid **7** can be directly protected as its boc derivative without isolation to afford intermediate **8**. Yields before isolation were close to 98% with 100% EE.

Reductive amination was also conducted using cell extracts from *E. coli* strain SC16496 expressing PDHmod and cloned FDH from *Pichia pastoris*. Cell extracts after polyethyleneamine treatment, clarification, and concentration were used to complete the reaction in 30 h with greater than 96% yield and more than 99.9% EE of product **7**. This process has now been used to prepare several hundred kilograms of boc-protected amino acid **8** to support the development of Saxagliptin [40].

16.3.2   ENZYMATIC AMMONOLYSIS OF (5S)-4,5-DIHYDRO-1H-PYRROLE-1,5-DICARBOXYLIC ACID, 1-(1,1-DIMETHYLETHYL)-5-ETHYL ESTER

The synthesis of DPP-IV inhibitor Saxagliptin **5** also required (5S)-5-amino-carbonyl-4,5-dihydro-1H-pyrrole-1-carboxylic acid, 1-(1,1-dimethylethyl)ester **10** (Figure 16.3C). Direct chemical ammonolyses were hindered by the requirement for aggressive reaction conditions, which resulted in unacceptable levels of amide race-mization and side-product formation, while milder two-step hydrolysis-condensation protocols using coupling agents such as 4-(4,6-dimethoxy-1,3,5-triazin-2-yl)-4-methylmorpholinium chloride (DMT-MM) [41] were compromised by reduced over-all yields. To address this issue, a biocatalytic procedure was developed based on the *Candida antartica* lipase B (CALB)-mediated ammonolysis of (5S)-4,5-dihydro-1H-pyrrole-1,5-dicarboxylic acid, 1-(1,1-dimethylethyl)-5-ethyl ester **9** with ammonium carbamate to furnish **10** without racemization and with low levels of side-product formation.

Screening experiments utilized process stream ester feed, which consisted of about 22% w/v (0.91M) of the ester in toluene. Since toluene precluded the use of free ammonia due to its low solubility in toluene, solid ammonium carbamate was employed. Reactions were performed using a mixture of neat process feed, ammo-nium carbamate (71 g/L, 2 mol eq of ammonia), and biocatalyst (25 g/L), shaken at 400 rpm, 50°C. Under these conditions, CALB and its immobilized forms Novozym 435 and Chirazyme L2 provided racemization-free amide with yields of 69%, 43%, and 40%, together with 21%, 18%, and 22% of side products (by HPLC), respectively, while all other biocatalysts (lipases) furnished less than 5% of the desired product [42]. The ammonolysis reaction with free CALB was then optimized with regard to the temperature and the CALB and ammonium carbamate loads to increase yield from 56% to 71%, with side products varying from 7% to 19%.

The inclusion of various additives was investigated with the aim of ameliorat-ing potential inhibitory phenomena, shifting the equilibrium toward amide synthesis and reducing side-product formation. Drying agents such as calcium chloride gave significant improvement (79% amide and 13% side products). The calcium chloride is known to complex alcohols as well as act as a desiccant, and its presumed binding of ethanol released during the course of amide formation may have served to miti-gate any deleterious effects of this alcohol on CALB catalysis. A dramatic increase in amide yield to 84% and 95% was achieved by including Sodalime and Ascarite, respectively, at 200 g/L in the reaction headspace, this presumably by way of adsorp-tion of carbon dioxide liberated from the decomposition of ammonium carbamate. A further increase in yield to 98% was attained via the combined use of 100 g/L of calcium chloride and 200 g/L of Ascarite.

A prep-scale reaction with the process ester feed was used. Ester (220 g/L) was reacted with 90 g/L (1.25 mol eq) of ammonium carbamate, 33 g/L (15% w/w of ester input) of CALB, 110 g/L calcium chloride, and 216 g/L of Ascarite (in the head-space), run at 50°C for 3 days. Complete conversion of ester was achieved, with the formation of 96% (182 g/L) of amide **10** and 4% of side products; after workup, 98% potency amide of greater than 99.9% EE was isolated in 81% yield.

16.4 ANTIVIRAL DRUG (HIV PROTEASE INHIBITOR)

16.4.1 ENZYMATIC PREPARATION OF (1S,2R)-[3-CHLORO-2-HYDROXY-1-(PHENYLMETHYL) PROPYL]-CARBAMIC ACID, 1,1-DIMETHYL-ETHYL ESTER

Atazanavir **11** (Figure 16.4A) is an acyclic aza-peptidomimetic, a potent HIV protease inhibitor [43,44] approved recently by the U.S. Food and Drug Adminstration for treatment of autoimmune diseases (e.g., AIDS). An enzymatic process has been developed for the preparation of (1S,2R)-[3-chloro-2-hydroxy-1-(phenylmethyl)propyl]carbamic acid, 1,1-dimethylethyl ester **13** (Figure 16.4B), a key chiral intermediate

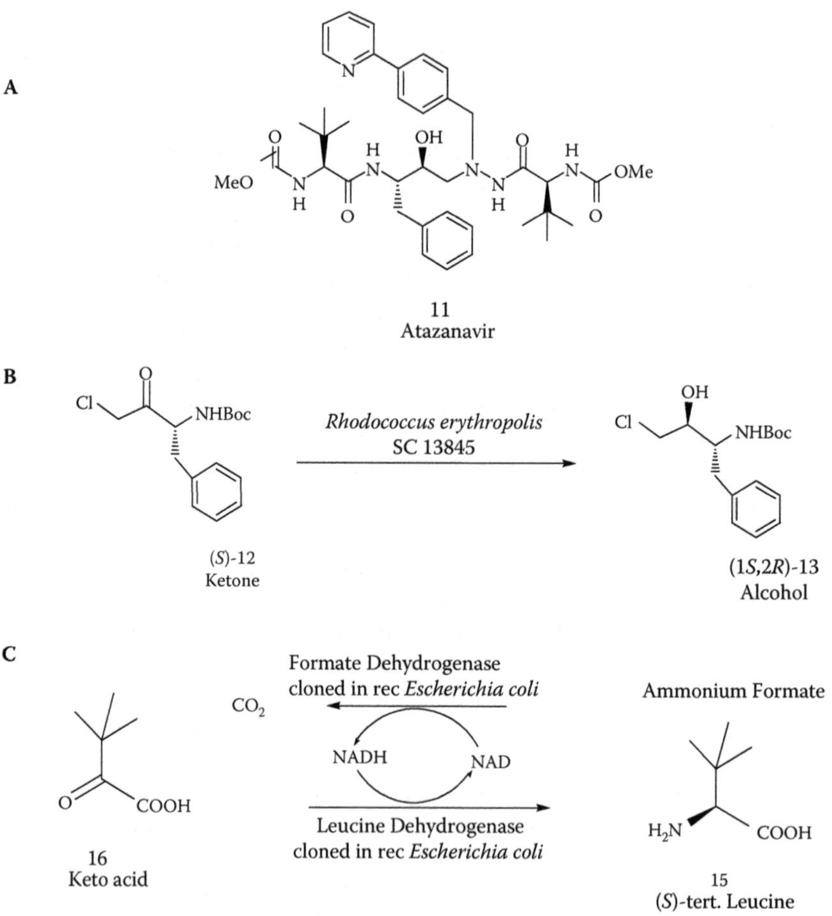

FIGURE 16.4 (A) HIV protease inhibitor atazanavir **11**. (B) Diastereoselective enzymatic reduction of (1S)-[3-chloro-2-oxo-1-(phenylmethyl)propyl] carbamic acid, 1,1-dimethylethyl ester **12**. (C) Enzymatic synthesis of (S)-tertiary-leucine **15**.

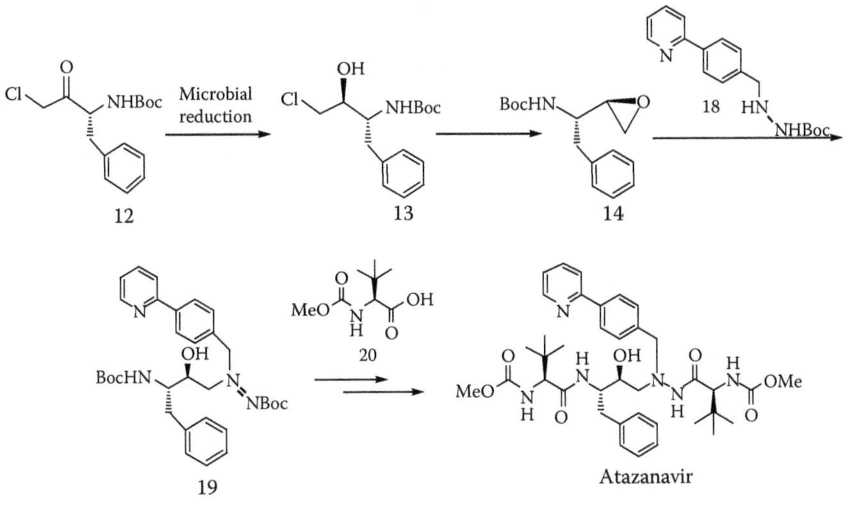

FIGURE 16.5 Chemoenzymatic synthesis of atazanavir **11**.

required for the total synthesis of the HIV protease inhibitor atazanavir. The diastereoselective reduction of (1*S*)-[3-chloro-2-oxo-1-(phenylmethyl)propyl] carbamic acid, 1,1-dimethylethyl ester **12** was carried out using *Rhodococcus, Brevibacterium*, and *Hansenula* strains to provide **13**. Three strains of *Rhodococcus* gave greater than 90% yield with a diastereomeric purity of more than 98% and an EE of 99.4% [45]. An efficient single-stage fermentation-biotransformation process was developed for the reduction of ketone **12** with cells of *Rhodococcus erythropolis* SC 13845 to yield **13** in 95% with a diasteromeric purity of 98.2% and an EE of 99.4%. Chemical reduction of chloroketone **12** using NaBH$_4$ produces primarily the undesired chlorohydrin diastereomer [46]. The (1*S*,2*R*)-**13** was converted to epoxide **14** (Figure 16.5) and used in the synthesis of atazanavir [46].

16.4.2 Enzymatic Synthesis of (S)-Tertiary-Leucine

Synthesis of atazanavir **11** also required (*S*)-tertiary leucine **15** (Figure 16.4C). An enzymatic reductive amination of ketoacid **16** to amino acid **17** by recombinant *Escherichia coli* expressing leucine dehydrogenase from *Thermoactinimyces intermedius* has been demonstrated. The reaction required ammonia and NADH as a cofactor. NAD produced during the reaction was converted back to NADH using recombinant *Escherichia coli* expressing formate dehydrogenase from *Pichia pastoris*. A reaction yield of more than 95% with an EE of more than 99.5% was obtained for **17** at 100 g/L substrate input [47].

Chiral epoxide **14** (Figure 16.5) can be coupled to hydrazinocarbamate **18** to prepared compound **19**. Chemical coupling of *N*-methoxycabonyl-(*S*)-tert, leucine **20**, and compound **19** can afford atazanavir **11**.

16.5 ANTIVIRAL DRUG (LOBUCAVIR)

16.5.1 REGIOSELECTIVE ENZYMATIC AMINOACYLATION

Lobucavir **21** (Figure 16.6) is a cyclobutyl guanine nucleoside analog under development as an antiviral agent for the treatment of herpesvirus and hepatitis B [48]. A prodrug in which one of the two hydroxyls is coupled to valine, **22**, has also been considered for development. Regioselective aminoacylation is difficult to achieve by chemical procedures but appeared to be suitable for an enzymatic approach [49]. Synthesis of the lobucavir L-valine prodrug **22** requires regioselective coupling of one of the two hydroxyl groups of lobucavir (**21**) with valine. Enzymatic processes were developed for aminoacylation of either hydroxyl group of lobucavir [49]. The selective hydrolysis of the di-cbz-valine ester **23** with lipase M gave **25** in 83% yield.

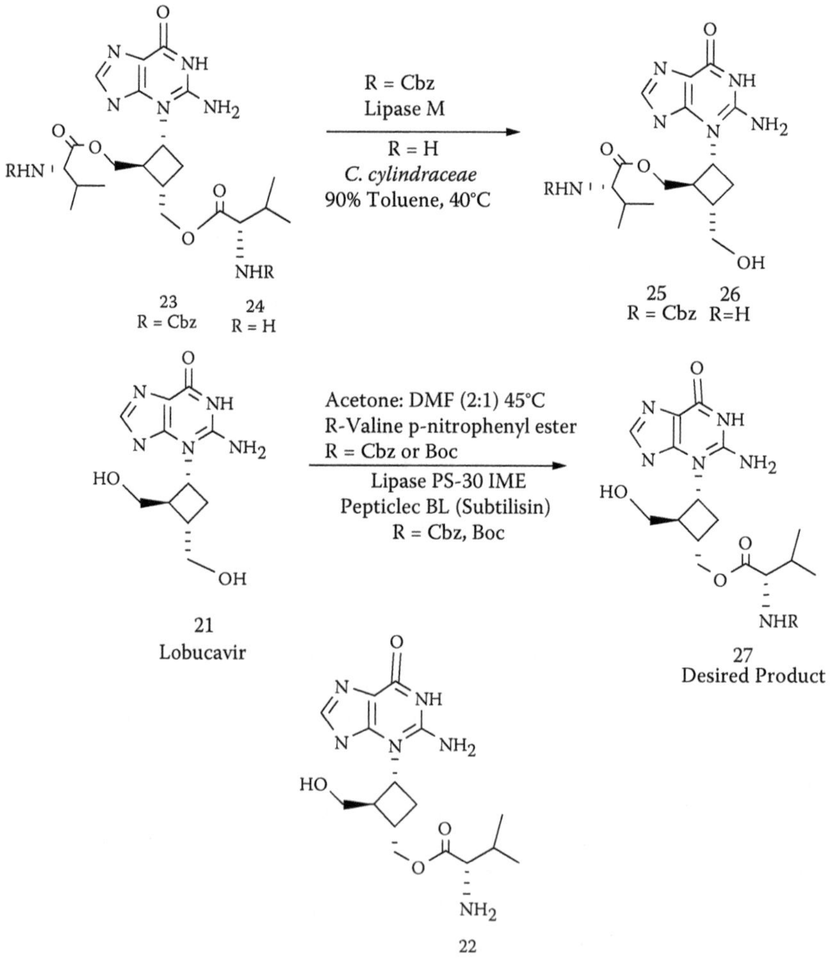

FIGURE 16.6 Antiviral drug (lobucavir) prodrug **22**. Regioselective enzymatic aminoacylation and hydrolytic reactions.

When the divaline ester of **24** dihydrochloride was hydrolyzed with lipase from *Candida cylindraceae*, **26** was obtained in 87% yield. The final intermediates for lobucavir prodrug, the methyl ester of **27** could be obtained by transesterification of lobucavir using ChiroCLEC™ BL (61% yield) or more selectively using lipase from *Pseudomonas cepacia* (84% yield).

16.6 ANTIVIRAL DRUG (HEPATITIS B VIRAL INHIBITOR)

16.6.1 ENZYMATIC ASYMMETRIC HYDROLYSIS AND ACETYLATION

Chiral monoacetate esters **29** and **31** are key intermediates for total chemical synthesis of **32** (Baraclude), a potential drug for hepatitis B virus (HBV) infection [50–52]. Baraclude is a carboxylic analog of 2′-deoxyguanosine in which the furanose oxygen is replaced with an exocyclic double bond; it was approved by the U.S. Food and Drug Administration (FDA) for treatment of HBV infection.

Enzymatic hydrolysis of (1α,2β,3α)-2-[(benzyloxy)methyl]-4-cyclopenten-1,3-diol diacetate has been demonstrated by Griffith and Danishefsky to obtain the corresponding monoester using acetylcholine esterase from electric eel [53,54]. We have described the enantioselective asymmetric hydrolysis of (1α,2β,3α)-2-[(benzyloxy)methyl]-4-cyclopenten-1,3-diol diacetate **28** (Figure 16.7) to the corresponding (+)-monoacetate **29** by lipase PS-30 from *Pseudomonas cepacia* and pancreatin. A

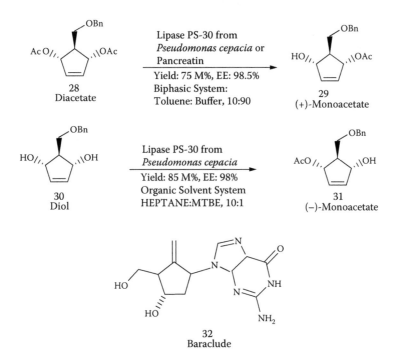

FIGURE 16.7 Antiviral drug Baraclude **32**: Enzymatic asymmetric hydrolysis of diacetate **28** and acetylation of diol **30**.

reaction yield of 85 mol% and an EE of 98% were obtained using lipase PS-30. Using pancreatin, a reaction yield of 75 mol% and an EE of 98.5% were obtained. We also demonstrated the enzymatic asymmetric acetylation of (1α,2β,3α)-2-[(benzyloxy) methyl]-4-cyclopenten-1,3-diol **30** to the corresponding (–)-monoacetate **31** using lipase PS-30 [55].

16.7 ANTICANCER DRUGS

16.7.1 PACLITAXEL SEMISYNTHETIC PROCESS

Among the antimitotic agents, paclitaxel (Taxol®) **33** (Figure 16.8), a complex, poly-cyclic diterpene, exhibits a unique mode of action on microtubule proteins respon-sible for the formation of the spindle during cell division. Paclitaxel is known to inhibit the depolymerization process of microtubulin [56]. Various types of cancers have been treated with paclitaxel, and the results of treatment of ovarian cancer and metastatic breast cancer are very promising and approved by the FDA. Paclitaxel was originally isolated from the bark of the yew *Taxus brevifolia* and has also been found in other *Taxus* species. Paclitaxel was obtained from *T. brevifolia* bark in very low (0.07%) yield, and cumbersome purification from other related taxanes was required. It is estimated that about 20,000 pounds of yew bark (equivalent to about 3,000 trees) are needed to produce 1 kg of purified paclitaxel [57]. The develop-ment of a semisynthetic process for the production of paclitaxel from baccatin III **34** (paclitaxel without the C-13 side chain) or 10-deacetylbaccatin III **35** (10-DAB; paclitaxel without the C-13 side chain and the C-10 acetate) and C-13 paclitaxel side chain (2R,3S)-**37** or acetate®-**39** was a very promising approach (Figure 16.8).

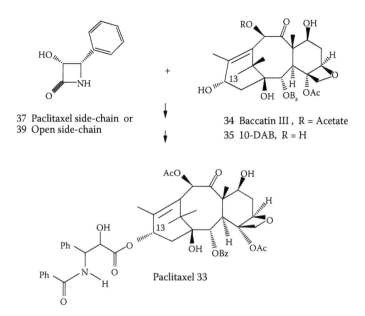

37 Paclitaxel side-chain or
39 Open side-chain

34 Baccatin III , R = Acetate
35 10-DAB, R = H

Paclitaxel 33

FIGURE 16.8   Anticancer drug paclitaxel **33**: Paclitaxel semisynthetic process.

Taxanes, baccatin III, and 10-DAB can be derived from renewable resources such as the needles, shoots, and young *Taxus* cultivars [58]. Thus, preparation of paclitaxel by a semisynthetic process would eliminate the cutting of yew trees.

Using selective enrichment techniques, two strains of *Nocardioides* were isolated from soil samples that contained the novel enzymes C-13 taxolase and C-10 deacetylase [59,60]. The extracellular C-13 taxolase, derived from the filtrate of the fermentation broth of *Nocardioides albus* SC 13911, catalyzed the cleavage of the C-13 side chain from paclitaxel and related taxanes such as Taxol C, cephalomannine, 7-β-xylosyltaxol, 7-β-xylosyl-10-deacetyltaxol, and 10-deacetyltaxol (Figure 16.9A). The intracellular C-10 deacetylase derived from fermentation of *Nocardioides luteus* SC 13912 catalyzed the cleavage of the C-10 acetate from paclitaxel, related taxanes, and baccatin III to yield 10-DAB (Figure 16.9B) The C-7 xylosidase derived from fermentation of *Moraxella* sp. (Figure 16.9C) catalyzed the cleavage of the C-7 xylosyl group [61] from various taxanes.

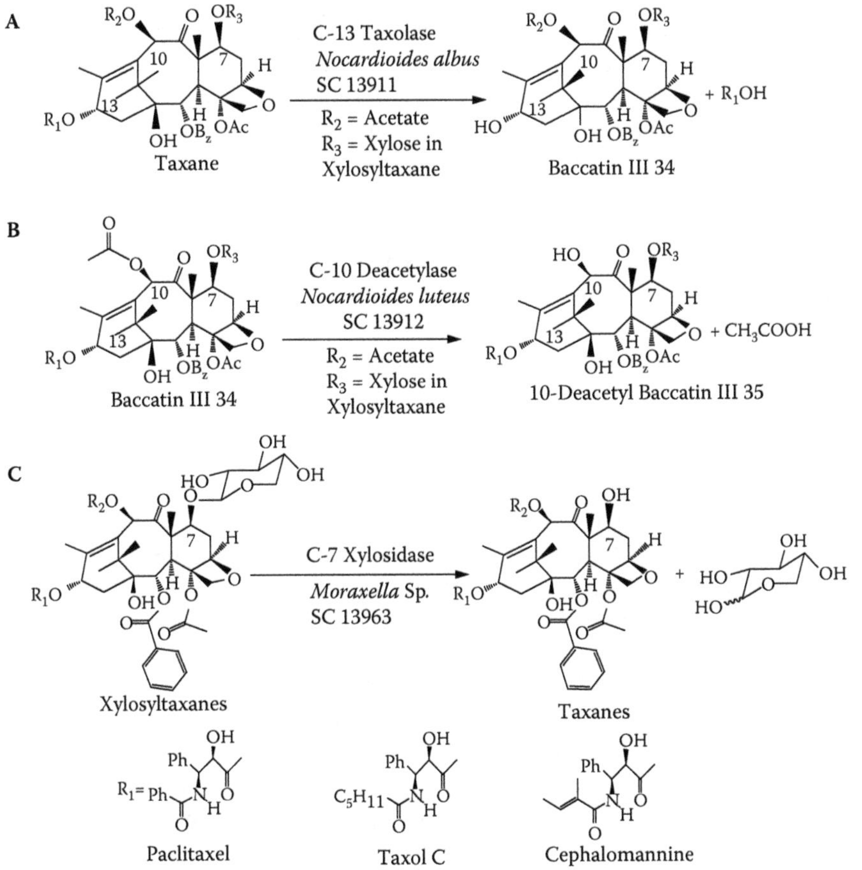

FIGURE 16.9 Anticancer drug paclitaxel **33**. (A) Enzymatic cleavage of the C-13 side chain from taxanes. (B) Enzymatic cleavage of the C-10 acetate from taxanes. (C) Enzymatic cleavage of the C-7 xylosyl groups from taxanes.

Fermentation processes were developed for growth of *Nocardioides albus* SC 13911 and *Nocardioides luteus* SC 13912 to produce C-13 taxolase and C-10 deacetylase, respectively, in 5,000-L batches, and a bioconversion process was demonstrated for the conversion of paclitaxel and related taxanes in extracts of *Taxus* cultivars to the single compound 10-DAB using both enzymes. In the bioconversion process, ethanolic extracts of the whole young plant of five different cultivars of *Taxus* were first treated with a crude preparation of the C-13 taxolase to give complete conversion of measured taxanes to baccatin III and 10-DAB in 6 h. *Nocardioides luteus* SC 13192 whole cells were then added to the reaction mixture to give complete conversion of baccatin III to 10-DAB. The concentration of 10-DAB was increased by 5.5 to 24 by treatment with the two enzymes. The bioconversion process was also applied to extracts of the bark of *T. bravifolia* to give a 12-fold increase in 10-DAB concentration. Enhancement of the 10-DAB concentration in yew extracts was useful in increasing the amount and ease of purification of this key precursor for the paclitaxel semisynthetic process using renewable resources.

Another key precursor for the paclitaxel semisynthetic process is the chiral C-13 paclitaxel side chain. Two different enantioselective enzymatic processes were developed for the preparation of the chiral C-13 paclitaxel side-chain synthon [62,63]. In one process, the enantioselective microbial reduction of 2-keto-3-(*N*-benzoylamino)-3-phenyl propionic acid ethyl ester **36** (Figure 16.10A) to yield (2*R*,3*S*)-*N*-benzoyl-3-phenyl isoserine ethyl ester **37** has been demonstrated using two strains of *Hansenula* [62]. Preparative-scale bioreduction of ketone **36** was demonstrated using cell suspensions of *Hansenula polymorpha* SC 13865 and *Hansenula fabianii* SC 13894 in independent experiments. In both batches, a reaction yield of more than 80% and EEs greater than 94% were obtained for (2*R*,3*S*)-**37**. In a single-stage bioreduction process, cells of *H. fabianii* were grown in a 15-L fermentor for 48 h, then the bioreduction process was initiated by addition of 30 g of substrate and 250 g of glucose and continued for 72 h. A reaction yield of 88% with an EE of 95% was obtained for (2*R*,3*S*)-**37**.

In an alternate process for the preparation of the C-13 paclitaxel side chain, the enantioselective enzymatic hydrolysis of racemic acetate *cis*-3-(acetyloxy)-4-phenyl-2-azetidinone **38** (Figure 16.10B), to the corresponding (*S*)-alcohol **39** and the unreacted desired (*R*)-acetate **38** was demonstrated [63] using lipase PS-30 from *Pseudomonas cepacia* (Amano International Enzyme Company) and BMS lipase (extracellular lipase derived from the fermentation of *Pseudomonas* sp. SC 13856). Reaction yields of more than 48% (theoretical maximum yield 50%) with EEs greater than 99.5% were obtained for the (*R*)-**38**. BMS lipase and lipase PS-30 were immobilized on Accurel polypropylene (PP), and the immobilized lipases were reused (10 cycles) without loss of enzyme activity, productivity, or the EE of the product (*R*)-**38**. The enzymatic process was scaled up to 250 L (2.5 kg substrate input) using immobilized BMS lipase and lipase PS-30. From each reaction batch, *R*-acetate **38** was isolated in 45 mol% yield (theoretical maximum yield 50%) and 99.5% EE. The (*R*)-acetate was chemically converted to (*R*)-alcohol **39**. The C-13 paclitaxel side-chain synthon (2*R*,3*S*-**37** or *R*-**39**) produced by either the reductive or resolution process could be coupled to bacattin III **34** after protection and deprotection to prepare paclitaxel by a semisynthetic process [64].

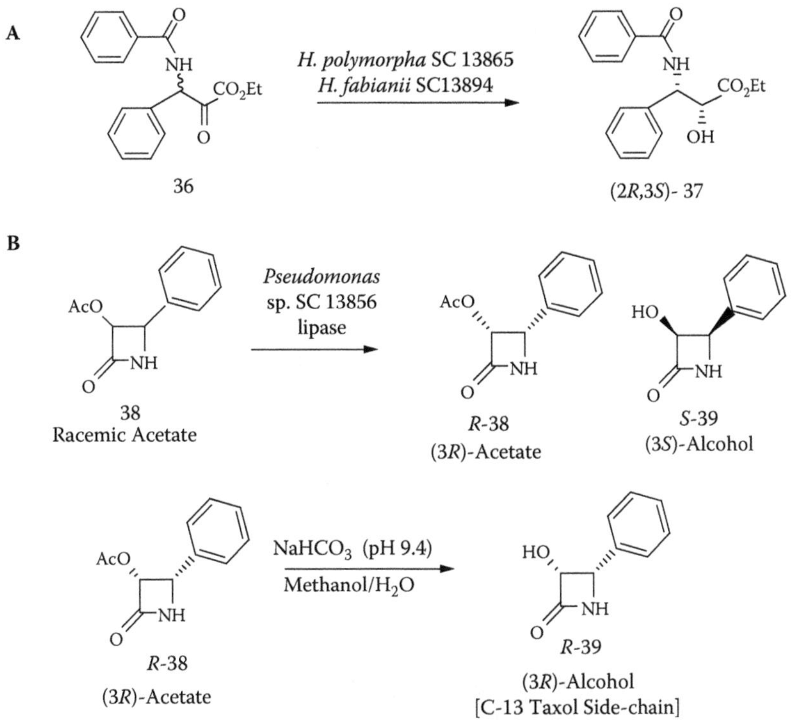

FIGURE 16.10 Preparation of the chiral C-13 paclitaxel side-chain synthon. (A) Enantio-selective microbial reduction of 2-keto-3-(*N*-benzoylamino)-3-phenyl propionic acid ethyl ester **36**. (B) Enantioselective enzymatic hydrolysis of *cis*-3-(acetyloxy)-4-phenyl-2-azetid-inone **38**.

16.7.2 Water-Soluble Taxane Derivatives

Due to the poor solubility of paclitaxel, various groups are involved in the development of water-soluble taxane analogs [65,66]. Taxane **40** (Figure 16.11A) is a water-soluble taxane derivative that, when given orally, was as effective as intravenous paclitaxel in five tumor models (murine M109 lung and C3H mammary 16/C cancer, human A2780 ovarian cancer cells [grown in mice and rats], and HCT/pk colon cancer) [67].

The chiral intermediate (3R-*cis*)-3-acetyloxy-4-(1,1-dimethylethyl)-2-azetidinone **41** (Figure 16.11B) was prepared for the semisynthesis of the new orally active taxane **40**. The enantioselective enzymatic hydrolysis of racemic *cis*-3-acetyloxy-4-(1,1-dimethylethyl)-2-azetidinone **41** to the corresponding undesired (*S*)-alcohol **42** and unreacted desired (*R*)-acetate **41** was carried out using immobilized lipase PS-30 (Amano) or BMS lipase (extracellular lipase derived from the fermentation of *Pseudomonas* sp. SC 13856). Reaction yields greater than 48% (theoretical maximum yield 50%) with EEs of more than 99% were obtained for the (*R*)-acetate **41**. Acetoxy β-lactam **41** was converted to *R*-hydroxy β-lactam **42** for use in the semisynthesis of **40** [68].

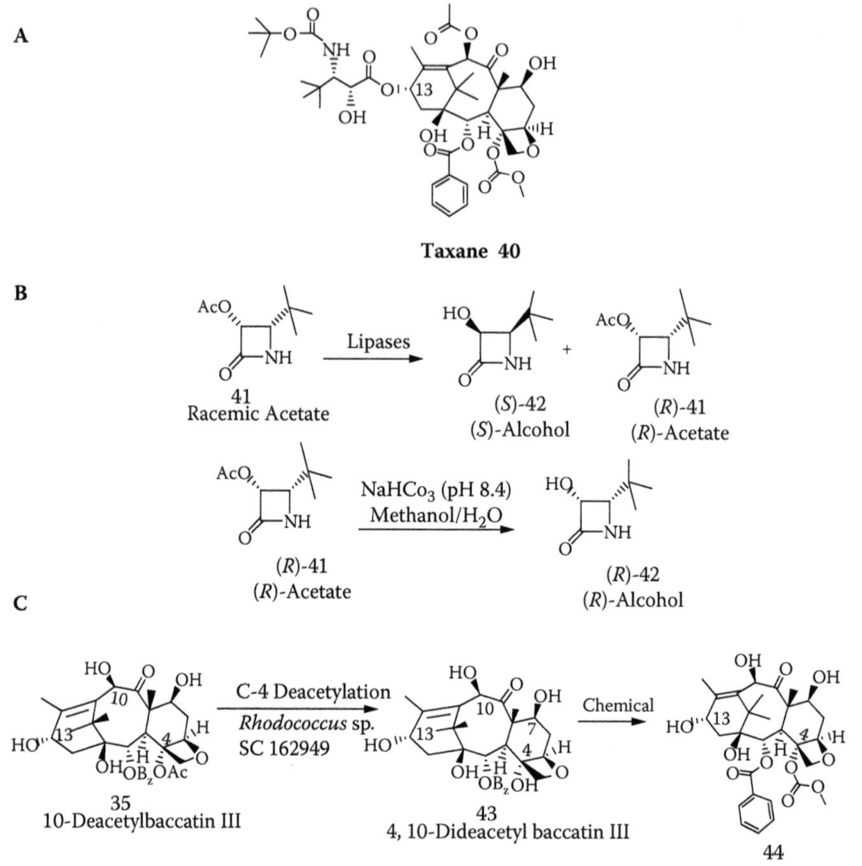

A

Taxane 40

B

41
Racemic Acetate

Lipases

(S)-42
(S)-Alcohol

(R)-41
(R)-Acetate

(R)-41
(R)-Acetate

NaHCo₃ (pH 8.4)
Methanol/H₂O

(R)-42
(R)-Alcohol

C

35
10-Deacetylbaccatin III

C-4 Deacetylation
Rhodococcus sp.
SC 162949

43
4, 10-Dideacetyl baccatin III

Chemical

44

FIGURE 16.11   (A) Water-soluble taxane **40**. (B) Enantioselective enzymatic hydrolysis of
cis-3-acetyloxy-4-(1,1-dimethylethyl)-2-azetidinone **41**. (C) Enzymatic C-4 deacetylation of
10-deactylbaccatin III **35**.

The synthesis of oral taxane **40** also required 4,10-dideacetyl baccatin **43**
(Figure 16.11C) as starting material for the chemical synthesis of the C-4 methylcarbonate derivative of 10-deacetyl baccatin III **44**. A microbial process was developed for deacetylation of 10-deacetylbaccatin III **35** to 4,10-dideacetylbaccatin III
43 using a *Rhodococcus* sp. SC 162949 isolated from soil using culture enrichment
techniques [69]. Chemically, **43** is converted to **44**, a C-4 methylcarbonate derivative
10 deactylbacattin III.

16.8   EPOTHILONES

16.8.1   MICROBIAL HYDROXYLATION OF EPOTHILONE B TO EPOTHILONE F

The clinical success of paclitaxel has stimulated research into compounds with
similar modes of activity in an effort to emulate its antineoplastic efficacy while
minimizing its less-desirable aspects, which include nonsolubility in water, difficult

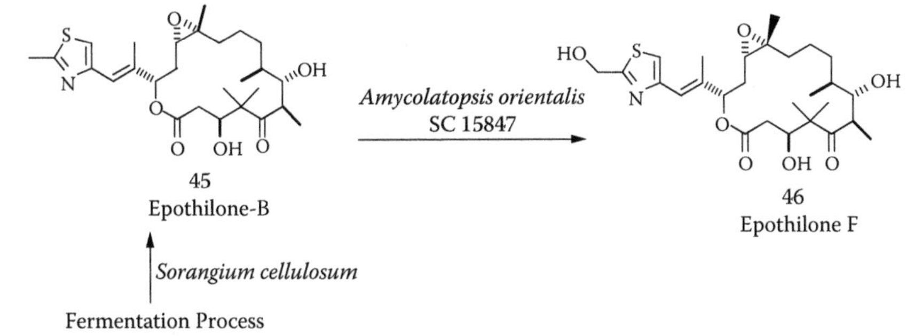

FIGURE 16.12 Microbial hydroxylation of epothilone B to epothilone F.

synthesis, and emerging resistance. The epothilones are a novel class of natural product cytotoxic compounds derived from the fermentation of the myxobacterium *Sorangium cellulosum* that are nontaxane microtubule-stabilizing compounds that trigger apoptosis [70,71]. The natural product epothilone B **45** (Figure 16.12) has demonstrated broad-spectrum antitumor activity in vitro and in vivo, including tumors with paclitaxel resistance [72]. The role of **45** as a potential paclitaxel successor has initiated interest in its synthesis, resulting in several total syntheses of **45** and various derivatives thereof [73,74]. The epothilone analogs were synthesized in an effort to optimize the water solubility, in vivo metabolic stability, and antitumor efficacy of this class of antineoplasic agents [75,76].

A fermentation process was developed for the production of epothilone B, and the titer of epothilone B was optimized and increased by a continuous feed of sodium propionate during fermentation. The inclusion of XAD-16 resin during fermentation to adsorb epothilone B and to carry out volume reduction made the recovery of product very simple [77]. A microbial hydroxylation process was developed for conversion of epothilone B **45** to epothilone F **46** by *Amycolatopsis orientalis* SC 15847. A bioconversion yield of 37%–47% was obtained when the process was scaled up to 100–250 L (R. N. Patel et al., unpublished results). Recently, the epothilone B hydroxylase along with the ferredoxin gene has been cloned and expressed in *Streptomyces rimosus* from *Amycolatopsis orientalis* SC 15847 by our colleagues at Bristol-Myers Squibb. Mutants and variants of this cloned enzyme have been used in the hydroxylation of epothilone B to epothilone F to obtain even higher yields of product [78].

16.9 INSULIN-LIKE GROWTH FACTOR 1 RECEPTOR INHIBITOR

16.9.1 ENZYMATIC PREPARATION OF (S)-2-CHLORO-1-(3-CHLOROPHENYL)ETHANOL

The synthesis of the leading candidate compound **47** in an anticancer program [79,80] required (S)-2-chloro-1-(3-chlorophenyl)ethanol **48** (Figure 16.13) as an intermediate. Other possible candidate compounds used analogs of the (S)-alcohol. From microbial screening of reduction of ketone **49** to the (S)-alcohol **48**, two cultures, namely, *Hansenula polymorpha* SC13824 (73.8% EE) and *Rhodococcus globerulus*

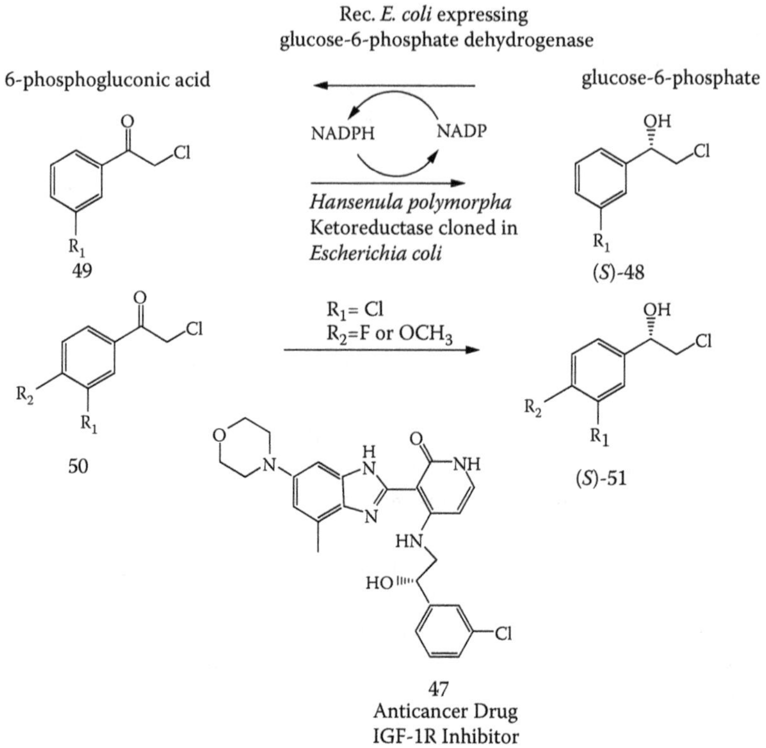

FIGURE 16.13 Anticancer drug **47**. Enzymatic preparation of (*S*)-2-chloro-1-(3-chloro-phenyl)ethanol **48**. IGF, insulin-like growth factor.

SC SC16305 (71.8% EE), were identified that had the highest enantioselectivity. A ketoreductase from *Hansenula polymorpha*, after purification to homogeneity, gave (*S*)-alcohol **48** with 100% EE. Amino acid sequences from the purified enzyme were used to design PCR primers for cloning the ketoreductase. The ketoreductase was cloned and expressed in *E. coli* together with a glucose-6-phosphate dehydrogenase from *Saccharomyces cerevisiae* to allow regeneration of the NADPH required for the reduction process. An extract of *E. coli* containing the two recombinant enzymes was used to reduce 2-chloro-1-(3-chloro-4-fluorophenyl)-ethanone **49** and related ketones **50** to the corresponding (*S*)-alcohols **48** and **51**. Intact *E. coli* cells provided with glucose were used to prepare (*S*)-2-chloro-1-(3-chloro-4-fluorophenyl)ethanol **48** at an 89% yield with 100% EE [81].

16.10   RETINOIC ACID RECEPTOR AGONIST

16.10.1   ENZYMATIC PREPARATION OF 2-(*R*)-HYDROXY-2-(1′,2′,3′,4′-TETRAHYDRO-1′,1′,4′,4′-TETRAMETHYL-6′-NAPHTHALENYL)ACETATE

Retinoic acid and its natural and synthetic analogs (retinoids) exert a wide variety of biological effects by binding to or activating a specific receptor or sets of

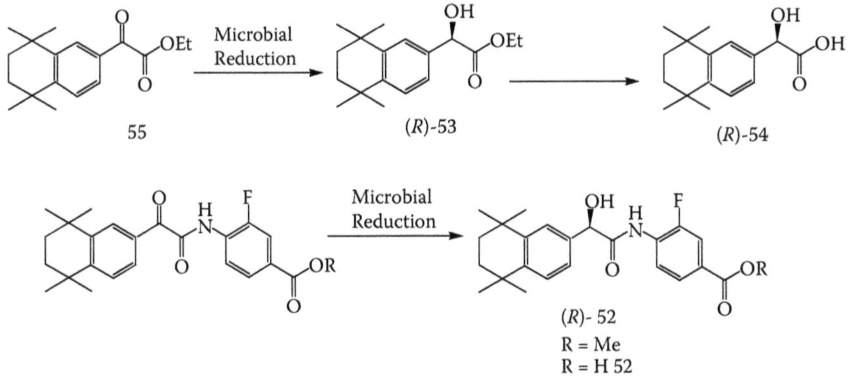

FIGURE 16.14 Retinoid acid receptor agonist. Enzymatic preparation of 2-(R)-hydroxy-2-(1′,2′,3′,4′-tetrahydro-1′,1′,4′,4′-tetramethyl-6′-naphthalenyl)acetate **54**.

receptors [82]. They have been shown to affect cellular growth and differentiation and are promising drugs for the treatment of cancers [83]. A few retinoids are already in clinical use for the treatment of dermatological diseases such as acne and psoriasis [84]. (R)-3-Fluoro-4-[[hydroxy-(5,6,7,8-tetrahydro-5,5,8,8-tetramethyl-2-naphthalenyl)-acetyl]amino]benzoic acid **52** (Figure 16.14) is a retinoic acid receptor gamma-specific agonist potentially useful as a dermatological and anticancer drug [85].

Ethyl-2-(R)-hydroxy-2-(1′,2′,3′,4′-tetrahydro-1′,1′,4′,4′-tetramethyl-6′-naphthalenyl) acetate **53** (Figure 16.14) and the corresponding acid **54** were prepared as intermediates in the synthesis of the retinoic acid receptor gamma-specific agonist [86]. Enantioselective microbial reduction of ethyl 2-oxo-2-(1′,2′,3′,4′-tetrahydro-1′,1′,4′,4′-tetramethyl-6-naphthalenyl) acetate **55** to alcohol **53** was carried out using *Aureobasidium pullulans* SC 13849 at a 98% yield and with an EE of 96%. At the end of the reaction, hydroxyester **53** was adsorbed onto XAD-16 resin and, after filtration, recovered in 94% yield from the resin with acetonitrile extraction. The recovered (R)-hydroxyester **53** was treated with Chirazyme L-2 or pig liver esterase to convert it to the corresponding (R)-hydroxyacid **54** in quantitative yield. The enantioselective microbial reduction of ketoamide **55** to the corresponding (R)-hydroxyamide **52** by *A. pullulans* SC 13849 has also been demonstrated [86].

16.11 TRYPTASE INHIBITORS

16.11.1 Enzymatic Preparation of (S)-N(tert-Butoxycarbonyl)-3-Hydroxymethyl Piperidine

(S)-N(tert-butoxycarbonyl)-3-hydroxymethylpiperidine **56** (Figure 16.15) is a key intermediate in the synthesis of a potent tryptase inhibitor **57** [87]. (S)-**56** was made by lipase PS-30 (*Pseudomonas cepacia*)-catalyzed resolution of R,S-N-(tert-butoxycarbonyl)-3-hydroxymethylpiperidine **56**. (S)-**56** was obtained at a 16% yield

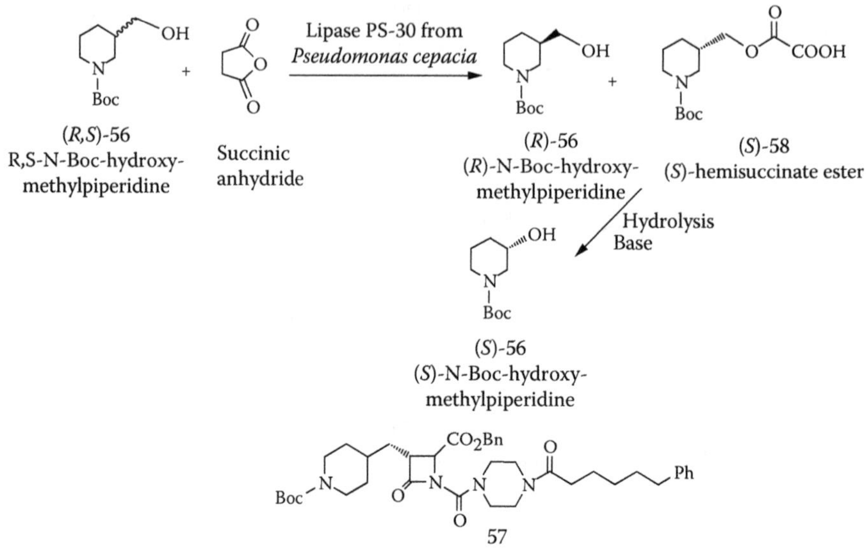

FIGURE 16.15   Tryptase inhibitor 57. Enzymatic preparation of (*S*)-*N*-(tert-butoxycarbonyl)-
3-hydroxymethyl piperidine **56**.

and more than 95% EE. Lipase PS also catalyzed esterification of the (*R*)-*S*-*N*-
(tert-butoxycarbonyl)-3-hydroxy methylpiperidine **56** with succinic anhydride
to yield (*R*)-*N*-(tert-butoxycarbonyl)-3-hydroxy methylpiperidine **56** and the
(*S*)-hemisuccinate ester **58**, which could be easily separated and hydrolyzed by base
to the (*S*)-**56**. The yield and EE could be improved greatly by repetition of the process. Using the repeated esterification procedure, (*S*)-**56** was obtained at a 32% yield
and 98.9% EE [88].

16.12   ANTICHOLESTEROL DRUGS: HYDROXYMETHYL
GLUTARYL CoA REDUCTASE INHIBITORS

16.12.1   PREPARATION OF (*S*)-4-CHLORO-3-HYDROXYBUTANOIC
ACID METHYL ESTER

(*S*)-4-Chloro-3-hydroxybutanoic acid methyl ester **59** (Figure 16.16) is a key chiral intermediate in the total chemical synthesis of **60**, a cholesterol antagonist that
acts by inhibiting hydroxymethyl glutaryl coenzyme A (HMG-CoA) reductase
[89,90]. The reduction of 4-chloro-3-oxobutanoic acid methyl ester **61** to (*S*)-**59** has
been demonstrated by cell suspensions of *Geotrichum candidum* SC 5469. In the
biotransformation process, a reaction yield of 95% and EE of 96% were obtained
for (*S*)-**59** at 10 g/L substrate input. The EE of (*S*)-**59** was increased to 98% by heat
treatment of cell suspensions (55°C for 30 min) prior to conducting the bioreduction
of **61** [91].

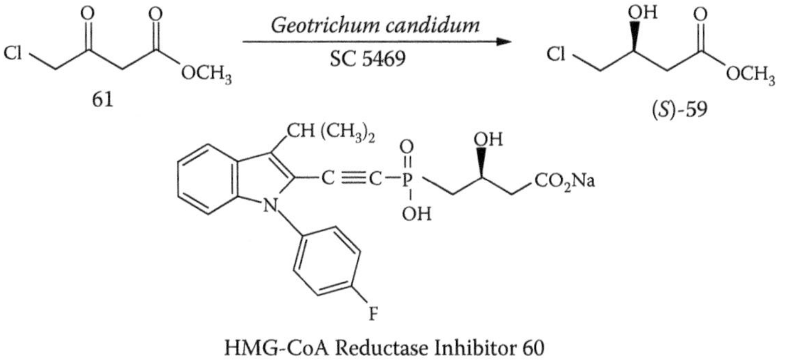

HMG-CoA Reductase Inhibitor 60

FIGURE 16.16 Anticholesterol drug **60**. Preparation of (*S*)-4-chloro-3-hydroxybutanoic acid methyl ester **59**.

16.12.2 ENZYMATIC REDUCTION OF 3,5-DIOXO-6-(BENZYLOXY) HEXANOIC ACID, ETHYL ESTER

The enantioselective reduction of a diketone 3,5-dioxo-6-(benzyloxy) hexanoic acid, ethyl ester **62** to (3*S*,5*R*)-dihydroxy-6-(benzyloxy) hexanoic acid, ethyl ester **63** has been demonstrated by cell suspensions of *Acinetobacter calcoaceticus* SC 13876 [92]. Compound **63** is a key chiral intermediate required for the chemical synthesis of [4-[4a, 6β(E)]]-6-[4,4-bis [4-fluorophenyl]-3-(1-methyl-1H-tetrazol-5-yl)-1,3-butadienyl]-tetrahydro-4-hydroxy-2H-pyren-2-one **64** (Figure 16.17), an anticholesterol drug that acts by inhibition of HMG-CoA reductase [93]. A reaction yield of 85% and an EE of 97% were obtained. Cell extracts of *A. calcoaceticus* SC 13876 in the presence of NAD⁺, glucose, and glucose dehydrogenase reduced **62** to the corresponding monohydroxy compounds [3-hydroxy-5-oxo-6-(benzyloxy) hexanoic acid ethyl ester **65** and 5-hydroxy-3-oxo-6-(benzyloxy) hexanoic acid ethyl ester **66**]. Both **65** and **66** were further reduced to the (3*S*,4*R*)-dihydroxy compound **63** in 92% yield and 99% EE by cell extracts. (3*S*,5*R*)-**63** was converted to **67**, a key chiral intermediate for the synthesis of **64**. Three different ketoreductases were purified to homogeneity from cell extracts, and their biochemical properties were compared. Reductase I only catalyzes the reduction of ethyl diketoester **62** to its monohydroxy products, whereas reductase II catalyzes the formation of dihydroxy products from monohydroxy substrates. A third reductase (III) was identified that catalyzes the reduction of diketoester **62** to syn-(3*R*,5*S*)-dihydroxy ester **63** [94], which now has been cloned and expressed in *E. coli*.

16.13 ANTI-ALZHEIMER'S DRUGS

16.13.1 ENANTIOSELECTIVE ENZYMATIC REDUCTION OF 5-OXOHEXANOATE AND 5-OXOHEXANENITRILE

Ethyl-(*S*)-5-hydroxyhexanoate **68** and (*S*)-5-hydroxyhexanenitrile **69** (Figure 16.18) are key chiral intermediates in the synthesis of anti-Alzheimer's drugs **70** [95].

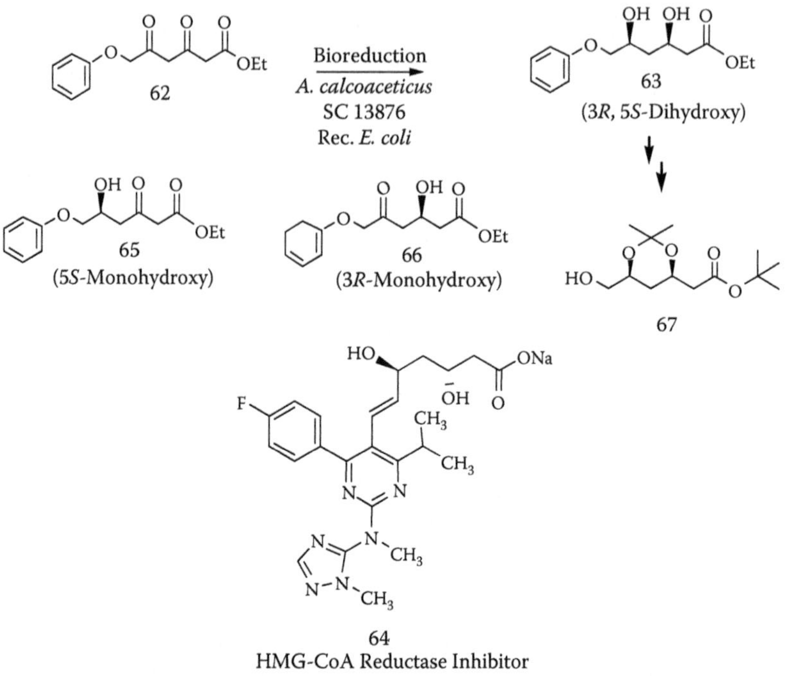

FIGURE 16.17 Anticholesterol drug **64**. Diastereoselective enzymatic reduction of 3,5-dioxo-6-(benzyloxy) hexanoic acid, ethyl ester **62**.

Both chiral compounds have been prepared by enantioselective reduction of ethyl-5-oxohexanoate **71** and 5-oxohexanenitrile **72** by *Pichia methanolica* SC 16116. Reaction yields of 80%–90% and more than 95% EEs were obtained for each chiral compound. In an alternate approach, the enzymatic resolution of racemic 5-hydroxy-hexane nitrile **73** by enzymatic succinylation was demonstrated using immobilized lipase PS-30 to obtain (*S*)-5-hydroxyhexanenitrile **69** in 35% yield (maximum yield is 50%). (*S*)-5-Acetoxy-hexanenitrile **74** was prepared by enantioselective enzymatic hydrolysis of racemic 5-acetoxyhexanenitrile **75** by *Candida antarctica* lipase. A reaction yield of 42% and an EE of more than 99% were obtained [96].

16.13.2 ENANTIOSELECTIVE MICROBIAL REDUCTION OF SUBSTITUTED ACETOPHENONE

The chiral intermediates (*S*)-1-(2′-bromo-4′-fluorophenyl)ethanol **76** and (*S*)-methyl 4-(2′-acetyl-5′-fluorophenyl)-butanol **77** are potential intermediates for the synthesis of several potential anti-Alzheimer's drugs, such as **78** [97,98]. The chiral intermediate (*S*)-1-(2′-bromo-4′-fluoro phenyl)ethanol **76** (Figure 16.19A) was prepared by the enantioselective microbial reduction of 2-bromo-4-fluoro acetophenone **79** [99]. Organisms from genuses *Candida*, *Hansenula*, *Pichia*, *Rhodotorula*, *Saccharomyces*, and *Sphingomonas* and baker's yeast reduced **79** to **76** in more than 90% yield and 99% EE.

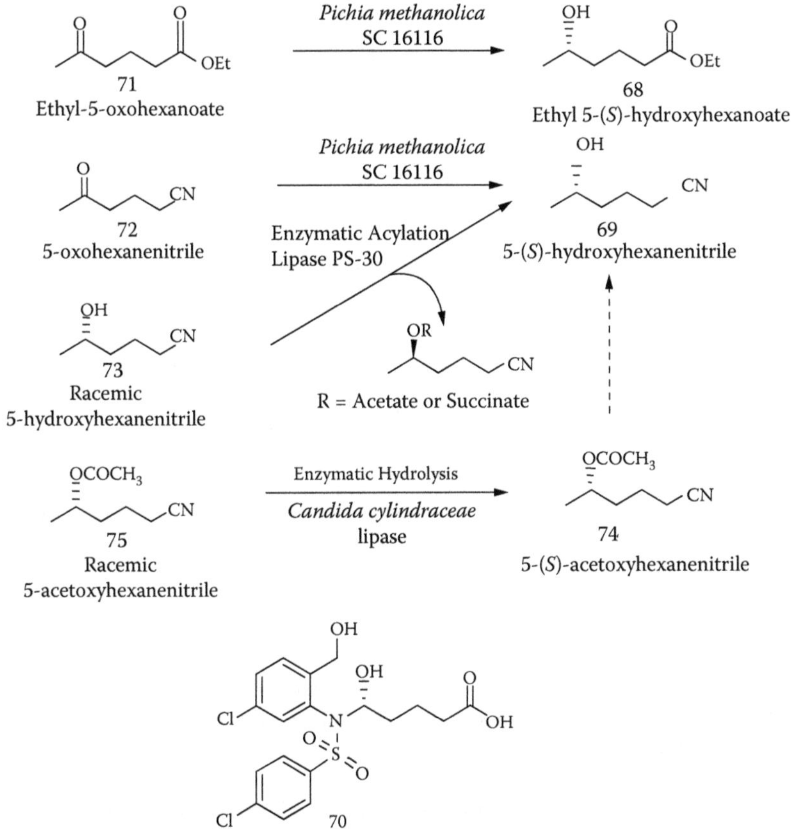

FIGURE 16.18 Anti-Alzheimer's drug **70**. Enantioselective enzymatic reduction of 5-oxo-hexanoate **71** and 5-oxohexanenitrile **72**.

In an alternate approach, the enantioselective microbial reduction of methyl-4-(2'-acetyl-5'-fluorophenyl) butanoates **80** (Figure 16.19B) was demonstrated using strains of *Candida* and *Pichia*. Reaction yields of 40%–53% and EEs of 90%–99% were obtained for the corresponding (*S*)-hydroxy esters **77**. The reductase that catalyzed the enantioselective reduction of ketoesters was purified to homogeneity from cell extracts of *Pichia methanolica* SC 13825. It was cloned and expressed in *E. coli*, and recombinant cultures were used for the enantioselective reduction of the keto-methyl ester **80** to the corresponding (*S*)-hydroxy methyl ester **77**. On a preparative scale, a reaction yield of 98% with an EE of 99% was obtained [99].

16.14 ANTI-INFECTIVE DRUGS

16.14.1 MICROBIAL HYDROXYLATION OF PLEUROMUTILIN OR MUTILIN

Pleuromutilin **81** (Figure 16.20) is an antibiotic from *Pleurotus* or *Clitopilus* basidiomycetes strains that kills mainly gram-positive bacteria and mycoplasms. A more

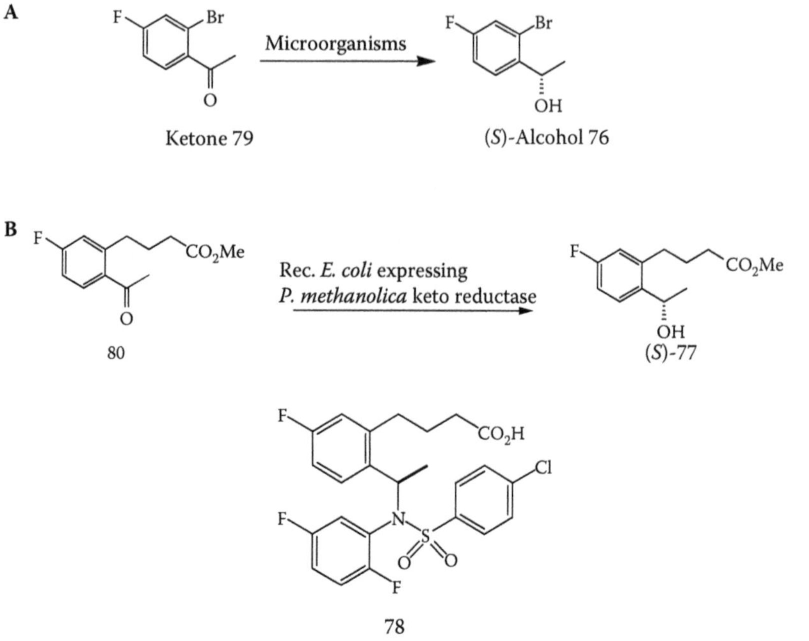

FIGURE 16.19 Anti-Alzheimer's drug **78**. (A) Enantioselective microbial reduction of substituted acetophenone **79**. (B) Enantioselective microbial reduction of methyl-4-(2′-acetyl-5′-fluorophenyl) butanoates **80**.

active semisynthetic analog, tiamulin, has been developed for the treatment of animals and poultry infection and has been shown to bind to prokaryotic ribosomes and inhibit protein synthesis [100]. Metabolism of pleuromutilin derivatives results in hydroxylation by microsomal cytochrome P-450 at the 2- or 8-position and inactivates the antibiotics [101]. Modification of the 8-position of pleuromutilin and analogs is of interest as a means of preventing the metabolic hydroxylation. Microbial hydroxylation of pleuromutilin **81** or mutilin **82** would provide a functional group at this position to allow further modification. The target analogs would maintain the biological activity of the parent compounds but not be susceptible to metabolic inactivation.

Biotransformation of mutilin and pleuromutilin by microbial cultures has been investigated to provide a source of 8-hydroxymutilin or 8-hydroxypleuromutilin [102]. *Streptomyces griseus* strains SC 1754 and SC 13971 (ATCC 13273) hydroxylated mutilin to (8*S*)-hydroxymutilin **83**, (7*S*)-hydroxymutilin **84**, and (2*S*)-hydroxymutilin **85**. *Cunninghamella echinulata* SC 16162 (NRRL 3655) gave (2*S*)-hydroxymutilin or (2*R*)-hydroxypleuromutilin **86** from biotransformation of mutilin or pleuromutilin, respectively. The biotransformation of mutilin by the *S. griseus* strain SC 1754 was scaled up in 15-, 60-, and 100-L fermentations to produce a total of 49 g of (8*S*)-hydroxymutilin, 17 g of (7*S*)-hydroxymutilin, and 13 g of (2*S*)-hydroxymutilin from 162 g of mutilin [102].

A C-8 ketopleuromutilin **87** derivative has been synthesized from the biotransformation product 8-hydroxymutilin [103]. A key step in the process was the selective

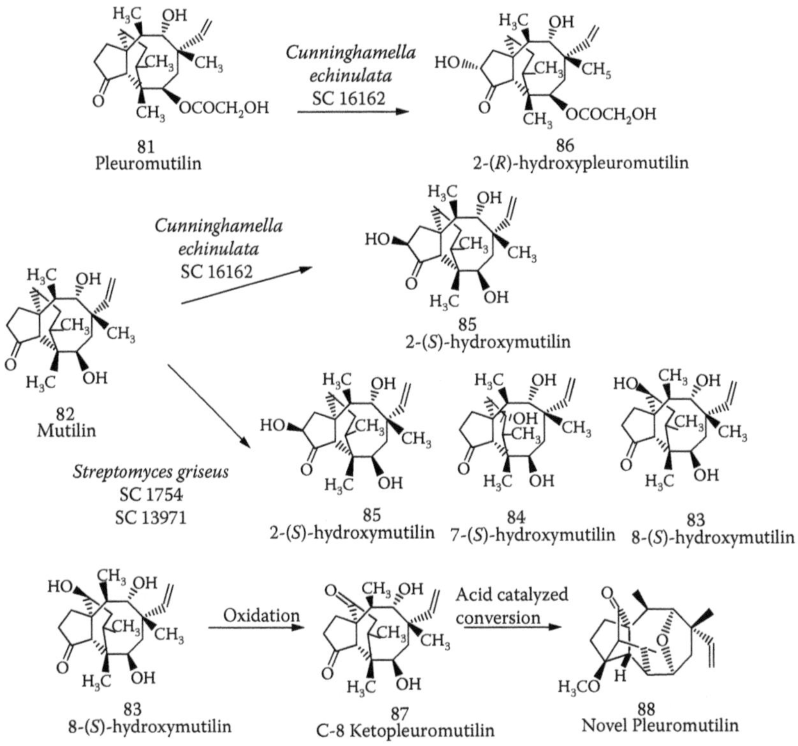

FIGURE 16.20 Anti-infective drugs. Microbial hydroxylation of pleuromutilin **81** or mutilin **82**.

oxidation at C-8 of 8-hydroxymutilin using tetrapropylammonium perruthenate. The presence of the C-8 keto group precipitated interesting intramolecular chemistry to afford **88** with a novel pleuromutilin-derived ring system by acid-catalyzed conversion of C-8 ketopleuromutilin.

16.14.2   ENZYMATIC SYNTHESIS OF L-β-HYDROXYVALINE

The asymmetric synthesis of β-hydroxy-α-amino acids by various methods has been demonstrated [104–106] because of their utility as starting materials for the total synthesis of monobactam antibiotics.

L-β-Hydroxyvaline **89** (Figure 16.21) is a key chiral intermediate required for the total synthesis of orally active monobactam [107], tigemonam **90**. The synthesis of L-β-hydroxyvaline **89** from α-keto-β-hydroxyisovalerate **91** by reductive amination using leucine dehydrogenase from *Bacillus sphaericus* ATCC 4525 has been demonstrated [107]. The NADH required for this reaction was regenerated by either formate dehydrogenase from *Candida boidinii* or glucose dehydrogenase from *Bacillus megaterium*. The required substrate **91** was generated either from α-keto-β-bromoisovalerate or its ethyl esters by hydrolysis with sodium hydroxide in situ.

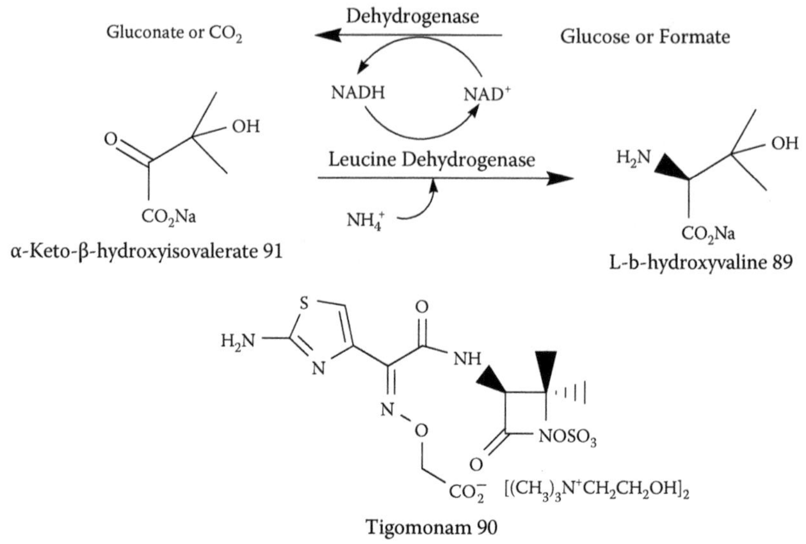

FIGURE 16.21 Anti-infective drug tigemonam **90**. Enzymatic synthesis of L-β-hydroxy-valine **89**.

In an alternate approach, the substrate was also generated from methyl-2-chloro-3,3-dimethyloxiran carboxylate and the corresponding isopropyl and 1,1-dimethyl ethyl ester by treatment with sodium bicarbonate and sodium hydroxide. In this process, an overall reaction yield of 98% and an EE of 99.8% were obtained for the L-β-hydroxyvaline **89**.

16.15 MELATONIN RECEPTOR AGONIST

16.15.1 ENANTIOSELECTIVE ENZYMATIC HYDROLYSIS OF RACEMIC 1-{2′,3′-DIHYDRO BENZO[B]FURAN-4′-YL}-1,2-OXIRANE

Epoxide hydrolase catalyzes the enantioselective hydrolysis of an epoxide to the corresponding enantiomerically enriched diol and unreacted epoxide [108,109]. The (S)-epoxide **92** (Figure 16.22A) is a key intermediate in the synthesis of a melatonin receptor agonist **93** [110]. The enantiospecific hydrolysis of the racemic 1-{2′, 3′-dihydro benzo[b]furan-4′-yl}-1,2-oxirane **92** to the corresponding (R)-diol **94** and unreacted (S)-epoxide **92** has been demonstrated [111]. Two A. niger strains (SC 16310, SC 16311) and Rhodotorula glutinis SC 16293 selectively hydrolyzed the (R)-epoxide, leaving behind the (S)-epoxide **92** in more than 95% EE and 45% yield (theoretical maximum yield is 50%). Several solvents at 10% v/v were evaluated in an attempt to improve the EE and yield and were found to have significant effects on both the extent of hydrolysis and the EE of unreacted (S)-epoxide **92**. Most solvents gave a lower EE of product and slower reaction rate than that of reactions without any solvent supplement, although MTBE gave a reaction yield of 45% and an EE of 99.9% for unreacted (S)-epoxide **92**.

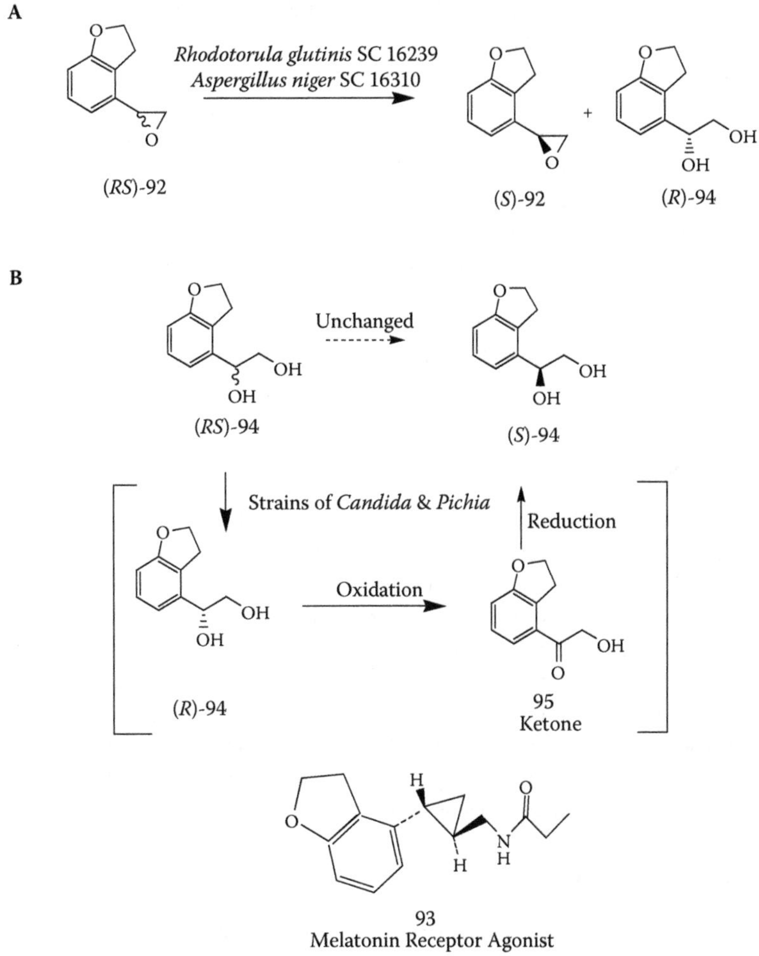

FIGURE 16.22 (A) Melatonin receptor agonist **93**. Enantioselective enzymatic hydrolysis of racemic 1-{2′,3′-dihydro benzo[b]furan-4′-yl}-1,2-oxirane **92**. (B) Dynamic kinetic resolution of (R,S)-1-{2′,3′-dihydrobenzo[b]furan-4′-yl}-ethane-1,2-diol **94**.

16.15.2 Biocatalytic Dynamic Kinetic Resolution of (R,S)-1-{2′,3′-Dihydrobenzo[b]furan-4′-yl}-ethane-1,2-diol

Most commonly used biocatalytic kinetic resolutions of racemates often provide compounds with high EE, but the maximum theoretical yield of product is only 50%. The reaction mixture contains about a 50:50 mixture of reactant and product that possess only slight differences in physical properties (e.g., a hydrophobic alcohol and its acetate), and thus separation may be very difficult. These issues with kinetic resolutions can be addressed by employing a "dynamic kinetic resolution" process involving a biocatalyst or biocatalyst with metal-catalyzed in situ racemization [112–114].

S-1-(2′,3′-Dihydrobenzo[b]furan-4′-yl)ethane-1,2-diol **94** (Figure 16.22B) is a potential precursor of (*S*)-epoxide **92** [115]. The dynamic kinetic resolution of the racemic diol **94** to the (*S*)-enantiomer **94** has been demonstrated [115]. Seven cultures (*Candida boidinii* SC 13821, SC 13822, SC 16115; *Pichia methanolica* SC 13825, SC 13860; and *Hansenula polymorpha* SC 13895, SC 13896) were found to be promising, providing (*S*)-diol **94** in 87%–100% EEs and 60%–75% yields. A new compound was formed during these biotransformations and was identified as the hydroxy ketone **95** by liquid chromatography-mass spectrometry (LC-MS). The area of the HPLC peak for hydroxy ketone **95** first increased with time, reached a maximum, and then decreased, as expected for the proposed dynamic kinetic resolution pathway. *Candida boidinii* SC 13822, *C. boidinii* SC 16115, and *P. methanolica* SC 13860 transformed the racemic diol **94** in 3–4 days to (*S*)-diol **94** in greater than 70% yield and 90%–100% EE.

16.16 CONCLUSION

The production of single enantiomers of drug intermediates is increasingly important in the pharmaceutical industry. Organic synthesis is one approach to the synthesis of single enantiomers, and biocatalysis provides an alternate opportunity to prepare pharmaceutically useful chiral compounds. The advantages of biocatalysis over chemical catalysis are that enzyme-catalyzed reactions are stereoselective and regioselective and can be carried out at ambient temperature and atmospheric pressure. In the course of the last decade, progress in biochemistry, protein chemistry, molecular cloning, random and site-directed mutagenesis, directed evolution of biocatalysts, and fermentation technology has opened up unlimited access to a variety of enzymes and microbial cultures as tools in organic synthesis.

ACKNOWLEDGMENTS

I would like to acknowledge Ronald Hanson, Animesh Goswami, Amit Banerjee, Venkata Nanduri, Jeffrey Howell, Steven Goldeberg, Robert Johnston, Mary-Jo Donovan, Dana Cazzulino, Thomas Tully, Thomas LaPorte, Lawrence Parker, John Wasylyk, Michael Montana, Ronald Eiring, Rapheal Ko, Linda Chu, Clyde McNamee, David Kronenthal, Michael Montana, and Richard Mueller for research collaboration and Robert Waltermire for reviewing the manuscript and making valuable suggestions.

REFERENCES

1. U.S. Food and Drug Administration. FDA's statement for the development of new stereoisomeric drugs. *Chirality* 4, 338, 1992.
2. Oliver M, Voigt CA, Arnold FH. Enzyme engineering by directed evolution. In: *Enzyme Catalysis in Organic Synthesis* Drauz, K. and Waldmann, H. (Eds.), Wiley-VCH, Gmbh, Weinheim. 1, 95, 2002.
3. Kazlauskas RJ. Enhancing catalytic promiscuity for biocatalysis. *Current Opinion in Chemical Biology* 9(2), 195, 2005.

4. Schmidt M, Baumann M, Henke E, et al. Directed evolution of lipases and esterases. *Methods in Enzymology (Protein Engineering)* 388, 199, 2004.

5. Reetz MT, Torre C, Eipper A, et al. Enhancing the enantioselectivity of an epoxide hydrolase by directed evolution. *Organic Letters* 6(2), 177, 2004.

6. Otey CR, Bandara G, Lalonde J, et al. Preparation of human metabolites of propranolol using laboratory-evolved bacterial cytochromes P450. *Biotechnology and Bioengineering* 93(3), 494, 2006.

7. Huisman GW, Lalonde JJ. Enzyme evolution for chemical process applications. In: *Biocatalysis in the Pharmaceutical and Biotechnology Industries*, Ed. Patel RN. CRC Press, Boca Raton, FL, 2007, p. 717.

8. Huisman GW, Gray D. Towards novel processes for the fine-chemical and pharmaceutical industries. *Current Opinion in Biotechnology* 13(4), 352, 2002.

9. Alphand V, Carrea G, Wohlgemuth R, et al. Towards large-scale synthetic applications of Baeyer-Villiger monooxygenases. *Trends in Biotechnology* 21(7), 318, 2003.

10. DiCosimo R. Nitrilases and nitrile hydratases. In: *Biocatalysis in the Pharmaceutical and Biotechnology Industries*, Ed. Patel RN. CRC Press, Boca Raton, FL, 2007, p. 1.

11. Patel RN. Synthesis of chiral pharmaceutical intermediates by biocatalysis. *Coordination Chemistry Reviews* 252(5–7), 659, 2009.

12. Simeo Y, Kroutil W, Faber K. Biocatalytic deracemization: dynamic resolution, stereoinversion, enantioconvergent processes, and cyclic deracemization. In: *Biocatalysis in the Pharmaceutical and Biotechnology Industries*, Ed. Patel RN. CRC Press, Boca Raton, FL, 2007, p. 27.

13. Simons C, Hanefeld U, Arends IWCE, et al. Towards catalytic cascade reactions: asymmetric synthesis using combined chemo-enzymatic catalysts. *Topics in Catalysis* 40(1–4), 35, 2006.

14. Turner NJ. Enzyme catalyzed deracemization and dynamic kinetic resolution reactions. *Current Opinion in Chemical Biology* 8(2), 114, 2004.

15. Robertson DE, Bornscheuer UT. Biocatalysis and biotransformation new technologies, enzymes and challenges. *Current Opinion in Chemical Biology* 9(2), 164, 2005.

16. Ishige T, Honda K, Shimizu S. Whole organism biocatalysis. *Current Opinion in Chemical Biology* 9(2), 174, 2005.

17. Stewart JD. Green chemical manufacturing with biocatalysis. In: *Environmental Catalysis*. Grassian, V.H. (Ed.), CRC Press, Boca Raton, FL, 2005, p. 649.

18. Fessner W-D, Jennewein S. Biotechnological applications of aldolases. In: *Biocatalysis in the Pharmaceutical and Biotechnology Industries*, Ed. Patel RN. CRC Press, Boca Raton, FL, 2007, p. 363.

19. Boyd DR, Bugg TDH. Arene *cis*-dihydrodiol formation: from biology to application. *Organic and Biomolecular Chemistry* 4(2), 181, 2006.

20. Pollard DJ, Woodley JM. Biocatalysis for pharmaceutical intermediates: the future is now. *Trends in Biotechnology* 25(2), 66, 2007.

21. Pohl M, Liese A. Industrial processes using lyases for C–C, C–N, and C–C bond formation. In: *Biocatalysis in the Pharmaceutical and Biotechnology Industries*, Ed. Patel RN. CRC Press, Boca Raton, FL, 2007, p. 661.

22. Fulton B, Goa KL. Olanzapine. A review of its pharmacological properties and therapeutic efficacy in the management of schizophrenia and related psychoses. *CNS Drugs* 53(2), 281, 1997.

23. Heiser JF, Wilcox CS. Serotonin 5-HT1A receptor agonists as antidepressants: pharmacological rationale and evidence for efficacy. *CNS Drugs* 10(5), 343, 1998.

24. Jajoo HK, Mayol RF, LaBudde JA, et al. Metabolism of the antianxiety drug buspirone in human subjects. *Drug Metabolism and Disposition* 17(6), 634, 1989.

25. Mayol RF. Buspirone metabolite for the alleviation of anxiety. U.S. Patent 6150365, 2000.

26. Yevich JP, New JS, Lobeck WG, et al. Synthesis and biological characterization of α-(4-fluorophenyl)-4-(5-fluoro-2-pyrimidinyl)-1-piperazinebutanol and analogs as potential atypical antipsychotic agents. *Journal of Medicinal Chemistry* 35(24), 4516, 1992.

27. Yevich JP, Mayol RF, Li J, et al. (*S*)-6-Hydroxy-buspirone for treatment of anxiety, depression and related disorders. U.S. Patent 2003022899, 2003.

28. Hanson RL, Parker WL, Brzozowski DB, et al. Preparation of (*R*)- and (*S*)-6-hydroxy-buspirone by enzymatic resolution or hydroxylation. *Tetrahedron: Asymmetry* 16(16), 2711, 2005.

29. Patel RN, Chu L, Nanduri V, et al. Enantioselective microbial reduction of 6-oxo-8-[4-[4-(2-pyrimidinyl)-1-piperazinyl]butyl]-8-azaspiro[4.5]decane-7,9-dione. *Tetrahedron: Asymmetry* 16(16), 2778, 2005.

30. Goldberg SL, Nanduri VB, Chu L, et al. Enantioselective microbial reduction of 6-oxo-8-[4-[4-(2-pyrimidinyl)-1-piperazinyl]butyl]-8-azaspiro[4.5]decane-7,9-dione: cloning and expression of reductases. *Enzyme and Microbial Technology* 39(7), 1441, 2006.

31. Gallwitz B. DDP IV inhibitors for the treatment of type 2 diabetes. *Diabetes Frontier* 18(6), 636, 2207.

32. Nielsen LL. Incretin mimetics and DPP-IV inhibitors for the treatment of type 2 diabetes. *Drug Discovery Today* 10(10), 703, 2005.

33. Sinclair EM, Drucker DJ. Glucagon-like peptide 1 receptor agonists and dipeptidyl peptidase IV inhibitors: new therapeutic agents for the treatment of type 2 diabetes. *Current Opinion in Endocrinology and Diabetes* 12(2), 146, 2005.

34. Weber AE, Thornberry N. Case history: JANUVIA (sitagliptin), a selective dipeptidyl peptidase IV inhibitor for the treatment of type 2 diabetes. *Annual Reports in Medicinal Chemistry* 42, 95, 2007.

35. Augeri DJ, Robl JA, Betebenner DA, Magnin DR, et al. Discovery and preclinical profile of saxagliptin (BMS-477118): a highly potent, long-acting, orally active dipeptidyl peptidase IV inhibitor for the treatment of type 2 diabetes. *Journal of Medicinal Chemistry* 48, 5025, 2005.

36. Hanson RL, Singh J, Kissick TP, et al. Synthesis of L-β-hydroxyvaline from α-keto-β-hydroxyisovalerate using leucine dehydrogenase from *Bacillus* species. *Bioorganic Chemistry* 18, 116, 1990.

37. Hanson RL, Schwinden MD, Banerjee A, et al. Enzymatic synthesis of L-6-hydroxynorleucine. *Bioorganic and Medicinal Chemistry* 7, 2247, 1999.

38. Hanson RL, Howell JM, LaPorte TL, et al. Synthesis of allyllysine ethylene acetal using phenylalanine dehydrogenase from *Thermoactinomyces intermedius*. *Enzyme and Microbial Technology* 26, 348, 2000.

39. Vu TC, Brzozowski DB, Fox R, et al. Preparation of cyclopropyl-fused pyrrolidine-based inhibitors of dipeptidyl peptidase IV. Bristol-Myers Squibb Company, *PCT Int. Appl.*, WO 2004/052850 A2, 2004.

40. Hanson RL, Goldberg SL, Brzozowski DB, et al. Preparation of an amino acid intermediate for the dipeptidyl peptidase IV inhibitor, saxagliptin, using a modified phenylalanine dehydrogenase. *Advanced Synthesis and Catalysis* 349(8+9), 1369, 2007.

41. Kunishima M, Kawachi C, Hioki K, et al. Formation of carboxamides by direct condensation of carboxylic acids and amines in alcohols using a new alcohol- and water-soluble condensing agent: DMT-MM. *Tetrahedron* 57(8), 1551, 2001.

42. Gill I, Patel RN. Biocatalytic ammonolysis of (5S)-4,5-dihydro-1H-pyrrole-1,5-dicarboxylic acid, 1-(1,1-dimethylethyl)-5-ethyl ester: preparation of an intermediate to the dipeptidyl peptidase IV inhibitor saxagliptin. *Bioorganic and Medicinal Chemistry Letters* 16(3), 705, 2006.

43. Bold G, Faessler A, Capraro H-G, et al. New aza-dipeptide analogs as potent and orally absorbed HIV-1 protease inhibitors: candidates for clinical development. *Journal of Medicinal Chemistry* 41(8), 3387, 1998.

44. Robinson BS, Riccardi KA, Gong YF, et al. BMS-232632, a highly potent human immunodeficiency virus protease inhibitor that can be used in combination with other available antiretroviral agents. *Antimicrobial Agents and Chemotherapy* 44(8), 2093, 2000.

45. Patel RN, Chu L, Mueller RH. Diastereoselective microbial reduction of (S)-[3-chloro-2-oxo-1-(phenylmethyl)propyl]carbamic acid, 1,1-dimethylethyl ester. *Tetrahedron: Asymmetry* 14(20), 3105, 2003.

46. Xu Z, Singh J, Schwinden MD, et al. Process research and development for an efficient synthesis of the HIV protease inhibitor BMS-232632. *Organic Process Research and Development* 6(3), 323, 2002.

47. Hanson R, Goldberg S, Patel RN. unpublished results.

48. Ireland C, Leeson PA, Castaner J. Lobucavir: antiviral. *Drugs Future* 22, 359, 1997.

49. Hanson R, Shi Z, Brzozowski D, et al. Regioselective enzymatic aminoacylation of lobucavir to give an intermediate for lobucavir prodrug. *Bioorganic and Medicinal Chemistry* 8(12), 2681, 2000.

50. Innaimo SF, Seifer M, Bisacchi GS, et al. Identification of BMS-200475 as a potent and selective inhibitor of hepatitis B virus. *Antimicrobial Agents and Chemotherapy*, 41, 1444, 1997.

51. Genovesi EV, Lamb L, Medina I, et al. Efficacy of the carbocyclic 2′-deoxyguanosine nucleoside BMS-200475 in the woodchuck model of hepatitis B virus infection. *Journal of Antimicrobial Agents and Chemotherapy* 43(3), 726, 1999.

52. Pendri YR, Chen C-P, Patel S, et al. Process for preparing the antiviral agent entecavir. PCT Int. Appl., WO 2004052310 A2 20040624 CAN 141:38813 AN, 2004.

53. Griffith DA, Danishefsky SJ. The total synthesis of allosamidin. Expansions of the methodology of aza-glycosidation pursuant to the total synthesis of allosamidin. A surprising enantiotopic sense for a lipase-induced deacetylation. *Journal of the American Chemical Society* 118, 9526, 1996.

54. Danishefsky SJ, Cabal MP, Chow K. Novel stereospecific silyl group transfer reactions: practical routes to the prostaglandins. *Journal of the American Chemical Society* 111, 3456, 1989.

55. Patel RN, Banerjee A, Pendri YR, et al. Preparation of a chiral synthon for an HBV inhibitor: enzymatic asymmetric hydrolysis of (1α,2β,3α)-2-(benzyloxymethyl)cyclopent-4-ene-1,3-diol diacetate and enzymatic asymmetric acetylation of (1α,2β,3α)-2-(benzyloxymethyl)cyclopent-4-ene-1,3-diol. *Tetrahedron: Asymmetry* 17(2), 175, 2006.

56. Holton R, Biediger R, Joatman P. Semisynthesis of Taxol and Taxotere. In: *Taxol: Science and Application*, Ed. Suffness M. CRC Press, New York, 1995, p. 97.

57. Kingston D. Natural taxoids: Structure and chemistry. In: *Taxol: Science and Application*, Ed. Suffness M. CRC Press, New York, 1995, p. 287.

58. Suffness M, Wall ME. Discovery and development of taxol. In: *Taxol: Science and Application,* Ed. Suffness M. CRC Press, New York, p. 3, 1995.

59. Hanson RL, Wasylyk JM, Nanduri VB, Cazzulino DL, Patel RN, Szarka LJ. Site-specific enzymatic hydrolysis of taxanes at C-10 and C-13. *Journal of Biological Chemistry*, 269, 22145, 1994.

60. Nanduri B, Hanson RL, LaPorte TL, et al. Fermentation and isolation of C-10 deacetylase for the production of 10-DAB from baccatin III. *Biotechnology and Bioengineering* 48, 547, 1995.

61. Hanson RL, Howell JM, Brzozowski DB, et al. Enzymic hydrolysis of 7-xylosyltaxanes by xylosidase from *Moraxella* sp. *Biotechnology and Applied Biochemistry* 26(3),153,1997.

62. Patel RN, Banerjee A, Howell JM, et al. Stereoselective microbial reduction of 2-keto-3-(*N*-benzoylamino)-3-pheny propionic acid ethyl ester. Synthesis of taxol side-chain synthon. *Tetrahedron: Asymmetry* 4, 2069, 1993.

63. Patel RN, Banerjee A, Ko RY, et al. Enzymic preparation of (3R-*cis*)-3-(acetyloxy)-4-phenyl-2-azetidinone: a taxol side-chain synthon. *Biotechnology and Applied Biochemistry* 20, 23, 1994.

64. Patel RN. Tour de paclitaxel: biocatalysis for semisynthesis. *Annual Review of Microbiology* 98, 361, 1995.

65. Baloglu E, Kingston D. A new semisynthesis of paclitaxel from baccatin III. *Journal of Natural Products* 62 (7), 1068, 1999.

66. Rose W, Long B, Fairchild C, et al. Preclinical pharmacology of BMS-275183, an orally active taxane. *Clinical Cancer Research* 7, 2016, 2001.

67. Mastalerz H, Cook D, Fairchild CR, et al. The discovery of BMS-275183: an orally efficacious novel taxane. *Bioorganic and Medicinal Chemistry* 11(2), 4315, 2003.

68. Patel RN, Howell JM, Chidambaram R, et al. Enzymatic preparation of (3R)-*cis*-3-acetyloxy-4-(1,1-dimethylethyl)-2-azetidinone: a side-chain synthon for an orally active taxane. *Tetrahedron: Asymmetry* 14, 3673, 2003.

69. Hanson RL, Parker WL, Patel RN. Enzymatic C-4 deacetylation of 10-deacetylbaccatin III. *Biotechnology and Applied Biochemistry*, 45(2), 81, 2006.

70. Gerth K, Pradella S, Perlova O, et al. Myxobacteria: proficient producers of novel natural products with various biological activities-past and future biotechnological aspects with the focus on the genus *Sorangium*. *Journal of Biotechnology* 106(2–3), 233, 2003.

71. Goodin S, Kane MP, Rubin EH. Epothilones: mechanism of action and biologic activity. *Journal of Clinical Oncology* 22(1), 2015, 2004.

72. Nicolaou KC, Roschangar F, Vourloumis D. Chemical biology of epothilones. *Angewandte Chemie, International Edition* 37(15), 2014, 1998.

73. Altmann K-H. The merger of natural product synthesis and medicinal chemistry: on the chemistry and chemical biology of epothilones. *Organic and Biomolecular Chemistry* 2(15), 2137, 2004.

74. Boddy CN, Hotta K, Tse ML, et al. Precursor-directed biosynthesis of epothilone in *Escherichia coli*. *Journal of the American Chemical Society* 126(24), 7436, 2004.

75. Lin N, Brakora K, Seiden M. BMS-247550 (Bristol-Myers Squibb/GBF). *Current Opinion in Investigational Drugs* 4(6), 746, 2003.

76. Kolman A. BMS—310705 Bristol-Myers Squibb/GBF. *Current Opinion in Investigational Drugs* 5(12), 1292, 2004.

77. Benigni D, Stankavage R, Chiang S-J, et al. Methods for the preparation, isolation and purification of epothilone B, and X-ray crystal structures of epothilone B. WO 2004026254, 2004.

78. Basch J, Chiang S-H. Cloning and expression of a cytochrome P450 hydroxylase gene from *Amycolatopsis orientalis*: hydroxylation of epothilone B for the production of epothilone F. *Journal of Industrial Microbiology and Biotechnology* 34(2), 171, 2007.

79. Wittman M, Carboni J, Attar R, et al. Discovery of a 1H-benzoimidazol-2-yl-1H-pyridin-2-one (BMS-536924) inhibitor of insulin-like growth factor I receptor kinase with in vivo antitumor activity. *Journal of Medicinal Chemistry* 48(18), 5639, 2005.

80. Wittman MD, Balasubramanian B, Stoffan K, et al. Novel 1H-(benzimidazol-2-yl)-1H-pyridin-2-one inhibitors of insulin-like growth factor I (IGF-1R) kinase. *Bioorganic and Medicinal Chemistry Letters* 17(4), 974, 2007.

81. Hanson RL, Goldberg S, Goswami A, et al. Purification and cloning of a ketoreductase used for the preparation of chiral alcohols. *Advanced Synthesis and Catalysis* 347(7+8), 1073, 2004.

82. Kagechika H, Kawachi E, Hashimoto Y, et al. Retinobenzoic acids. I. Structure-activity relationships of aromatic amides with retinoidal activity. *Journal of Medicinal Chemistry* 32(12), 2583, 1989.
83. Kagechika H, Shudo K. Retinoids. Vitamin A for clinical applications. *Farumashia* 26, 35, 1990.
84. Morriss-Kay GM. Retinoids: their physiological function and therapeutic potential. In: *Advances in Organ Biology*, Vol. 3, Ed. Bittar EE. JAI Press, Greenwich, CT, 1997, p. 79.
85. Moon RC, Mehta RG. Anticarcinogenic effects of retinoids in animals. *Advances in Experimental Medicine and Biology* 206, 399, 1986.
86. Patel RN, Chu L, Chidambaram R, et al. Enantioselective microbial reduction of 2-oxo-2-(1′,2′,3′,4′-tetrahydro-1′,1′,4′,4′-tetramethyl-6′-naphthalenyl)acetic acid and its ethyl ester. *Tetrahedron: Asymmetry* 13(4), 349, 2002.
87. Bisacchi G, Slusarchyk WA, Treuner U, et al. Preparation of amidino and guanidine azetidinone compounds as tryptase inhibitors. WO 9967215, 1999.
88. Goswami A, Howell JM, Hua EY, et al. Chemical and enzymatic resolution of (*R,S*)-N-(tert-butoxycarbonyl)-3-hydroxymethylpiperidine. *Organic Process Research and Development* 5(4), 415, 2001.
89. Jagoda E, Stouffer B, Ogan M, et al. Radioimmunoassay for hydroxyphosphinyl-3-hydroxybutanoic acid (SQ 33,600), a hypocholesterolemia agent. *Therapeutic Drug Monitoring* 15(3), 213, 1993.
90. Karanewsky DS, Badia MC, Ciosek CP, et al. Phosphorus-containing inhibitors of HMG-CoA reductase. I. 4-[(2-Arylethyl)hydroxyphosphinyl]-3-hydroxy-butanoic acids: a new class of cell-selective inhibitors of cholesterol biosynthesis. *Journal of Medicinal Chemistry* 33(11), 2952, 1990.
91. Patel RN, McNamee C, Banerjee A, et al. Stereoselective reduction of β-keto esters by *Geotrichum candidum*. *Enzyme and Microbial Technology* 14(9), 731, 1992.
92. Patel RN, Banerjee A, McNamee C, et al. Enantioselective microbial reduction of 3,5-dioxo-6-(benzyloxy)hexanoic acid, ethyl ester. *Enzyme and Microbial Technology* 15, 1014, 1993.
93. Sit S, Parker R, Motoc I, et al. Synthesis, biological profile, and quantitative structure-activity relationship of a series of novel 3-hydroxy-3-methylglutaryl coenzyme A reductase inhibitors. *Journal of Medicinal Chemistry* 33, 2982, 1990.
94. Guo Z, Chen Y, Goswami A, et al. Synthesis of ethyl and t-butyl (3R,5S)-dihydroxy-6-benzyloxyhexanoates via diastereo- and enantioselective microbial reduction. *Tetrahedron: Asymmetry*, 17(10), 1589, 2006.
95. Prasad CVC, Wallace OB, Noonan JW, et al. Hydroxytriamides as potent γ-secretase inhibitors. *Bioorganic and Medicinal Chemistry Letters*, 14(12), 3361, 2004.
96. Nanduri VB, Banerjee A, Howell JM, et al. Purification of a stereospecific 2-ketoreductase from *Gluconobacter oxydans*. *Journal of Industrial Microbiology and Biotechnology*, 25, 171, 2000.
97. Prasad CVC, Vig S, Smith DW, et al. 2,3-Benzodiazepin-1,4-diones as peptidomimetic inhibitors of γ-secretase. *Bioorganic and Medicinal Chemistry Letters* 14(13), 3535, 2004.
98. Schenk D, Games D, Seubert P. Potential treatment opportunities for Alzheimer's disease through inhibition of secretases and Aβ immunization. *Journal of Molecular Neuroscience* 17, 259, 2001.
99. Patel RN, Goswami A, Chu L, et al. Enantioselective microbial reduction of substituted acetophenones. *Tetrahedron: Asymmetry*, 15, 1247, 2004.
100. Hoegenauer G. Mechanism of action of antibacterial agents. Tiamulin and pleuromutilin. *Antibiotics* (New York) 5(1), 344, 1979.
101. Berner H, Vyplel H, Schulz G, et al. Chemistry of pleuromutilins. V. Photoisomerization of AB-trans-anellated mutilan-11-one. *Tetrahedron* 39(10), 1745, 1983.

102. Hanson RL, Matson JA, Brzozowski DB, et al. Hydroxylation of mutilin by *Streptomyces griseus* and *Cunninghamella echinulata*. *Organic Process Research and Development* 6(4), 482, 2002.
103. Springer DM, Sorenson ME, Huang S, et al. Synthesis and activity of a C-8 keto pleuromutilin derivative. *Bioorganic and Medicinal Chemistry Letters* 13(10), 1751, 2003.
104. Wandrey C, Wichmann R, Leuchtenberger W, et al. Continuous enzymic transformation of water-soluble α-keto carboxylic acids into the corresponding amino acids. Eur. Pat. Appl. 1981.
105. Brunhuber NM, Blanchard JS. The biochemistry and enzymology of amino acid dehydrogenases. *Critical Reviews in Biochemistry and Molecular Biology* 29(6), 415, 1994.
106. Bommarius AS. Reduction of C=N bonds. In: *Enzyme Catalysis in Organic Synthesis,* 2nd ed., Drauz, K. and Waldman, H. (Eds.), Wiley-VCH, Gmbh, Weinheim, Vol. 3, 2003, p. 1047.
107. Hanson RL, Singh J, Kissick TP, et al. Synthesis of L-β-hydroxyvaline from α-keto-β-hydroxyisovalerate using leucine dehydrogenase from *Bacillus* species. *Bioorganic Chemistry* 18(2), 116, 1990.
108. Archelas A, Furstoss R. Biocatalytic approaches for the synthesis of enantiopure epoxides. *Topics in Current Chemistry*, 200, 159. 1999.
109. Steinreiber A, Faber K. Microbial epoxide hydrolases for preparative biotransformations. *Current Opinion in Biotechnology* 12(6), 552, 2001.
110. Catt JD, Johnson G, Keavy DJ, et al. Preparation of benzofuran and dihydrobenzofuran melatonergic agents. United States, 1999. U.S. 5856529 A 19990105, CAN 130:110151 AN 1999:34508.
111. Goswami A, Totleben MJ, Singh AK, et al. Stereospecific enzymatic hydrolysis of racemic epoxide: a process for making chiral epoxide. *Tetrahedron: Asymmetry* 10(16), 3167, 1999.
112. Schnell B, Faber K, Kroutil W. Enzymatic racemization and its application to synthetic biotransformations. *Advanced Synthesis and Catalysis* 345(6+7), 653, 2003.
113. Stecher H, Faber K. Biocatalytic deracemization techniques. Dynamic resolutions and stereoinversions. *Synthesis* (1), 1, 1997.
114. Martin-Matute J, Baeckvall JE. Dynamic kinetic resolution catalyzed by enzymes and metals. *Current Opinion in Chemical Biology* 11(2), 226, 2007.
115. Goswami A, Mirfakhrae KD, Patel RN. Deracemization of racemic 1,2-diol by biocatalytic stereoinversion. *Tetrahedron: Asymmetry*, 10(21), 4239, 1999.

17 Induction of Phenolics and Terpenoids in Edible Plants Using Plant Stress Responses

Hyun-Jin Kim, Dae Young Kwon, and Suk Hoo Yoon

CONTENTS

Key Words: Carotenoid; chitosan; methyl jasmonate; phenolic compound; salinity; secondary metabolites; stress response; terpenoid.

17.1 INTRODUCTION

Epidemiological studies of the relationship between dietary habits and disease risk have shown that food has a direct impact on health [1]. In particular, it is generally accepted that plant-derived foods exert some beneficial effects on human health, such as anticancer, antioxidant, anti-inflammation, antidiabetic, and antihypertension effects [1–4]. With increasing knowledge of health beneficial effects of plant-derived foods, the consumption of plant-derived foods with potential health benefits has been continuously growing at a rate of 5%–10% per year [5], and several health organizations around the world have recommended an increase in the intake of plant-derived foods to improve our health status and to delay the development of chronic diseases [1]. The capacity of some plant-derived foods for reducing the risk of some diseases has been associated with plant secondary metabolites called phytochemicals having various functional properties [1,6]. However, it is true that these metabolites have low potency as bioactive compounds compared to pharmaceutical drugs, but since they are ingested regularly and in significant amounts as part of the diet, they may have a noticeable long-term physiological effect [1]. Due to a high interest in phytochemicals in plant-derived foods, there have been many investigations to increase their amounts in plants with various techniques, including gene modification [7,8]. Although gene modification is widely and powerfully used to induce the content of phytochemicals in plants, genetically modified crops face public concerns about safety [9]. Thus, alternative methods to induce secondary metabolites in plants have been suggested; among them, a tool using plant stress response is considered crucial.

Under biotic and abiotic stresses such as pathogen attack, physical wounding, and environmental changes, plants induce various defense responses to adapt and survive, and it was clearly elucidated that secondary metabolites that play important roles in plant defense responses can be stimulated by biotic and abiotic stresses [10–13].

In this review, we describe how to induce secondary metabolites with various biological functionalities for humans in edible plants using methyl jasmonate (MeJA), chitosan, and salt stress.

17.2   PLANT STRESS RESPONSES AND SECONDARY METABOLITES

Plants in the evolutionary process are capable of improving tolerance to adapt and survive in enormous biotic and abiotic environmental stresses, including insect attack, pathogen infection, physical damage, ultraviolet (UV) radiation, and changes of temperature, water capacity, and nutrients [14]. Under such stresses, plants induce networks of defense responses such as proteinase inhibitors, antifungal agents, and lignification; it is clear that most plant defense responses are associated with stress-induced secondary metabolites [10–14].

Unlike the primary metabolites, secondary metabolites have been viewed in the past as the waste products from mistakes of the primary metabolism because they are not involved in respiration, growth/development, and photosynthesis of plants. Now, it is known through much investigation that these compounds play important roles in the interaction of the cell (organism) with its environment, ensuring the survival of the organism in its ecosystem [15–17]. In addition, it has been found that many secondary metabolites may have various health-beneficial properties, such as antioxidant, anticancer, and anti-inflammatory activities.

Plant secondary metabolites biosynthesized from primary metabolites in different ways, including shikimate, phenylpropanoid, malonate, mevalonate, and methylerythriol phosphate (MEP) pathways (Figure 17.1), are generally classified into three chemically distinct groups: phenolic compounds, terpenoids, and nitrogen-containing compounds [18]. Some of them act as direct defenses toxic to potential herbivores, help attenuate the amount of light reaching the photosynthetic cells, and serve as precursors for the synthesis of lignin, suberin, and other polyphenolic barriers against pathogen attack. In addition to playing a role in plant defense responses, they have an enormous value from an economical point of view [15]. First, quite a few are used as specialty chemicals such as drugs, flavors, fragrances, insecticides, and dyes. Of all drugs used in Western medicine, about 25% are derived from plants, either as a pure compound or as derived from a natural synthon [15]. There is also an enormous

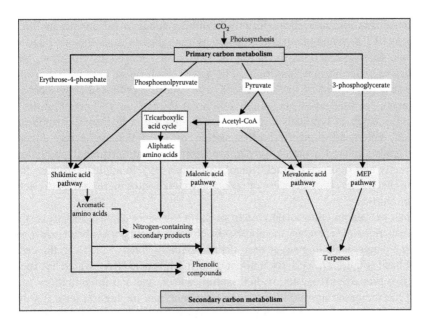

FIGURE 17.1 A simplified view of the major pathways of secondary metabolite biosynthesis and their interrelationship with primary metabolism. (From [18].)

potential value for new drug development. With a total world market for medicines of about US$250 billion per year, it is obvious that natural products from plants are a valuable commodity [15]. A much more difficult group to assess in terms of economical terms is that of the medicinal plants. It is estimated that about 80% of the world's population depends on traditional medicinal plants for their primary health care [15]. Traditional medicine has served as a lead for many important drugs, and probably the activity is in many cases due to a combination of secondary metabolites present in these plants [15]. In addition, secondary metabolites are important factors determining the appearance and nutritional quality of plant-derived foods and edible plants, and presently there are many interesting health-promoting effects of secondary metabolites in food because of their biological functional properties [1,6].

Secondary metabolites that show economical importance in many areas as well as play important roles in plant defense responses can also be altered in unstressed healthy plants by various modulators such as elicitors, intermediates of plant stress responses, and salt without negative changes in the appearance of plants.

17.3 MODULATORS OF SECONDARY METABOLITE PRODUCTION

17.3.1 CHITOSAN

Chitosan, an important structural component of several fungi cell walls, is a liner β-(1,4)-glucosamine polymer produced by deacetylation of chitin and was reported to be the most active ingredient contained within fungal cell walls [19]. It was convincingly demonstrated that chitosan had a potential dual role: inducing a defense response as well as directly inhibiting fungi in various plants, including pea [19,20]. When plants are attacked by fungal pathogens, they respond with multiple defense reactions, including activation of endogenous chitinase, which is inactive under normal conditions. The activated chitinase may be important in deconstructing chitin of the fungal cell wall, and the produced oligosaccharide fragments are used as stress signals (elicitors) to increase various plant defense responses, such as synthesis of callose and elicitation of hydrogen peroxide production, enhanced transcription/translation of plant defense genes encoding pathogenesis-related proteins, proteinase inhibitor, and antifungal phytoalexins to directly and indirectly attack fungal pathogens [19]. In addition, it was clearly shown that the levels of a variety of secondary metabolites increased by the activation of chitinase.

Thus, exogenous chitosan derived from chitin can be used as a biotic stress signal, and an increase of secondary metabolites by chitosan treatment has been investigated in some plants, including sweet basil [9] and *Lupinus luteus* L. [21]. In the case of sweet basil (Figure 17.2) [9], the amount of rosmarinic acid (RA), which has various bioactive properties (e.g., antioxidant, antimicrobial, and anti-inflammatory activities) [22], increased up to 0.1% chitosan treatment, and the highest level (2.0 mg/g fresh tissue) was 2.5 times higher than the control. This value was close to the accumulated amount of RA in *Coleus blumei* treated by MeJA and *Pythium* elicitor [23]. Essential oil was also accumulated in sweet basil by chitosan treatment, and at 0.5% chitosan treatment, the amounts of eugenol and L-linalool in sweet basil were twice as much as those in the control. It was similar to the reports that UV-B treatment or color mulching increased essential oil content in basil, but mechanical wounding did not induce essential oil content except for linalool [24–26].

17.3.2 METHYL JASMONATE

In plants, responses to many biotic and abiotic stresses are, at least in part, mediated by the jasmonate family, including jasmonate (JA) and its methyl ester MeJA, which are synthesized via the octadecanoid pathway, beginning with phospholipase A to release linolenic acid from chloroplast membrane [27,28]. They play a central role in regulating the biosynthesis of many secondary metabolites, and induction of plant secondary metabolite accumulation by the JA family is not limited to certain types of metabolites in most plants [29], as shown in Table 17.1.

The signaling property of the JA family is interesting not only for elucidation of plant defense response mechanisms, but also for practical applications in the production of economically useful compounds [30]. For example, the exogenous JA family significantly induces the production of the anticancer drug Taxol in cell suspension cultures of *Taxus* [31]. In particular, unlike other JA family members, MeJA can be released with various organic volatile compounds from stressed plants to the air to transfer a stress signal to healthy neighboring (receiver) plants, and the receiver plants induce defense systems without being subjected to any stress factor [32,33]. Due to this characteristic of MeJA, the application of exogenous MeJA to accumulate a variety of secondary metabolites in many plants, including sweet basil [27], tomato [30], apples [34], *Silybum marianum* [35], and raspberries [36], has been

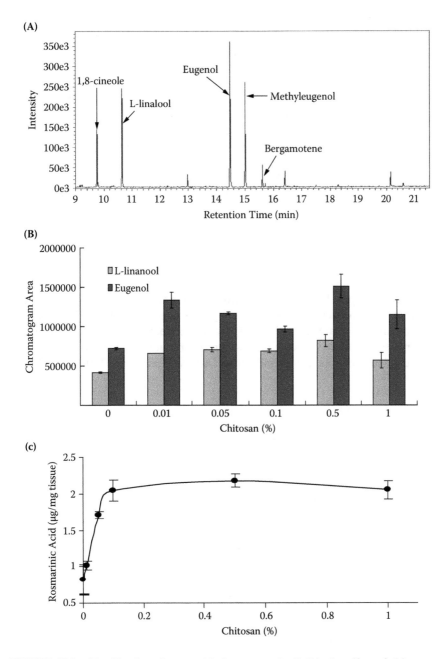

FIGURE 17.2 Identification of terpenoids from sweet basil (A); the effect of chitosan on terpenoids (B) and rosmarinic acid (C) in sweet basil. (From [9].)

reported. Among various secondary metabolites, phenolic compounds and terpenoids showing antioxdant activity have been mainly accumulated by MeJA application. For example, MeJA elicited the accumulation of RA, eugenol, and caffeic acid

TABLE 17.1

Methyl Jasmonate-Induced Secondary Metabolites in Various Plants and Plant Cell Suspensions

Plants	Elicitor	Secondary Metabolites	Reference
Cupressus lusitanica	MeJA	β-Thujaplicin	37
Glycyrrhiza glabra	MeJA	Soyasaponin 5-deoxyflavonoid	38
Taxus cell suspension	MeJA	Taxol, paclitaxel, baccatin III	39
Tomato	MeJA	Caffeoylputrescine	30
Silybum marianum	MeJA	Silymarin	35
Apple	MeJA	β-Carotene	34
Sweet basil	MeJA	Rosmarinic acid, eugenol, linalool, caffeic acid	27
Raspberries	MeJA	Flavon oids	36

in sweet basil (Figure 17.3) [27], phenylpropanoids in radish sprout [17], antioxidant flavonoids in raspberries [36], β-carotene in golden delicious apples [34], and caffeic acid derivative in romaine lettuce [32]. These results clearly indicated that MeJA can be a very useful tool for inducing plant secondary metabolites, but the possibility of the accumulation of toxic compounds in MeJA-treated plants was little studied.

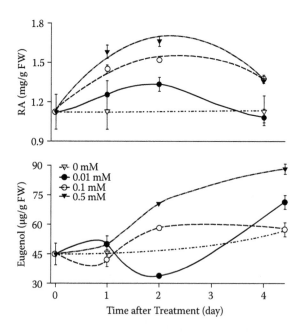

FIGURE 17.3 Time course of the effect of MeJA on rosmarinic acid (RA) and eugenol in sweet basil. (From [27].)

17.3.3 SALINITY STRESS

The salinity of soil and irrigation water is one of the major environmental factors decreasing crop yield because the leaf surface expansion, plant growth, and primary carbon metabolism of many plants are negatively affected by nutritional imbalance, osmotic stress, water deficit, and oxidative stress under high-salinity stress [40–43]. Moreover, although it is critically dependent on the salt sensitivity of plants, salinity-induced oxidative stress modulates the content of secondary metabolites. To prevent further water loss from the leaves in response to decreased soil water potential, salt-affected plants induce the production of abscisic acid, which is a plant hormone responding to water deficiency and is actually synthesized via the carotenoid biosynthesis pathway (Figure 17.4) [43]. Thus, during the abscisic acid biosynthesis many intermediate carotenoids can be accumulated by salt stress, and the induced carotenoids are of greater importance in the prevention of salt-induced oxidative damage and osmotic balance. The total amount of carotenoids in the romaine lettuce increased with irrigation containing NaCl, but at high-salt concentration, decrease of plant growth was observed. The carotenoid content in the 5 mM treated romaine

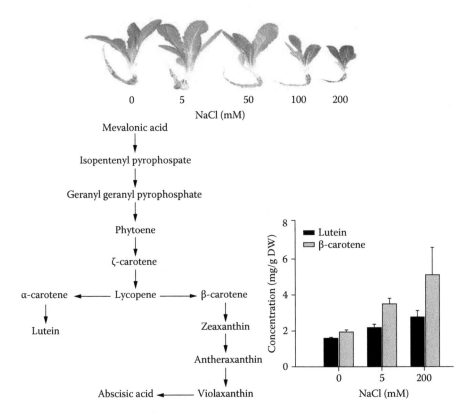

FIGURE 17.4 Growth change of salt-treated romaine lettuce (top), carotenoid biosynthesis pathway (left), and the contents of lutein and β-carotene in salt-treated romaine lettuce (right). (From [43].)

lettuce (13.9 mg CE/g DW) without any appearance change was 70% higher than that of the untreated control (8.3 mg CE/g DW) (Figure 17.4). Among various carotenoids, the contents of lutein (known as an eye health promoter, anticolon cancer agent, and antioxidant [44,45]) and β-carotene (known as an important antioxidant [46]) increased by 37% and 80%, respectively, whereas phenolic content in lettuces was not changed by salinity stress. However, the opposite result has been reported: The carotenoid content of salt-sensitive sugarcane decreased with increased salt concentration, whereas the content of phenolic compounds increased [33].

17.4 FUTURE STUDIES

Since it was pharmacologically and physiologically demonstrated that many plant secondary metabolites (called phytochemicals) had various health-functional properties, many tools (e.g., gene modification technique, cell culture technique, and breeding) to stimulate them in plants have been introduced, but studies on enhancing phytochemical contents in plant-derived foods or whole edible plants including vegetables, herbal plants, and medicinal plants are few. In this chapter, we introduced just three tools (i.e., MeJA, chitosan, and salinity) to increase phytochemicals in edible plants without appearance change, and there is no place for doubt that they can be useful tools for improving the nutritional quality of plant-derived foods or whole edible plants. Before they are actually applied to food and agricultural industries, however, the safety of edible plants subjected to plant stress responses will be guaranteed, and postharvesting or food-processing techniques to maintain the amounts of phytochemicals induced by stress responses in edible plants and processed products from farm to table must be developed.

REFERENCES

1. Espín, J.C., García-Conesa, M.T., and Tomás-Barberán, F.A. Nutraceuticals: facts and fiction. *Phytochemistry* 68, 2986–3008, 2007.
2. Jun, M., Jeong, W.-S., and Ho, C.-T. Health promoting properties of natural flavor substances. *Food Sci. Biotechnol.* 15(3), 329–338, 2006.
3. Yeo, E.-J., Kim, K.-T., Han, Y.S., Nah, S.-Y., and Paik, H.-D. Antimicrobial, anti-inflammatory, and anti-oxidative activities of *Scilla scilloides* (Lindl.) Druce root extract. *Food Sci. Biotechnol.* 15(4), 639–642, 2006.
4. Block, G., Patterson, B., and Subar, A. Fruit, vegetables, and cancer prevention: a review of the epidemiological evidence. *Nutr. Cancer* 18, 1–29, 1992.
5. Foley, C.M., and Kratz, A.M. Resources and guidelines on buying and using nutraceuticals. In: Roberts, A.J., O'Brien, M.E., and Subak-Sharpe, G., Eds., *Nutraceuticals—The Complete Encyclopedia of Supplements, Herbs, Vitamins, and Healing Foods*, Perigree, New York, 2000, p. 635.
6. Hounsome, N., Hounsome, B., Tomos, D., and Edwards-Jones, G. Plant metabolites and nutritional quality of vegetables. *J. Food Sci.* 73, R48–R65, 2008.
7. DellaPenna, D. Nutritional genomics: manipulating plant micronutrients to improve human health. *Science* 285, 375–379, 1999.
8. Oksman-Caldentey, K.-M., and Inzé, D. Plant cell factories in the post-genomic era: new ways to produce designer secondary metabolites. *Trends Plant Sci.* 9, 433–440, 2004.

9. Kim, H.-J., Chen, F., Wang, X., and Rajapakse, N.C. Effect of chitosan on the biological properties of sweet basil (*Ocimum basilicum* L.). *J. Agric. Food Chem.* 53, 3696–3701, 2005.
10. Bennett, R.N., and Wallsgrove, R.M. Secondary metabolites in plant defense mechanisms. *New Phytol.* 127, 617–633, 1994.
11. Kessler, A., and Baldwin, I.T. Plant responses to insect herbivory: the emerging molecular analysis. *Annu. Rev. Plant Biol.* 53, 299–328, 2002.
12. Kliebenstein, D.K. Secondary metabolites and plant/environment interactions: a view through *Arabidopsis thaliana* tinged glasses. *Plant Cell Environ.* 27, 675–684, 2004.
13. de Bruxelles, G.L., and Roberts, M.R. Signals regulating multiple responses to wounding and herbivores. *Crit. Rev. Plant Sci.* 20, 487–521, 2001.
14. Dixon, R.A., and Paiva, N.L. Stress-induced phenylpropanoid metabolism. *Plant Cell* 7, 1085–1097, 1995.
15. Verpoorte, R. Secondary metabolism. In: Verpoorte, R. and Alfermann, A.W., Eds., *Metabolic Engineering of Plant Secondary Metabolism*, Kluwer Academic, Dordrecht, Netherlands, 2000, pp. 1–29.
16. Hartman, T. From waste products to ecochemicals: fifty years research of plant secondary metabolism. *Phytochemistry* 68, 2831–2846, 2007.
17. Kim, H.-J., Chen, F., Wang, X., and Choi, J.-H. Effect of methyl jasmonate on phenolic, isothiocyanate, enzymes in radish sprout (*Raphanus sativus* L.). *J. Agric. Food Chem.* 54, 7263–7269, 2006.
18. Taiz, L., and Zeiger, E. *Plant Physiology*, 3rd ed. Sinauer, Sunderland, MA, 2002, p. 286.
19. Agrawal, G.K., Rakwal, R., Tamogami, S., Yonekura, M., Kubo, A., and Saji, H. Chitosan activates defense/stress response(s) in the leaves of *Oryza sativa* seedlings. *Plant Physiol. Biochem.* 40, 1061–1069, 2002.
20. Hadwiger, L.A., and Beckman, J.M. Chitosan as a component of pea *Fusarium solani* interactions. *Plant Physiol.* 66, 205–211, 1980.
21. Kneer, R., Poulev, A.A., Olesinski, A., and Raskin, I. Characterization of the elicitor-induced biosynthesis and secretion of genistein from roots of *Lupinus luteus* L. *J. Exp. Bot.* 50, 1553–1559, 1999.
22. Jayasinghe, C., Gotoh, N., Aoki, T., and Wada, S. Phenolics composition and antioxidant activity of sweet basil (*Ocimum basilicum* L.). *J. Agric. Food Chem.* 51, 4442–4449, 2003.
23. Szabo, E., Thelen, A., and Petersen, M. Fungal elicitor preparations and methyl jasmonate enhance rosmarinic acid accumulation in suspension cultures of Coleus blumei. *Plant Cell Rep.* 18, 485–489, 1999.
24. Johnson, C.B., Kirby, J., Naxakis, G., and Pearson, S. Substantial UV-B-mediated induction of essential oils in sweet basil (*Ocimum basilicum* L.). *Phytochemistry* 51, 507–510, 1999.
25. Loughrin, J. H., and Kasperbauer, M. J. Light reflected from colored mulches affects aroma and phenol content of sweet basil (*Ocimum basilicum* L.) leaves. *J. Agric. Food Chem.* 49, 1331–1335, 2001.
26. Zabaras, D., and Wyllie, S.G. The effect of mechanical wounding on the composition of essential oil from *Ocimum basilicum* L. leaves. *Molecules* 6, 79–86, 2001.
27. Kim, H.-J., Chen, F., Wang, X., and Rajapakse, N.C. Effect of methyl jasmonate on secondary metabolites of sweet basil (*Ocimum basilicum* L.). *J. Agric. Food Chem.* 54, 2327–2332, 2006.
28. Turner, J.G., Ellis, C., and Devoto, A. The jasmonate signal pathway. *Plant Cell* 14(Suppl.), 153–164, 2002.
29. Zhao, J., Davis, L.C., and Verpoorte, R. Elicitor signal transduction leading to production of plant secondary metabolites. *Biotechnol. Adv.* 23, 283–333, 2005.

30. Chen, H., Jones, A.D., and Howe, G.A. Constitutive activation of the jasmonate signaling pathway enhances the production of secondary metabolites in tomato. *FEBS Lett.* 580, 2540–2546, 2006.

31. Ketchum, R.E., Rithner, C.D., Qiu, D., Kim, Y.S., Williams, R.M., and Croteau, R.B. Taxus metabolomics: methyl jasmonate preferentially induces production of taxoids oxygenated at C-13 in *Taxus media* cell cultures. *Phytochemistry* 62, 901–909, 2003.

32. Kim, H.-J., Fonseca, J.M., Choi, J.-H., and Kubota, C. Effect of methyl jasmonate on phenolic compounds and carotenoids of romaine lettuce (*Lactuca sativa* L.). *J. Agric. Food Chem.* 55, 10366–10372, 2007.

33. Wahid, A., and Ghazanfar, A. Possible involvement of some secondary metabolites in salt tolerance of sugarcane. *J. Plant Physiol.* 163, 723–730, 2006.

34. Perez, A., Sanz, C., Richardson, D., and Olias, J. Methyl jasmonate promotes β-carotene synthesis and chlorophyll degradation in golden delicious apple peel. *J. Plant Growth Regul.* 12, 163–167, 1993.

35. Sánchez-Sampedro, M.A., Fernández-Tárrango, J., and Corchete, P. Yeast extract and methyl jasmonate-induced silymarin production in cell cultures of *Silybum marianum* (L.) Gaertn. *J. Biotechnol.* 119, 60–69, 2005.

36. Wang, S.Y., and Zheng, W. Preharvest application of methyl jasmonate increases fruit quality and antioxidant capacity in raspberries. *Int. J. Food Sci. Technol.* 40, 187–195, 2005.

37. Zhao, J., Fujita, K., Yamada, J., and Sakai, K. Improved beta-thuaplicin production in *Cupresus lusitanica* suspension cultures by fungal elicitor and methyl jasmonate. *Appl. Microbiol. Biotechnol.* 55, 301–305, 2001.

38. Hayashi, H., Huang, P., and Inoue, K. Up-regulation of soyasaponin biosynthesis by methyl jasmonate in cultured cells of *Glycyrhiza glabra*. *Plant Cell Physiol.* 44, 404–411, 2003.

39. Yukimune, Y., Tabata, H., and Hara, Y. Methyl jasmonate-induced overproduction of paclitaxel and baccatin III in Taxus cell suspension cultures. *Nature Biotechnol.* 14, 1129–1132, 1996.

40. Parida, A.K., and Das, A.B. Salt tolerance and salinity effects on plants: a review. *Ecotoxicol. Environ. Safety* 60, 324–349, 2005.

41. Hu, Y., Burucs, Z., von Tucher, S., and Schmidhalter, U. Short-term effects of drought and salinity on mineral nutrient distribution along growing leaves of maize seedlings. *Environ. Exp. Bot.* 60, 268–275, 2007.

42. Shalhevet, L., and Hsiao, T.C. Salinity and drought—a comparison of their effects on osmotic adjustment, assimilation, transpiration and growth. *Irrig. Sci.* 7, 249–264, 1986.

43. Kim, H.-J., Fonseca, J.M., Choi, J.-H., and Kubota, C. Salt in irrigation water affects nutritional and visual properties of romaine lettuce (*Lactuca sativa* L.). *J. Agric. Food Chem.* 56, 3772–3776, 2008.

44. Matsuda, F., Morino, K., Ano, R., Kuzawa, M., Wakasa, K., and Miyagawa, H. Metabolic flux analysis of the phenylpropanoid pathway in elicitor-treated potato tuber tissue. *Plant Cell Physiol.* 46, 454–466, 2005.

45. Cisneros-Zevallos, L. The use of controlled postharvest abiotic stresses as a tool for enhancing the nutraceutical content and adding-value of fresh fruits and vegetables. *J. Food Sci.* 68, 1560–1565, 2003.

46. Tomas-Barberan, F.A., Loaiza-Velarde, J., Bonfanti, A., and Saltveit, M.E. Early wound- and ethylene-induced changes in phenylpropanoid metabolism in harvested lettuce. *J. Am. Soc. Hort. Sci.* 122, 399–404, 1997.

18 Branched-Chain Fatty Acid as a Functional Lipid

Hirosuke Oku and Teruyoshi Yanagita

CONTENTS

18.1 INTRODUCTION

Branched-chain fatty acids (BCFAs) are a group of fatty acids with methyl branching at variable positions on their fatty acyl chains. BCFAs are not normally encountered in the internal tissues of mammals and are detectable only to a variable extent in the skin surface lipid of animals [1]. For this reason, the physiological significance of BCFAs in diets has been poorly understood, especially in terms of biological activities of these fatty acids.

BCFAs make up the major proportion of fatty acids in the lipid extract from certain bacteria, such as *bacilli* [2]. Biosynthesis of BCFAs occurs with the branched-chain amino acids as primary precursors and malonyl-CoA (coenzyme A) as the chain extender (Figure 18.1). These BCFAs biosynthesized by bacteria and included in fermented food may contribute to the food's regulation of cell biology or metabolism [3–5]. A saturated BCFA, 13-methyltetradecanoic acid (13-MTD), was purified from a fermented soy product as an antitumor compound [6]. A BCFA was also found to induce apoptotic cell death in human cancer cells [6]. In this review, we describe the biological activities of BCFAs, with special reference to 13-MTD.

18.2 ANTICANCER ACTIVITY OF BCFAS

Since 1985, cancer patients at different clinical stages have used a fermented soy product as a nutritional and therapeutic supplement. Results in these patients indicated that the product might improve clinical condition as well as survival rate,

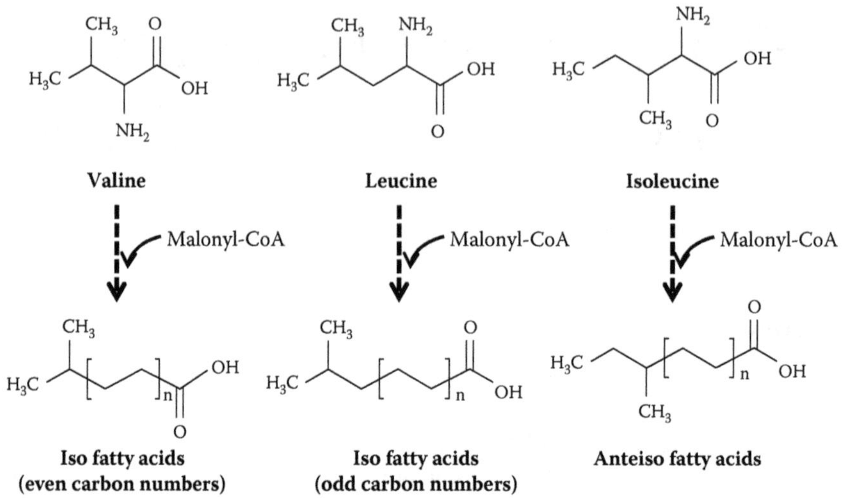

FIGURE 18.1 Biosynthetic pathway of branched-chain fatty acids.

and laboratory studies have shown inhibition of tumor cell growth by this product. Several compounds responsible for the anticancer activity are BCFAs (C15 to C21). The most abundant component in this fermented soy product, 13-MTD, was evaluated for anticancer activity [6]. 13-MTD caused cell death through apoptosis in several human tumor cell lines. Furthermore, oral administration of 13-MTD inhibited growth of tumors orthotopically implanted in nude mice. These *in vitro* and *in vivo* data demonstrate that 13-MTD inhibits tumor cell growth by inducing apoptosis and is a potential agent for cancer chemotherapy.

Although these studies found BCFAs to be antitumor agents, no information on the structural characteristics relevant to their biological activity has been available. Our group examined the relationship between chain length and antitumor activity and showed that the cytotoxicities of iso-BCFAs vary according to chain length [7]. The highest activity was observed with iso-16:0.

Two major types of BCFAs occur in nature: iso series and anteiso series [2]. Anteiso-15:0, having methyl branching at antepenultimate positions (anteiso), showed cytotoxicity comparable to that of iso-15:0 [7]. Thus, the D_{50} of both iso-15:0 and anteiso-15:0 were lower than those of corresponding straight-chain fatty acids.

Several studies have shown that certain polyunsaturated fatty acids have a selective cytotoxic effect on tumor cells and minimal or no effect on normal cells [8]. By association, selective cytotoxicity of a BCFA was studied with normal and tumor lung cells. 13-MTD was toxic to both normal and tumor cells, with higher cytotoxicity than the straight-chain fatty acid n-C14:0.

Conjugated linoleic acids (CLAs) have been reported to be antitumoral fatty acids [9,10]. Two types of biologically active CLAs are known: the *cis*-9,*trans*-11 isomer and the *trans*-10,*cis*-12 isomer. The former is the principal dietary form of CLA and was used in our experiment as a reference. Comparisons of cytotoxicity to breast cancer cells revealed that the cytotoxicity of 13-MTD was almost equivalent to that of CLA.

18.3 ANTILIPOGENIC ACTIVITY OF BCFAS

The regulation of cell growth is a homeostatic balance between stimulatory and inhibitory signals. The negative growth control by induction of apoptosis has attracted the attention of many investigators and has provided a strategy for treatment of malignancies [11–16]. Complex signal transduction systems are involved in induction of apoptosis [17,18], and the mechanisms of apoptosis induction by BCFA remain unknown.

Malignant tissues express high levels of fatty acid synthase (FAS) [19–25], leading to the notion that FAS is a target for anticancer drug development [15]. It has been proposed that inhibition of FAS triggers the signal transduction flux toward induction of apoptosis and causes cell death [12,14]. Fatty acid might exert its biological activity by affecting the fatty acid or lipid metabolism [26]. It thus can be expected that fatty acids exert their biological effects by modulation of fatty acid metabolism, including the pathways of biosynthesis and degradation. We therefore examined the effect of 13-MTD on lipid biosynthesis from [^{14}C]acetate.

13-MTD significantly decreased incorporation of [^{14}C]acetate into triacylglycerol and free fatty acid compared with myristic acid [7]. Decreased [^{14}C]acetate incorporation was also noted with CLA, but to a lower extent compared with 13-MTD. Myristic acid enhanced transformation of free fatty acid into triacylglycerols more than did 13-MTD or CLA. 13-MTD decreased cholesterogenesis in breast cancer cells and decreased fatty acid biosynthesis.

The *in vitro* effect of 13-MTD on the enzymes involved in fatty acid biosynthesis was studied. Five enzyme activities were assayed: FAS, acetyl-CoA carboxylase, glucose-6-phosphate dehydrogenase (G6PDH), adenosine triphosphate (ATP)-citrate lyase, and malic enzyme. Of the enzymes involved in fatty acid biosynthesis, FAS, acetyl-CoA carboxylase, and G6PD from rat liver were inhibited *in vitro* by 13-MTD [7].

G6PD is the first enzyme in the hexose monophosphate shunt, which provides important precursors for both fatty acid and nucleotide synthesis. Thus, 13-MTD might synthetically decrease fatty acid biosynthesis by reducing the substrate nicotinamide adenine dinucleotide phosphate (NADPH) in addition to directly inhibiting FAS. Furthermore, G6PD is an interface enzyme between the glycolytic pathway and the hexose monophosphate shunt. Inhibition of G6PD therefore decreases substrates for both fatty acid and nucleotide biosynthesis. BCFAs such as 13-MTD may affect cell dynamics by reducing substrates for both DNA duplication and membrane biogenesis.

13-MTD was shown to be toxic against both cancer and normal lung cells *in vitro* [7]. However, this observation does not necessarily indicate that 13-MTD damages both tumor and normal tissues *in vivo*. 13-MTD effectively inhibited *in vivo* growth of various cancer cells without significant side effects [6]. Cancer cells express higher levels of FAS and are more dependent on fatty acid biosynthesis for survival than are normal cells [16]. It is thus accepted that inhibition of FAS is selectively cytotoxic to human cancer cells *in vivo* [13,16]. Rapidly growing normal and cancer cells may require increased fatty acid biosynthesis. However, normal tissue does not need to undergo rapid regeneration under normal situations. Thus, the demand of normal tissues for fatty acid synthesis may be smaller than that of cancer cells, which may explain in part the selective cytotoxicity of FAS inhibition against cancer cells. However, rapidly regenerating tissues such as intestinal epithelium or bone marrow may need more fatty acid synthesis than other tissues, even under normal conditions. For this reason, care should be taken when referring to the effect of FAS inhibition on rapidly proliferating tissues.

18.4 STRUCTURE–CYTOTOXICITY RELATIONSHIP OF UNUSUAL FATTY ACIDS

Unusual fatty acids represented by BCFAs are incorporated into cellular lipid and may impair lipid metabolism, leading to apoptotic cell death. We studied the incorporation of unusual fatty acids (see Figure 18.2) into cellular lipids of breast cancer cells and its relevance to cytotoxicity [27]. Fatty acids added to culture medium were preferentially incorporated into triacylglycerols rather than into phospholipids. Acyl chain length appeared to be the critical factor for incorporation into cellular lipids. Incorporation of BCFAs was greater than that of cycloalkyl and almost comparable to that of hydroxy fatty acids. Cytotoxicities of BCFAs were stronger than those of cycloalkyl or hydroxy fatty acids. Thus, of the fatty acids studied, only BCFAs exhibited cytotoxicity against breast cancer cells. These observations suggest that the branched acyl chain structure has some relation to cytotoxicity.

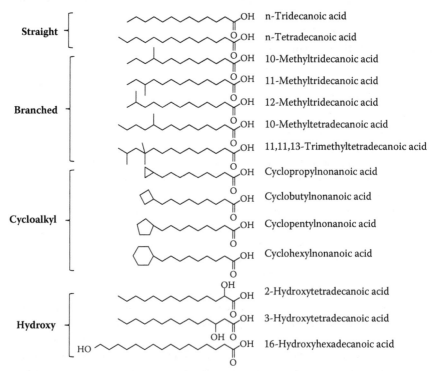

FIGURE 18.2   Chemical structures of fatty acids used in authors' study [27].

18.5   INCORPORATION OF BCFAS INTO CELLULAR LIPIDS

Epidemiological studies have indicated that the fatty acid composition of the diet correlates with risk of mammary cancer [28–30]. The level of stearic acid in tumor membrane phosphatidylcholine has been considered to be an independent tumor marker of breast cancer prognosis [31]. Furthermore, given the involvement of lipogenic enzymes in oncogenesis, membrane lipids could be factors in modulation of tumor growth [32,33]. Therefore, incorporation of 13-MTD into membrane or storage lipid might play a role in induction of apoptosis in cancer cells. We therefore studied the incorporation of 13-MTD into cellular glycerolipid and the positional distribution of 13-MTD in lipid molecules [34].

13-MTD was incorporated preferentially into triacylglycerols rather than into phospholipids. The proportion of 13-MTD in both phospholipids and triacylglycerols increased to saturation level in 6 h of incubation. The metabolic fate of fatty acids thus showed diversity with their chemical structures and resulted in uneven positional distribution in triacylglycerol or phospholipid molecules [35]. Triacylglycerols contained almost equal proportions of 13-MTD at the sn-2 position and at the sn-1,3 positions. 13-MTD was incorporated into phosphatidylcholine to a greater extent than into phosphatidylethanolamine, with preference for the sn-2 position.

In conclusion, we found that the 13-MTD was incorporated into membrane phospholipids even to a lesser extent than into triacylglycerol, and this could influence the membrane environment as the case of death stimulus.

18.6   CASPASE-INDEPENDENT INDUCTION
OF APOPTOSIS BY BCFAs

Caspases are involved in signal transduction leading to apoptotic cell death [36,37]. Caspase-3 acts at the final step in the caspase cascade to cleave proteins involved in cytoskeletal and nuclear structures, resulting in the cell shrinking and DNA fragmentation of late-stage apoptosis. 13-MTD, however, has no effect on the activity of caspase-3 in SKBR-3 cells [34]. In contrast to 13-MTD, CLA significantly elevated caspase-3 activity after 2 and 4 h of incubation. Despite the lack of caspase-3 activation, 13-MTD reduced cell viability to the same extent that CLA did [34].

To confirm the noninvolvement of the caspase pathway, cytotoxicity and apoptosis induction were studied in the presence of a caspase-3 inhibitor (Z-DEVD-FMK) and a negative control for the caspase-3 inhibitor (Z-FA-FMK). The negative control only inhibited cysteine protease and had no inhibitory effect on caspase-mediated apoptosis. The presence of a caspase-3 inhibitor, as well as a negative control, did not alleviate the cytotoxicity of 13-MTD and did not prevent induction of apoptosis.

The effect of 13-MTD on expression of several regulatory genes involved in apoptosis induction was studied (unpublished observation). 13-MTD and myristic acid decreased the expression of the p53 gene to the same extent, with no significant difference between treatments. CLA increased the messenger RNA (mRNA) level of caspase-3 and increased its activity. 13-MTD had no effect on expression of caspase-8, apoptosis-inducing factor (AIF), Bax, Bad, and Bcl-X_L.

Hydrogen peroxide mediates induction of apoptosis in response to several external stimuli [38,39], and supplementation of medium with catalase prevents apoptosis induction [40,41]. Adding catalase to culture medium did not protect or rescue cells from cytotoxicity of 13-MTD.

Several series of studies have demonstrated that short-chain or medium-chain fatty acids increase cellular Ca^{2+} mobilization via G-protein-coupled orphan receptors [42,43]. Furthermore, excess loading of mitochondria with Ca^{2+} results in abnormal mitochondrial metabolism, which triggers programmed cell death [44]. No mobilization of Ca^{2+} occurred when cells were loaded with 13-MTD. Oleic acid, included as a positive control, stimulated cellular Ca^{2+} mobilization, as did CLA.

For induction of apoptosis, mitochondria play a pivotal role in signal transduction pathways [45]. To examine mitochondrial membrane integrity, we used MitoCapture reagent (BioVision Research Product, Mountain View, CA) to stain apoptotic cells after treatment of cells with 13-MTD. After 4 h of treatment with 13-MTD, the MitoCapture dye fluoresced predominantly green, indicating disruption of mitochondrial integrity. Myristic acid slightly increased the green fluorescence compared with control vehicle buffer, but to a much lower extent compared with 13-MTD. CLA treatment for 4 h liberated the cells from the bottom of the culture dishes into medium and resulted in green fluorescence.

The most notorious apoptotic factors released from permeabilized mitochondria are cytochrome C and AIF. However, the results of the foregoing studies may rule out involvement of cytochrome C in the death pathway of 13-MTD-dependent apoptosis. AIF is therefore the most plausible candidate for death signal transducer for apoptosis induction by 13-MTD. The percentage of AIF-punctuated apoptotic nuclei was significantly increased by 13-MTD treatment. This percentage was increased to some extent by treatment with myristic acid. However, the difference between control and myristic acid treatment was statistically insignificant. Treatment of cells with vehicle buffer (control) resulted in little nuclear staining of AIF.

18.7 SUMMARY

In summary, treatment with 13-MTD induced no change in activity and expression of caspase-3. Furthermore, 13-MTD induced no mobilization of cellular calcium. However, 13-MTD altered mitochondrial transmembrane potential and induced AIF translocation from the mitochondria to the nucleus after 4 h of treatment. These results support the view that incorporation of 13-MTD into cellular lipids triggers apoptosis via a caspase-independent pathway.

REFERENCES

1. Nicolaides N. (1974) Skin lipids: their biochemical uniqueness. *Science* 186: 19–26.
2. Kaneda T. (1991) Iso- and anteiso-fatty acids in bacteria: biosynthesis, function, and taxonomic significance. *Microbiol Rev* 55: 288–302.

3. Alonso L, Fontecha J, Lozada L, Fraga MJ, Juárez M. (1999) Fatty acid composition of caprine milk: major, branched-chain, and trans fatty acids. *J Dairy Sci* 82: 878–884.
4. DePooter H, Decloedt M, Schamp N. (1981) Composition and variability of the branched-chain fatty acid fraction in the milk of goats and cows. *Lipids* 16: 286–292.
5. Dewhurst RJ, Moorby JM, Vlaeminck B, Fievez V. (2007) Apparent recovery of duodenal odd- and branched-chain fatty acids in milk of dairy cows. *J Dairy Sci* 90: 1775–1780.
6. Yang Z, Liu S, Chen S, Chen H, Haung M, Zheng J. (2000) Induction of apoptotic cell death and *in vivo* growth inhibition of human cancer cells by a saturated branched-chain fatty acid, 13-methyltetradecanoic acid. *Cancer Res* 60: 505–509.
7. Wongtangtintharn S, Oku H, Iwasaki H, Toda T. (2004) Effect of branched-chain fatty acids on fatty acid biosynthesis of human breast cancer cells. *J Nutr Sci Vitaminol (Tokyo)* 50: 137–143.
8. Diggle CP. (2002) *In vitro* studies on the relationship between polyunsaturated fatty acids and cancer: tumour or tissue specific effects? *Prog Lipid Res* 41: 240–253.
9. Pariza MW, Park Y, Cook ME. (2001) The biologically active isomers of conjugated linoleic acid. *Prog Lipid Res* 40: 283–298.
10. Palombo JD, Ganguly A, Bistrian BR, Menard MP. (2002) The antiproliferative effects of biologically active isomers of conjugated linoleic acid on human colorectal and prostatic cancer cells. *Cancer Lett* 177: 163–172.
11. Pizer ES, Wood FD, Pasternack GR, Kuhajda FP. (1996) Fatty acid synthase. (FAS): a target for cytotoxic antimetabolites in HL60 promyelocytic leukemia cells. *Cancer Res* 56: 745–751.
12. Pizer ES, Jackisch C, Wood FD, Pasternack GR, Davidson NE, Kuhajda FP. (1996) Inhibition of fatty acid synthesis induces programmed cell death in human breast cancer cells. *Cancer Res* 56: 2745–2747.
13. Pizer ES, Wood FD, Heine HS, Romantsev FE, Pasternack GR, Kuhajda FP. (1996) Inhibition of fatty acid synthesis delays disease progression in a xenograft model of ovarian cancer. *Cancer Res* 56: 1189–1193.
14. Pizer ES, Chrest FJ, DiGiuseppe JA, Han WF. (1998) Pharmacological inhibitors of mammalian fatty acid synthase suppress DNA replication and induce apoptosis in tumor cell lines. *Cancer Res* 58: 4611–4615.
15. Kuhajda FP, Pizer ES, Li JN, Mani NS, Frehywot GL, Townsend CA. (2000) Synthesis and antitumor activity of an inhibitor of fatty acid synthase. *Proc Natl Acad Sci USA* 97: 3450–3454.
16. Kuhajda FP. (2000) Fatty-acid synthase and human cancer: new perspective on its role in tumor biology. *Nutrition* 16: 202–208.
17. Beere H. (2001) Apoptosis hots up from the cold. *Trends Biochem Sci* 26: 278–280.
18. Zimmermann KC, Pinkoski MJ. (2001) In the company of killers. *Trends Mol Med* 7: 195–197.
19. Alò PL, Visca P, Marci A, Mangoni A, Botti C, Di Tondo U. (1996) Expression of fatty acid synthase (FAS) as a predictor of recurrence in stage I breast carcinoma patients. *Cancer* 77: 474–482.
20. Jensen V, Ladekarl M, Holm-Nielsen P, Melsen F, Soerensen FB. (1995) The prognostic value of oncogenic antigen 519 (OA-519) expression and proliferative activity detected by antibody MIB-1 in node-negative breast cancer. *J Pathol* 176: 343–352.
21. Alò PL, Visca P, Trombetta G, Mangoni A, Lenti L, Monaco S, Botti C, Serpieri DE, Di Tondo U. (1999) Fatty acid synthase (FAS) predictive strength in poorly differentiated early breast carcinomas. *Tumori* 85: 35–40.

22. Epstein JI, Carmichael M, Partin AW. (1995) OA-519 (fatty acid synthase) as an independent predictor of pathologic stage in adenocarcinoma of the prostate. *Urology* 45: 81–86.

23. Shurbaji MS, Kalbfleisch JH, Thurmond TS. (1996) Immunohistochemical detection of a fatty acid synthase (OA-519) as a predictor of progression of prostate cancer. *Hum Pathol* 27: 917–921.

24. Rashid A, Pizer ES, Moga M, Miligraum LZ, Zahurak M, Pasternack GR, Kuhajda FP, Hamilton SR. (1997) Elevated expression of fatty acid synthase and fatty acid synthetic activity in colorectal neoplasia. *Am J Pathol* 150: 201–208.

25. Pizer ES, Lax SF, Kuhajda FP, Pasternack GR, Kurman RJ. (1998) Fatty acid synthase expression in endometrial carcinoma: correlation with cell proliferation and hormone receptors. *Cancer* 83: 528–537.

26. Igarashi M, Miyazawa T. (2001) The growth inhibitory effect of conjugated linoleic acid on a human hepatoma cell line, HepG2, is induced by a change in fatty acid metabolism, but not the facilitation of lipid peroxidation in the cells. *Biochim Biophys Acta* 1530: 162–171.

27. Wongtangtintharn S, Oku H, Inafuku M, Iwasaki H, Toda T. (2006) Effect of carbon chain structures of fatty acids on their cytotoxicity to breast cancer cells. *J Jpn Soc Nutr Food Sci* 59: 115–118.

28. Chan PC, Ferguson KA, Dao TL. (2002) Effects of different dietary fats on mammary carcinogenesis. *Cancer Res* 43:1079–1083.

29. Carroll KK, Hopkins GJ. (1979) Dietary polyunsaturated fat versus saturated fat in relation to mammary carcinogenesis. *Lipids* 14: 155–158.

30. Tinsley IJ, Schmitz JA, Pierce DA. (1981) Influence of dietary fatty acids on the incidence of mammary tumors in the C3H mouse. *Cancer Res* 41: 1460–1465.

31. Bougnoux P, Chajès V, Lanson M, Hacene K, Body G, Couet C, Le Floch O. (1992) Prognostic significance of tumor phosphatidylcholine stearic acid level in breast carcinoma. *Breast Cancer Res Treat* 20: 185–194.

32. Chajès V, Lanson M, Fetissof F, Lhuillery C, Bougnoux P. (1995) Membrane fatty acids of breast carcinoma: contribution of host fatty acids and tumor properties. *Int J Cancer* 63: 169–175.

33. Fonck K, Scherphof GL, Konings AW. (1982) Control of fatty acid incorporation in membrane phospholipids. X-ray-induced changes in fatty acid uptake by tumor cells. *Biochim Biophys Acta* 692: 406–414.

34. Wongtangtintharn S, Oku H, Iwasaki H, Inafuku M, Toda T, Yanagita T. (2005) Incorporation of branched-chain fatty acid into cellular lipids and caspase-independent apoptosis in human breast cancer cell line, SKBR-3. *Lipids Health Dis* 4: 29 (doi: 10.1186/1476-511X-4-29).

35. Ramirez M, Amate L, Gil A. (2001) Absorption and distribution of dietary fatty acids from different sources. *Early Hum Dev* 65: S95–S101.

36. Philchenkov A. (2004) Caspases: potential targets for regulating cell death. *J Cell Mol Med* 8: 432–444.

37. Fulda S, Debatin KM. (2004) Apoptosis signaling in tumor therapy. *Ann NY Acad Sci* 1028: 150–126.

38. Kim DK, Cho ES, Um HD. (2000) Caspase-dependent and -independent events in apoptosis induced by hydrogen peroxide. *Exp Cell Res* 257: 82–88.

39. Hancock JT, Desikan R, Neill SJ. (2001) Role of reactive oxygen species in cell signalling pathways. *Biochem Soc Trans* 29: 345–350.

40. Arimura T, Kojima-Yuasa A, Suzuki M, Kennedy DO, Matsui-Yuasa I. (2003) Caspase-independent apoptosis induced by evening primrose extract in Ehrlich ascites tumor cells. *Cancer Lett* 201: 9–16.

41. Yonezawa T, Katoh K, Obara Y. (2004) Existence of GPR40 functioning in a human breast cancer cell line, MCF-7. *Biochem Biophys Res Commun* 314: 805–809.

42. Briscoe CP, Tadayyon M, Andrews JL, Benson WG, Chambers JK, Eilert MM, Ellis C, Elshourbagy NA, Goetz AS, Minnick DT, Murdock PR, Sauls HR Jr, Shabon U, Spinage LD, Strum JC, Szekeres PG, Tan KB, Way JM, Ignar DM, Wilson S, Muir AI. (2003) The orphan G protein-coupled receptor GPR40 is activated by medium and long chain fatty acids. *J Biol Chem* 278: 11303–11311.
43. Berridge MJ, Bootman MD, Lipp P. (1998) Calcium—a life and death signal. *Nature* 395: 645–648.
44. Bras M, Queenan B, Susin SA. (2005) Programmed cell death via mitochondria: different modes of dying. *Biochemistry* 70: 231–239.
45. Oku H, Wongtangtintharn S, Iwasaki H, Toda T. (2003) Conjugated linoleic acid (CLA) inhibits fatty acid synthetase activity *in vitro*. *Biosci Biotechnol Biochem* 67: 1584–1586.

19 Retrobiosynthetic Production of 2′-Deoxyribonucleoside from Glucose, Acetaldehyde, and Nucleobase through Multistep Enzymatic Reactions

Jun Ogawa, Nobuyuki Horinouchi,
and Sakayu Shimizu

CONTENTS

19.1 INTRODUCTION

There will be a need for 2′-deoxyribonucleoside in the near future due to increasing demand in new medical and biotechnology fields. 2′-Deoxyribonucleoside is a building block of promising antisense drugs for cancer therapy. For some recently developed antiviral reagents, such as azidothymidine for treatment of human immunodeficiency virus (HIV) infections, 2′-deoxyribonucleoside is a synthesis intermediate. 2′-Deoxyribonucleoside is also a precursor of an indispensable material used for widespread polymerase chain reaction (PCR) applications, 2′-deoxyribonucleoside triphosphate. The current 2′-deoxyribonucleoside sources include hydrolyzed herring and salmon sperm DNA, which are not suitable sources for sudden high demands. Microbial/enzymatic processes for 2′-deoxyribonucleoside production could possibly remove this bottleneck of raw material supply [1].

We focused on the reactions involved in 2′-deoxyribonucleoside degradation for the synthesis of 2′-deoxyribonucleoside. All reactions in 2′-deoxyribonucleoside degradation are reversible, so the biochemical retrosynthesis of 2′-deoxyribonucleosides from their metabolites is possible. Some of the metabolites, such as triosephosphates, can be obtained from cheap sugar materials through the glycolytic pathway. Therefore, we proposed the multistep enzymatic process presented in Figure 19.1, with glucose, acetaldehyde, and a nucleobase as the starting materials. The enzymes involved in the process are (1) glycolytic enzymes, which generate D-glyceraldehyde 3-phosphate from glucose; (2) deoxyriboaldolase, which catalyzes the condensation of acetaldehyde and D-glyceraldehyde 3-phosphate (G3P) to yield 2-deoxyribose 5-phosphate (DR5P); (3) phosphopentomutase, which catalyzes the intermolecular transfer of phosphate from DR5P to 2-deoxyribose 1-phosphate (DR1P); and

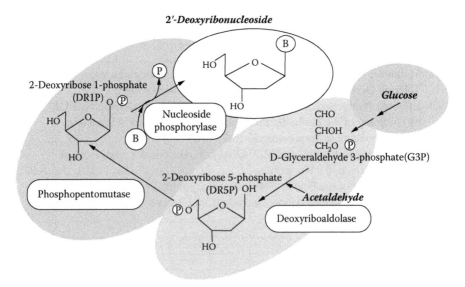

FIGURE 19.1 Multistep enzymatic process for the production of 2′-deoxyribonucleoside from glucose, acetaldehyde, and a nucleobase through the reverse reactions of 2′-deoxyribonucleoside degradation.

(4) nucleoside phosphorylase, which catalyzes nucleobase transfer to the pentosyl moiety to generate 2'-deoxyribonucleosides.

19.2   METABOLIC ANALYSIS OF *ESCHERICHIA COLI* EXPRESSING ACETALDEHYDE-TOLERANT DEOXYRIBOALDOLASE FROM *KLEBSIELLA PNEUMONIAE*

The key step of 2'-deoxyribonucleoside synthesis is generation of an important intermediate, 2-deoxyribose. We focused on the deoxyriboaldolase-catalyzing reaction for 2-deoxyribose synthesis. Deoxyriboaldolase catalyzes reversible cleavage of DR5P to G3P and acetaldehyde in degradation of 2'-deoxyribonucleoside. Screening for suitable catalysts for the production of DR5P through the reverse deoxyriboaldolase reaction was carried out. The screening reactions were done with a high acetaldehyde concentration to reveal acetaldehyde-tolerant enzymes because a high concentration of acetaldehyde is required to push the deoxyriboaldolase reaction in the direction of DR5P synthesis. Among about 200 strains tested, *Klebsiella pneumoniae* B-4-4 was selected as an acetaldehyde-tolerant deoxyriboaldolase producer for further investigation [2].

The gene encoding the acetaldehyde-tolerant deoxyriboaldolase was cloned from the chromosomal DNA of *K. pneumoniae* B-4-4. It contains an open reading frame consisting of 780 nucleotides corresponding to 259 amino acid residues. The predicted amino acid sequence exhibited 94.6% homology with that of deoxyriboaldolase from *Escherichia coli* [3]. The deoxyriboaldolase of *K. pneumoniae* was expressed in recombinant *E. coli* cells. A deoxyriboaldolase-expressing *E. coli* transformant, 10B5/pTS8, was constructed. Strain 10B5/pTS8 was superior for DR5P production because of a defect in alkaline phosphatase activity decomposing DR5P to 2-deoxyribose [3]. The specific activity of deoxyriboaldolase in the cell extracts of the *E. coli* transformant reached as high as 2.5 U/mg, which was three times higher than that in the *K. pneumoniae* cell extracts.

To determine whether the glycolytic pathway could provide G3P, various compounds involved in the glycolytic pathway were examined as precursors of G3P in the reactions with washed cells of the *E. coli* transformant and *K. pneumoniae* B-4-4 (Table 19.1). Dihydroxyacetone phosphate (DHAP) and fructose 1,6-diphosphate (FDP) were utilized well for G3P generation by both strains.

19.3   EFFICIENT PRODUCTION OF DR5P FROM GLUCOSE AND ACETALDEHYDE VIA FDP BY COUPLING OF ALCOHOLIC FERMENTATION SYSTEM OF BAKER'S YEAST AND DEOXYRIBOALDOLASE-EXPRESSING *E. COLI*

If FDP could be provided efficiently from glucose, production of DR5P from glucose and acetaldehyde by using a deoxyriboaldolase-expressing *E. coli* 10B5/pTS8 became possible. Temporary accumulation of FDP in the course of adenosine triphosphate (ATP) regeneration by baker's yeast was reported [4–8]. Based on this information, we designed a novel metabolic and enzymatic DR5P production process

TABLE 19.1

Evaluation of Glycolysis Intermediates as the Substrates for 2-Deoxyriboase 5-Phosphate Production by a Deoxyriboaldolase-Expressing *E. coli* 10B5/pTS8

	DR5P Production (mM)	
Substrate	*K. pneumoniae* B-4-4	*E. coli* 10B5/pTS8
Glyceraldehyde 3-phosphate	68.8	59.2
Glyceraldehyde	—	—
Dihydroxyacetone phosphate	64.5	48.9
Dihydroxyacetone	—	—
Glucose	—	0.3
Glucose 6-phosphate	—	3.9
Glucose 1,6-diphosphate	—	—
Fructose	—	—
Fructose 6-phosphate	—	—
Fructose 1,6-diphosphate	36.1	38.9

Note: The reactions were carried out with various glycolysis and glycerol metabolism intermediates (100 mM). —, not detected.

consisting of FDP synthesis through glucose fermentation and DR5P synthesis through deoxyriboaldolase-catalyzed aldol condensation with glucose and acetaldehyde as starting materials (Figure 19.2). In this process, toluene-treated baker's yeast generates FDP from glucose and inorganic phosphate, and then the strain 10B5/pTS8 converts the FDP into DR5P via G3P.

It is important that almost all the inorganic phosphate is consumed in the first step (FDP production from glucose by baker's yeast) because the second step (deoxyriboaldolase reaction by the *E. coli* transformant) is inhibited by inorganic phosphate. When the FDP production was carried out with 750 mM inorganic phosphate, almost all the inorganic phosphate was converted into FDP, so the subsequent DR5P production proceeded well. Addition of acetaldehyde was effective on FDP production. In the presence of a downstream intermediate of alcoholic fermentation, acetaldehyde, FDP degradation might be depressed, resulting in increase in FDP accumulation. A standing reaction with an almost fully filled Erlenmeyer flask, to maintain a low soluble oxygen concentration, was effective. Under the optimized conditions with toluene-treated yeast cells, 356 mM (121 g/L) FDP was produced from 1,111 mM glucose and 750 mM potassium phosphate buffer (pH 6.4) with a catalytic amount of adenosine monophosphate (AMP) in 5 h. The molar yields of FDP for glucose and inorganic phosphate were 32% and 95.9%, respectively, and the apparent AMP turnover was 47.5 [9].

For optimization of DR5P production from the enzymatically prepared FDP and acetaldehyde with deoxyriboaldolase-expressing *E. coli* 10B5/pTS8, surfactants xylene and polyoxyethylene laurylamine (PL) were added for improvement of the permeation of phosphorylated compounds [10]. Under the preparative reaction

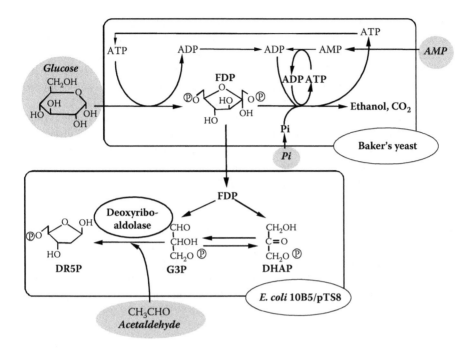

FIGURE 19.2   2-Deoxyribose 5-phosphate synthesis from glucose and acetaldehyde by baker's yeast and a deoxyriboaldolase-expressing *E. coli*. ADP, adenosine diphosphate.

conditions at 28°C with 178 mM enzymatically prepared FDP and 400 mM acetaldehyde as the substrates, 246 mM DR5P (52.6 g/L) was produced in 2 h. The molar yields of DR5P regarding FDP and acetaldehyde were 69.1% and 61.5%, respectively. The molar yield of glucose through the total two-step reaction was 22.1% [9].

19.4   BIOCHEMICAL RETROSYNTHESIS OF 2′-DEOXYRIBONUCLEOSIDES FROM GLUCOSE, ACETALDEHYDE, AND A NUCLEOBASE

As mentioned, DR5P was successfully produced through the two-step multienzyme-catalyzed process, i.e., the first step; FDP production from glucose, a high concentration of inorganic phosphate, and a catalytic amount of AMP by toluene-treated baker's yeast, and the second step; and DR5P production from acetaldehyde and enzymatically prepared FDP via G3P by deoxyriboaldolase-expressing *E. coli* 10B5/pTS8 (Figure 19.2). The DR5P enzymatically produced from glucose and acetaldehyde was further converted to 2′-deoxyribonucleosides through the reverse reactions of 2′-deoxyribonucleoside degradation. All reactions in 2′-deoxyribonucleoside degradation are reversible, so the biochemical retrosynthesis of 2′-deoxyribonucleosides from their metabolites is possible. DR5P is further converted to 2′-deoxyribonucleosides through phosphopentomutase-catalyzing isomerization of DR5P to DR1P and nucleoside phosphorylase-catalyzing nucleobase transfer to the pentosyl moiety to generate 2′-deoxyribonucleosides (the third step). *E. coli* transformants expressing phosphopentomutase (*E. coli* BL21/pTS17) and commercial nucleoside phosphorylase were used for the third step. Adenine was the best nucleobase for the third step reaction, however, the adenosine deaminase activity of *E. coli* BL21/pTS17 transformed 2′-deoxyadenosine produced to 2′-deoxyinosine. From 12.3 mM enzymatically prepared DR5P and 50 mM adenine, 9.9 mM 2′-deoxyinosine was produced [11]. The yield of glucose throughout the whole reaction process was 17.7%.

Next, we investigated a means of one-pot enzymatic synthesis of 2′-deoxyribonucleoside in which the concentrations of phosphate and phosphorylated glycolysis intermediates could be controlled more easily and attained much higher production of 2′-deoxyribonucleoside because phosphate and the phosphorylated glycolysis intermediates such as FDP, G3P, and DHAP generated through the first and the second steps severely inhibited the third step [11]. In this one-pot system, phosphate generated by nucleoside phosphorylase-catalyzed base transfer is reused well for ATP regeneration by the yeast glycolytic pathway (Figure 19.3).

The one-pot reaction mixture comprised 600 mM glucose, 333 mM acetaldehyde, 100 mM adenine, 26 mM MgSO$_4$ · 7H$_2$O, 33 mM potassium phosphate buffer (pH 7.0), 1.3 mM MnCl$_2$ · 4H$_2$O, 0.13 mM glucose 1,6-diphosphate, 0.53 % (v/v) PL, 1.3 % (v/v) xylene, 10 mM adenosine, 4% (w/v) acetone-dried cells, 6.6% (w/v) wet cells (approximately 0.7% w/v dry cells) of *E. coli* 10B5/pTS8, 26.6% (w/v) wet cells

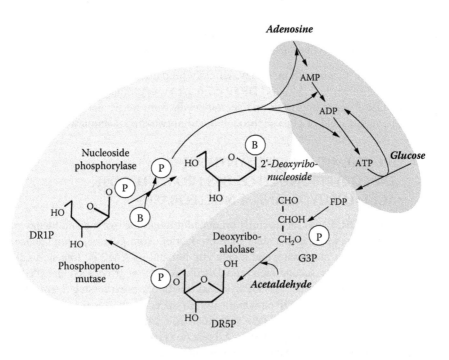

FIGURE 19.3 Multistep enzymatic process for 2′-deoxyribonucleoside production from glucose, acetaldehyde, and a nucleobase through glycolysis, reverse reactions of 2′-deoxyribonucleoside degradation, and ATP regeneration by yeast glycolytic pathway recycling the phosphate that was generated by nucleoside phosphorylase. ADP, adenosine diphosphate.

(approximately 2.7% w/v dry cells) of *E. coli* BL21/pTS17, and 5.3 U/mL commercial purine nucleoside phosphorylase. After 24 h reaction at 33°C with shaking (120 rpm), the 2'-deoxyribonucleoside (2'-deoxyinosine) production had reached 33.3 m*M*. This was almost 3.4 times higher than the production through the three-step process. The energy carrier (adenosine) turnover through the whole process was 3.3. The yields of glucose, acetaldehyde, and adenine were 2.8%, 10%, and 33.3%, respectively [12].

The one-pot process showed different features from those of the three-step process. Adenosine could be used as an energy carrier in the one-pot process, while AMP was required for the three-step process. High concentrations of phosphate (750 m*M*) and anaerobic conditions were required for the first step (FDP production through glycolysis) of the three-step process according to the Harden-Young equation [6,7]. On the other hand, 2'-deoxyribonucleoside production with the one-pot process proceeded well with a low concentration of phosphate (33 m*M*), which could prevent inhibition of phosphopentomutase- and nucleoside phosphorylase-catalyzed reactions by phosphorylated glycolysis intermediates and phosphate itself, resulting in an increase in 2'-deoxyribonucleoside production. Furthermore, 2'-deoxyribonucleoside production with the one-pot process increased with shaking, while a static condition (anaerobic condition) was required for the first step of the three-step process. Adenine, which is one of the substrates, exhibits low solubility. Accordingly, the agitation efficiency might contribute to the reactivity [12].

19.5 IMPROVEMENT OF THE ONE-POT MULTISTEP ENZYMATIC PROCESS FOR PRACTICAL PRODUCTION OF 2'-DEOXYRIBONUCLEOSIDE FROM GLUCOSE, ACETALDEHYDE, AND A NUCLEOBASE

We constructed deoxyriboaldolase-phosphopentomutase coexpressing *E. coli*, strain BL21/pACDR-pTS17, and optimized the one-pot 2'-deoxyribonucleoside production from glucose, acetaldehyde, and adenine using acetone-dried yeast, *E. coli* BL21/pACDR-pTS17, and commercial nucleoside phosphorylase. It was found that maintaining low adenine concentration in the reaction mixture was effective for improvement of the yield of 2'-deoxyribonucleoside to adenine. As for stability of catalysts, acetone-dried yeast could not keep its activity more than 20 h. Thus, by feeding a small amount of adenine, adenosine, and acetone-dried yeast into the reaction mixture, the amount of 2'-deoxyribonucleoside production was greatly increased to 75 m*M* (2'-deoxyinosine), and yield of 2'-deoxyribonucleoside to added base was considerably improved (83.3%). Once a 2'-deoxyribonucleoside was produced, the other 2'-deoxyribonucleosides could be synthesized through base exchange reactions catalyzed by various nucleoside phosphorylases [13,14].

19.6 CONCLUSIONS

Above all, we could demonstrate the practical production of 2'-deoxyribonucleoside from glucose, acetaldehyde, and a nucleobase by multistep enzymatic processes. In

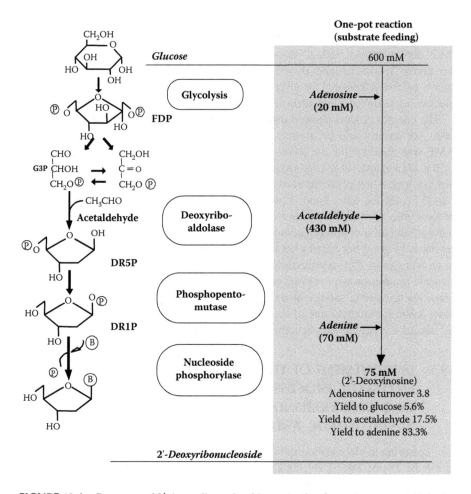

FIGURE 19.4 Summary of 2′-deoxyribonucleoside production from glucose, acetaldehyde, and a nucleobase in the one-pot (substrate-feeding) reaction.

the processes, especially in the one-pot process, successive reactions (i.e., glycolysis for FDP generation, reverse reactions of 2′-deoxyribonucleoside degradation, and ATP regeneration by yeast glycolytic pathway recycling the phosphate generated by nucleoside phosphorylase reaction) might be well coupled. The use of economical substrates (glucose, acetaldehyde, and a nucleobase), the high amount of product accumulation, and the high yield to the starting materials make the multistep enzymatic process suitable for industrial application (Figure 19.4). We hope our multistep enzymatic processes of 2′-deoxyribonucleoside and its synthesis intermediate production will contribute to the development of life science and various industries through the practical supply of 2′-deoxyribonucleosides.

ACKNOWLEDGMENTS

This work was partially supported by the Industrial Technology Research Grant Program (02A07001a to J. O.) of the New Energy and Industrial Technology Development Organization, Japan, and Grants-in-Aid for Scientific Research (16688004 to J. O.) and COE for Microbial-Process Development Pioneering Future Production Systems (to S. S.) from the Ministry of Education, Science, Sports, and Culture, Japan.

REFERENCES

1. Schmid, A., J.S. Dordick, B. Hauer, A. Kiener, M. Wubbolts, and B. Witholt. 2001. Industrial biocatalysis today and tomorrow. *Nature* 409: 258–268.
2. Ogawa, J., K. Saito, T. Sakai, N. Horinouchi, T. Kawano, S. Matsumoto, M. Sasaki, Y. Mikami, and S. Shimizu. 2003. Microbial production of 2-deoxyribose 5-phosphate from acetaldehyde and triosephosphate for the synthesis of 2'-deoxyribonucleosides. *Biosci. Biotechnol. Biochem.* 67: 933–936.
3. Horinouchi, N., J. Ogawa, T. Sakai, T. Kawano, S. Matsumoto, M. Sasaki, Y. Mikami, and S. Shimizu. 2003. Construction of deoxyriboaldolase-overexpressing *Escherichia coli* and its application to 2-deoxyribose 5-phosphate synthesis from glucose and acetaldehyde for 2'-deoxyribonucleoside production. *Appl. Environ. Microbiol.* 69: 3791–3797.
4. Tochikura, T., M. Kuwahara, S. Yagi, H. Okamoto, Y. Tominaga, T. Kano, and K. Ogata. 1967. Fermentation and metabolism of nucleic acid-related compounds in yeast. *J. Ferment. Technol.* 45: 511–529.
5. Yamamoto, K., H. Kawai, and T. Tochikura. 1981. Preparation of uridine diphosphate-*N*-acetylgalactosamine from uridine diphosphate-*N*-acetylglucosamine by using microbial enzymes. *Appl. Environ. Microbiol.* 41: 392–395.
6. Harden, A., and W. Young. 1905. The alcoholic fermentation of yeast-juice. *Proc. R. Soc. Lond. Ser. B.* 77: 405–420.
7. Wakisaka, S., Y. Ohshima, M. Ogawa, T. Tochikura, and T. Tachiki. 1988. Characteristics and efficiency of glutamine production by coupling of a bacterial glutamine synthetase reaction with the alcoholic fermentation system of baker's yeast. *Appl. Environ. Microbiol.* 64: 2953–2957.
8. Yamamoto, S., M. Wakayama, and T. Tachiki. 2005. Theanine production by coupled fermentation with energy transfer employing *Pseudomonas taetrolens* Y-30 glutamine synthetase and baker's yeast cells. *Biosci. Biotechnol. Biochem.* 69: 784–789.
9. Horinouchi, N., J. Ogawa, T. Kawano, T. Sakai, K. Saito, M. Matsumoto, M. Sasaki, Y. Mikami, and S. Shimizu. 2006. Efficient production of 2-deoxyribose 5-phosphate from glucose and acetaldehyde by coupling of the alcoholic fermentation system of baker's yeast and deoxyriboaldolase-expressing *Escherichia coli*. *Biosci. Biotechnol. Biochem.* 70: 1371–1378.
10. Fujio, T., and A. Maruyama. 1997. Enzymatic production of pyrimidine nucleotides using *Corynebacterium ammoniagenes* cells and recombinant *Escherichia coli* cells: enzymatic production of CDP-choline from orotic acid and choline chloride (part I). *Biosci. Biotechnol. Biochem.* 61: 956–959.
11. Horinouchi, N., J. Ogawa, T. Kawano, T. Sakai, K. Saito, M. Matsumoto, M. Sasaki, Y. Mikami, and S. Shimizu. 2006. Biochemical retrosynthesis of 2'-deoxyribonucleosides from glucose, acetaldehyde, and a nucleobase. *Appl. Microbiol. Biotechnol.* 71: 615–621.

12. Horinouchi, N., J. Ogawa, T. Kawano, T. Sakai, K. Saito, M. Matsumoto, M. Sasaki, Y. Mikami, and S. Shimizu. 2006. One-pot microbial synthesis of 2'-deoxyribonucleoside from glucose, acetaldehyde, and a nucleobase. *Biotechnol. Lett.* 28: 877–881.
13. Utagawa, T. 1999. Enzymatic production of nucleoside antibiotics. *J. Mol. Catal. B Enzym.* 6: 215–222.
14. Yokozeki, K., and T. Tsuji. 2000. A novel enzymatic method for the production of purine-2'-deoxyribonucleosides. *J. Mol. Catal. B Enzym.* 10: 207–213.

20 Bioconversion of Marine Phospholipid in Super Critical Carbon Dioxide to Produce Functional DHA-Enriched Lysophospholipid

Koretaro Takahashi, Hirohisa Naito,
Takeshi Ohkubo, and Mikako Takasugi

CONTENTS

20.1 INTRODUCTION

Health benefits of docosahexaenoic acid (DHA) itself are well known, but DHA is usually supplied in the form of triglyceride or in the form of ethyl esters. Ethyl ester is not allowed for food uses, and triglyceride requires emulsifiers when adding to foods. In contrast, phospholipid from DHA can be applied easily as food additives because it has both hydrophobic and hydrophilic groups in the same molecule. Another reason why the phospholipid form DHA is beneficial is because the phospholipid form is usually more stable against oxidation in aqueous media (Nara et al. 1997). Diacyl glycerophospholipids are known as a naturally occurring emulsifier, and lysophospholipids derived from diacyl glycerophospholipids are known as high-performance emulsifiers. For this reason, if we can prepare DHA-bound lysophospholipid, it should become a health beneficial high-performance emulsifier.

279

20.2 BIOCATALYTIC PARTIAL HYDROLYSIS OF PHOSPHOLIPID IN SUPERCRITICAL CARBON DIOXIDE

High-performance health beneficial emulsifier (i.e., the DHA-bound lysophospholipid) can be prepared from squid meal phospholipid because squid meal phospholipid contains DHA exclusively in position *sn*-2 (Hosokawa 1996). For this reason, if we eliminate the fatty acid moiety in the *sn*-1 position by applying position 1,3-specific lipase, or phospholipase A_1, we can obtain the *sn*-2 DHA-bound lysophospholipid. As a substrate for the desired reaction, squid meal phosphatidylcholine (PC) is the most useful so far.

Partial hydrolytic reactions were carried out by varying the conditions in the supercritical carbon dioxide ($SC\text{-}CO_2$) to obtain the *sn*-2 DHA-bound lysophosphatidylcholine (DHA-LPC). Lipozyme RMIM was first employed for the partial hydrolysis of squid PC (DHA-PC) in $SC\text{-}CO_2$ medium because it had already been successfully done in organic solvent system (Ono et al. 1997a, 1997b). However, when lipozyme RMIM-mediated partial hydrolysis occurred in the $SC\text{-}CO_2$ medium, the result was that the more moisture that was added, the more the reaction became impaired (Figure 20.1). In fact, maximum yield in this reaction remained 20%. By changing the enzyme to phospholipase A_1 and then examining the effect of moisture on the yield of the desired LPC by varying the amount of adding water, the LPC yield increased up to 200 μL water addition and then it decreased (Figure 20.2). The optimum water amount corresponded to 10 times the amount of lipozyme RMIM. Effect of $SC\text{-}CO_2$ pressure on the LPC yield under the phospholipase A_1-mediated reaction was examined. Basically, the pressure itself gave a small effect on the yield. However, among 10 MPa, 17.7 MPa, and 25.0 MPa, 10 MPa gave the best yield (Figure 20.3). The effect of $SC\text{-}CO_2$ flow rate on the yield of LPC was also small in the phospholipase A_1-mediated reaction system, as shown in Figure 20.4. The LPC yield was 40% to 45% despite the $SC\text{-}CO_2$ flow rate. Reaction temperature basically

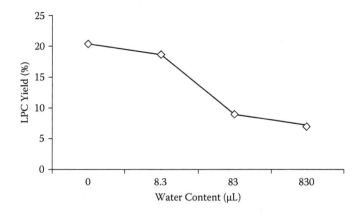

FIGURE 20.1 Effect of added moisture on the yield of DHA-bound lysophosphatidylcholine (DHA-LPC) after partial hydrolysis of DHA-bounded phosphatidylcholine (PC) mediated by lipozyme RMIM: 20 mg PC, 53 mg lipozyme RMIM, 17.7 MPa pressure, 33°C, 3.0 mL/min $SC\text{-}CO_2$ flow rate, 4 h reaction, 10 mL reaction volume.

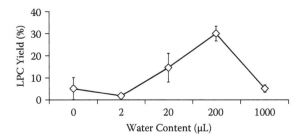

FIGURE 20.2   Effect of added moisture on the yield of DHA-LPC after partial hydrolysis of DHA-PC mediated by phospholipase A_1: 20 mg PC, 20 mg phospholipase A_1, 17.7 MPa pressure, 33°C, 3.0 mL/min SC-CO_2 flow rate, 4 h reaction.

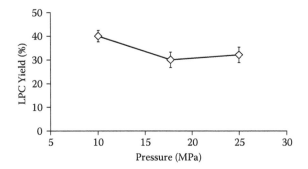

FIGURE 20.3   Effect of pressure on the yield of DHA-LPC after partial hydrolysis of DHA-PC mediated by phospholipase A_1. Conditions the same as in Figure 20.2 except that SC-CO_2 pressure varied.

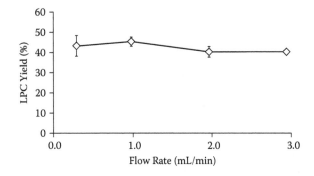

FIGURE 20.4   Effect of SC-CO_2 flow rate on the yield of DHA-LPC after partial hydrolysis of DHA-PC mediated by phospholipase A_1. Conditions same as in Figure 20.2 except SC-CO_2 flow rate varied.

did not have an effect on the yield of the desired LPC in phospholipase A_1-mediated system up to 60°C, giving around a 45% yield. This phenomenon is extraordinary for an enzymatic reaction. However, when the temperature exceeded 60°C, the yield

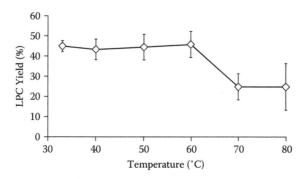

FIGURE 20.5   Effect of temperature on the yield of DHA-LPC after partial hydrolysis of DHA-PC mediated by phospholipase A_1. Condition same as in Figure 20.2 except reaction temperature varied.

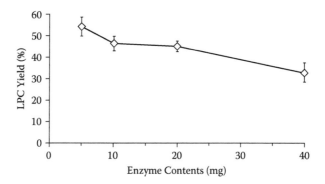

FIGURE 20.6   Effect of phospholipase A_1 amount on the yield of DHA-LPC after partial hydrolysis of DHA-PC. Condition same as in Figure 20.2 except phospholipase A_1 amount varied.

was drastically impaired, no doubt due to denaturation of the enzyme (Figure 20.5). We considered that 33°C is the most desirable condition among the examined temperatures regarding the denaturation of the enzyme.

Generally, enzyme amount simply affects reaction velocity. For this reason, it was predicted that the velocity should increase in proportion to the increase in phospholipase A_1 amount. However, as shown in Figure 20.6, under the same reaction time, the yield of the desired LPC was rather impaired when the amount of phospholipase A_1 was increased. This may be due to lack of essential water for the activation of phospholipase A_1 under the closed-batch condition employed. By varying the reaction time, it was borne out that there is an optimum reaction time that gives the maximum yield. When the reaction time was elongated, the yield of the desired LPC was impaired (Figure 20.7). Hydrolysis of the obtained LPC should be responsible for this. Throughout these experiments, it was concluded that the following partial hydrolytic condition gives the best yield in SC-CO_2 medium when the substrate PC amount is 20 mg in 10 mL reaction volume.

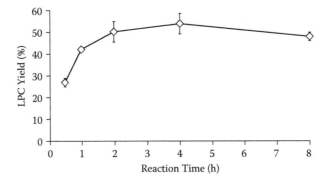

FIGURE 20.7   Effect of reaction time on the yield of DHA-LPC after partial hydrolysis of DHA-PC mediated by phospholipase A_1. Conditions same as in Figure 20.2 except reaction time varied.

Substrate PC amount: 20 mg in 10 mL reaction volume
Added water: 200 μL
Amount of phospholipase A_1: 5 mg
Pressure: 10 MPa
Temperature: 33°C
Fluid velocity: 1 mL/min of SC-CO_2
Reaction time: 4 h

Figure 20.8 is a view from the reaction cell window under the optimum condition to obtain the LPC. As can be seen, phospholipase A_1 stuck on the wall of the reaction cell and window. Under ordinary conditions for enzymatic reaction, both in aqueous and in nonaqueous media, occurrence of the adhered enzyme should seriously impair the efficiency of the enzymatic reaction. However, in

FIGURE 20.8   View from the reaction cell window under the optimum condition to obtain the DHA-LPC. Probe seen from the left wall is for temperature measurement; the rotating object near the bottom is a magnetic stirrer.

Table 20.1   Fatty Acid Composition of the Substrate Squid PC and the Obtained LPC after Partial Hydrolysis in SC-CO$_2$ Media

Fatty Acids	Substrate Squid PC (%)	Product LPC (%)
$C_{14:0}$	1.62	0.13
$C_{16:0}$	34.03	0.83
$C_{16:1}$	0.58	0.06
$C_{18:0}$	1.35	1.01
$C_{18:1}$	2.94	0.13
$C_{20:1}$	0.03	0.13
$C_{20:4}$	0.92	0.91
$C_{20:5}$ (EPA)	8.40	10.82
$C_{22:6}$ (DHA)	39.90	83.93

SC-CO$_2$ medium, there seemed to be no impairment by the adhesion. This should be owing to very strong permeability of the SC-CO$_2$ allowing the dissolved substrate to contact sufficiently with the phospholipase A$_1$. Under the aforementioned optimum reaction condition, the obtained LPC derived from squid PC contained approximately 73% DHA and 18% eicosapentaenoic acid (EPA), as shown in Table 20.1, whereas the original substrate squid PC contained 45% DHA and 7% EPA.

20.3   TRANSPORT OF THE OBTAINED LPC IN SMALL INTESTINAL EPITHELIAL CELL MODEL

LPC should be absorbed easily in the small intestine because it has both hydrophobic and hydrophilic groups; also, the molecular weight of LPC is much smaller than PC. For this reason, it was predicted that DHA-LPC should be transported much easier than DHA-PC through the small intestinal epithelial cells. For transport studies *in vitro*, a small intestinal epithelial cell model was constructed (Satsu et al. 2001, Hossain et al. 2006, Rieux et al. 2007). Briefly, Caco-2 cells were plated onto a polycarbonate transwell filter. The cells were grown in Dulbecco's modified Eagle's medium with glucose, glutamine, penicillin, streptomycin, nonessential amino acids, and heat-incubated fetal calf serum in a 5% CO$_2$ atmosphere at 37°C. These cells were given fresh Dulbecco's modified Eagle's medium at 3-day intervals until used. Cells were grown for 20 days (postconfluence) before the experiments. Before each experiment, cells were washed twice with phosphate-buffered saline and preequilibrated for 10 min in Hanks' balanced salt solution containing morpholinoethanesulfonic acid in the apical chamber and Hanks' balanced salt solution containing 2-[4-(2-hydroxyethyl)-1-piperazinyl] ethanesulfonic acid in the basolateral chamber. Flux of Lucifer Yellow (a kind of fluorescent reagent) and the transepithelial electrical resistance between the apical chamber and the basolateral chamber (transepithelial electrical resistance [TEER] value) were measured to verify

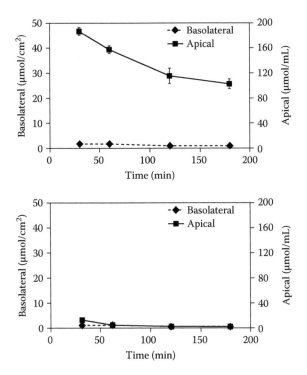

FIGURE 20.9   Direct incorporation of DHA-LPC (bottom) and DHA-PC (top) into M cell-expressed small intestinal epithelial cell model.

the completion of the small intestinal epithelial cell model. After the completion of the small intestinal epithelial cell model, DHA-LPC labeled with 1,6-diphenyl-1,3,5-hexatriene (DPH), a fluorescent reagent (Andrich and Vanderkooi 1976, Blitterswijk et al. 1981, Kumar and Misra 1991, Koba et al. 1993, Ohno et al. 1981, Cooper and Meddings 1991, Giannettini et al. 1991, Muller et al. 1990, Gareau et al. 1991) and, as comparison, DPH-labeled DHA-PC solutions were added to the apical chamber. As shown in Figure 20.9, LPC (i.e., the DHA-LPC) seemed to incorporate much faster than the DHA-PC. In fact, the fatty acid composition of the lipid obtained from the small intestinal epithelial cells of the model demonstrated the increase in DHA amount (Figure 20.10). Although there is no doubt that this should be the main transport route of the DHA-LPC, we monitored the TEER value of the small intestinal epithelial cell model to make sure that tight junction routes between the epithelial cells may not be responsible in transporting the DHA-LPC. As shown in Figure 20.11, the DHA-LPC itself decreased the TEER value, showing that the DHA-LPC tends to open the tight junction more than the DHA-PC does. However, as shown in Figure 20.12, when bile acid coexists, the TEER value did not decrease, implying that as long as bile acid coexists, the tight junction opening may not occur. We should say that direct incorporation into the epithelial cell is the most responsible transport route for the DHA-LPC.

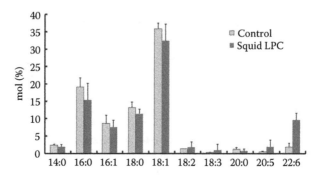

FIGURE 20.10 Fatty acid composition of the lipid obtained from the small intestinal epithelial cell. Dark-color bars designate the fatty acid composition after the incorporation of DHA-LPC.

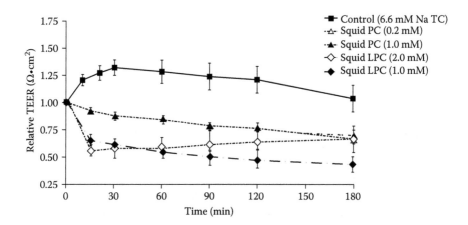

FIGURE 20.11 Effect of DHA-LPC itself on transepithelial electrical resistance (TEER) between the apical chamber and the basolateral chamber. When the TEER value decreases, it indicates that tight junctions between the epithelial cells are opened.

20.4 HEALTH BENEFITS OF THE DHA-LPC

LPC is known to cause hemolysis. This is because LPC incorporates very easily in the erythrocyte cell membrane and weakens the erythrocyte itself. However, this hemolysis depends on the level of LPC concentration. Under low-LPC concentration, hemolysis will not occur. We evaluated the deformability of the erythrocytes when treated with low-concentration DHA-LPC and the DHA-PC by measuring the flow speed of the erythrocytes going through a microchannel array called MC fan (Hosokawa et al. 1995, Nojima et al. 1995). As comparison, flow speed of soy PC- and LPC-incorporated erythrocytes was also measured.

The left figure in Figure 20.13 shows the flow speed of the erythrocytes after 3-h incubation with phospholipids, and the right shows that of 1-h incubation. As can be

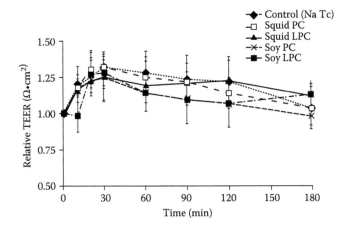

FIGURE 20.12   Effect of DHA-LPC under bile acid coexistence on TEER between the apical chamber and the basolateral chamber.

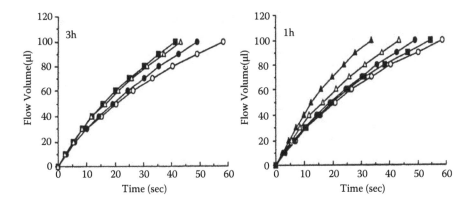

FIGURE 20.13   Flow curves of human erythrocytes treated with some PCs and DHA-LPC for 3 h (left) and 1 h (right). ○, Control; ●, soy PC 10 μ*M*; ■, DHA-PC 10 μ*M*; △, DHA-LPC 5 μ*M*; ▲, DHA-LPC 10 μ*M*.

seen, all of the DHA-containing phospholipid-incorporated erythrocytes showed a higher flow rate than the soy PC-incorporated erythrocytes. A notable feature is that the DHA-LPC improved the deformability of the human erythrocyte more rapidly than the DHA-PC did.

We also predicted that the DHA-LPC should suppress leukotriene B_4 (LTB_4) release from mast cells as other ω-3 fatty acid-containing lipids do. To measure the LTB_4 release from mast cells, mast cells were cultured with interleukin 3 (IL-3) and sodium alginate for 2 days, then rinsed with Tyrode buffer, and centrifuged to obtain a precipitated cell pellet. Calcium ionophore, calcium chloride, and lipid samples were added, then incubated for 20 min. Reactions were terminated with acetone/methanol that contains prostaglandin B_2 as an internal standard. LTB_4 itself was measured by

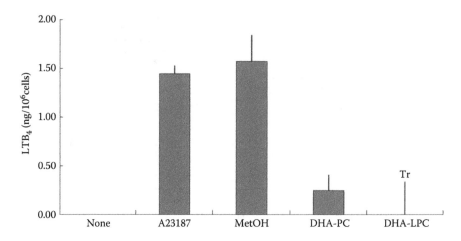

FIGURE 20.14 Effect of DHA-LPC and DHA-PC on LTB_4 release from mast cell lines. A23187 designates calcium ionophore, and MetOH designates methanol as controls.

high-performance liquid chromatography (HPLC). As shown in Figure 20.14, both the DHA-PC and the DHA-LPC suppressed the LTB_4 release. Between these two, the DHA-LPC seemed more effective on suppressing LTB_4. From this result, we may say that the DHA-LPC must be one of the most quick-acting DHA-bound lipid classes for allergy therapy.

Other than the health benefits discussed, the DHA-LPC should have many other health beneficial effects on humans that have to be exploited.

ACKNOWLEDGMENTS

We are grateful to Masashi Hosokawa and Shigeru Katayama of Hokkaido University, who advised us how to develop the small intestinal epithelial cell model.

REFERENCES

Andrich, M.P., and Vanderkooi, J.M., Temperature dependence of 1,6-diphenyl-1,3,5-hexatriene fluorescence in phospholipid artificial membranes, *Biochemistry*, 15, 1257, 1976.

Blitterswijk, W.J.V., Hoeven, R.P.V., and Dermeer, B.W.V., Lipid structural order parameters (reciprocal of fluidity) in biomembranes derived from steady-state fluorescence polarization measurements, *Biochem. Biophys. Acta*, 644, 323, 1981.

Cooper, P., and Meddings, J.B., Erythrocyte membrane fluidity in malignant hyperthermia, *Biochim. Biophys. Acta*, 1069, 151, 1991.

Gareau, R., Goulet, H., Chenard, C., Caron, C., and Brisson, R., Fluorescence studies on aged and young erythrocyte populations, *Cell. Mol. Biol.*, 37, 15, 1991.

Gianettini, J., Chauvet, M., Dell'Amico, M., Chautan, M., and Bourdeaux, M., Fluidity of human erythrocyte ghosts, *Biochem. Int.*, 24, 917, 1991.

Hosokawa, M., Enzymatic synthesis of phospholipids containing polyunsaturated fatty acids and their physiological functions [in Japanese], Ph.D. dissertation, Hokkaido University, Hakodate, Japan, 1996.

Hosokawa, M., Takahashi, K., Kikuchi, Y., and Hatano, M., Preparation of therapeutic phospholipids through porcine pancreatic phospholipase A$_2$ mediated esterification and lipozyme mediated acidolysis, *J. Am. Oil Chem. Soc.*, 72, 1287, 1995.

Hossain, Z., Kurihara, H., Hosokawa, M., and Takahashi, K., Docosahexaenoic acid and eicosapentaenoic acid-enriched phosphatidylcholine liposomes enhance the permeability, transportation and uptake of phospholipids in Caco-2 cells, *Mol. Cell. Biochem.*, 285, 155, 2006.

Koba, K., Wakamatsu., K., Obata, K., and Sugano, M., Effect of dietary proteins on linoleic acid desaturation and membrane fluidity in rat liver microsomes, *Lipids*, 28, 457, 1993.

Kumar, V., and Misra, U.K., Hepatic plasma membrane fluidity and dietary proteins, *Indian J. Biochem. Biophys.*, 28, 301, 1991.

Muller, S., Donner, M., and Stoltz, J.F., Fluorescent probe for membrane fluidity approach, *Ann. Int. Conf. IEEE Eng. Med. Biol. Soc.*, 12, 1624, 1990.

Nara, E., Miyashita, K., and Ota, T., Oxidative stability of liposomes prepared from soybean PC, chicken egg PC, and salmon egg PC, *Biosci. Biotechnol. Biochem.*, 61, 1736, 1997.

Nojima, M., Hosokawa, M., Takahashi, K., Hatano, M., and Kikuchi, Y., Effect of *in vitro* treatment with glycerophospholipids containing highly polyunsaturated fatty acids on deformability of human erythrocytes [in Japanese], *Nippon Suisan Gakkaishi*, 61, 197, 1995.

Ohno, H., Shimizu, N., Tsuchida, E., Sasakawa, S., and Honda, K., Fluorescence polarization study on the increase of membrane fluidity of human erythrocyte ghosts induced by synthetic water-soluble polymers, *Biochim. Biophys. Acta*, 649, 221, 1981.

Ono, M., Hosokawa, M., Inoue, Y., and Takahashi, K., Concentration of docosahexaenoic acid-containing phospholipid through lipozyme IM-mediated hydrolysis, *J. Jpn. Oil Chem. Soc.*, 46, 867, 1997a.

Ono, M., Hosokawa, M., Inoue, Y., and Takahashi, K., Water activity-adjusted enzymatic partial hydrolysis of phospholipids to concentrate polyunsaturated fatty acids, *J. Am. Oil Chem. Soc.*, 74, 1415, 1997b.

Rieux, A.D. , Fievez, V., Théate, I., Mast, J., Preat, V., and Chneider, Y.-J., An improved in vitro model of human intestinal follicle-associated epithelium to study nanoparticle transport by M cells, *Eur. J. Pharm. Sci.*, 30, 380, 2007.

Satsu, H., Yokoyama, T., Ogawa, N., Fujiwara-Hatano, Y., and Shimizu, M., The changes in the neuronal PC12 and the intestinal epithelial Caco-2 cells during the coculture. The functional analysis using an in vitro coculture system, *Cytotechnology*, 35, 73, 2001.

21 Regiospecific Quantification of Triacylglycerols by Mass Spectrometry and Its Use in Olive Oil Analysis

Jiann-Tsyh Lin

CONTENTS

Key Words: Mass spectrometry; olive oil; quantification; regiospecific; triacylglycerols.

21.1 INTRODUCTION

The structures of stereospecific (*sn*-1, *sn*-2, and *sn*-3) triacylglycerols (TAGs) are shown in Figure 21.1 using dioleoylpalmitoylglycerol as the example. The regiospecific isomers (*sn*-2 and *sn*-1,3) do not differentiate the *sn*-1 and *sn*-3 location isomers (Figure 21.1B and 21.1C). The location of acyl chains on the glycerol backbone of TAG from biological origins can be used in determining the acylation steps in the biosynthetic pathway. Regiospecific locations of acyl chains on the glycerol backbone affect the physical properties of the oils (TAG) for industrial uses (e.g., viscosity, pour point, melting point, heat of fusion, solubility, crystal structure, and polymorphism) [1]. It also affects the human absorption of oil in food and is thus valuable in nutritional value for food industry.

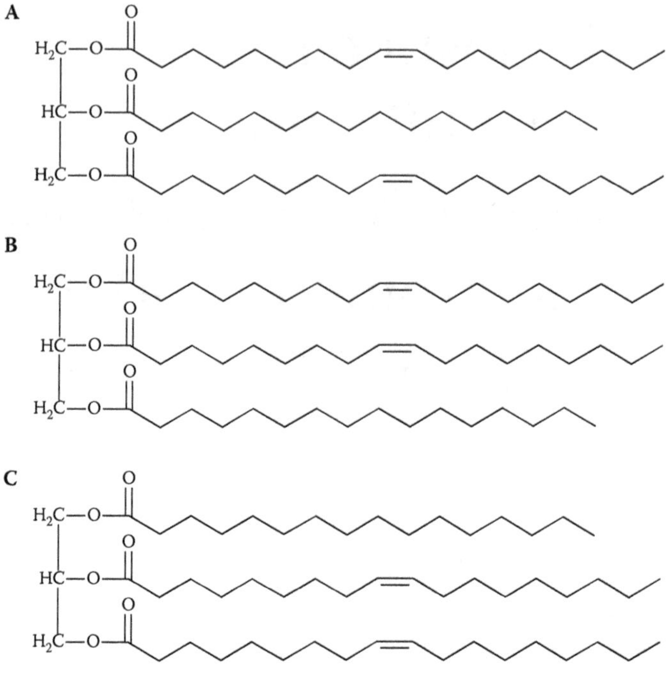

FIGURE 21.1 Structures of stereospecific dioleoylpalmitoylglycerols. (A) 1,3-Dioleoyl-2-palmitoyl-*sn*-glycerol. (B) 1,2-Dioleoyl-3-palmitoyl-*sn*-glycerol. (C) 2,3-Dioleoyl-1-palmitoyl-*sn*-glycerol.

Regioisomers of TAG in biological samples have been identified by atmospheric pressure chemical ionization-mass spectrometry (APCI-MS) [2,3] and by electro-spray ionization-mass spectrometry (ESI-MS) [4,5] based on the premise that the loss of the acyl chain from the *sn*-1 or *sn*-3 position is energetically favored over the loss from the *sn*-2 position. The quantification of the regioisomers, dilinoleoyloleoyl-glycerol (LOL, LLO), has been determined from a linear calibration curve of the ratio of [LL]+ and [LO]+ fragment ions derived from various concentrations of regio-specific LOL and LLO standards by APCI-MS [6,7]. Recently, calibration curves of LLO, LOO, POO, and PPO were used for quantification of oils and fat [8].

The regiospecific characterization of TAG as lithiated adducts by collisionally activated dissociation tandem mass spectrometry (CAD-MS^2) has been reported by Hsu and Turk [9]. The ions were reported from the loss of fatty acids as α,β-unsaturated fatty acid specific at the *sn*-2 position of TAG of lithiated adducts. We have used these ions to quantitate six regiospecific diricinoleoylacylglycerols (RRAcs) containing ricinoleate (a hydroxyl fatty acid) in castor oil [10].

Figure 21.1 shows the structures of *sn*-OPO (1,3-dioleoyl-2-palmitoyl-*sn*-glycerol), *sn*-OOP (1,2-dioleoyl-3-palmitoyl-*sn*-glycerol), and *sn*-POO (1-palmi-toyl-2,3-dioleoyl-*sn*-glycerol). Currently, TAGs of *sn*-AAB and *sn*-BAA are not commercially available and cannot be differentiated by MS. AAB shown here is the mixture of AAB and BAA as 1(3),2-dioleoyl-3(1)-palmitoyl-*sn*-glycerol unless specified. ABA shown here is stereospecific as *sn*-ABA. The regiospecific analysis of TAGs ABA and AAB containing only common fatty acids (nonhydroxyl fatty acids) is described here by ESI-MS³ of lithiated adducts. We have used this method to quantify the regiospecific ABA and AAB in extra virgin olive oil.

21.2   MASS SPECTROMETRY OF OPO

The MS method used was similar to that of our recent report [10]. Figure 21.2 shows the MS² spectrum of [OPO + Li]⁺ at *m/z* 865.5. The spectrum is simple and shows the precursor ion and the fragment ions of [OPO + Li − OCOOH]⁺ at *m/z* 583.4 and [OPO + Li − PCOOH]⁺ at *m/z* 609.4, reflecting the neutral losses of oleic acid (OCOOH) and palmitic acid (PCOOH), respectively.

Figure 21.3 shows the MS³ spectrum of [OPO + Li − OCOOH]⁺ at *m/z* 583.5. A regiospecific ion was derived from the loss of palmitate specific at the *sn*-2 position as α,β-unsaturated palmitate corresponding to the ion of [OPO + Li − OCOOH − P′CH=CHCOOH]⁺ at *m/z* 329.2. The ion of [OPO + Li − OCOOH − O′CH=CHCOOH]⁺ at *m/z* 303 was not detected because the oleoyl chain was not at the *sn*-2 position. The fragmentation pathway for the ions from the loss of fatty acids specific at the *sn*-2 position as α,β-unsaturated fatty acids from TAG-lithiated adducts was previously proposed [9]. An ion at *m/z* 527.3 (Figure 21.3) may be the acid anhydride of oleate and palmitate, [OCOOCOP + Li]⁺, and is the same as [OPO + Li − OCOOH − C₃H₄O]⁺. The loss of C₃H₄O was from the glycerol backbone. The fragmentation pathway for the formation of acid anhydride ion has been proposed [10]. Both postulated pathways involved the intermediate containing a 1,3-dioxolane five-membered ring with the two carbon atoms of the ring originating from the glycerol backbone.

Figure 21.4 shows the MS³ spectrum of [OPO + Li − PCOOH]⁺ at *m/z* 609.3. The major fragment ion [OPO + Li − PCOOH − O′CH=CHCOOH]⁺ at *m/z* 329.2 was apparently from the loss of fatty acid at the *sn*-1(3) position as α,β-unsaturated oleate because the precursor ion contained no oleoyl chain at the *sn*-2 position. As shown in Figure 21.3, the precursor ion containing an acyl chain at the *sn*-2 position produced a fragment ion from the loss of a fatty acid as α,β-unsaturated fatty acid at the *sn*-2 position only, not *sn*-1,3 positions. As shown in Figure 21.4, the precursor ion containing no acyl chain at the *sn*-2 position produced a fragment ion from the loss of a fatty acid as α,β-unsaturated acyl chain at the *sn*-1,3 position. The MS³ spectra of other ABA, POP, OSO, and SOS were similar to those of Figures 21.3 and 21.4: [ABA + Li − ACOOH − B′CH=CHCOOH]⁺ were detected and [ABA + Li − ACOOH − A′CH=CHCOOH]⁺ were not detected.

21.3   MASS SPECTROMETRY OF OOP

Figure 21.5 shows the MS³ spectrum of [OOP + Li − OCOOH]⁺ at *m/z* 583.4. This precursor ion was the mixture of 1,2-PO and 1,3-OP from OOP. The precursor ion, 1,3-OP, produced the fragment ions of both [OOP + Li − OCOOH − P′CH=CHCOOH]⁺ at *m/z* 329 and [OOP + Li − OCOOH − O′CH=CHCOOH]⁺ at *m/z* 303.1. The precursor ion, 1,2-PO, produced the fragment ion of [OOP + Li − OCOOH − O′CH=CHCOOH]⁺ at

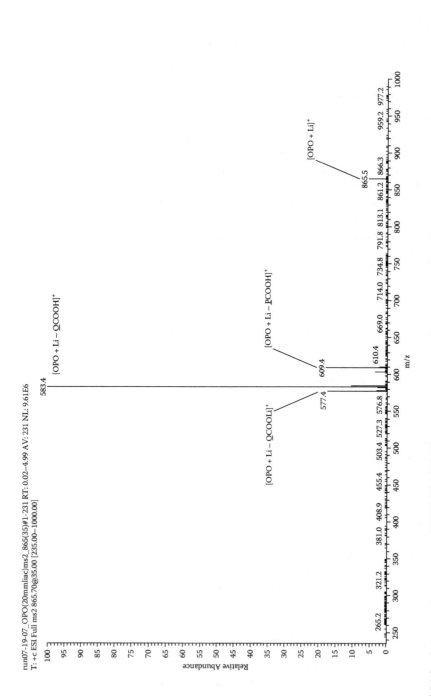

FIGURE 21.2   Ion trap mass spectrum of ESI-MS^2 of [OPO + Li]$^+$ ion at m/z 865.5. Both O and OCOOH are oleic acid. Both P and PCOOH are palmitic acid. OPO is 1,3-dioleoy-2-palmitoyl-sn-glycerol. OCOOLi is the lithium salt of oleic acid.

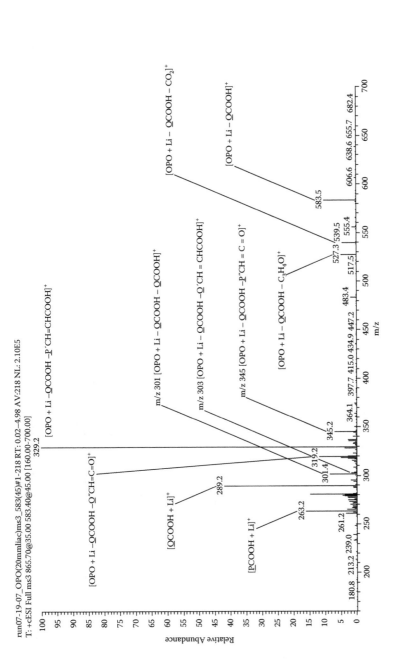

FIGURE 21.3 Ion trap mass spectrum of ESI-MS³ of [OPO + Li – QCOOH]⁺ at m/z 583.5. For abbreviations, see Figure 21.2. P′CH=CHCOOH is the α,β-unsaturated palmitic acid from the sn-2 position; Q′CH=CHCOOH is α,β-unsaturated oleic acid from the sn-1,3 position (not detected); C₃H₄O is the loss of glycerol backbone to form acid anhydride of two fatty acids (6); P″CH=C=O is palmitoyl ketene from the sn-2 position; and Q″CH=C=O is oleoyl ketene from the sn-1,3 position.

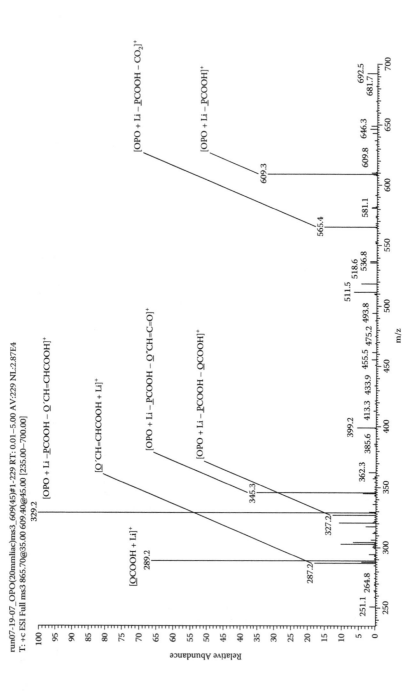

FIGURE 21.4 Ion trap mass spectrum of ESI-MS³ of [OPO + Li – PCOOH]⁺ at *m/z* 609.3. For abbreviations, see Figures 21.2 and 21.3. This spectrum is very much the same as that of ESI-MS³ of [OOP + Li – PCOOH]⁺ at *m/z* 609.3. OOP is the standard 1,2-dioleoyl-3-palmitoyl-*rac*-glycerol.

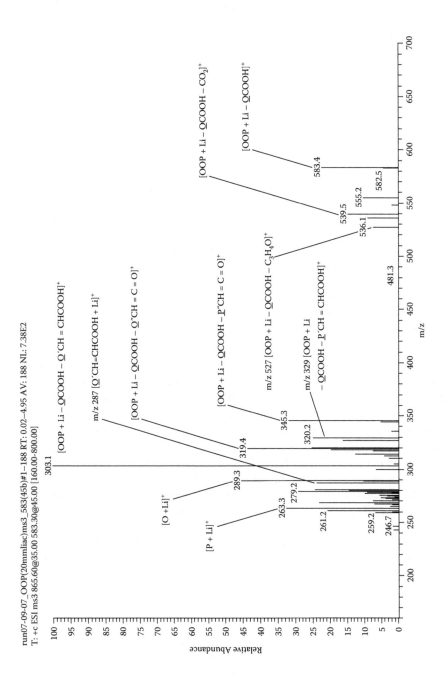

FIGURE 21.5 Ion trap mass spectrum of ESI-MS³ of [OOP + Li – $\underline{O}$COOH]⁺ at *m/z* 583.4. For abbreviations, see Figures 21.2, 21.3, and 21.4.

m/z 303.1, but not [OOP + Li − O̲COOH − P̲′CH=CHCOOH]⁺ at *m/z* 329. The ratio of the abundances of these two fragment ions, *m/z* 329 and *m/z* 303.1, was 0.25. The fragment ion of [OOP + Li − O̲COOH − O̲′CH=CHCOOH]⁺ at *m/z* 303.1 was the base peak while [OOP + Li − O̲COOH − P̲′CH=CHCOOH]⁺ at *m/z* 329 was a minor peak (25%). For the accurate quantification of the percentages of the regioisomers in the mixture of OPO and OOP, this minor peak from OOP must be considered because it interferes with the ion [OPO + Li − O̲COOH − P̲′CH=CHCOOH]⁺ at *m/z* 329.2 from OPO.

21.4　REGIOSPECIFIC ANALYSIS OF THE MIXTURE OF OPO AND OOP

Analysis of regiospecific OPO and OOP depends on the fragment ions from the losses of fatty acid as α,β-unsaturated fatty acids specific at the *sn*-2 position. Figure 21.3 from the OPO standard shows the major ion of [OPO + Li − O̲COOH − P̲′CH=CHCOOH]⁺ at *m/z* 329.2 from the loss of P̲′CH=CHCOOH at the *sn*-2 position, and the ion of [OOP + Li − O̲COOH − O̲′CH=CHCOOH]⁺ at *m/z* 303 was not detected. Figure 21.5 from the regiospecific OOP standard shows the ion of [OOP + Li − O̲COOH − O̲′CH=CHCOOH]⁺ at *m/z* 303.1 at the *sn*-2 position and at less abundance the ion of [OOP + Li − O̲COOH − P̲′CH=CHCOOH]⁺ at *m/z* 329 from the loss of P̲′CH=CHCOOH at the *sn*-1,3 position. The ion at *m/z* 303.1 from the loss of O̲′CH=CHCOOH at the *sn*-2 position can be used to detect a very small amount of OOP because OPO does not produce the ion at *m/z* 303 (Figure 21.3).

The MS³ spectrum of the sample containing both regiospecific OPO and OOP will be the mix of ions shown in Figures 21.3 and 21.5. The ratio of these two regioisomers can be estimated from the ratio of the abundances of the ions at *m/z* 329.2 and *m/z* 303.1 from the losses of fatty acids mostly at the *sn*-2 position as the α,β-unsaturated fatty acids in assuming a linear relationship [6]. However, the ion observed at *m/z* 329.2 resulted not only from the loss of P′CH=CHCOOH at the *sn*-2 position of OPO but also from the loss of P̲′CH=CHCOOH at the *sn*-1,3 position of OOP to a lesser extent. The assumption was that the abundances of the ions [OOP + Li − O̲COOH − O̲′CH=CHCOOH]⁺ at *m/z* 303.1 and [OOP + Li − O̲COOH − P̲′CH=CHCOOH]⁺ at *m/z* 329 from OOP combined and the abundance of the ion [OPO + Li − O̲COOH − P̲′CH=CHCOOH]⁺ at *m/z* 329.2 from OPO were proportional to the ratio of OOP to OPO. The abundance of the ion [OOP + Li − O̲COOH − P̲′CH=CHCOOH]⁺ at *m/z* 329 from OOP in a sample of regiomixture can be estimated from the abundance of the ion at *m/z* 303 multiplied by the ratio of the abundances of *m/z* 329 and *m/z* 303 from the regiospecifically pure OOP standard as shown in Figure 21.5 and Table 21.1, which was determined to be 0.25. The MS³ spectra of other AAB shown in Table 21.1 also contained the ions of [AAB + Li − A̲COOH − B̲′CH=CHCOOH]⁺ and [AAB + Li − A̲COOH − A̲′CH=CHCOOH]⁺ as Figure 21.5.

Table 21.1 shows the ratios of the peak abundances of [AAB + Li − A̲COOH − B̲′CH=CHCOOH]⁺ and [AAB + Li − A̲COOH − A̲′CH=CHCOOH]⁺ from the MS³ of various commercially available TAG standards, AAB. The ratio varied depending on the molecular species of AAB and MS operating conditions, including the

TABLE 21.1

Ratios of the Peak Abundances of [AAB + Li − ACOOH − B′CH=CHCOOH]⁺

and [AAB + Li − ACOOH − A′CH=CHCOOH]⁺ on the MS³ of Triacylglycerol

Standards, AAB, Lithium Adducts[a]

AAB[b]	AAB + Li	AAB + Li − ACOOH	CE[c] of MS²	CE[c] of MS³	Peak Mass[d]	Ratio
OOP	865.5	583.4	35%	45%	329/303	0.25
PPO	839.5	583.4	35%	45%	303/329	0.05
OOS	893.5	611.5	36%	44%	329/331	0.17
SSO	895.6	611.5	35%	42%	331/329	0.15
OOL	889.6	607.4	35%	44%	329/327	0.10
LLO	887.7	607.4	35%	45%	327/329	0.28
LLP	861.5	581.4	36%	46%	327/303	0.37
SSP	869.8	585.5	35%	45%	331/303	0.17

[a] A, ACOOH and B are fatty acids; B′CH=CHCOOH is α,β-unsaturated fatty acid from triacylglycerol AAB.

[b] O, oleic acid; P, palmitic acid; S, stearic acid; L, linoleic acid.

[c] Collision energy.

[d] The masses are [AAB + Li − ACOOH − B′CH = CHCOOH]⁺/[AAB + Li − ACOOH − A′CH = CHCOOH]⁺.

collision energies of MS² and MS³. The collision energies for MS² and MS³ fragmentation that were applied to obtain the MS³ spectra are also given in Table 21.1. On the analysis of biological samples, the MS operating conditions should be the same as those of the AAB standards. The data in Table 21.1 have been reproducible.

21.5 EARLIER ANALYSIS OF TAG IN OLIVE OIL

Identification and quantification of the molecular species of TAGs (nonregiospecific) in olive oil by MS have been reported [11–13]. The contents varied, and the latest report [11] showed the TAG (nonregiospecific) contents in olive oil (Aix-en-Provence, France) as follows: OOO (35%), OOP (22%), OOL (15%), PLO (7.7%), PPO (3.6%), OOS (3.2%), LLO (2.9%), LLP (1.1%), PPL (0.8%). The quantitative regiospecific analysis of AAB and ABA in olive oil has only been reported for regiospecific LOL as 0% among LLO, LOL, and OLL combined [6]. The presence of some regiospecific OOP, OOS, OLO, LLO, POP, and LLP in olive oil has also been reported [2].

21.6 REGIOSPECIFIC QUANTIFICATION OF TAG IN OLIVE OIL

The fractionation of the molecular species of TAG in olive oil was as shown in Figure 21.6 [14,15]. Figure 21.7 shows the ion trap mass spectrum of ESI-MS³ of [OOP + Li − OCOOH]⁺ and [OPO + Li − OCOOH]⁺ at *m/z* 583.6 from the high-performance liquid chromatographic (HPLC) fraction of olive oil containing OOO, OOP (OPO), and PPO (POP) (same partition numbers). The ratio of the abundances of *m/z*

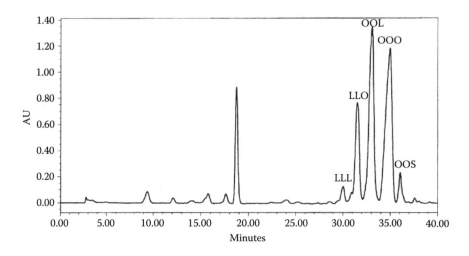

FIGURE 21.6 HPLC fractionation of extra virgin olive oil. Olive oil (1 mg) was chromatographed with a C_{18} analytical column, a linear gradient from 100% methanol to 100% 2-propanol in 40 min, 1-mL/min flow rate, 0.5 min/fraction, and detected at 205 nm. Triacylglycerols with the same partition times were eluted and collected together, such as fractions 63–64 (containing LLO, LLP), fractions 65–68 (OOL, POL, PPL), fractions 69–71 (OOO, OOP, PPO), and fractions 72–74 (OOS, POS).

329.3 and 303.2 was 31% from the mixture of OOP and OPO in olive oil (Figure 21.7). The ion m/z 303.2 was [OOP + Li − $\underline{O}$COOH − $\underline{O}'$CH=CHCOOH]$^+$ alone, and ion m/z 329.3 was from both [OOP + Li − $\underline{O}$COOH − $\underline{P}'$CH=CHCOOH]$^+$ and [OPO + Li − $\underline{O}$COOH − $\underline{P}'$CH=CHCOOH]$^+$. According to the ratio of OOP in Table 21.1 as 25%, the abundances of ions [OOP + Li − $\underline{O}$COOH − $\underline{O}'$CH=CHCOOH]$^+$ at m/z 303.2 and [OOP + Li − $\underline{O}$COOH − $\underline{P}'$CH=CHCOOH]$^+$ at m/z 329.3 from OOP combined was 100% + 25% = 125%. The abundances of these three ions combined was 100% + 31% = 131%. The content of OOP (and POO) among the three stereoisomers (OOP, OPO, and POO) combined was 125% divided by 131% and was 95%.

We have used the current method to estimate approximately the contents of regioisomers of ABA and AAB in the three stereoisomers (AAB, ABA, and BAA) combined in olive oil as follows: OOP (95%), POP (98%), OOL (61%), OOS (94%), PLP (100%), LLO (67%), and LLP (74%). AAB actually includes BAA. The saturated fatty acids palmitic acid and stearic acid were mostly located at the *sn*-1,3 positions, and unsaturated fatty acids oleic acid and linoleic acid were mostly located at the *sn*-2 position. On the plant TAG biosynthetic pathway, unsaturated fatty acids are incorporated mostly at the *sn*-2 position.

21.7 REGIOSPECIFIC QUANTIFICATION OF TAG USING CALIBRATION CURVES

Calibration curves have been used for the quantification of regioisomers [6,8]. Figure 21.8 shows the calibration curve using different ratios of OOP and OPO,

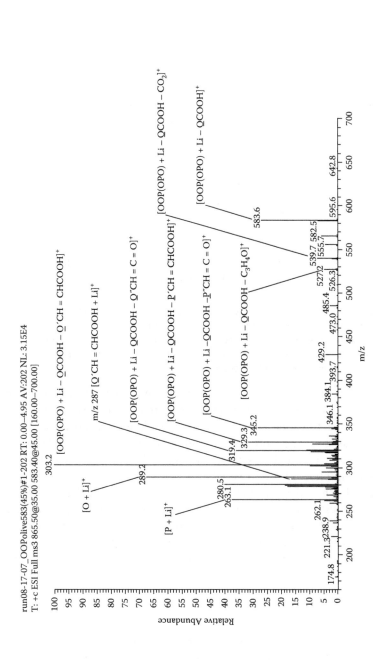

FIGURE 21.7 Ion trap mass spectrum of ESI-MS³ of [OOP(OPO) + Li − OCOOH]⁺ at m/z 583.6 from the HPLC fraction of olive oil containing OOO (triolein) as the major component in this fraction. For abbreviations, see Figures 21.2, 21.3, and 21.4.

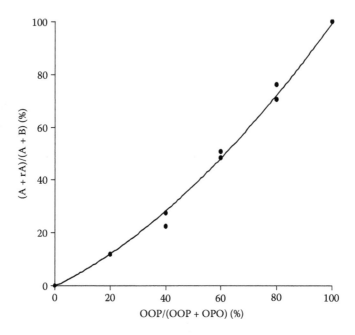

FIGURE 21.8 Calibration curve of the content of regioisomer (OOP) estimated at various contents of OOP in the mixture of OOP and OPO. A, the abundance of m/z at 303.1 from the mixture of OOP and OPO; B, the abundance of m/z at 329.2 from the mixture of OOP and OPO; r, the ratio of the abundances of m/z 329 and m/z 303.1 from OOP standard, see Figure 21.5; OOP, content of OOP; OPO, content of OPO.

which was nearly linear. The reason that this was not a straight line might be because this was an MS³ (two fragmentations) study. The current method is applicable for quantification; however, using calibration curves from each regioisomeric pair could be more accurate but more time consuming. Using the calibration curve of Figure 21.8 for estimation, OOP content among OOP and OPO combined in olive oil would be 97%, which was not much different from 95% estimated by the current method. However, at the worst deviation part (see Figure 21.8), 40% estimated by the current method (linear relationship) was actually 51% by the calibration curve (Figure 21.8). The ratios shown in Table 21.1 could vary if MS conditions change. Ideally, it is best to obtain the ratio values or calibration curves on the same day or within a few days of when the TAG samples are analyzed.

21.8 CONCLUSION

The MS³ method was developed to estimate the contents of regioisomers of TAG, AAB and ABA, from biological origins. The content of AAB in (AAB + ABA) is estimated as $(A + rA)/(A + B)$. A is the abundance of the ion [AAB + Li – ACOOH – ACH=CHCOOH]⁺ from AAB alone. A minor amount of AAB can be detected by this method because ABA could not contribute to this ion. B is the abundances of the ions [ABA + Li − $\underline{A}$COOH − $\underline{B}'$CH=CHCOOH]$^+$ and [AAB + Li − $\underline{A}$COOH − $\underline{B}'$CH=CHCOOH]$^+$ combined at the same m/z value. The r is the ratio of the abundances of the ions [AAB + Li − $\underline{A}$COOH − $\underline{B}'$CH=CHCOOH]$^+$/ [AAB + Li − $\underline{A}$COOH − $\underline{A}'$CH=CHCOOH]$^+$ from AAB (Table 21.1). The method can target the specific m/z of ABA and AAB in a biological sample and can thus avoid the contamination of ions from other TAG molecular species.

REFERENCES

1. Foubert, I., Dewettinck, K., Van de Walle, D., Dijkstra, A.J., and Quinn, P.J., Physical properties: Structural and physical characteristics. In *The Lipid Handbook* (3rd ed.), Gunstone, F.D., Harwood, J.L., and Dijkstra, A.J., Eds., CRC Press, Taylor & Francis Group, Boca Raton, FL, 2007, pp. 535–590.
2. Mottram, H.R., Woodbury, S.E., and Evershed, R.P., Identification of triacylglycerol positional isomers present in vegetable oils by high performance liquid chromatography/ atmospheric pressure chemical ionization mass spectrometry, *Rapid Commun. Mass Spectrom.*, 11, 1240–1252, 1997.
3. Mottram, H.R., Crossman, Z.M., and Evershed, R.P., Regiospecific characterization of the triacylglycerols in animal fats using high performance liquid chromatography-atmospheric pressure chemical ionization mass spectrometry, *Analyst*, 126, 1018–1024, 2001.
4. Kalo, P., Kemppinen, A., Ollilainen, V., and Kuksis, A., Regiospecific determination of short-chain triacylglycerols in butterfat by normal-phase HPLC with on-line electrospray-tandem mass spectrometry, *Lipids*, 39, 915–928, 2004.
5. Marzilli, L.A., Fay, L.B., Dionisi, F., and Vouros, P., Structural characterization of tri-acylglycerols using electrospray ionization-MS^n ion-trap MS, *J. Am. Oil Chem. Soc.*, 80, 195–202, 2003.
6. Jakab, A., Jablonkai, I., and Forgacs, E., Quantification of the ratio of positional iso-mer dilinoleoyl-oleoyl glycerols in vegetable oils, *Rapid Commun. Mass Spectrom.*, 17, 2295–2302, 2003.
7. Byrdwell, W.C., The bottom-up solution to the triacylglycerol lipidome using atmo-spheric pressure chemical ionization mass spectrometry, *Lipids*, 40, 383–417, 2005.
8. Leskinen, H., Suomela, J.P., and Kallio, H., Quantification of triacylglycerol regio-isomers in oils and fat using different mass spectrometric and liquid chromatographic methods, *Rapid Commun. Mass Spectrom.*, 21, 2361–2373, 2007.
9. Hsu, F.-F., and Turk, J., Structural characterization of triacylglycerols as lithiated adducts by electrospray ionization mass spectrometry using low-energy collisionally activated dissociation on a triple stage quadrupole instrument, *J. Am. Soc. Mass Spectrom.*, 10, 587–599, 1999.
10. Lin, J.T., and Arcinas, A., Regiospecific analysis of diricinoleoylacylglycerols in castor (*Ricinus communis* L.) oil by electrospray ionization-mass spectrometry, *J. Agric. Food Chem.*, 55, 2209–2216, 2007.
11. Ollivier, D., Artaud, J., Pinatel, C., Durbec, J., and Guerere, M., Differentiation of French virgin oil RDOs by sensory characteristics, fatty acids and triacylglycerol compositions and chemometrics, *Food Chem.*, 97, 382–393, 2006.
12. Jakab, A., Heberger, K., and Forgacs, E., Comparative analysis of different plant oils by high-performance liquid chromatography–atmospheric pressure chemical ionization mass spectrometry, *J. Chromatogr. A*, 976, 255–263, 2002.

13. Gomez-Ariza, J.L., Arias-Borrego, A., Garcia-Barrera, T., and Beltran, R., Comparative study of electrospray ionization sources coupled to quadrupole time-of-flight mass spectrometer for olive oil authentication, *Talanta*, 70, 859–869, 2006.

14. Lin, J.T., Woodruff, C.L., and McKeon, T.A., Non-aqueous reversed-phase high-performance liquid chromatography of synthetic triacylglycerols and diacylglycerols, *J. Chromatogr. A*, 782, 41–48, 1997.

15. Lin, J.T., and McKeon, T.A., Separation of the molecular species of acylglycerols by HPLC. In *HPLC of Acyl Lipids*, Lin, J.T., and McKeon, T.A., Eds., HNB, New York, 2005, pp. 199–220.

22 Utilization of Membrane Separation Technology for Purification and Concentration of Anserine and Carnosine Extracted from Chicken Meat

*Hiroshi Nabetani, Shoji Hagiwara,
Nobuya Yanai, Shigenobu Shiotani, Joosh
Baljinnyam, and Mitsutoshi Nakajima*

CONTENTS

22.1 INTRODUCTION

Membrane separation technology is a fine filtration technology that can separate molecules according to their molecular size. It requires low initial cost and little energy while keeping the quality of products high because it is a simple process that requires no heat treatment [1].

On the other hand, approximately 200,000 tons of egg-laying hens are discarded annually in Japan because the quality of their meat is low. However, chicken meat is rich in anserine and carnosine, which are dipeptides having unique and strong antioxidant functionality.

Anserine and carnosine consisting of β-alanine and L-histidine (Figure 22.1) are strong antioxidants against hypochlorite radical (ClO•), and they are antioxidants that reduce lipid oxidation, which affects flavor, aroma, texture, color, and nutritional composition [2,3]. Also, anserine and carnosine have immunoresponse modulation, blood fat reduction, and enhanced wound-healing functions *in vivo* [4–6]. Figure 22.2 shows an example of the physiological functionality of anserine and carnosine [6]. Anserine and carnosine remarkably reduced an oxidative stress in normal human volunteers in combination with vitamin C and ferulic acid.

$$NH_2-CH_2-CH_2-CO-NH-\underset{\overset{|}{COOH}}{CH}-CH_2$$

Anserine
(β-alanyl-1-methyl-L-histidine)

$$NH_2-CH_2-CH_2-CO-NH-\underset{\overset{|}{COOH}}{CH}-CH_2$$

Carnosine
(β-alanyl-L-histidine)

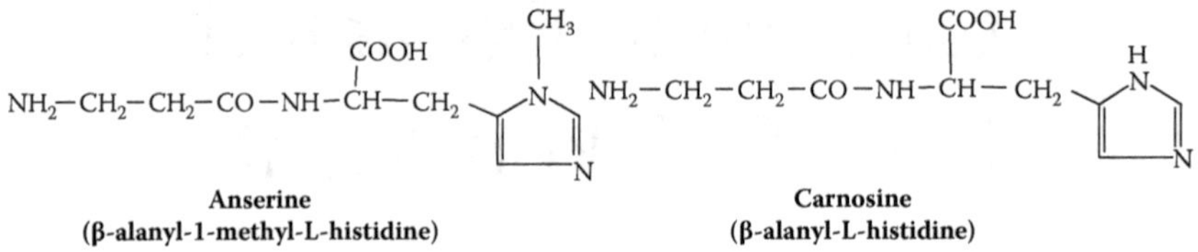

FIGURE 22.1 Rational formula of anserine and carnosine (AC).

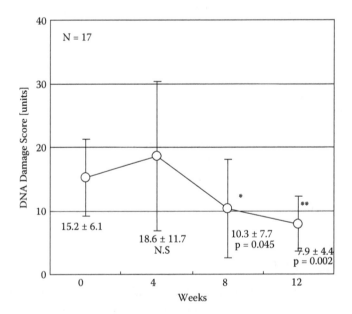

FIGURE 22.2 Reducing effect of AC (anserine and carnosine) in combination with vitamin C and ferulic acid on oxidative injury of lymphocyte DNA in normal human volunteers. Symbols indicate mean Comet assay score [7] and vertical bars indicate ± standard deviation; *$p < .05$, **$p < .01$.

If anserine and carnosine contained in the chicken meat can be purified at low cost using membrane separation technology, they can be promising components of a wide variety of functional foods. In this study, to add extra value to the discarded hen meat and utilize it, a membrane separation process that can purify anserine and carnosine contained in extract from the chicken meat was developed and its efficiency was demonstrated.

22.2 MATERIALS AND METHODS

Chicken extract was obtained from whole chicken carcasses by heating the carcasses with water (Figure 22.3). Then, the extract was treated with ion exchange resin to remove acidic and neutral amino acids and proteins. The chicken extract after the treatment with the ion exchange resin contained 6.79 g/L of anserine and carnosine, and their purity was 60%–70%. Impurities contained in the extract after the treatment with the ion exchange resin were creatinine and sodium chloride. Concentrations of these impurities were 2.30 g/L and 0.85 g/L, respectively. The extract after the treatment with the ion exchange resin was used as material for membrane separation experiments. A bench-scale membrane separation unit supplied by DSS (Danish Separation System) was used in this study.

To purify anserine and carnosine contained in the chicken extract, impurities (mainly creatinine and sodium chloride) need to be removed. Molecular weights of these impurities are 113 and 58, respectively, while average molecular weight of anserine and carnosine is 234. Therefore, 13 different kinds of nanofiltration membranes that had NaCl rejection values of 10% to 60% or molecular weight cutoff values of 700 to 2,500 were chosen. Membranes tested in this study are listed in Table 22.1.

At first, total circulation experiments were conducted with the membrane unit (Figure 22.4). In this experiment, composition of the feed solution was kept constant by returning the permeate from the membranes to the feed tank, and the effect

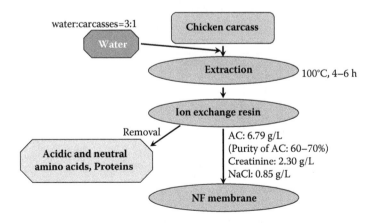

FIGURE 22.3 Procedure for preparation of chicken extract.

TABLE 22.1

Nanofiltration Membranes Tested in This Study

Membrane Type	NaCl Rejection[a]	Manufacturer	Material
NFT50	55	DSS	Polypiperazine/polyamide
DRA4510	45	DAISEN	Polyamide
Desal DL	15	Desalination	Polyamide (aromatic)
Desal DK	50	Desalination	Polyamide (aromatic)
NTR7430	30	NITTO DENKO	Sulfonated polyether sulfone
NTR7450	50	NITTO DENKO	Sulfonated polyether sulfone
NTR7250	60	NITTO DENKO	Polyvinyl alcohol
MPF34	35	Abcor	Polysulfone
MPF36	10	Abcor	Polysulfone
MPF44	25	Abcor	Polyacrylonitrile (PAN)
MPF50	700[a]	Abcor	Polyacrylonitrile (PAN)
G-5	1,000[a]	Desalination	Polyamide
G-10	2,500[a]	Desalination	Polyamide

[a] MW cutoff.

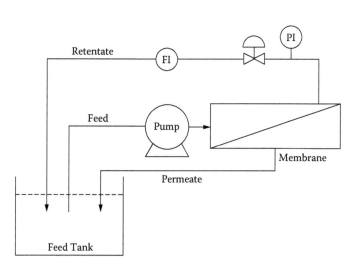

Feed flow rate: 5.8–11.3 L/min
Operating Pressure: 1–6 MPa
Temperature: 25°C

FIGURE 22.4 Schematic flow diagram of an experimental apparatus for total circulation experiment.

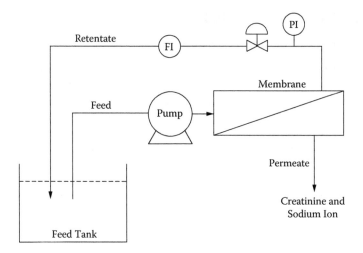

Area of membrane: 360 cm^2
Temperature: 25°C
Initial Volume of Feed: 11.3 L
Final Volume of Feed: 3.0 L

FIGURE 22.5 Schematic flow diagram of an experimental apparatus for batchwise concentration experiment.

of operating conditions on separation efficiency of each membrane was evaluated under different operating conditions (feed flow rate 5.8–11.3 L/min; operating pressure 1–6 MPa; temperature 25°C). Based on the experimental results obtained in the total circulation experiment, suitable membranes and operating conditions for purification of anserine and carnosine were selected.

Then, batchwise concentration experiments were performed with the selected membranes (360 cm²) and conditions (Figure 22.5). In this experiment, impurities such as creatinine and sodium chloride were taken out from the system together with permeate, and anserine and carnosine were purified and concentrated. Initial and final volumes of the feed were 11.3 and 3.0 L, respectively.

During all the membrane separation experiments, feed and permeate were sampled periodically. Concentration of anserine and carnosine was analyzed with high-performance liquid chromatography (HPLC). Concentration of creatinine and sodium ion was determined with HPLC and ICP-AES (inductively coupled plasma-atomic emission spectroscopy), respectively. Then, rejection ability of a membrane against each component was calculated with Eq. 22.1:

$$R_i = 1 - C_{p,i}/C_{f,i} \tag{22.1}$$

where R_i, $C_{p,i}$, and $C_{f,i}$ are observed rejection against component i and concentrations of component i in permeate and in feed, respectively.

22.3   RESULTS AND DISCUSSION

22.3.1   TOTAL CIRCULATION EXPERIMENTS

Effects of operating pressure on permeate flux value and rejection value during the total circulation experiments are shown in Figure 22.6. Operating pressure had little effect on the rejection value. Permeate flux value increased linearly with operating pressure up to 4 MPa. Once the operating pressure exceeded 4 MPa, permeate flux did not increase linearly. Therefore, suitable operating pressure for separation of anserine and carnosine with the nanofiltration membranes was determined to be 4 MPa. Similar results were obtained with other membranes.

Effects of feed flow rate on permeate flux value and rejection value during the total circulation experiments are shown in Figure 22.7. Feed flow rate showed almost no effect on rejection value. In the case of membranes that showed low permeate flux value, feed flow rate had almost no effect on permeate flux value. However, in the case of membranes with high permeate flux value such as NFT-50, permeate flux increased with feed flow rate up to 10 L/min as shown in Figure 22.7. Therefore, 10 L/min was chosen as a suitable feed flow rate for separation of anserine and carnosine with the nanofiltration membranes.

A summary of results obtained in the total circulation experiments with the 13 different nanofiltration membranes is listed in Table 22.2. The purpose of the nanofiltration treatment is to improve the purity of anserine and carnosine contained in the feed solution by removing impurities such as creatinine and sodium chloride into the permeate (Figure 22.8). Therefore, a membrane that shows higher rejection ability against anserine and carnosine and low rejection ability against creatinine and sodium ion is preferred for this purpose. Furthermore, higher permeate flux value implies that a process with the membrane will require smaller membrane area and lower initial cost. Based on these criteria, four membranes (NFT-50, DRA-4510, Desal DK, and Desal DL) were chosen as suitable membranes for purification and concentration of anserine and carnosine from chicken extract.

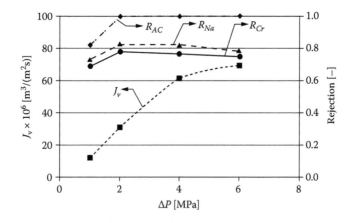

FIGURE 22.6   Effect of operating pressure ΔP on permeate flux J_v and observed rejection for anserine-carnosine R_{AC}, creatinine R_{Cr}, and sodium ion R_{Na} with NFT-50 membrane.

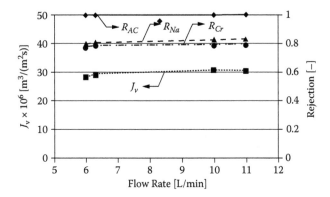

FIGURE 22.7 Effect of flow rate on permeate flux J_v and observed rejection for anserine-carnosine R_{AC}, creatinine R_{Cr}, and sodium ion R_{Na} with NFT-50 membrane.

TABLE 22.2
Summary of Results of the Total Circulation Experiments

Membrane	$J_v \times 10^6$ [m³/(m² s)]	R_{AC} [–]	R_{Cr} [–]	R_{NaCl} [–]
NFT-50	61.1	0.998	0.765	0.811
DRA-4510	54.9	0.994	0.813	0.835
Desal DK	42.4	0.992	0.713	0.733
Desal DL	36.8	0.997	0.439	0.446
MPF36	34.7	0.751	0.490	0.257
NTR7250	29.2	0.888	0.564	0.234
MPF50	28.5	0.017	0.035	—
NTR7430	27.8	0.925	0.600	0.719
NTR7450	13.9	0.941	0.704	0.842
MPF34	11.8	1.000	0.990	0.980
G-10	8.8	0.453	0.214	0.588
MPF44	6.3	0.940	0.886	0.757
G-5	4.6	0.406	0.070	0.593

Note: AC, anserine and carnosine.

22.3.2 BATCHWISE CONCENTRATION EXPERIMENTS

Batchwise concentration experiments were conducted with the four kinds of membranes under the selected operating conditions. Experimental results obtained in the batchwise concentration experiments are shown in Figure 22.9, where feed flow rate and operating pressure were 10 L/min and 4 MPa, respectively. Figure 22.9 shows changes in yield of each component with *concentration factor*, which is defined as the ratio of initial feed volume to feed volume. Yield of creatinine and sodium decreased with increase in volume reduction factor, while that of anserine and carnosine was

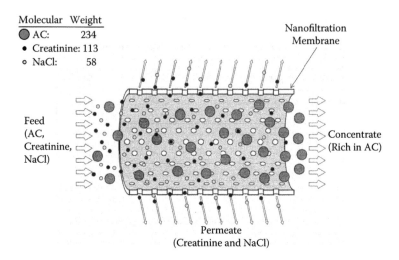

FIGURE 22.8 Principle of purification and concentration process for AC (anserine and carnosine) using a nanofiltration membrane.

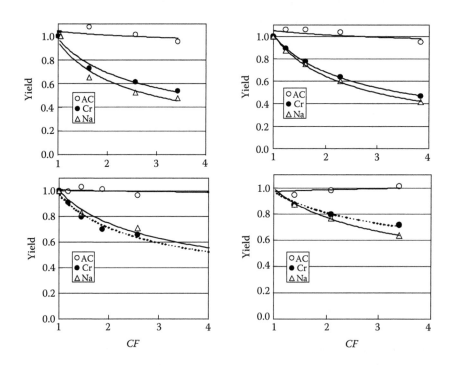

FIGURE 22.9 Changes in yield of each component with concentration factor (CF) during batchwise concentration experiments, where CF is defined as the ratio of initial feed volume $V_{f,0}$ to feed volume V_f. Flow rate: 10 L/min; pressure: 4 MPa. AC, anserine and carnosine.

almost constant at unity. These results imply that anserine and carnosine were concentrated and their purity was increased during the batchwise concentration treatments.

22.3.3 PROPOSED MATHEMATICAL MODEL

Based on the experimental results obtained in the batchwise concentration experiments, a mathematical model that can express efficiency of the purification and concentration process with nanofiltration membranes was proposed. Figure 22.10 shows the proposed model. By solving this model, change in concentration of each component with time can be calculated.

Changes in purity and yield of anserine-carnosine, concentrations of each component, and permeate flux value with processing time calculated with the mathematical model are shown in Figure 22.11 together with experimental results. The lines show calculated values, and the plots show experimental values. The calculated values are in good agreement with experimental values, and it was confirmed that the efficiency of the membrane purification process could be predicted with this model precisely.

22.3.4 PROCESS DESIGN

Then, by applying the mathematical model, a practical size membrane process was designed. The minimum purity and yield of anserine and carnosine required in the final product were set at 90% and 95%, respectively. The NFT-50 membrane was chosen as the nanofiltration membrane. A flow diagram of the designed process is shown in Figure 22.12. The simulation result showed that 5.3 m² of nanofiltration membrane

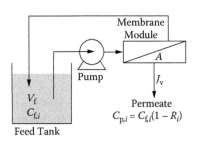

Nomenclature

A: membrane area [m²]
C: concentration [kg/m³]
CF: concentration factor [−]
J_v: permeate flux [m³/(m²s)]
k: mass transfer coefficient [m/s]
R: rejection value (= $1 - C_p/C_f$) [−]
t: time [s]
V: volume [m³]
α: constant [−]

\<Subscripts\>
f: feed solution
i: component i
p: permeate

$$
\begin{cases}
dV_f/dt = -AJ_v \\
d(C_{f,i}V_f)/dt = -AJ_vC_{f,i}(1 - R_i) \\
J_v = k\ln(\alpha/CF) \\
CF = V_{f,0}/V_f
\end{cases}
$$

FIGURE 22.10 A mathematical model to describe efficiency of a purification-and-concentration process with a nanofiltration membrane.

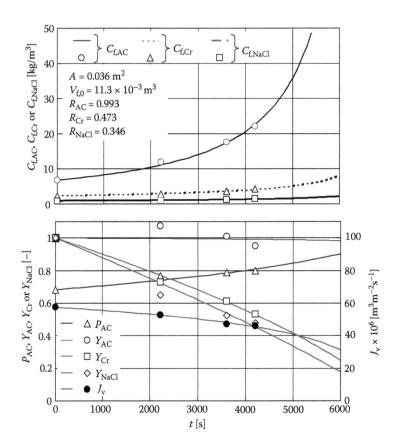

FIGURE 22.11 Changes in purity P and yield Y of anserine-carnosine, concentration of each component, and permeate flux value during treatment with NFT-50 membrane (experimental values are plots, and calculated values are lines).

was needed for the process, which could process 3.6 t of chicken carcasses in a day, and 7.4 kg of purified anserine and carnosine could be obtained with this process.

Applicability of this purification-and-concentration process with a nanofiltration membrane might be high because it is a simple process that requires low cost and low energy consumption. We are planning to apply this process to purification and concentration of value-added components contained in a wide variety of natural resources.

22.4 CONCLUSIONS

To add extra value to discarded chicken carcasses and utilize them, efficiency of a membrane process for purification and concentration of antioxidative dipeptides contained in chicken extract was investigated.

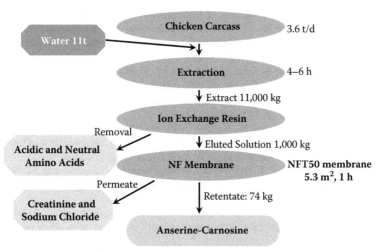

(conc.: 90 kg/m^3, purity: 90%, yield: 98%, AC Powder: 7.4 kg)

FIGURE 22.12 Schematic flow diagram of the designed process that can process 3.6 t of chicken carcasses in a day.

1. Thirteen different kinds of nanofiltration membranes were tested, and suitable membranes and operating conditions for purification and concentration of anserine and carnosine were selected.
2. By applying the selected membranes and conditions, anserine and carnosine were purified and concentrated with a pilot-scale unit.
3. Based on the experimental results, a mathematical model that can express efficiency of a nanofiltration process was proposed, and an industrial-scale nanofiltration process that could process 3.6 t of chicken carcasses in a day was designed by applying the model.

ACKNOWLEDGMENT

This work was supported in part by a grant-in-aid ("Development of Evaluation and Management Methods for Supply of Safe, Reliable and Functional Food and Farm Produce") from the Ministry of Agriculture, Forestry, and Fisheries of Japan.

REFERENCES

1. Cheryan, M. 1998. *Ultrafiltration and Microfiltration Handbook*. Lancaster, PA: Technomic.
2. Yanai, N., S. Shiotani, M. Mizuno, H. Nabetani, and M. Nakajima. 2004. A simple and rapid HPLC method for quantification of histidine-containing dipeptides, anserine and carnosine, in animal extracts. *Journal of the Japanese Society for Food Science and Technology* 51: 87–91.

3. Yanai, N., S. Shiotani, M. Mizuno, H. Nabetani, and M. Nakajima. 2004. Characteristics of anti-oxidative activity of carnosine and anserine mixture isolated from chicken extract: comparison with other botanical antioxidants. *Journal of the Japanese Society for Food Science and Technology* 51: 238–246.

4. Yanai, N., S. Shiotani, S. Kanazawa, M. Mizuno, H. Nabetani, and M. Nakajima. 2005. Purification and characterization of antioxidative dipeptides, anserine and carnosine, contained in chicken extract. Proceedings of the 34th Annual Meeting of UJNR Food and Agriculture Panel, p. 125, Susono, Japan.

5. Nabetani, H., N. Yanai, and M. Mizuno. 2006. Separation and purification of antioxidative dipeptides, anserine and carnosine, in chicken extract, and evaluation of their physiological functionality. *Japan Journal of Food Engineering* 7: 15–23.

6. Yanai, N., S. Shiotani, J. Baljinnyam, S. Hagiwara, H. Nabetani, and M. Nakajima. 2007. Purification and clinical application of antioxidative dipeptides obtained from chicken extract. *Membrane* 32: 197–202.

7. Collins, A.R., A.G. Ma, and S.J. Duthie. 1995. The kinetics of repair of oxidative DNA damage (strand breaks and oxidized pyrimidines) in human cells. *Mutation Research* 336: 69–77.

23 Application of Phospholipases for Highly Functional Phospholipid Preparation

Yukihiro Yamamoto, Masashi Hosokawa, and Kazuo Miyashita

CONTENTS

23.1 INTRODUCTION

Phospholipids (PLs) are attractive molecules because they contain both hydrophobic and hydrophilic residues in their molecular structure. This unique structure gives amphiphilic properties that play an important role in biomembrane organization and

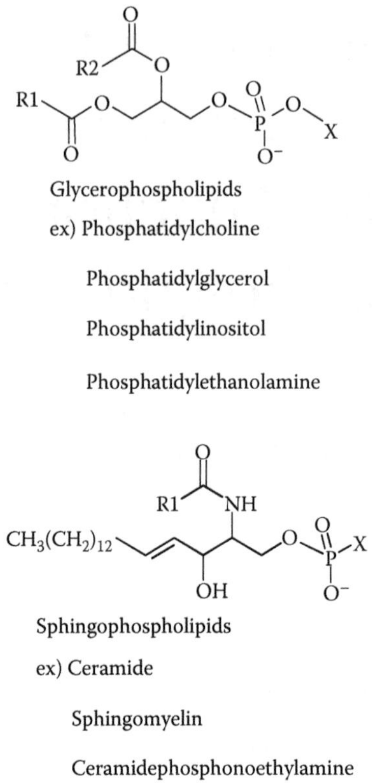

Glycerophospholipids

ex) Phosphatidylcholine

Phosphatidylglycerol

Phosphatidylinositol

Phosphatidylethanolamine

Sphingophospholipids

ex) Ceramide

Sphingomyelin

Ceramidephosphonoethylamine

FIGURE 23.1   Two main classes of phospholipids. R, alkyl or acyl residue; X, −OH, choline, serine, ethanolamine, glycerol, and so on.

lipoprotein formation. For this reason, PLs are ubiquitous and essential molecules in all organisms and are also precursors of several chemical mediators in the body.

PLs are classified into two main classes depending on their backbone structure and fatty acid-bonding types: glycerophospholipids and sphingophospholipids having glycerol and sphingosine backbones, respectively (Figure 23.1). Their biological and chemical functions are determined by their structures. The amphiphilic properties and biocompatibility of PLs are very beneficial for their application in the fields of food, cosmetics, and medicine. For example, lecithin (whose main component is phosphatidylcholine [PC]) is used in a variety of foods, such as dressing, mayonnaise, bread, and chocolate, as an emulsifier. In cosmetics, PLs are used not only as emulsifiers, but also as enhancers for the absorption of active ingredients because of their high affinity for cell membranes. Highly purified PC is used as a drug that regulates metabolism in the liver. Many drug delivery systems (DDS), which have recently received good attention in the medical field, are founded on the property of PLs to form liposomes as nanoparticles [1].

The fatty acid compositions of PLs are very different depending on their origins and lipid class. PC extracted from soybean and egg yolk binds linoleic acid and palmitic acid as its main fatty acids. PLs rich in n-3 polyunsaturated fatty acid (PUFA), such as eicosapentaenoic acid (EPA) and docosahexaenoic acid (DHA), are available from marine products such as salmon roe and squid meal. Phosphatidylethanolamine (PE) and phosphatidylserine (PS) have a higher level of PUFA compared to PC.

The selection of the proper PL molecule is critical for its effective use in a given application. In addition, for application in the medical and fine chemical fields, it is important to prepare the defined PL molecule at high purity. Therefore, tailor-made syntheses of PLs are desirable and essential. Compared to chemical methods, enzymatic methods possess greater selectivity and specificity for modification of PLs. Enzymatic reactions are also easier and simpler, and can often be conducted under mild conditions, protecting PLs from oxidation.

Modification of PLs mediated by PL-related enzymes has been well documented in previous reports [2–4]. In this chapter, we summarize the phospholipase A_2-(PLA$_2$-) and phospholipase D- (PLD-) catalyzed methodology for synthesis of new or less-abundant PLs as well as natural-type PLs and describe the functions and potential uses of synthetic PLs.

23.2 PHOSPHOLIPASE A_2 (SN-2 POSITION-SPECIFIC REACTION)

Phospholipase A_2 (EC 3.1.1.4) is known to hydrolyze L-α-phosphatidylcholine (PC) to produce L-α-lysophosphatidylcholine and free fatty acid (FFA). PLA$_2$ is also a valuable tool in the synthesis of glycerophospholipids with defined fatty acids in the sn-2 position. Lipase is the most common enzyme used to modify acyl groups on glycerophospholipids. Some lipases exhibit sn-1 positional specificity, and others are nonspecific. On the other hand, PLA$_2$ catalyzes sn-2 position-specific reactions [5]. PLA$_2$ belongs to three main families: secretory PLA$_2$ (sPLA$_2$), cytosolic PLA$_2$ (cPLA$_2$), and Ca^{2+}-independent PLA$_2$ (iPLA$_2$); however, only sPLA$_2$-catalyzed reactions have been reported for modification of fatty acid of PLs. To date, two main strategies have been used for PL synthesis: esterification and transesterification (acidolysis) (Figure 23.2).

23.2.1 Esterification

Esterification begins with lyso-PL and FFA (Figure 23.2A). Table 23.1 summarizes the existing literature on PLA$_2$-catalyzed esterification [6–14].

This esterification reaction was first reported by Pernas et al. [6] and Na et al. [7] with very low yields. Because PLA$_2$ preferentially catalyzes hydrolysis reactions over esterification, it is necessary but difficult to regulate the water content in the reaction mixture to obtain high yields. By precisely controlling water activity a$_w$, the reaction equilibrium shifted from hydrolysis to esterification, and the PC yield was successfully increased to 49%–60% with a variety of FFAs [11,12]. Water activity was controlled at a higher level (a$_w$ = 0.43) during the initial stage of the reaction to promote the reaction rate and then decreased stepwise to a$_w$ = 0.11. Furthermore, the application of formamide as a "water mimic" is worth some attention. A water mimic can activate enzymes in the place of water due to its high dielectric constant and ability to form multipoint hydrogen bonds with proteins [15,16]. By using

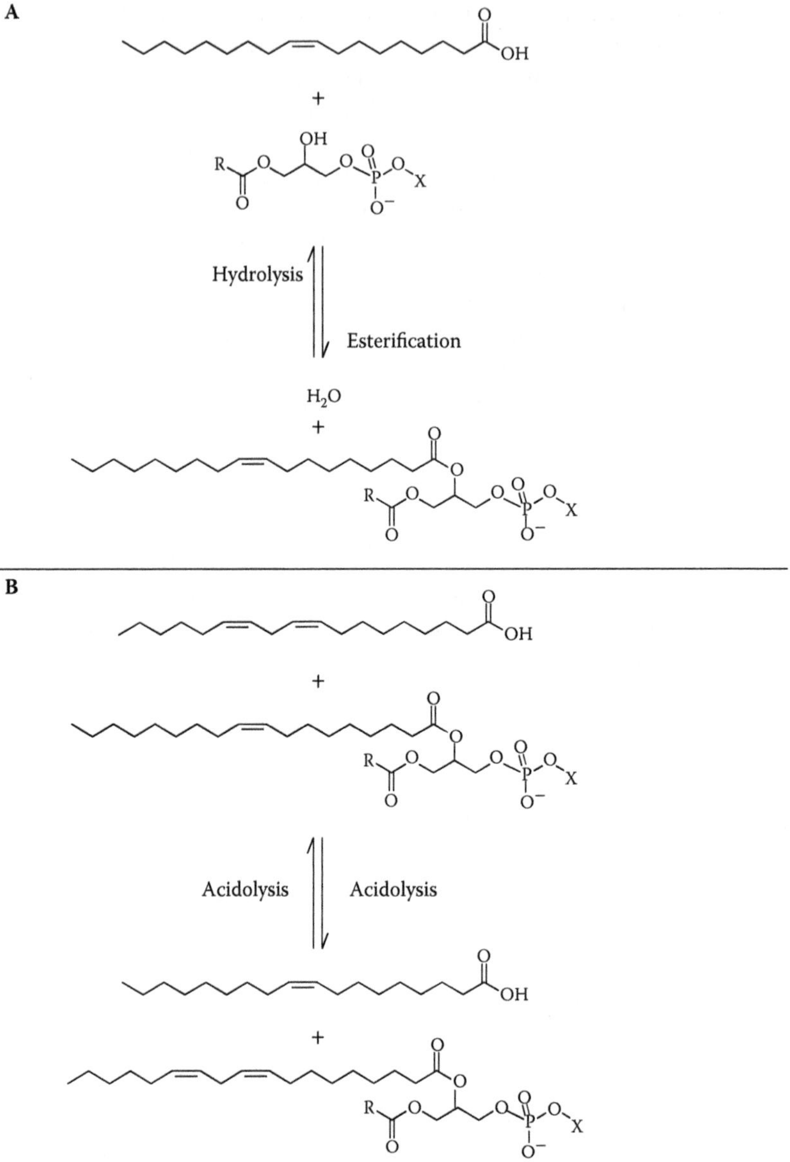

FIGURE 23.2 Reaction scheme of PLA$_2$-catalyzed esterification (A) and transesterification (B). R, alkyl residue; X, −OH, choline, serine, ethanolamine, glycerol, and so on.

formamide, esterification can be efficiently conducted since the hydrolysis reaction is suppressed. At the optimum conditions of 110 mg lysophosphatidylcholine (LPC), 180 mg EPA, 5,500 mg glycerol, 500 μL formamide, 23 mg PLA$_2$, and 3 μmol CaCl$_2$ at 25°C for 48 h, the reaction yield of EPA-PC reached 60 mol% [10]. This reaction system is also effective in synthesizing PLs containing conjugated linoleic acids,

Table 23.1 Esterification of Lysophospholipid and Fatty Acid by Phospholipase A_2

PLA_2 sources	Substrate PL	Acyl donor	Optimal reaction condition	Yield (%)	Reference
Naja naja venom Bee venom Porcine pancreas Bovine pancreas *Streptomyces violaceoruber*	LPC, LPE and egg yolk PC	C18:2	70 mM LPC and 700 mM oleic acid were dispersed in the 150 ml dehydrated toluene followed by adding lyophilized 4.5 U PLA_2 at 37°C for 24 h.	6.5	P. Pernas et al. (1990) ref. 6.
Porcine pancreas	LPC (C16:0 68%, C18:0 26.2% C18:1 5.3% C18:2 0.5%)	PUFA from fish oil, DHA, C12:0, C16:0, C18:0, C18:2	Microemulsion system; 1.2% of 0.1 M borate-HCl buffer (pH 8.2), 3.4% AOT, 4.0% LPC, 4.0% PUFA, 87.4% isooctane and PLA_2 (2.5 x 10^4 U/g LPC) were mixed under N_2 at 30°C for 16 h.	6	A. Na et al. (1990) ref. 7.
Porcine pancreas (PLA_2 was immobilized to porous glass beads)	LPC from egg yolk	C4:0, C10:0, C14:0, C16:0, C18:0, C18:1, C18:2, C18:3	0.2 mg PLA_2 was suspended in 1 ml of anhydrous benzene containing 7 mM LPC and 500 mM oleic acid at 25°C for 8 days.	35	I. Mingarro et al. (1994) ref. 8.
Porcine pancreas (PLA_2 was immobilized to polymer carrier; Deloxan®)	PC from soybeans or PC from egg yolk were hydrolyzed to LPC with PLA_2	PUFA from fish oil (mainly contains DHA and EPA)	Reaction unit (reactor); LPC was dissolved in 91% PUFA and 9% propane at 45°C for 20 h.	26	M. Härröd and I. Elfman. (1995) ref. 9.

Abbreviations: PC, phosphatidylcholine; LPC, lysophosphatidylcholine; LPE, lysophosphatidylethanolamine; CLA, conjugated linoleic acid; DHA, docosahexaenoic acid; AOT, sodium bis (2-ethylhexyl) sulfosuccinate.

Table 23.1 (continued)   Esterification of Lysophospholipid and Fatty Acid by Phospholipase A$_2$

PLA$_2$ sources	Substrate PL	Acyl donor	Optimal reaction condition	Yield (%)	Reference
Porcine pancreas	LPC from soybean PC	EPA	110 mg LPC, 180 mg EPA, 5500 mg glycerol, 23 mg PLA$_2$, 500 l formamide and 3 mol CaCl$_2$. Reaction was conducted at 25°C for 48 h.	60	M. Hosokawa et al. (1995) ref. 10.
Porcine pancreas (PLA$_2$ was immobilized to XAD-8)	LPC from egg yolk	C6:0, C8:0, C10:0, C12:0, C14:0, C16:0, C18:1, C18:2, C18:3	10 mM LPC and 1.8 M oleic acid were dispersed in the 3 ml dehydrated toluene followed by adding 50 mg of immobilized PLA$_2$ at 25°C for 360 h. a$_w$ = 0.43 –> 0.11.	60	D. Egger et al. (1997) ref. 11.
Pig pancreas (PLA$_2$ was immobilized to XAD-8)	PC from egg yolk (PC was hydrolyzed by PLA$_2$ to yield LPC)	C6:0	860 mg LPC and 1.2 M hexanoic acid in toluene were dispersed followed by adding 11 g of immobilized PLA$_2$ for 96 h. 25°C - > 40°C a$_w$ = 0.43 –> 0.11.	49	D. Adlercreutz et al. (2004) ref. 12.
Porcine pancreas	LPC from egg yolk	CLA (9c,11t, 10t,12c 9t,11t, 9c,11c, and mixture of 9c,11t *and* 10t,12c)	11 mg LPC, 18 mg CLA, 550 mg glycerol, 50 l formamide, 3.3 x 10^4 U PLA$_2$ and 0.3 mol CaCl$_2$ at 37°C for 6 h.	65	Y. Yamamoto et al. (2006) ref. 13.
Porcine pancreas	Soy LPC	DHA	110 mg Soy LPC, 180 mg DHA, 5500 mg glycerol, 60 mg PLA$_2$, 3 mmol CaCl$_2$, 0.5 ml formamide at 40°C for 48 h under decompression (132–165 Torr).	90	S. Awano et al. (2006) ref. 14.

Abbreviations: PC, phosphatidylcholine; LPC, lysophosphatidylcholine; CLA, conjugated linoleic acid; DHA, docosahexaenoic acid; EPA, eicosapentaenoic acid.

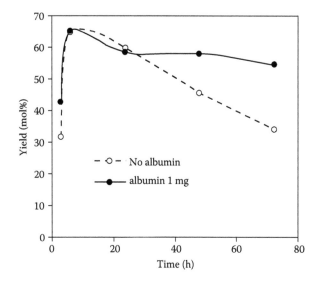

FIGURE 23.3 Time course of CLA-PC synthesis by PLA_2 in the presence of albumin [13]. Reaction mixture: 11 mg LPC, 18 mg CLA-mixture, 550 mg glycerol, 3.3×10^4 U PLA_2, 50 mL formamide, 0.3 mmol $CaCl_2$, and 1 mg albumin. The reaction was conducted at 37°C.

which are known to exhibit an antiobesity effect. The addition of a protein such as albumin to the reaction mixture could retain the high yield (Figure 23.3) [13]. Recently, the yield of PC containing DHA reached 90% through the combination of regulating water activity and the use of a water mimic [14].

Porcine pancreatic PLA_2 is the most widely used enzyme in hydrolysis and in synthesis. Utilization of PLA_2 from venom (snake or bee) or microbial PLA_2 (*Streptomyces violaceoruber*) for esterification of LPC and FFA has been reported in only one study [6]. In addition, lysophosphatidylethanolamine (LPE) was used as a substrate for PE synthesis [10], although bee venom PLA_2 did not promote esterification.

For the lyso-PL substrate, LPC is the most available because it is prepared as a food emulsifier on the industrial scale. To prepare structured PLs such as fatty acid-defined PE, PS, or phosphatidylglycerol (PG), it would be better to first synthesize fatty acid-defined PC by PLA_2, followed by PLD-mediated transesterification as described in Section 23.3 on PLD.

23.2.2 Transesterification (Acidolysis)

Transesterification (acidolysis) is the reaction between diacyl-PL and FFA. This reaction can be performed in a single step (Figure 23.2B), while esterification requires hydrolysis to prepare lyso-PL. To date, few studies on transesterification mediated by PLA_2 have been reported (Table 23.2 [17–21]).

Water mimics are effective in increasing the incorporation of the desired fatty acids in PC by PLA_2-mediated acidolysis. By employing glycerol as a water mimic, 35% and 34% incorporation of EPA and DHA into soy PC were obtained, respectively,

Table 23.2   Transesterificaiton of Phospholipid and Fatty Acid by Lipase and Phospholipase A$_2$

PLA$_2$ sources	Substrate PL	Acyl donor	Optimal reaction condition	Incorporation (%)	Reference
Porcine pancreas (PLA$_2$ was immobilized to celite)	Soy PC	C12:0	Solvent free system: The water content was adjusted to 1 % (vol/wt) of the weight of a substrate mixture (PC/FFA, 1/10), and 40% PLA$_2$ was added at 60°C for 24 h.	7	A.M. Aura et al. (1995) ref. 17.
Porcine pancreas	Soy PC (> 95%)	EPA, DHA, EPA-EE, DHA-EE	14 μmol Soy PC was combined with 60 μmol of each acyl donors and 0.55 g glycerol. 10 mg PLA$_2$ was dissolved in distilled water (50ml) containing 3 μmol CaCl$_2$. Reaction was carried at 25°C for 48 h.	EPA, 35 DHA, 34	M. Hosokawa et al. (1998) ref. 18.
Porcine pancreas	PC	EPA-EE	PC and EPAEE was dissolved in 5 ml toluene at the molar ratio of 1/10 (PC/EPAEE) with initial a$_w$of 0.25 at 50° C for 48 h.	14.3	C.W. Park et al. (2000) ref. 19.
Porcine pancreas (PLA$_2$ was immobilized to celite)	Soybean PL, (PC ≅ 94%)	CLA (mixture of 9c,11t and 10t,12c)	20 mg PLA$_2$ was added to the reaction mixture of 100 mg PL and 185 mg CLA mixtures dissolved in hexane. An amount of 8 ml Tris-HCl buffer (pH 8) containing 20 mM CaCl$_2$ was sprayed over the PLA$_2$before adding to the reaction mixture. Reaction was carried at 50° C for 72 h.	4.2	M. Hossen and E. Hernandez (2006) ref. 20.
Porcine pancreas (PLA$_2$was immobilized to Amberlite XAD7, Superlite DAX8, Celite 545, Dowex 50W, Lewatit VPOC1600, Duolite A568 or Accurel EP 100)	Soybean PC (PC, 93%)	C8:0, CLA (mixture of 9c,11t and 10t,12c), DHA	Solvent-free system: Reaction was initiated by the addition of 300 mg immobilized PLA$_2$ (XAD7, 72 mg PLA$_2$/g carrier). Substrate ratio; 9 (3 for CLA and DHA) mol/mol C8:0/PC, water addition; 2%, enzyme 30wt%, at 45° C for 48 h.	C8:0, 29 CLA, 30 DHA, 20	A.F. Vikbjerg et al. (2007) ref. 21.

Abbreviations: PC, phosphatidylcholine; FFA, free fatty acid; EPA, eicosapentaenoic acid; DHA, docosahexaenoic acid; EPA-EE, eicosapentaenoic acid ethyl ester; DHA-EE, docosahexaenoic acid ethyl ester; CLA, conjugated linoleic acid..

while the ethyl esters of EPA and DHA were not substrates [18]. The fatty acid composition at the *sn*-2 position of PL has an important influence on its nutritional and medical functions [22]. In this reaction, EPA was successfully incorporated into the *sn*-2 position of soy PC at 70% of total fatty acids at *sn*-2, indicating high specificity of the PLA$_2$ enzyme.

Along with PLA$_2$, lipase is widely used to exchange the *sn*-1 position of PL. Acidolysis by lipase for *sn*-1 positional specificity is also important to prepare defined PL molecular species. To complicate matters, PL for use in the food industry should be synthesized in the absence of toxic organic solvents. To this end, a solvent-free system was developed by Aura et al. [17] and Vikbjerg et al. [21]. Enzyme immobilization is an important technique to provide enzyme operational stability, dispersibility, and selectivity [23], all of which affect the yield or incorporation of fatty acids to PS. Several immobilization supports, including Amberlite XAD7 (Sigma), Superlite DAX8 (Supelco), Duolite A568 (Rohm and Haas), Dowex 50W (Dow Chemical Company), Celite 545 (BHD), Accural EP 100 (Akzo), and Lewatite VPOC 1600 (Lanxess AG), were screened. Amberlite XAD7, a nonionic, weakly polar macroreticular resin (matrix: acylic ester), was shown to be superior in both activity per immobilized particle and specific activity. Immobilized enzymes have also been applied to esterification and PLD-catalyzed transphosphatidylation (see Section 23.3).

Porcine pancreatic PLA$_2$ has seven disulfide bonds [24] in its enzyme protein, whereas microbial PLA$_2$ (e.g., *Streptomyces violaceoruber* A-2688) has only two disulfide bonds [25]. Microbial PLA$_2$ may show different characteristics on esterification and acidolysis.

23.3 PHOSPHOLIPASE D (MODIFICATION OF POLAR HEAD GROUPS)

Phospholipase D (EC 3.1.4.4) is a lipolytic enzyme that hydrolyzes the terminal phosphodiester bond on PLs. Due to its ability to transfer the phosphatidyl moiety of glycerophospholipids to various alcohols (transphosphatidylation), PLD is also used to synthesize PLs with desired head groups that are poorly accessible via the chemical route (Figure 23.4). This ability has been utilized for the synthesis of natural PLs that are rare in nature, such as PG and PS. Novel types of PLs (phosphatidyl-X) have also been synthesized via PLD-mediated transphosphatidylation to add the amphiphilic properties of PLs to the acceptor compounds. These reactions are typically carried out in biphasic systems with water (containing PLD or a hydrophilic alcohol acceptor) and an organic solvent such as chloroform, ether, ethyl acetate, benzene, or toluene.

23.3.1 SYNTHESIS OF NATURAL PLs

Juneja et al. reported transphosphatidylation mediated by PLD in detail. PG, PE, and PS were synthesized from lecithin or PC by PLD [26–30]. PG, PE, and PS were prepared in quantity by chemoenzymatic synthesis under optimized conditions by

Pisch's group [31]. Phosphatidylinositol (PI) was synthesized from dioleoyl phos-phatidylcholine (DOPC) using mutant PLD [32], although the yield was only 2.5% of PI isomers. While some PLD with plant origins were reported to catalyze PI synthesis, these enzymes were not easily available for industrial purposes [33]. The synthesis of cardiolipin (diphosphatidylglycerol) was also reported by Piazza et al. [34].

These reaction methods were applied to the preparation of PS containing PUFAs from squid skin lecithin (PC 44.2%, PE 29.4%, PS 3.3%, LPE 12.2%) and serine [35]. Lysophosphatidylglycerol was also synthesized from egg LPC and glycerol by PLD [36]. Toxic organic solvents should be avoided in the production of PLs for use as human food. Therefore, PG and PS have been synthesized in aqueous media, with 72% and 80% yields, respectively [37,38].

23.3.2 Synthesis of Novel PLs

PLD-catalyzed transphosphatidylation is a very effective reaction for the prepara-tion of novel PLs with functional moieties. Many compounds were converted to PL derivatives via transphosphatidylation by PLD. Table 23.3 gives the reaction system of novel glycerophospholipids [39–53]; the structures of acceptor compounds (alco-hols) are presented in Figure 23.5.

23.3.2.1 Priority of Hydroxyl Group

The priority of the hydroxyl group is as follows: primary hydroxyl group > phenolic hydroxyl group > secondary hydroxyl group. Phosphatidylation occurs preferentially at the primary hydroxyl group. A variety of structures, including sterically bulky compounds such as nucleosides or genipins, can be high-yielding acceptors. In the synthesis of phosphatidyl terpenes, we found that the yield decreased with the chain length of substrate terpenes. A saturated terpene (phytol) showed a lower yield than an unsaturated one (geranylgeraniol) [52].

Phosphatidylation of phenolic compounds has also been reported [46,49], although the yields were low. In particular, catenol, which has a neighboring phenolic hydroxyl group in the molecule, showed a very low yield (less than 10%) [46].

Secondary hydroxyl groups can also be acceptors. Hossen and Hernandez [50] synthesized phosphatidyl-sitosterol in a biphasic system of chloroform and 0.2M sodium acetate buffer (pH 5.6). However, again, the yield was low (30%).

In cases of compounds with both primary and secondary hydroxyl groups, the primary hydroxyl groups are selectively transphosphatidylated. Nucleoside [39], L-ascorbic acid [40], saccharide [41], kojic acid and arbutin [43], and genipin [44], which have secondary hydroxyl groups or phenolic hydroxyl groups in the molecule, are selectively phosphatidylated at the position of the primary hydroxyl group.

23.3.2.2 Stereoisomeric Recognition of PLD from *Streptomyces* sp.

Stereoisomeric recognition of PLD from *Streptomyces* sp. was examined by Wang et al. [42]. In the reaction with both (*S*)- and (*R*)-prolinol, the yield was almost the

Table 23.3   Preparation of Novel Phospholipids by Phospholipase D

Acceptor alcohol	PLD sources	Typical reaction condition	Reference
Several nucleosides (primary alkanols)	Streptomyces sp. AA 586 Cabbage	5 mmol nucleoside and 550 U PLD were dissolved in 5 ml appropriate buffer. Then, 20 ml $CHCl_3$ solution of 0.5 mmol several PC was added, and the mixture was stirred at 45°C for 6 h.	S. Shuto et al. (1987) ref. 39.
L-ascorbic acid	Streptomyces lydicus (Honen) Streptomyces sp. Streptomyces sp. Cabbage	1 ml of 10 mM egg yolk PC in diethyl ether, 0.8 ml of 1 M ascorbic acid, 0.05 mL PLD solution (90 U, lydicus) in 1 ml of 10 mM sodium acetate buffer (pH 5.1), and 0.15 ml of deionized water. 37°C for 24 h.	A. Nagao et al. (1991) ref. 40.
Saccarides	Actinomadura sp. 362	0.1 g of egg lecithin, 1 ml diethyl ether, 1 g of saccharide, 40 mg of NaCl, 0.5 ml of water, and 0.5 ml of PLD solution (10 U) were put into a 25-ml glass test tube, 30°C for 24 h.	Y. Kokusho et al. (1993) ref. 41.
Azosugars Nucleosides Peptides	Streptomyces sp.	To 0.7 ml of 1.6 M alcohol in 100 mM NaOAc and 50 mM $CaCl_2$ buffer (pH 6.5) were added PLD (180 U) and 0.026 mmol DMPC in chloroform. 30°C, 4 h.	P. Wang et al. (1993) ref. 42.
Kojic acid Arbutin	Streptomyces sp.	0.68 mmol DPPC and 6.62 mmol alcohol were added to a mixture of 20 ml of ethyl acetate and 20 ml of 0.2 M sodium acetate buffer (pH 5.6, 4% $CaCl_2$) containing 50 U PLD, and incubated 35°C for 4 h.	M. Takami et al. (1994) ref. 43.
Genipin	Streptomyces sp.	1.36 mmol DPPC and 4.42 mmol genipin were added to a mixture of 30 ml of ethyl acetate and 30 ml of 0.2 M sodium acetate buffer (pH 5.6, 4% $CaCl_2$) containing 50 U PLD, and incubated 35°C for 3 h.	M. Takami et al. (1994) ref. 44.
Dihydroxyacetone	Streptomyces sp.	8 mol DPPC and 200 mol dihydroxyaceton dimer were incubated with PLD (2.5 U) in a biphasic system of 0.5 ml ethyl acetate and 0.5 ml of 0.2 M sodium acetate buffer (pH 5.6) at 20°C for 6 h.	M. Takami et al. (1994) ref. 45.

Abbreviations: PC, phosphatidylcholine; DMPC, dimyristoyphosphatidylcholine; DPPC, dipalmitoyphosphatidlcholine.

Table 23.3 (continued)   Preparation of Novel Phospholipids by Phospholipase D

Acceptor alcohol	PLD sources	Typical reaction condition	Reference
Aromatic compounds	*Streptomyces* sp.	0.68 mmol DPPC and 8.1 mmol alcohol were mixed with 50 ml of benzene and 50 ml of 0.2 M sodium acetate buffer (pH 5.6) containing PLD (100 U) and incubated at room temperature for 5 h.	M. Takami et al. (1994) ref. 46.
Vitamin E derivative	*Streptomyces hydicus*	2.5 ml of 25 mM egg yolk PC and 25 mM Toc-Et in diethyl ether and 25 ml of PLD (10 U) and 10 mM CaCl₂ in 10 mM acetate buffer (pH 5.1) at 37°C for 2 h.	T. Koga et al. (1994) ref. 47.
1,8-Octandiol	*Streptomyces* sp.	To a solution of 30 mmol 1,8-octanediol and 10 mmol DSPC in 200 ml chloroform were added 30 ml NaOAc buffer (pH 5.6) and 3 ml of 0.4 M aqueous Ca(OAc)₂. PLD (370 U in 3 ml of 0.4 M NaOAc buffer) was added. 35°C, 50 h.	M. Koketsu et al. (1997) ref. 48.
Several compounds	*Streptomyces* sp.	20 mM DPPC and 20 mM alcohol were dissolved in 1 ml of dry chloroform containing 20 mg PLD. The mixture contained 100 mg ion-exchanged resin (IRC-50). 45°C.	J.O. Rich and Y.L. Khmelnitsky et al. (2001) ref. 49.
Sitosterol	*Streptomyces* sp.	20 mg PC (soybean) and 104 mg β-sitosterol (1:10 mol ratio) were dissolved in 2 ml chloroform and 100 ml if 0.2 M sodium acetate buffer (pH 5.6) containing 40 mM CaCl₂ and 20 U PLD, 40°C for 12 h.	M. Hossen and E. Hernandez et al. (2004) ref. 50.
N- or C2-substituted derivatives of ethanolamine	*Streptomyces* sp. Cabbage	38.2 mmol DOPC in 18.8 ml of diethyl ether and 2.4 ml of reaction buffer containing 2.88 mmol of acceptor alcohol and PLD. 30°C for 5 (*Streptomyces*) or 8 (Cabbage) h.	M. Dippe et al. (2008) ref. 51.
Terpenes	*Streptomyces* sp.	50 mmol SoyPC, 2,000 mmol alcohol, 1.6 U PLD, 0.8 ml of 0.2 M sodium acetate buffer (pH 5.6), 37°C, 24 h.	Y. Yamamoto et al. (2008) ref. 52,53.

Abbreviations: PC; phosphatidylcholine, DMPC; dimyristoylphosphatidylcholine, DPPC; dipalmitoylphosphatidlcholine, Toc-Et; 2,5,7,8-tetramethyl-l-6-hydroxy-2-(hydroxyethyl) chronanol, DSPC; distearoylphosphatidylcholine, DOPC; dioleoylphosphatidylcholine.

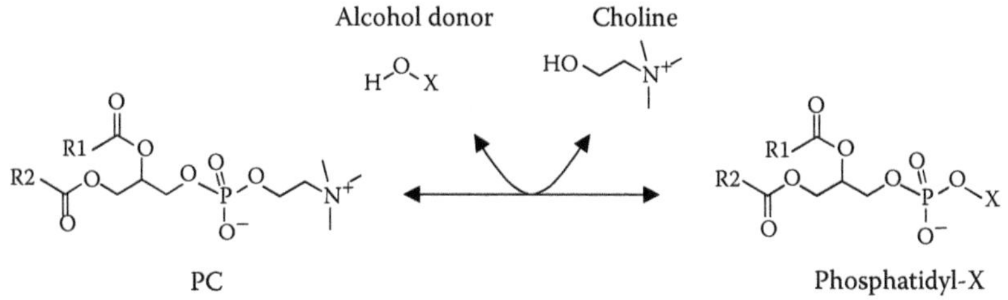

FIGURE 23.4 Reaction scheme of PLD-catalyzed transphosphatidylation.

same (84% and 85%, respectively). When lower concentrations of the alcohol were used, however, PLD showed a slight preference for the R enantiomer (S 52% and R 68%). In the reaction with serinol, the enzyme prefers the pro-R hydroxyl group, as indicated in a separate reaction with (S)-1,1-dideuteroserinol to give a 4:1 mixture of **3** in part 14 of Figure 23.5. In contrast, when (S)-1,1-dideuteroglycerol was used, no preference was found. Similarly, **12** in part 15 of Figure 23.5 was prepared in 37% yield as a 1:1 mixture of diastereomers. In our report on the synthesis of phosphatidyl-geraniol and -nerol, the yields were 90% and 91%, respectively, which indicates that PLD from *Streptomyces* sp. does not distinguish between the Z/E configurations of monoterpene [53].

23.3.2.3 PLD Source

In almost all of the transphosphatidylation reactions used in the synthesis of novel PLs, PLD from *Streptomyces* sp. has been used. Several reports showed PLD from cabbage, peanuts, and *Streptomyces chromofuscus* have low activity for transphosphatidylation [39,40,42,52], although they are able to transphosphatidylate PC to PE and PG. In these few reports, PLD from cabbage catalyzed transphosphatidylation with PC and ethanolamine derivatives [51]. Compared to PLD from *Streptomyces* sp., PLD from cabbage may possess a more rigid substrate specificity.

23.3.2.4 Substituent Effects

The effect of substituents of glycerophospholipids was studied by Takami et al. [46] using various PLs with a series of *p*-substituted phenols. There was a correlation between the reaction efficiency and the Hammets' sigma constants σ_p of the substituents. Phenols with more electron-donating substituents (lower σ_p) were more effective acceptors, with the exception of −OH.

PAF-acether (1-O-alkyl-2-acethyl-*sn*-glycero-3-phosphocholine), a PL mediator implicated in numerous biological situations, such as platelet activation, bronchoconstriction, and neutrophil activation [54], or alkylphosphate esters, which are characterized by a PL-like amphiphilic structure lacking the glycerol backbone, have attracted attention as anticancer drugs [55]. These could act as substrates for PL in PLD-catalyzed transphosphatidylation.

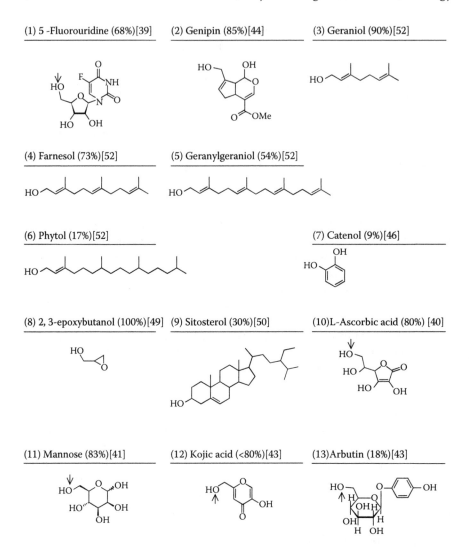

FIGURE 23.5 Structures of alcohol acceptor for PLD-catalyzed transphosphatidylation that has been used for novel PL synthesis. Numbers in parentheses give the yield. Arrows indicate the hydroxyl group that binds phosphatidyl residue.

23.4 FUNCTIONS OF NOVEL PHOSPHOLIPIDS

A number of functional compounds have been used for phosphatidylation. Functional properties of synthesized PLs have also been evaluated.

23.4.1 PHOSPHATIDYL-NUCLEOSIDES

Nucleoside and nucleobase derivatives are widely used for chemotherapy, but they have several disadvantages: The antitumor spectra are not broad enough; normal

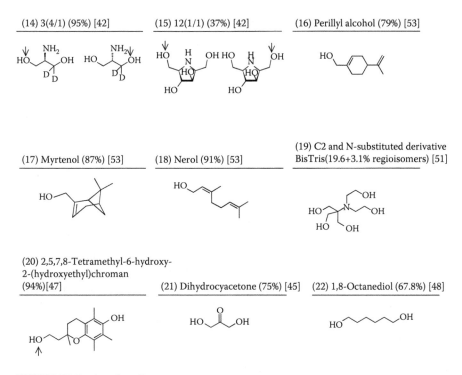

(14) 3(4/1) (95%) [42]   (15) 12(1/1) (37%) [42]   (16) Perillyl alcohol (79%) [53]

(17) Myrtenol (87%) [53]   (18) Nerol (91%) [53]   (19) C2 and N-substituted derivative BisTris(19.6+3.1% regioisomers) [51]

(20) 2,5,7,8-Tetramethyl-6-hydroxy-2-(hydroxyethyl)chroman (94%)[47]   (21) Dihydrocyacetone (75%) [45]   (22) 1,8-Octanediol (67.8%) [48]

FIGURE 23.5   (continued)

cells are affected, causing toxic side effects; tumors can develop resistance to these drugs; and the drugs are often inactive against metastases, as pointed out by Shuto et al. [56]. To solve these problems, various 5-nucleosides were phosphatidylated, and attractive antitumor effects were obtained *in vivo*. Several structures, such as alkyl ester backbones or no glycerol backbone, were investigated. Antitumor effects of 5′-phosphatidylnucleosides were found to be dependent on their recognition as diacylphospholipids by the body, indicating that diacylglycerol structure is important. Recently, 5′-*O*-dipalmitoylphosphatidyl-2′-*C*-cyano-2′-deoxy-1-β-D-*arabino*-pentofuranosylcytosine (DPP-CNDAC) was phosphatidylated by PLD, and liposomal DPP-CNDAC showed promise for cancer therapy [57,58].

23.4.2   PHOSPHATIDYL-L-ASCORBIC ACID

L-Ascorbic acid is a well-known hydrophilic antioxidant. To prevent active oxygen radicals from attacking membrane lipids, 6-phosphatidyl-L-ascorbic acid was synthesized from egg yolk PC [40]. The attack of active oxygen radicals originating in the aqueous phase is believed to be one of the initiation steps of membrane lipid peroxidation. When 6-phosphatidyl-L-ascorbic acid was included in multilamellar liposomes of egg yolk PC, it could more effectively retard the aqueous peroxyl-induced peroxidation of PC than L-ascorbic acid [59].

23.4.3 PHOSPHATIDYL-KOJIC ACID AND -ARBUTIN

Kojic acid and arbutin are tyrosinase inhibitors. Kojic acid inhibits tyrosinase activity by chelation and as an antioxidant, while arbutin is a competitive inhibitor of tyrosinase. To impart hydrophobicity to these compounds to prevent degradation, phosphatidyl-kojic acid and phosphatidyl-arbutin were synthesized from dipalmitic-PC (DPPC) [43]. Their inhibition of L-DOPA (3,4-dihydroxyphenylalanine) to dopachrome (precursor of melanin), catalyzed by tyrosinase *in vitro*, was of a similar level to the parent compounds. These phosphatidyl derivatives show promise for application in the cosmetics industry.

23.4.4 PHOSPHATIDYL-GENIPIN

Genipin is a component of Chinese medicine that exhibits a variety of pharmacological activities, including antitumor effects. However, genipin is a water-soluble compound that may present disadvantages in pharmaceutical use [44]. To address this problem, genipin was phosphatidylated from DPPC, and enhanced cytotoxity to cancer cells was shown *in vitro* [44]. This result shows that the phosphatidyl residue works as a good carrier into cells. Phosphatidylated genipin is suggested to improve affinity toward cell membranes by penetrating the lipid bilayers, then liberating genipin by the action of PLD on the inner side of the bilayers.

23.4.5 PHOSPHATIDYL-VITAMIN E DERIVATIVE

The chromanol ring is the active moiety of α-tocopherol for radical-scavenging activity [47]. Phosphatidyl-chromanol (PCh) suppressed autoxidation of lard more effectively than the original compound [60]. The explanation for this phenomenon is that PCh forms reverse micelles in the oil, trapping the residual water. This water contains dissolved trace metal ions, which initiate the oxidation. In further studies of the antioxidative activity in Fe(III)/ascorbic acid-catalyzed peroxidation of a fish oil emulsion and the autoxidation of a rat brain homogenate using vitamin E, 2,5,7,8-tetramethyl-6-hydroxy-2-(hydroxyethyl) chroman, PCh, and 2-(2′,5′,7′,8′,-tetramethyl-6′-hydroxychromanyl) ethylphosphate (Ch-P) (Figure 23.6), Ch-P showed the highest activity [61]. It is speculated that this is due to the high water solubility of Ch-P.

23.4.6 PHOSPHATIDYL-TERPENE ALCOHOL

Terpenes are functional compounds with an isoprenoid structure, found in the essential oils of plants such as lemon grass, cherry, mandarin, and citronella. To utilize functional terpenes, especially in the pharmaceutical field, geraniol, farnesol, geranylgeraniol, phytol, perillyl alcohol, myrtenol, and nerol were phosphatidylated, and their antiproliferative effects on cancer cells were investigated *in vitro* [52,53]. Among these seven compounds, phosphatidyl-monoterpenes showed a markedly antiproliferative effect on human prostate PC-3 and human leukemia HL-60 cells,

Vitamin E [47]

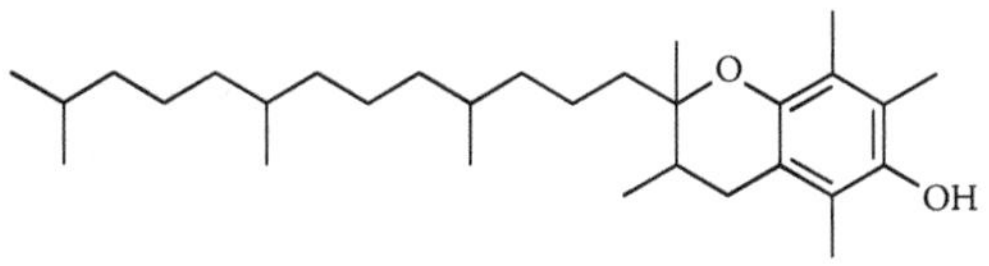

2, 5, 7, 8, -Tetramethyl-6-hydroxy-2-(hydroxyethyl) chromanol [47]

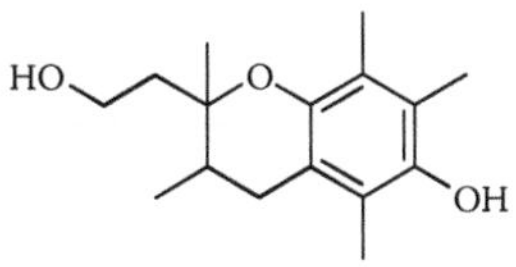

PCh: 1, 2-Diacyl-sn-glycero-3-phopho-2'-hydroxyethyl-2',5',7',8',-tetramethyl-6'-hydroxychroman[47]

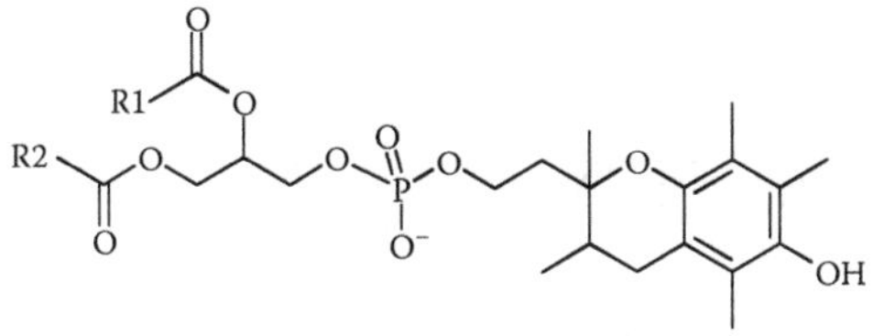

Ch-P: 2-(2',5',7',8',-Tetramethyl-6'-hydroxychromanyl)ethylphosphate [47]

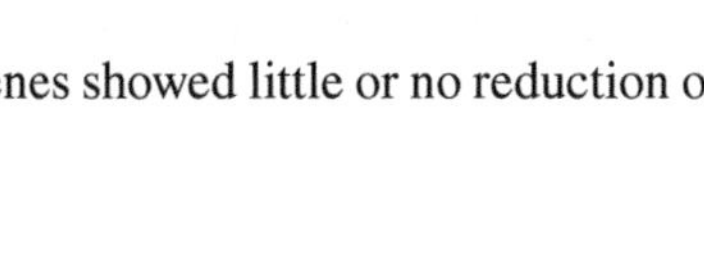

FIGURE 23.6 Vitamin E, 2,5,7,8,-tetramethyl-6-hydroxy-2-(hydroxyethyl)chromanol, and its phospholipid derivatives.

while the precursor monoterpenes showed little or no reduction of viability of cancer cells (Figure 23.7).

23.5 CONCLUSION AND OTHER REMARKS

Enzymatic strategies for PL modification and novel PL synthesis using PLA$_2$ and PLD were described in this chapter. The promising outlook for novel PLs in the pharmaceutical and medical fields was also discussed.

Enzymatic production of PLs has many advantages, such as reaction specificity, mild conditions, and reduced environmental pollution. A limitation to the industrial application of enzymatic approaches is the high cost of the enzymes; however, the use of enzymatic immobilization or microbial production (in a reactor system) to lower the cost is promising.

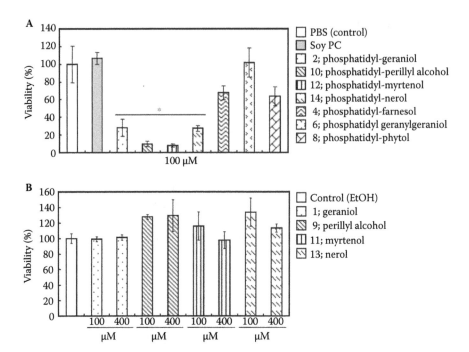

FIGURE 23.7　Viability of PC-3 cells treated with phosphatidylated terpene alcohols. (A) PC-3 cells were incubated for 72 h in culture media containing phosphatidylated terpene alcohols. (B) PC-3 cells were incubated for 72 h in the culture media containing terpene alcohols. *$p <$ .01 versus control. Similar results were obtained in the WST-1 assay of HL-60 cells.

Recently, PI synthesis was accomplished by mutated PLD from *Streptomyces antibioticus* [32]. Mutated PLD has a flexible binding substrate pocket that can recognize sterically hindered *myo*-inositol as a substrate. This successful result is supported by the evaluation of the catalytic molecular mechanism of PLD from *Streptomyces* sp. strain PMF [62]. Application of mutagenesis to PLA$_2$ holds considerable promise in the synthesis of novel types of PL containing various functional compounds.

REFERENCES

1. A. Samad, Y. Sultana, and M. Aqil: Liposomal drug delivery system: an update review. *Current Drug Delivery* 4 (2007) 297–305.
2. P. D'Arrigo and S. Servi: Using phospholipids for phospholipid modification. *Trends in Biotechnology* 15 (1997) 90–96.
3. Y. Iwasaki and T. Yamane: Phospholipases in enzyme engineering of phospholipids for food, cosmetics and medical applications. In *Lipid Biotechnology*, ed. by T.M. Kuo and H.W. Gardner, Dekker, New York, 2002, pp. 417–431.
4. Z. Guo, A.F. Vikbjerg, and X. Xu: Enzymatic modification of phospholipids for functional applications and human nutrition. *Biotechnology Advances* 23 (2005) 203–205.
5. P. Adlercreutz, A.M. Lyberg, and D. Adlercreutz: Enzymatic fatty acid exchange in glycerophospholipids. *European Journal of Lipid Science and Technology* 105 (2003) 638–645.

6. P. Pernas, J.L. Olivier, M.D. Legoy, and G. Bereziat: Phospholipid synthesis by extra cellular phospholipase A_2 in organic solvents. *Biochemical and Biophysical Research Communications* 168 (1990) 655–650.
7. A. Na, C. Eriksson, S.G. Eriksson, E. Österberg, and K. Holmberg: Synthesis of phosphatidylcholine with (n-3) fatty acids by phospholipase A_2 in microemulsion. *Journal of the American Oil Chemists' Society* 67 (1990) 766–770.
8. I. Mingarro, C. Abad, and L. Braco: Characterization of acylating and deacylating activities of an extracellular phospholipase A_2 in a water-restricted environment. *Biochemistry* 33 (1994) 4652–4660.
9. M. Härröd and I. Elfman: Enzymatic synthesis of phosphatidylcholine with fatty acid, isooctane, carbon dioxide, and propane as solvents. *Journal of the American Oil Chemists' Society* 72 (1995) 641–646.
10. M. Hosokawa, K. Takahashi, Y. Kikuchi, and M. Hatano: Preparation of therapeutic phospholipids through porcine pancreatic phospholipase A_2-mediated esterification and lipozyme-mediated acydolysis. *Journal of the American Oil Chemists' Society* 72 (1995) 1287–1291.
11. D. Egger, E. Wehtje, and P. Adlercreutz: Characterization and optimization of phospholipase A_2 catalyzed synthesis of phosphatidylcholine. *Biochimica et Biophysica Acta* 1343 (1997) 76–84.
12. D. Adlercreutz and E. Wehtje: An enzymatic method for the synthesis of mixed-acid phosphatidylcholine. *Journal of the American Oil Chemists' Society* 81 (2004) 553–557.
13. Y. Yamamoto, M. Hosokawa, and K. Miyashita: Production of phosphatidylcholine containing conjugated linoleic acid mediated by phospholipase A_2. *Journal of Molecular Catalysis B: Enzymatic* 41 (2006) 92–96.
14. S. Awano, K. Miyamoto, M. Hosokawa, M. Mankura, and K. Takahashi: Production of docosahexaenoic acid bounded phospholipids via phospholipase A_2 mediated bioconversion. *Fisheries Science* 72 (2006) 909–911.
15. H. Kitaguchi and A.M. Klibanov: Enzymatic peptide synthesis via segment condensation in the presence of water mimics. *Journal of the American Chemical Society* 111 (1989) 9272–9273.
16. M. Reslow, P. Adlercreutz, and B. Mattiasson: Modification of the microenvironment of enzymes in organic solvents. Substitution of water by polar solvents. *Biocatalysis* 6 (1992) 307–318.
17. A.M. Aura, P. Forssell, A. Mustranta, and K. Poutanen: Transesterification of soy lecithin by lipase and phospholipase. *Journal of the American Oil Chemists' Society* 72 (1995) 1375–1379.
18. M. Hosokawa, M. Ito, and K. Takahashi: Preparation of highly unsaturated fatty acid-containing phosphatidylcholine by transesterification with phospholipase A_2. *Biotechnology Techniques* 12 (1998) 583–586.
19. C.W. Park, S.J. Kwon, J.J. Han, and J.S. Rhee: Transesterification of phosphatidylcholine with eicosapentaenoic acid ethyl ester using phospholipase A_2 in organic solvent. *Biotechnology Letters* 22 (2000) 147–150.
20. M. Hossen and E. Hemandez: Enzyme-catalyzed synthesis of structured phospholipids with conjugated linoleic acid. *European Journal of Lipid and Science and Technology* 107 (2005) 730–736.
21. A.F. Vikbjerg, H. Mu, and X. Xu: Synthesis of structured phospholipids by immobilized phospholipase A_2 catalyzed acidolysis. *Journal of Biotechnology* 128 (2007) 545–554.
22. K. Takahashi and M. Hosokawa: Production of tailor-made polyunsaturated phospholipids through bioconversions. *Journal of Liposome Research* 11 (2001) 343–353.
23. I. Chibata and T. Tosa: Industrial applications of immobilized enzymes and immobilized microbial cells. *Applied Biochemistry and Bioengineering* 1 (1976) 329–357.

24. M. Sugiyama, K. Ohtani, M. Izuhara, T. Koike, K. Suzuki, S. Imamura, and H. Misaki: A novel prokaryotic phospholipase A_2; characterization, gene cloning, and solution structure. *Journal of Biological Chemistry* 277 (2002) 20051–20058.

25. W.C. Puijik, H.M. Verheij, and G.H. De Haas: The primary structure of phospholipase A_2 from porcine pancreas. *Biochimica et Biophysica Acta* 492 (1977) 254–259.

26. L.R. Juneja, N. Hibi, N. Inagaki, T. Yamane, and S. Shimizu: Comparative study on conversion of phosphatidylcholine to phosphatidylglycerol by cabbage phospholipase D in micelle and emulsion system. *Enzyme and Microbial Technology* 9 (1987) 350–354.

27. L.R. Juneja, N. Hibi, T. Yamane, and S. Shimizu: Repeated batch and continuous operations for phosphatidylglycerol synthesis from phosphatidylcholine with immobilized phospholipase D. *Applied Microbiology and Biotechnology* 27 (1987) 146–151.

28. L.R. Juneja, T. Kazuoka, T. Yamane, and S. Shimizu: Kinetic evaluation of conversion of phosphatidylcholine to phosphatidylethanolamine by phospholipase D from different sources. *Biochimica et Biophysica Acta* 960 (1988) 334–341.

29. L.R. Juneja, T. Kazuoka, N. Goto, T. Yamane, and S. Shimizu: Conversion of phosphatidylcholine to phosphatidylserine by various phospholipase D in the presence of L- or D-serine. *Biochimica et Biophysica Acta* 1003 (1989) 277–283.

30. L.R. Juneja, E. Taniguchi, S. Shimizu, and T. Yamane: Increasing productivity by removing choline in conversion of phosphatidylcholine to phosphatidylserine by phospholipase D. *Journal of Fermentation and Bioengineering* 73 (1992) 357–361.

31. S. Pisch, U.T. Bornscheuer, H.H. Meyer, and R.D. Schmid: Properties of unusual phospholipids IV: chemoenzymatic synthesis of phospholipids bearing acetylenic fatty acids. *Tetrahedron* 53 (1997) 14627–14634.

32. A. Masayama, T. Takahashi, K. Tsukada, S. Nishikawa, R. Takahashi, M. Adachi, K. Koga, A. Suzuki, T. Yamane, H. Nakano, and Y. Iwasaki: *Streptomyces* phospholipase D mutants with altered substrate specificity capable of phosphatidylinositol synthesis. *Chembiochem* 9 (2008) 974–981.

33. S.B. Mandal, P.C. Sen, and P. Chakrabarti: In vitro synthesis of phosphatidylinositol and phosphatidylcholine by phospholipase D. *Phytochemistry* 19 (1980) 1661–1663.

34. G.J. Piazza and W.N. Marmer: Conversion of phosphatidylcholine to phosphatidylglycerol with phospholipase D and glycerol. *Journal of the American Oil Chemists' Society* 84 (2007) 645–651.

35. M. Hosokawa, T. Shimatani, T. Kanda, Y. Inoue, and K. Takahashi: Conversion to docosahexaenoic acid-containing phosphatidylserine from squid skin lecithin by phospholipase D-mediated transphosphatidylation. *Journal of Agricultural and Food Chemistry* 48 (2000) 4550–4554.

36. C. Vitro, I. Svensson, and P. Adlercreutz: Hydrolytic and transphosphatidylation activities of phospholipase D from savoy cabbage towards lysophosphatidylcholine. *Chemistry and Physics of Lipids* 106 (2000) 41–51.

37. N. Dittrich and R.U. Hofmann: Transphosphatidylation by immobilized phospholipase D in aqueous media. *Biotechnology and Applied Biochemistry* 34 (2001) 189–194.

38. Y. Iwasaki, Y. Mizumoto, T. Okada, T. Yamamoto, K. Tsutsumi, and T. Yamane: An aqueous suspension system for phospholipase D-mediated synthesis of PS without toxic organic solvent. *Journal of the American Oil Chemists' Society* 80 (2003) 653–657.

39. S. Shuto, S. Ueda, S. Imamura, K. Fukukawa, A, Matsuda, and T. Ueda: A facile one-step synthesis of 5'-phosphatidylnucleosides by an enzymatic two-phase reaction. *Tetrahedron Letters* 28 (1987) 199–202.

40. A. Nagao, N. Ishida, and J. Terao: Synthesis of 6-phosphatidyl-L-ascorbic acid by phospholipase D. *Lipids* 26 (1991) 390–394.

41. Y. Kokusho, A. Tsunoda, S. Kato, H. Machida, and S. Iwasaki: Production of various phosphatidylsaccharides by phospholipase D from *Actinomadura* sp. strain no. 362. *Bioscience, Biotechnology, and Biochemistry* 57 (1993) 1302–1305.

42. P. Wang, M. Schuster, Y.F. Wang, and C.H. Wong: Synthesis of phospholipid-inhibitor conjugates by enzymatic transphosphatidylation with phospholipase D. *Journal of the American Chemical Society* 115 (1993) 10487–10491.

43. M. Takami, N. Hidaka, S. Miki, and Y. Suzuki: Enzymatic synthesis of novel phosphatidylkojic acid and phosphatidylarbutin, and their inhibitory effects on tyrosinase activity. *Bioscience, Biotechnology, and Biochemistry* 58 (1994) 1716–1717.

44. M. Takami and Y. Suzuki: Enzymatic synthesis of novel phosphatidylgenipin, and its enhanced cytotoxity. *Bioscience, Biotechnology, and Biochemistry* 58 (1994) 1897–1898.

45. M. Takami and Y. Suzuki: Synthesis of novel phosphatidyldihydroxyacetone *via* transphosphatidylation reaction by phospholipase D. *Bioscience, Biotechnology, and Biochemistry* 58 (1994) 2136–2139.

46. M. Takami, N. Hidaka, and Y. Suzuki: Phospholipase D-catalyzed synthesis of phosphatidyl aromatic compounds. *Bioscience, Biotechnology, and Biochemistry* 58 (1994) 2140–2144.

47. T. Koga, A. Nagao, J. Terao, K. Sawada, and K. Mukai: Synthesis of a phosphatidyl derivative of vitamin E and its antioxidant activity in phospholipid bilayers. *Lipids* 29 (1994) 83–89.

48. M. Koketsu, T. Nitoda, H. Sugino, L.R. Juneja, M. Kim, T. Yamamoto, N. Abe, T. Kajimoto, and C.H. Wong: Synthesis of a novel sialic acid derivative (sialylphospholipid) as an antirotaviral agent. *Journal of Medicinal Chemistry* 40 (1997) 3332–3335.

49. J.O. Rich and Y.L. Khmelnitsky: Phospholipase D-catalyzed transphosphatidylation in anhydrous organic solvent. *Biotechnology and Bioengineering* 72 (2001) 374–377.

50. M. Hossen and E. Hernandez: Phospholipase D-catalyzed synthesis of novel phospholipid-phytosterol conjugates. *Lipids* 39 (2004) 777–782.

51. M. Dippe, C.M. Klaus, A. Schierhorn, and R.U. Hofmann: Phospholipase D-catalyzed synthesis of new phospholipids with polar head groups. *Chemistry and Physics of Lipids* 152 (2008) 71–77.

52. Y. Yamamoto, M. Hosokawa, H. Kurihara, and K. Miyashita: Preparation of phosphatidylated terpenes via phospholipase D-mediated transphosphatidylation. *Journal of the American Oil Chemists' Society* 85 (2008) 313–320.

53. Y. Yamamoto, M. Hosokawa, H. Kurihara, T. Maoka, and K. Miyashita: Synthesis of phosphatidylated-monoterpene alcohols catalyzed by phospholipase D and their antiproliferative effects on human cancer cells. *Bioorganic and Medicinal Chemistry Letters* 18 (2008) 4044–4046.

54. V.T. Lamant, B. Archaimbault, J. Durand, and M. Rigaud: Enzymatic synthesis of structural analogs of PAF-acether by phospholipase D-catalyzed transphosphatidylation. *Biochimica et Biophysica Acta* 1123 (1992) 347–350.

55. I. Aurich, P. Dürrschmidt, A. Schierhorn, and R.U. Hofmann: Production of octadecylphospho-L-serine by phospholipase D. *Biotechnology Letters* 24 (2002) 585–590.

56. S. Shuto, H. Itoh, A. Sakai, K. Nakagami, S. Imamura, and A. Matsuda: Antitumor phospholipids with 5-fluorouridine as a cytotoxic polar-head: synthesis of 5'-phosphatidyl-5-fluorouridines by phospholipase D-catalyzed transphosphatidylation. *Bioorganic and Medicinal Chemistry* 3 (1995) 235–243.

57. S. Shuto, H. Awano, N. Shimazaki, K. Hanaoka, and A. Matsuda: Enzymatic synthesis of 5'-phosphatidyl derivatives of 1-(2-*c*-cyano-2-deoxy-β-D-arabino-pentofuranosyl) cytosine (CNDAC) and their notable antitumor effects in mice. *Bioorganic and Medicinal Chemistry Letters* 6 (1996) 1021–1024.

58. T. Asai, K. Shimazu, M. Kondo, K. Kuromi, K. Watanabe, K. Ogino, T. Taki, S. Shuto, A. Matsuda, and N. Oku: Anti-neovascular therapy by liposomal DPP-CNDAC targeted to angiogenic vessel. *FEBS Letters* 520 (2002) 167–170.

59. A. Nagao and J. Terao: Antioxidant activity of 6-phosphatidyl-L-ascorbic acid. *Biochemical and Biophysical Research Communications* 172 (1990) 385–389.

60. T. Koga and J. Terao: Antioxidant activity of a novel phosphatidyl derivative of vitamin E in lard and its model system. *Journal of Agricultural and Food Chemistry* 42 (1994) 1291–1294.

61. S. Miyamoto, T. Koga, and J. Terao: Synthesis of a novel phosphate ester of a vitamin E derivative and its antioxidant activity. *Bioscience, Biotechnology, and Biochemistry* 62 (1998) 2463–2466.

62. I. Leiros, S. McSweeney, and E. Hough: The reaction mechanism of phospholipase D from *Streptomyces* sp. strain PMF. Snapshots along the reaction pathway reveal a penta-coordinate reaction intermediate and an unexpected final product. *Journal of Molecular Biology* 339 (2004) 805–820.

24 Synthesis of Useful Glycosides by Cyclodextrin Glucanotransferases and Glycosidases

Hirofumi Nakano and Taro Kiso

CONTENTS

24.1 INTRODUCTION

Glycosidic compounds constructed from glycosyl (glycone) and nonglycosyl (aglycone) compounds are generally termed *glycosides*. Most natural glycosides are synthesized by glycosyl transferases. Such enzymes, however, are not necessarily suitable for practical or industrial uses; the enzymes require expensive nucleotide donors as well as cofactors. Their inflexible selectivity for acceptors often hinders the production of glycosides of desired structures. The limited availability of cheap enzymes in large quantities and instability in the soluble state are also shackling their application. Therefore, other classes of carbohydrate-active enzymes are utilized to produce glycosides of vitamins, polyphenols, and various hydroxyl compounds, aiming at improving low aqueous solubility, liability to oxidation and to light, unfavorable tastes, and toxicity of aglycone compounds.

Cyclodextrin glucanotransferase (CGTase) is a potent enzyme, particularly for practical glycosylation due to high synthetic efficiency, availability of starch as a cheap glycosyl donor substrate, a broad tolerance for acceptors, as well as availability of commercial enzymes [1]. Glycosidases are also attractive especially because of large entries of cheap and well-characterized enzymes [2]. Glycosidases catalyze not only transglycosylation but also condensation (reverse hydrolysis). In addition, a number of studies have been conducted with phosphorylases irrespective of limited enzyme availability [3]. Phosphorylases generally show high substrate selectivity and positional specificity compared to glycosidases.

This chapter focuses on the synthesis of glycosides by CGTases and glycosidases, especially the following glycosides of potential usefulness: glycosides of glycerol [4], a sweet natural glycoside with improved taste [5], and thioglycosides [6–8].

24.2 REACTIONS FOR THE SYNTHESIS OF GLYCOSIDES

24.2.1 TRANSGLYCOSYLATION

There are two basic methods of glycoside synthesis: transglycosylation and condensation (reverse hydrolysis) (Figure 24.1) [1,2]. The reactions are also termed the kinetically controlled method and the equilibrium-controlled method, respectively.

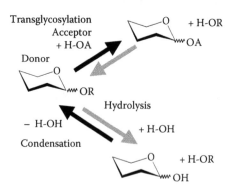

FIGURE 24.1 Two synthetic reactions for glycosides.

Transglycosylation is used for the production of various glycosides and oligosaccharides, in which glycosyl residues of appropriate donor substrates are transferred to acceptors (agylcone compounds) to generate new glycosidic linkages. Most glycosyl transferases hardly catalyze hydrolysis, and the synthesis proceeds efficiently, although particular enzymes (e.g., CGTases) that are also classified as transferases show hydrolytic action to certain limited degrees. This can be understood as the transfer of a glycosyl residue to a water molecule. Anomer-retaining hydrolases, glycosidases and glycanases that yield the same anomeric products as that of the substrates, catalyze transglycosylation. However, anomer-inverting hydrolases that yield inverted anomeric products show virtually no transglycosylation activity. In hydrolase-catalyzed reactions, a transfer product also serves as a substrate for hydrolysis. Therefore, it is necessary to optimize the reaction condition regarding such things as incubation time and enzyme/substrate concentrations. In glycosyl transferase-catalyzed reactions, no considerable degradation of transfer products occurs even after long-term incubation.

24.2.2 Condensation

Condensation is catalyzed by glycosyl hydrolases; if the reaction equilibrium shifts in the direction of synthesis by adopting high concentrations of substrates (e.g., monosaccharides and aglycone compounds), there are low water concentrations, or products are removed from the reaction system [2]. This reaction seems to be advantageous when appropriate oligosaccharides and glycosides are unavailable as donor substrates in transglycosylation. Condensation is catalyzed by anomer-inverting as well as by -retaining hydrolases; for instance, glucoamylase, an anomer-inverting enzyme, catalyzes not transglycosylation, but condensation. Yields in condensation, however, are lower compared to those of transglycosylation, which is attributable to the unfavorable thermodynamic equilibrium in aqueous reaction systems. In addition, longer reaction time and higher enzyme activity are required to obtain sufficient yields. Therefore, this methodology has not been fully utilized for the industrial production of glycosides so far, except for a few oligosaccharides.

24.3 PRODUCTION OF GLYCOSYL GLYCEROLS

24.3.1 Occurrence and Production Methods

Glycosylated glycerol exists as a hydrophilic moiety of glyceroglycolipids in plants, algae, and bacterial membranes. O-α-D-Glucosyl glycerol (Glc-GL; Figure 24.2) was found in traditional Japanese fermented foods such as *sake* (rice wine), *miso* (fermented soybean paste), and *mirin* (alcoholic seasoning made from Koji mash) approximately at 0.5%–1% [9,10]. Glc-GL exhibits about half the sweetness, high thermal stability, low Maillard reactivity, high water-holding activity, noncariogenicity, and low digestibility, and possible uses as ingredients for foods and cosmetics are expected. Glc-GL is produced by a Koji-mold α-glucosidase by fermentation. Therefore, a production method using *Aspergillus niger* α-glucosidase was previously developed [9]. In this section, an alternative method is presented that produces

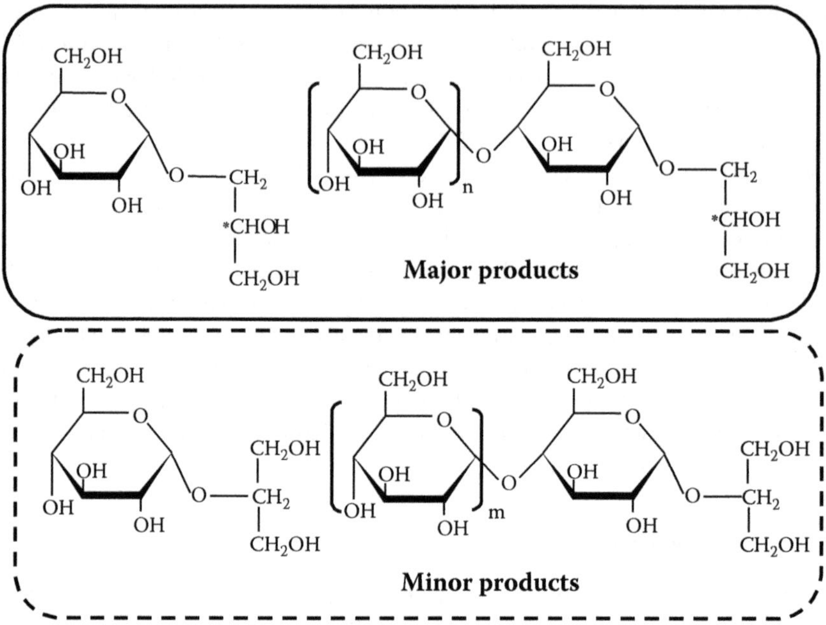

FIGURE 24.2 Glycosyl GLs produced by the CGTase reaction. The asterisks suggest an asymmetric carbon of the GL moiety. The products are distributed mainly from mono- to tetraglucosides (n, m = 1–4).

not only Glc-GL but also a series of maltooligosyl GLs (collectively termed glycosyl GLs) from starch and GL using CGTases [4].

24.3.2 PRODUCTS AND REACTION CONDITIONS
OF CGTASE-CATALYZED SYNTHESIS

Glycosyl GLs are produced by the intermolecular transglycosylation action of CGTases, in which α-1,4-linked glycosyl residues of starch are transferred to GL. Some products are shown in Figure 24.2: O-α-D-glucosyl-(1→1)-GL and O-α-D-glucosyl-(1→2)-GL are identified as major (90%–95%) and minor (5%–10%) components of the monoglucosides (Glc-GL), respectively. In addition, two diasteromeric isomers were possible for O-α-D-glucosyl-(1→1)-GL; theoretically, binding between the C1-OH of GL and the Glc residue makes C2 of GL a new asymmetric center. Insufficient experimental evidence, however, was obtained for these isomers. The main diglucoside was O-α-D-glucosyl-(1→4)-O-α-D-glucosyl-(1→1)-GL. Based on the above evidence together with the well-studied action of CGTases, larger products were reasonably estimated to have maltooligosyl residues, with which the primary OH group (C1 or C3) of the GL moiety is mainly substituted.

CGTases from *Geobacillus stearothermophilus* and *Thermoanaerobacter* sp. gave higher yields compared to those from *Bacillus macerans* and *B. circulans* (Figure 24.3). High thermal stability of the former enzymes allowed relatively high

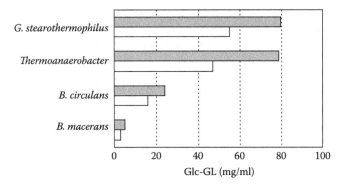

FIGURE 24.3 Transglycosylation of GL by various CGTases. GL (10% and 20%) and soluble starch (10%) were reacted with CGTases from the indicated origins at pH 6.0 and 50°C for 72 h. GL concentrations: open bars, 10%; shaded bars, 20%.

TABLE 24.1
Transglycosylation Products by CGTase and α-Glucosidase

Component	Saccharides (mg/mL)	
	CGTase	**α-Glucosidase**
Glucosyl GL	69.1	75.6
Maltosyl GL	36.6	7.8
Maltotriosyl GL	13.7	0
Maltotetraosyl GL	3.7	0
Glucose	2.4	128
Maltose	1.1	18.6
Unknown[a]	61.3	4.7
Total	188	238

Note: Soluble starch (20%) and maltose (20%) were used as donor substrates for *Thermoanaerobacter* CGTase and *A. niger* α-glucosidase in the reaction with 25% GL, respectively.

[a] Nonseparable or unidentified saccharides including cyclodextrins expressed as milligrams Glc/milliliter.

reaction temperatures, at 60°C–80°C. The reaction of 30% (w/v) GL and 20% (w/v) starch was established as the most efficient condition.

Table 24.1 shows the products formed by CGTase and α-glucosidase when soluble starch and maltose were used as donors, respectively. The yield of whole products using the CGTase method [4] was comparable to or rather higher than that using the conventional method [9,10]. Monoglucoside was the main product in the latter method, while maltooligosyl GLs were produced much more in the former. Low production of reducing sugars such as glucose and maltose by the CGTase may be of significance. Reducing sugars reduce usefulness, such as low heat colorability and low cariogenicity, unless they are separated from glycosyl GLs.

24.3.3 INHIBITION OF α-AMYLASE BY GLYCOSYL GLYCEROLS

Glc-GL is reported to function as an inhibitor for rat intestinal disaccharide hydrolases [10], which suppress the digestion and consequently blood sugar level. Expecting similar physiological effects, if any, inhibition against porcine pancreas α-amylase was measured *in vitro* (Table 24.2). A large IC_{50} (median inhibition concentration) value of Glc-GL suggested weak inhibition. On the other hand, maltosyl GL and maltotriosyl GL, which are particular products of CGTase but not of α-glucosidase, were found to be more potent inhibitors. A few Glc residues may be required for sufficient affinity with the amylase, which favors glucans as natural substrates. Three glycosides were not hydrolyzed by amylase. Like Glc-GL, maltooligosylated GL showed low Maillard reactivity, high heat stability, noncariogenicity, high moisturizing power, low sweetness, and the stabilizing effect of ascorbic acid.

TABLE 24.2
Inhibition of Porcine Pancreas α-Amylase by Glycosyl GLs

Glycoside	IC_{50} (mM)
Glucosyl GL	26.4
Maltosyl GL	0.245
Maltotriosyl GL	0.065

24.3.4 ENZYMATIC PRODUCTION OF OTHER GLYCOSYL GLYCEROLS AND REACTIVITY IN ESTERIFICATION

Glycosyl GLs were synthesized by the condensation action of several glycosidases: α-glucosidases, β-glucosidases, β-galactosidases, and α-mannosidases were reacted with GL and monosaccharides of the respective substrates, D-glucose, D-galactose, and D-mannose, at elevated concentrations. The major products were O-α-D-glucosyl-(1→1)-GL, O-β-D-glucosyl-(1→1)-GL, O-β-D-galactosyl-(1→1)-GL, and O-α-D-mannosyl-(1→1)-GL, respectively, in which the glycosyl residues were bound with one of the primary OH groups of the GL moiety. The reaction proceeded well at pH 4.5–5.0 and relatively high temperatures, 60°C–65°C. The yields increased with substrate concentrations up to approximately $2M$, although more concentrated substrates were inhibitory. An approximate molar ratio of 1:2 was appropriate for the monosaccharides and GL.

GL also serves as a preferable acceptor in transglycosylation by various glycosylases and glycanases. It has been long known that *Escherichia coli* β-galactosidase mainly produces O-β-D-galactosyl-(1→1)-GL in the reaction of lactose and GL. Contrarily, an endogalactanase from *Penicillium* glycosylated the secondary OH group of GL almost exclusively to produce O-β-D-galactosyl-(1→2)-GL and O-β-D-galactosyl-(1→4)-O-β-D-galactosyl-(1→2)-GL, using β-1,4-galactan as a donor [11]. An exogalactanase from *B. subtilis* transferred digalactosyl residue to

GL to produce O-β-D-galactosyl-(1→4)-O-β-D-galactosyl-(1→2)-GL and O-β-D-galactosyl-(1→4)-O-β-D-galactosyl-(1→1)-GL at an approximate ratio of 2:1 [12].

Some glycosyl GLs and glycosides with other polyol moieties were derived to glycolipids by the esterification with lipases in an aqueous system comprised mostly of the substrate and the lipase in addition to a minimum amount of water [13–16]. Monoglycosides served as better substrates than diglycosides in the reaction with middle-chain fatty acids such as lauric acid and long-chain fatty acids such as oleic acid. The products were mono-, di-, and triesters, which indicated that the esterification occurred at the GL moiety as well as at the sugar residues. The yields of esters were almost the same with different substitute positions of GL (the primary and secondary OH groups), the monosaccharide residues (glucose, galactose, fructose), and the linkage types (α and β).

24.4 TRANSGLYCOSYLATION OF MOGROSIDE V BY CGTASES

24.4.1 NATURAL HIGHLY SWEET GLYCOSIDES

Stevioside is a highly sweet glycoside extracted from the leaves of a Chinese medical plant, *Stevia rebaudiana* Bertoni. The glycoside has approximately 140-fold sweetness of sucrose and is useful as a low-calorie sweetener. In spite of such intensive sweetness, however, the quality of taste, especially its bitter taste and aftertaste, has limited the applications as ingredients for foods and beverages. Glycosylated stevioside of improved taste was thus produced by CGTase [1,17]. Several related compounds such as rebaudioside A (steviol-tetraglucoside), rubusoside (steviol-bisglucoside), and glycyrrhritin were also glycosylated by CGTase and β-fructofuranosidase to improve the taste as well as the solubility.

Mogroside V (MgV) is a natural sweet glycoside that has approximately 350 times the relative sweetness of sucrose (Figure 24.4) [18]. MgV is contained in extract from the fruit of *luo-han-guo* (*Siraita grosvenori* Swingle), a perennial herb cultured in

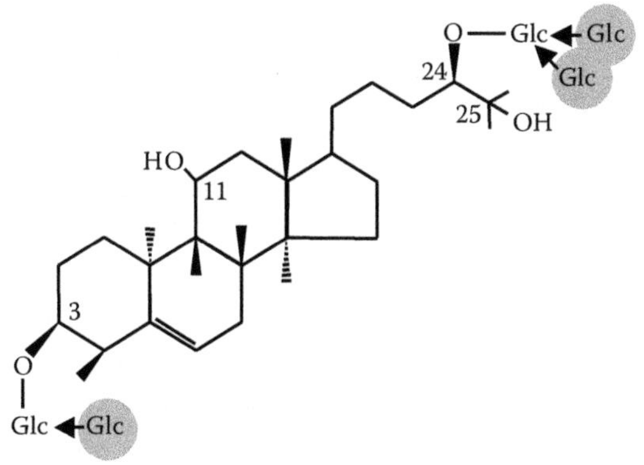

FIGURE 24.4 Structure of MgV. Shaded Glc residues indicate possible glycosylation sites by CGTases.

China. Traditionally, the extract has been used not only as a sweetener but also as a folk medicine for remediation of fever and cough. In recent years, the free radical-eliminating activity and antitumor activity of MgV and related glycosides have been elucidated [19]. In Japan, an extract comprised mainly of MgV is used as an ingredient for a commercialized sweetener with low caloric value. This section deals with the further glycosylation of MgV to improve the quality of sweetness.

24.4.2 Reaction Conditions and Products

Several CGTases glycosylated MgV to give considerable yields of more than 80% when 6% MgV, 10% tapioca starch, and the enzymes were incubated at 50°C and pH 6.5 (Figure 24.5). It is known that CGTases specifically glycosylate C4-OH of the nonreducing end Glc residue of acceptor molecules. In the glycosylation of stevioside, α-1,4-glycosylation occurred at the C4-OH groups for both the 13-O-β-sophorosyl and 19-O-β-glucosyl residues. MgV has three possible Glc residues with free C4-OH groups: two Glc residues linking at C24 through β-1,6- and β-1,2-linkages and β-1,6-linked Glc residue at C3 (Figure 24.4). Nuclear magnetic resonance (NMR) analysis of the product having one additional Glc residue (1Glc-MgV) suggested that these three possible residues were involved in the glycosylation. In other words, 1Glc-MgV contained three glycosides with different glycosylation positions. Because of the difficulty of separating these products, however, it has not been determined whether glycosylation occurred equally at these residues or preferentially at a certain residue.

24.4.3 Sweetness and Tastes of Glycosylation Products

Three glycosylated products (1Glc-MgV, 2Glc-MgV, and 3Glc-MgV) were separated in reverse-phase (ODS) chromatography performed using gradient elution with ethanol. Liquid chromatography-mass spectrometric (LC-MS) analysis suggested that they had an additional 1–3 Glc residue in addition to the five intrinsic Glc residues,

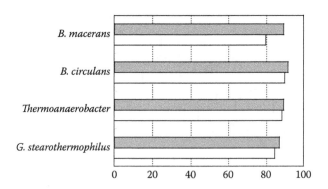

FIGURE 24.5 Transglycosylation of MgV by various CGTases. MgV (6%) and tapioca starch (10%) were reacted with CGTases from the indicated origins at pH 6.5 and 50°C. Reaction time: open bars, 24h; shaded bars, 48 h. *G.*, *Geobacillus*; *B.*, *Bacillus*.

TABLE 24.3

Sensory Evaluations of Transglycosylation Products of MgV

| Glycosides | Sweetness | Quality of Sweetness | | |
		Bitterness	Aftertaste	Peculiarity
Sucrose	1	1.00	1.00	1.00
MgV	378	1.50	2.50	1.81
1Glc-MgV	191	1.36	1.91	1.36
2Glc-MgV	97	1.23	1.69	1.46
3Glc-MgV	53	1.30	1.40	1.40

Note: In the evaluation of the quality of sweetness, the score of sucrose was expressed as 1.0; a score closer to 1.0 signifies a more favorable taste.

respectively. For each product, the quality of sweetness such as bitterness, after-taste, and peculiarity together with relative intensity of sweetness was compared with those of MgV through sensory evaluation using sucrose as a standard sweetener (Table 24.3). The products were judged to have improved quality, particularly with respect to the aftertaste. The intensity of sweetness considerably diminished with an increasing number of the Glc residues; the addition of one Glc residue resulted in a reduction in sweetness almost by half, although the smaller products, especially 1Glc-MgV, still retained sufficient sweetness that was almost comparable to that of stevioside (150-fold). In the reaction mixture, the intensity of sweetness decreased to approximately 30% that of MgV. However, the sweetness was recovered to approximately 80% by partial hydrolysis with an α-amylase of a high-saccharifying activity type. This suggests that partial hydrolysis of larger products may afford both sufficient intensity and better qualities of sweetness.

24.5 SYNTHESIS OF THIOGLYCOSIDES

24.5.1 Glycosides with Various Linkages

Several types of glycosides occur in nature: *O*-, *N*-, *C*-, and *S*-glycosides. *O*-Glycosides are widely distributed, especially in the second metabolites of plants. *N*-Glycosides are represented by nucleotides. *C*-Glycosides are found, though rarely, as antibiotics and natural color substances. *S*-Glycosides are also rare, although they still exist as antibiotics and plant glycosides. In addition, some glycosides contain glycosyl ester linkages. So far, however, enzyme synthesis has been focused on the glycosylation of the OH groups in alcoholic or phenolic compounds to obtain *O*-glycosides. It has been demonstrated that a sucrose phosphorylase from *Streptococcus mutans* glycosylates the carboxyl groups, for example, of acetic acid and benzoic acid to give acetyl and benzoyl glucose, respectively [20].

The following section describes the versatility of the usual glycosidases in providing thio(*S*)-glycosides [6,7]. Thiol (SH) compounds have general problems such as characteristic smell, high volatility, low solubility, and liability to oxidation. Glycosylation is expected to improve such inexpedience. Two reactions were utilized for this purpose: condensation by β-galactosidases and transglycosylation by β-fructofuranosidases.

24.5.2 β-GALACTOSYLATION

Thio (S)-glucosidase (sinigrinase, EC 3.2.3.1) is known to hydrolyze natural β-S-glucosides, sinigrin and sinalbin. A S-glucosidase from a plant origin (white mustard), however, gave no transfer and condensation products. Instead, it was found that several β-galactosidases from microbial origins showed hydrolyzing activity on a synthetic glycoside, p-nitrophenyl S-β-D-galactoside, even if not as high compared to those on usual O-galactosides. Moreover, the enzymes glycosylated not only the OH group of 2-mercaptoethanol (ME) but also the SH group in condensation with D-galactose (Gal). The products were isolated to be confirmed as 2-mercaptoethyl O-β-D-galactoside (Gal-O-ME) and 2-hydroxyethyl S-β-D-galactoside (Gal-S-ME), respectively (Figure 24.6) [6]. The ratio of these products differed greatly depending on the enzymes and the reaction conditions. The enzymes from *Aspergillus* and *Penicillium* produced Gal-S-ME significantly in addition to Gal-O-ME, while the other enzymes predominantly produced the usual Gal-O-ME (Figure 24.7). Effective synthesis of Gal-S-ME was attained in the reactions of $1.25M$ ME and $0.8M$–$2.8M$ Gal. More concentrated substrates caused enzyme inhibition by Gal and inactivation by ME. Gal-O-ME was produced to a maximum (equilibrium) level at an early stage as shown in Figure 24.8. This is reasonable because the enzyme has much higher O-galactosidase activity than S-galactosidase activity. In spite of the lower production rate, however, Gal-S-ME accumulated considerably during a long-term reaction. In general, the S-glycosidic bond does not cleave easily with an acid catalyst. A sulfur atom shows the smaller electronegativity or softer basicity than an oxygen atom. This feature weakens interaction with substrates as acid-base catalysis. The high equilibrium concentration of Gal-S-ME is attributable to such high chemical stability of the S-glycosidic linkage. Gal-S-ME also showed considerable resistance to enzyme hydrolysis compared to Gal-O-ME in diluted aqueous solutions.

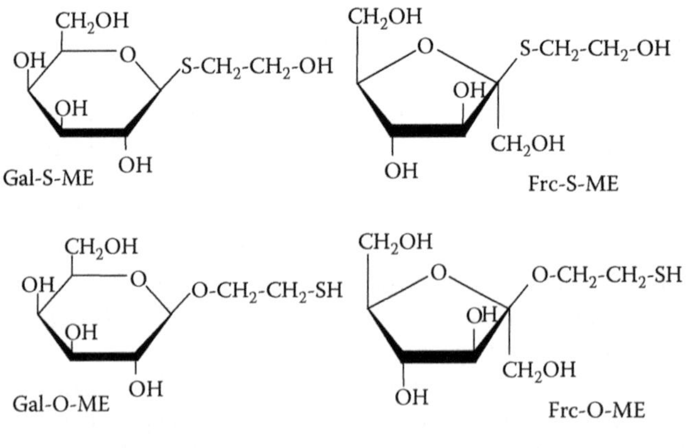

FIGURE 24.6   Glycosylation products of ME.

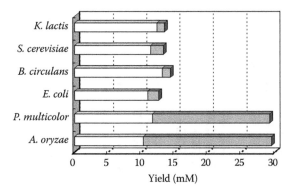

FIGURE 24.7 Galactosylation of ME by various β-galactosidases. ME (0.25*M*) and Gal (2.0*M*) were reacted with enzymes from the indicated origins at pH 4.5–7.0 (appropriately selected for each enzyme) and 37°C for 24 h. Open bars, Gal-*O*-ME; shaded bars, Gal-*S*-ME. *K.*, *Klyuveromyces*; *S.*, *Saccharomyces*; *E.*, *Escherichia*; *B.*, *Bacillus*; *P.*, *Penicillium*; *A.*, *Aspergillus*.

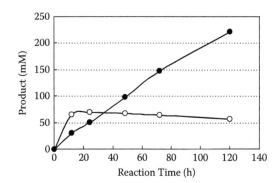

FIGURE 24.8 Time course of formation of Gal-*O*-ME and Gal-*S*-ME by condensation of ME and Gal with β-galactosidase. Gal (2.0*M*) and ME (1.0*M*) were reacted with *A. oryzae* β-galactosidase at pH 5.5 and 37°C. Open circle, Gal-*O*-ME; closed circle, Gal-*S*-ME.

24.5.3 β-Fructofuranosylation

Transfructofuranosylation of ME was performed using sucrose as a donor substrate [7]. Two products, 2-hydroxyethyl *S*-β-D-fructofuranoside (Frc-*S*-ME) and 2-mercaptoethyl *O*-β-D-fructofuranoside (Frc-*O*-ME), were produced concomitantly (Figure 24.6 and Figure 24.9). The product ratios were also different in the enzymes as well as the reaction time (Figure 24.9); the two *Candida* enzymes were effective for synthesizing Frc-*S*-ME, while the *Arthrobacter* enzyme gave no Frc-*S*-ME but only Frc-*O*-ME in spite of broad acceptor specificity in the usual synthesis of oligosaccharides and glycosides [21]. Frc-*O*-ME tended to be produced more from the initial stage, although it disappeared by hydrolysis on prolonged incubation.

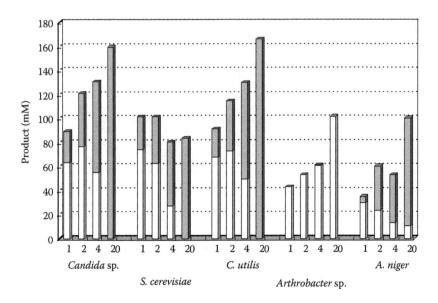

FIGURE 24.9   Transfructosylation of ME by various β-fructofuranosidases. ME (1.0*M*) and sucrose (1.5*M*) were reacted with enzymes from the indicated origins at pH 6.0–7.0 (appropriately selected for each enzyme) and 40°C. Open bars, Frc-O-ME; shaded bars, Frc-S-ME. Reaction time (h) is indicated in the figure. *S.*, *Saccharomyces*; *C.*, *Candida*; *A.*, *Aspergillus*.

Contrarily, Frc-S-ME accumulated continuously, probably without hydrolysis. This is also attributable to the high chemical stability of the *S*-glycosidic linkage and to high resistance to the enzyme hydrolysis. The most efficient synthesis of Frc-S-ME was accomplished approximately at 1.9*M* sucrose and 3.0*M* ME. The specific accumulation of Frc-S-ME is favorable because difficulties in purification, especially the separation of Frc-S-ME and Frc-O-ME, are resolved.

24.5.4   INHIBITION OF SEVERAL GLYCOSIDASES BY *S*-FRUCTOSIDE

Specific inhibitors for glycosidases have been studied for several reasons. Acarbose, an α-glucosidase inhibitor, is useful as a therapeutic agent for diabetes due to inhibition of the intestinal digestion of carbohydrates [22]. Small-molecule glycosidase inhibitors have been used as investigation tools in the reaction mechanism and active center structures. Some inhibitors are utilized as ligands for affinity chromatography. Although many synthetic inhibitors for various glycosidases have been developed, there are few for β-fructofuranosidases. Frc-S-ME is a small compound that shows high resistance to the hydrolysis, and the β-fructofuranosyl moiety seemed to exhibit corresponding affinity.

As expected, Frc-S-ME was hardly hydrolyzed by several β-fructofuranosidases [8]. The inhibition experiment (Table 24.4) revealed that *S*-fructoside inhibited the enzymes competitively, except the *Arthrobacter* enzyme, which did not synthesize

TABLE 24.4

Inhibition Activity of Frc-S-ME Against β-Fructosidases and α-Glucosidases

Enzyme	Origin	Substrate	Ki (mM)	Type of Inhibition
β-Fructosidase	A. niger	Sucrose	46	Competitive
	Candida sp.		1.7	Competitive
	S. cerevisiae		5.5	Competitive
	Arthrobacter sp.		No inhibition	—
α-Glucosidase	Rat intestine	Sucrose	2.4	Noncompetitive
		Maltose	2.4	Noncompetitive
	Saccharomyces sp.	Sucrose	3.5	Noncompetitive
		Maltose	7.4	Noncompetitive

Frc-S-ME [7]. Respective Ki values suggested that the affinity was almost comparable to those with sucrose, which were evaluated with respective Km values of 17, 7.2, and 30 mM. It was also found that sucrase activity of the rat intestinal sucrase/isomaltase complex, which belongs not to β-fructosidases but to α-glucosidases, was strongly suppressed by Frc-S-ME. In addition, Frc-S-ME as well as an α-glucosidase from *Sacchromyces* inhibited maltase activity of the intestinal glycosidase complex. The inhibition type against the α-glucosidases was noncompetitive in the hydrolysis of sucrose and maltose. Frc-S-ME exhibited weak inhibition of other α-glucosyl hydrolases such as the *Bacillus* α-glucosidase, the *Rhizopus* glucoamylase, and the porcine kidney trehalase. Although the inhibition activity of Frc-S-ME against the intestinal glycosidases does not seem to be sufficient for pharmaceutical application, substitution with various aglycone moieties may afford more potent inhibition.

24.6 CONCLUDING REMARKS

Glycosylation is effective for the structural modification of biologically active compounds. Enzymatic glycosylation is advantageous because troublesome protection and deprotection steps in the chemical synthesis can be avoided. In addition, the reaction proceeds under mild conditions. Effective and specific glycosylation is attained by selecting appropriate enzymes and reaction conditions. Some ordinary and cheap carbohydrate-active enzymes show remarkable ability and versatility and can be potent catalytic tools in synthesizing glycosides. Glycosylation occasionally increases or changes biological activities as suggested here and in other studies [23]. It is hoped, therefore, that enzyme glycosylation will be extensively studied to alter not only the physicochemical properties but also the biological activity of various compounds to enhance its usefulness, especially in food, cosmetic, and pharmaceutical ingredients.

REFERENCES

1. Nakano, H., and Kitahata, S. 2005. Application of cyclodextrin glucanotransferases to the synthesis of oligosaccharides and glycosides. In *Handbook of Industrial Biocatalysis*, ed. C.T. Hou. Boca Raton, FL: CRC Press, Taylor & Francis Group, pp. 22-1-15.
2. Fujimoto, H., Isomura, M., and Ajisaka, K. 1996. Synthesis of oligosaccharide components of glycoconjugates using glycosidases. *J. Appl. Glycosci.* 43: 265–272.
3. Kitao, S., and Sekine, H. 1994. α-D-Glucosyl transfer to phenolic compounds by sucrose phosphorylase from *Leuconostoc mesenteroides* and production of α-arbutin. *Biosci. Biotechnol. Biochem.* 58: 38–42.
4. Nakano, H., Kiso, T., Okamoto, K., Tomita, T., Manan, M.B.A., and Kitahata, S. 2003. Synthesis of glycosyl glycerol by cyclodextrin glucanotransferases. *J. Biosci. Bioeng.* 95: 583–588.
5. Yoshikawa, S., Murata, Y., Sugiura, M., Kiso, T., Shizuma, M., Kitahata, S., and Nakano, H. 2005. Transglycosylation of mogroside V, a triterpene glycoside in *Siraitia grosvenori*, by cyclodextrin glucanotransferase and improvement of the quality of sweetness. *J. Appl. Glycosci.* 52: 247–252.
6. Nakano, H., Shizuma, M., Kiso, T., and Kitahata, S. 2000. Galactosylation of thiol group by β-galactosidase. *Biosci. Biotechnol. Biochem.* 64: 735–740.
7. Nakano, H., Murakami, H., Shizuma, M., Kiso, T., Araujo, T.L., and Kitahata, S. 2000. Transfructosylation of thiol group by β-fructofuranosidases. *Biosci. Biotechnol. Biochem.* 64: 1472–1476.
8. Kiso, T., Hamayasu, K., Fujita, K., Hara, K., Kitahata, S., and Nakano, H. 2003. Inhibition of β-fructofranosidases and α-glucosidases by synthetic thio-fructofuranoside. *Biosci. Biotechnol. Biochem.* 67: 1719–1724.
9. Takenaka, F., Uchiyama, H., and Imamura, T. 2000. Identification of α-D-glucosylglycerol in sake. *Biosci. Biotechnol. Biochem.* 64: 378–385.
10. Takenaka, F., and Uchiyama, H. 2001. Effect of α-D-glucosylglycerol on the *in vitro* digestion of disacchrides by rat intestinal enzymes. *Biosci. Biotechnol. Biochem.* 65: 1458–1463.
11. Nakano, H., Takenishi, S., and Watanabe, Y. 1988. Formation of transfer products from soybean arabionogalactan and glycerol by galactanase from *Penicillium citrinum*. *Agric. Biol. Chem.* 52: 1912–1921.
12. Nakano, H., Kitahata, S., Kinugasa, H., Watanabe, Y., Fujimoto, H., Ajisaka, K., and Takenishi, S. 1991. Transfer reaction catalyzed by exo-β-1,4-galactanase from *Bacillus subtilis*. *Agric. Biol. Chem.* 55: 2075–2082.
13. Nakano, H., Kitahata, S., Tominaga, Y., and Takenishi, S. 1991. Esterification of glycosides with glycerol and trimethylolpropane moieties by *Candida cylindracea* lipase. *Agric. Biol. Chem.* 55: 2083–2089.
14. Nakano, H., Kitahata, S., Tominaga, Y., Kiku, Y., Ando, K., Kawashima, Y., and Takenishi, S. 1992. Synthesis of glucosides with trimethylopropane and two related polyol moieties by cyclodextrin glucanotransferase and their esterification with lipase. *J. Ferment. Bioeng.* 73: 237–238.
15. Nakano, H., Kiku, Y., Ando, K., Kawashima, Y., Kitahata, S., Tominaga, Y., and Takenishi, S. 1994. Lipase-catalyzed esterification of mono- and oligoglycosides with fatty acids. *J. Ferment. Bioeng.* 78: 237–238.
16. Nakano, H., Kitahata, S., Shimada, Y., Nakamura, M., Tominaga, Y., and Takenishi, S. 1995. Esterification of glycosides by a mono- and diacylglycrol lipase from *Penicillium camembertii* and comparison of the products with *Candida cylindracea* lipase. *J. Ferment. Bioeng.* 80: 24–29.

17. Fukunaga, Y., Miyata, T., Nakayasu, N., Mizutani, K., Kasai, R., and Tanaka, O. 1989. Enzymic transglycosylation products of stevioside: Separation and sweetness-evaluation. *Agric. Biol. Chem.* 53: 1603–1607.
18. Kasai, R., Nie, R.L., Nashi, K., Ohtani, K., Zhou, J., Tao, G.D., and Tanaka, O. 1989. Sweet cucurbitane glycosides from the fruits of *Siraitia siamensis* (chi-zi luo-han-guo), a Chinese folk medicine. *Agric. Biol. Chem.* 53: 3347–3349.
19. Takasaki, M., Konoshima, T., Murata, Y., Sugiura, M., Nishino, H., Tokuda, H., Matsumoto, K., Kasai, R., and Yamasaki, K. 2003. Anticarcinogenic activity of natural sweeteners, cucurbitane glycosides, from *Momordica grosvenori*. *Cancer Lett.* 198: 37–42.
20. Sugimoto, K., Nomura, K., Nishiura, H., Ohdan, K., Nishimura, T., Hayashi, H., and Kuriki, T. 2007. Novel transglucosylation reaction of sucrose phosphorylase to carboxylic compounds such as benzoic acid. *J. Biosci. Bioeng.* 104: 22–29.
21. Fujita, K., Hara, K., Hashimoto, H., and Kitahata, S. 1990. Transfructosylation catalyzed by β-fructofuranosidase I from *Arthrobacter* sp. K-1. *Agric. Biol. Chem.* 54: 2655–2661.
22. Hanozet, G., Pircher, H.-P., Vanni, P., Oesch, B., and Semenza, G. 1981. An example of enzyme hysteresis. *J. Biol. Chem.* 256: 3703–3711.
23. Sugimoto, K., Nomura, K., Nishimura, T., Kiso, T., Sugimoto, K., and Kuriki, T. 2005. Synthesis of α-arbutin-α-glucoside and their inhibitory effects on human tyrosinase. *J. Biosci. Bioeng.* 99: 272–276.

25 Biotechnological Production and Properties of Carotenoid Pigments

Milan Čertík, Vladimíra Hanusová, Emília Breierová, Ivana Márová, and Peter Rapta

CONTENTS

Key Words: Biotechnology; carotene; microorganism; pigment; production; properties.

25.1 INTRODUCTION

Carotenoids represent one of the broadest groups of natural antioxidants (over 600 characterized structurally) with significant biological effects and numerous industrial applications. Lycopene is a typical acyclic carotene that serves as a starting metabolite for formation of carotenoid derivatives via specific routes (β-carotene, torulene, etc.). Xanthophylls include hydroxy-, methoxy- oxo-, epoxy-, carboxy-, and aldehydic groups (torularhodin, zeaxanthin, astaxanthin, etc.), which results in a broad structural variety of carotenoid compounds.

Commercially, carotenoids are used as food colorants and nutritional supplements, with an estimated global market of some \$935 million by 2005 (Fraser and Bramley 2004). They are present in all photosynthetic organisms and responsible for most of the yellow-to-red colors of fruits and flowers. Because animals lack the ability to synthesize carotenoids, the characteristic colors of many birds, insects, and marine invertebrates are due to the presence of carotenoids that originate in the diet. Carotenoid pigments have also been found in various microorganisms, including bacteria, algae, yeasts, and fungi.

There is an increased interest in carotenoids as natural antioxidants and free-radical scavengers for their ability to reduce and alleviate chronic diseases, various pathological stages, and aging. However, the application of chemical synthetic methods to prepare carotenoid compounds as food additives has been strictly regulated in recent years. Therefore, attention is paid to finding suitable natural methods for their production. One possibility lies in biotechnological techniques potentially employing microorganisms that are able to convert various substrates into carotenoid pigments.

25.2 CAROTENOID-PRODUCING MICROORGANISMS

Despite the availability of a variety of natural and synthetic carotenoids, there is currently renewed interest in microbial sources. Microorganisms accumulate several types of carotenoids as a part of their response to various environmental stresses. Microbial production of carotenoids, when compared with extraction from vegetables or chemical synthesis, seems to be of paramount interest mainly because of the problems of seasonal and geographic variability in the production and marketing of several of the colorants of plant origin.

25.2.1 BACTERIA

Carotenoid pigments are synthesized in all of the photosynthetic and nonphotosynthetic bacteria (*Streptomyces, Flavobacterium, Mycobacterium, Brevibacterium, Rhodomicrobium, and Erwinia*) and cyanobacteria. Gene clusters responsible for the synthesis of carotenoids have been isolated from various carotenogenic bacteria, including the epiphytic bacteria *Erwinia* species and the marine bacterium *Agrobacterium aurantiacum.* Carotenoids in cyanobacteria have two main functions: They serve as light-harvesting pigments in photosynthesis, and they protect against photooxidative damage (Bryant 1996). In bacteria that cannot naturally synthesize carotenoids (e.g., *Escherichia coli*), their carotenoid biosynthesis has been achieved by the introduction of the carotenogenic genes from other sources (Sandmann 2003, Schmidt-Dannert 2000, Wang et al. 2000).

25.2.2 YEASTS

A number of red yeast species *(Rhodotorula, Rhodosporidium, Sporidiobolus, Sporobolomyces, Cystofilobasidium, Kockovaella,* and *Phaffia)* are known as producers of carotene pigments. Studies with yeast mutants or carotenoid biosynthesis inhibitors have shown that carotenoid-deficient yeast strains are sensitive to free oxygen radicals or oxidizing environment, and that this sensitivity can be relieved by the addition of exogenous carotenoids (Davoli et al. 2004). The major yeast pigments are β-carotene, γ-carotene, torulene, torularhodin, and astaxanthin (Dufosse 2006). Among yeasts, *Rhodotorula* species are one of main carotenoid-forming microorganisms with predominant synthesis of β-carotene, torulene, and torularhodin (Davoli et al. 2004, Libkind and van Broock 2006, Maldonade et al. 2008). *Cystofilobasidium* and *Dioszegia* were also found to synthesize these three pigments. Some yeast carotenoids are modified with oxygen-containing functional groups. For example, astaxanthin is almost exclusively formed by *Phaffia rhodozyma* (Čertík et al. 2005, Lukacs et al. 2006). Nevertheless, although there are many strategies for stimulation of carotene biosynthetic machinery in yeasts, attention is still focused on unexplored yeasts' habitats for selection of hyperproducing strains, which is the important step toward the design and optimization of a biotechnological process for pigment formation (Libkind and van Broock 2006, Maldonade et al. 2008).

25.2.3 FUNGI

Studies of a number of fungi, including *Neurospora crassa, Blakeslea trispora, Mucor hiemalis, M. circinelloides,* and *Phycomyces blakesleeanus* (oleaginous fungi with carotene-rich oil), have been published over the last 20 years (Feofilova et al. 1995, Bhosale 2004, Dufosse 2006). Fungal carotenoid content is relatively simple with dominant levels of β-carotene. Recent work with dimorphic fungal mutants *M. circinelloides* (Iturriaga et al. 2005) and *B. trispora* (Cerda-Olmedo 2001) showed that these strains could be useful in biotechnological production of carotenoids in the usual fermentors. On the other hand, *Phycomyces* bearing its full carotenogenic potential on both solid substrates and in liquid media is also considered for large-scale fermentations (Dufosse 2006). Apart from β-carotene, genetically modified fungus *Fusarium sporotrichioides* synthesized a sufficient amount of lycopene from cheap corn fiber material (Jones et al. 2004).

25.2.4 ALGAE

Astaxanthin is a high-value carotenoid (Higuera-Ciapara et al. 2006) and has been identified in several algae (*Haematococcus pluvialis* and *Chlorella zofingiensis*)

(Guerin et al. 2003, Johnson and Schroeder 1995). The algae *Chlorococcum* is another promising commercial source of ketocarotenoids due to its relatively fast growth rate, ease of cultivation in outdoor systems, and high tolerance to extreme pH and high temperature (Zhang and Lee 1999). Moreover, the esters of astaxanthin and adonixanthin and free cantaxanthin were also detected as the major carotenoids in the *Chlorococcum* cells. Interestingly, Yuan et al. (2002) described *Chlorococcum* algae as a good producer of lutein and β-carotene with only a small amount of ketocarotenoids.

25.3   BIOSYNTHESIS OF MICROBIAL CAROTENOID PIGMENTS

Thorough biochemical analysis of carotenoid biosynthesis, classical genetics, and more recently molecular genetics resulted in the elucidation of the main routes for the synthesis of acyclic and cyclic carotenoids at a molecular level (Sandmann 2001). Little is known, however, about the biosynthesis of carotenoids containing additional modifications of the end groups, the polyene chain, the methyl groups, or molecular rearrangements that contribute to the tremendous structural diversity of carotenoids. At present, hundreds of individual carotenoids have been characterized (Britton et al. 1998), and novel carotenoids continue to be isolated. All carotenoids are derived from the isoprenoid or terpenoid pathway.

25.3.1   ISOPRENOID BIOSYNTHESIS

Starting with the simple compounds acetyl-CoA (coenzyme A), glyceraldehyde-3-phosphate, and pyruvate, which arise via the central pathways of metabolism, the key intermediate isopentenyl diphosphate is formed by two independent routes. It is then converted by bacteria, yeasts, fungi, plants, and animals into thousands of different naturally occurring products.

In fungi, carotenoids are derived by sequence reactions via the mevalonate biosynthetic pathway. The main product 3-hydroxy-3-methylglutaryl-CoA (HMG-CoA) is finally reduced to the mevalonic acid. This two-step reduction of HMG-CoA to mevalonate is highly controlled and is also a major control factor of sterol synthesis (Metzler 2003).

In bacteria and plastids of plants, formation of prenyl pyrophosphates, which are the precursors of carotenoids, proceeds via an alternative—the glyceraldehyde-3-phosphate:pyruvate pathway. The pathway is apparently the sole source of isoprenoid compounds for the unicellular algae *Scenedesmus* (Metzler 2003). Initially, pyruvate is decarboxylated by a 1-deoxyxylulose-5-phosphate synthase and condensed with D-glyceraldehyde 3-phosphate to form 1-deoxyxylulose 5-phosphate. This product may be the branching point for independent routes to isopentyl pyrophosphate (IPP) and dimethylallyl pyrophosphate (DMAPP).

Before polyprenyl formation begins, one molecule of IPP must be isomerized to DMAPP. However, there is evidence that the pathways of prenyl diphosphate synthesis differ in prokaryotes and eukaryotes (Sandmann 2001). In *Synechocystis* and in *E. coli,* parallel synthesis of IPP and DMAPP has been found.

25.3.2 Carotene Biosynthesis

Carotenoids are synthesized in bacteria, yeasts, fungi, algae, and green plants. In addition to a very few bacterial carotenoids with 30, 45, or 50 carbon atoms, C40 carotenoids represent the majority of the more than 600 known structures. Especially, bacterial carotenoids are most diverse. Hydroxy groups at the ionone ring may be glycosylated or carry a glycoside fatty acid ester moiety. Furthermore, carotenoids with aromatic rings or acyclic structures with different polyene chains and typically 1-methoxy groups can also be found. Typical fungal carotenoids possess 4-keto groups, may be monocyclic, or possess 13 conjugated double bonds. 3-Hydroxy α- and β- as well as 5,6-epoxy β-carotene derivatives are abundant in the chloroplasts of some algal groups and green plants (Britton et al. 1998).

All carotenoids are derived from the isoprenoid or terpenoid pathway. From prenyl diphosphates of different chain lengths, specific routes branch off into various terpenoid end products. The prenyl diphosphates are formed by different prenyl transferases after isomerization of IPP to DMAPP by successive 1′-4 condensations with IPP molecules. Condensation of one molecule of dimethylallyl diphosphate (DMADP) and three molecules of isopentyl diphosphate (IDP) produces the diterpene geranylgeranyl diphosphate (GGDP) that forms one-half of all C40 carotenoids. The head-to-head condensation of two GGDP molecules results in the first colorless carotenoid, phytoene. Phytoene synthesis is the first committed step in C40 carotenoid biosynthesis (Britton et al. 1998, Sandmann 2001).

Subsequent desaturation reactions lengthen the conjugated double-bond system to produce neurosporene or lycopene. Two completely unrelated types of phytoene desaturases exist. The enzyme found in bacteria (except cyanobacteria) and in fungi catalyzes the entire four-step desaturation process of phytoene to lycopene (Schmidt-Dannert 2000). The plant-type phytoene desaturase from cyanobacteria, algae, and plants carries out a two-step desaturation reaction with different ζ-carotene stereoisomers as reaction products (Britton et al. 1998, Schmidt-Dannert 2000).

Following desaturation, carotenoid biosynthesis branches into routes for acyclic and cyclic carotenoids. In phototrophic bacteria, additional desaturation at C3,4, introduction of hydroxy and methoxy groups at C1(C1′) of neurosporene and lycopene and, under aerobic conditions, a keto-group at C2 of neurosporene lead to the synthesis of the acyclic xanthophylls spheroidene or spheroidenone and spirilloxanthin, respectively. Synthesis of cyclic carotenoids involves cyclization of one or both end groups of lycopene or neurosporene. Typically, β-rings are introduced, but formation of ϵ-rings is common in higher plants, and carotenoids with γ-rings are found, for example, in certain fungi. Most cyclic carotenoids contain at least one oxygen function at one of the ring carbon atoms. Cyclic carotenoids with keto groups at C4(C4′) or hydroxy groups at C3(C3′) (e.g., zeaxanthin, astaxanthin, echinenone, and lutein) are widespread in microorganisms and plants. The leaf xanthophylls violaxanthin and neoxanthin and algae carotenoids such as fucoxanthin contain β-rings with a C(5,6)-epoxy group. The C(5,6)-epoxy β-end group serves as a key intermediate in the formation of a variety of other end groups (Schmidt-Dannert 2000).

25.3.3 Genes Coding Biosynthesis of Carotenoids

Since 1988, the biochemistry of carotenogenesis, especially the cloning of carotenogenic genes, has made considerable progress. At present, more than 150 genes, encoding 24 different crt enzymes, have been isolated from bacteria, yeasts, fungi, algae, and plants. All enzymes involved in carotenoid biosynthesis are membrane associated or integrated into membranes. These genes can be used to engineer a variety of diverse carotenoids in recombinant microorganisms. Complete carotenoid biosynthesis pathways have been cloned from a number of bacteria where the *crt* genes are clustered. The availability of these genes may help in the characterization of further enzymes of carotenogenic pathways and their regulation. The following paragraphs summarize the main types of carotenogenic genes (Sandmann 2001, 2003, Lee and Schmidt-Dannert 2002, Fraser and Bramley 2004).

Phytoene synthase is encoded by the closely related bacterial *crtB* and eukaryotic *psy* genes. The phytoene formed in the enzymatic reaction was present in both a 15-*cis* and all-*trans* isomeric configuration. The essential cofactors required were adenosine triphosphate (ATP) in combinations with either Mn^{2+} or Mg^{2+}. Phytoene synthesis was inhibited by phosphate ions and squalestatin.

Several phytoene desaturases are known that differ in the number of desaturation steps and in their structures. Among them, the bacterial/fungal type is encoded by *crtI,* and the cyanobacterial/algal/plant type is encoded by *crtP* or *pds*. Phytoene desaturase converts phytoene to ζ-carotene with phytofluene as an intermediate. The reaction is stimulated by NAD, NADP, and oxygen.

ζ-Carotene desaturase catalyzes the final two desaturation steps to lycopene by introducing two double bonds at positions 7,8 of ζ-carotene and 7′,8′ of neurosporene. From cyanobacteria, two structurally unrelated genes for ζ-carotene desaturase have been cloned. One *crtQ* (formerly called *zds*) from *Anabaena* is related to the bacterial *crtI* gene, whereas the second *crtQb* (also called *crtQ-2*) from *Synechocystis* is quite similar to *crtP*.

Two completely unrelated types of lycopene cyclases have been identified in bacteria, fungi, and plants. One encodes the classical monomeric bacterial β-cyclase gene (*crtY*), which may be an ancestor of *crtL* (the gene for a β-cyclase in cyanobacteria). The plant β- and ε-cyclase genes were developed from the *crtL* gene. In gram-positive bacteria, two genes, *crtYc* and *crtYd,* are present. From this type of lycopene cyclase genes, the fungal lycopene cyclase/phytoene synthase fusion gene *crtYB* has evolved. In addition to its two-step reaction in which both sides of the lycopene molecule are cyclized to β-ionone rings with the monocyclic γ-carotene as an intermediate, neurosporene as well as 1-hydroxylycopene were cyclized to β-zeacarotene and hydroxy γ-carotene, respectively. The cofactors involved in the reaction are either NADH (nicotinamide adenine dinucleotide) or NADPH (nicotinamide adenine dinucleotide phosphate).

Bacterial β-carotene hydroxylases, *CrtZ*, were characterized with respect to substrate specificity and reaction mechanism after expression in *E. coli*. The reactions depend on oxygen and are stimulated by 2-oxoglutarate, ascorbate, and Fe^{2+}, which is typical for a dioxygenase reaction. Similar mechanistic properties were also found for bacterial ketolases encoded by *crtW*. These enzymes catalyze the introduction of a keto group at C(4) of a β-ring. The most commonly occurring ketocarotenoids are the keto derivatives of β,β-carotene, namely, echinenone, cantaxanthin, adonirubin, and astaxanthin. However, the accumulation of a range of products in microbial systems indicates that the order of introduction of the hydroxy and keto groups into one end group or both end groups in the molecule is not rigidly controlled. In addition, it is apparent that the β-hydroxylases and β-C(4)-oxygenases probably operate independently and are versatile in their substrate requirements.

Other types of genes involved in carotenoid biosynthesis in microorganisms have also been characterized. For example, crtA, B, C, D, E, F, and I genes in phototrophic bacteria *Rhodobacter capsulatus* and crtD, I, B, and C genes in *Rhodobacter sphaeroides* have been involved in conversion of neurosporene to various types of acyclic xanthophylls (Britton et al. 1998). The availability of a large number of carotenogenic genes opens new possibilities to modify and engineer the carotenoid biosynthetic pathways in microorganisms and plants for production of tailor-made carotenoid pigments.

25.4   REGULATION OF MICROBIAL CAROTENOID OVERPRODUCTION

To improve the yield of carotenoid pigments and subsequently decrease the cost of this biotechnological process, diverse studies have been performed by optimizing the culture conditions, including nutritional and physical factors. Factors such as nature and concentration of carbon and nitrogen sources, minerals, vitamins, pH, aeration, temperature, light, and stress have a major influence on cell growth and yield of carotenoids. Because carotenoid biosynthesis is governed by the levels and activities of enzymes employed for the total carbon flux through the carotenoid synthesizing system, the efficient formation of carotenoids can also be achieved by construction of hyperproducing strains with mutagenesis and genetic/metabolic engineering.

As mentioned in section 25.3.2, the flow of carbon source to carotenoids comprises several metabolic steps. Efficiency of that flow depends on the cooperation of individual pathways engaged in this process, and which pathway is suppressed or activated varies with the growth medium composition, cultivation conditions, and microbial species and their developmental stage. Because overall yield of carotenoids is directly related to the total biomass yield, keeping both high growth rates and high flow carbon efficiency to carotenoids by optimal cultivation conditions is essential to achieve the maximal pigment productivity.

25.4.1   NUTRITIONAL AND PHYSIOLOGICAL REGULATION

25.4.1.1   Carbon and Nitrogen Sources

The efficiency of the carbon source conversion into pigments and the optimization of the growth medium with respect to its availability and price have been the subject of intensive studies. Numerous sources, including pentoses and hexoses, various disaccharides, glycerol, ethanol, methanol, oils, *n*-alkanes, or a wide variety of wastes derived from agricultural production (e.g., molasses, bananas, starch wastes, grape must, sugar cane molasses, whey permeate, hydrolysates of wood and peat, raw coconut milk, vegetable oils), have been considered as potential carbon sources for biotechnological production of carotenoids (Bhosale and Gadre 2001, Buzzini 2000, Buzzini 2001, Buzzini and Martini 1999, Frengova et al. 1994, Squina et al. 2002, Dominguez-Bocanegra and Torres-Munoz 2004, Lukacs et al. 2005, 2006). In all experiments with *Rhodotorula* strains, the major carotenoids torularhodin, torulene, β-carotene, and γ-carotene were quantified in different proportions depending on the medium composition. On the other hand, astaxanthin was almost synthesized as the main pigment in *Phaffia* strains. The work of Tinoi at al. (2005) demonstrated the effectiveness of using a widely available agroindustrial waste product as substrate and the importance of the sequential simplex optimization method in obtaining high carotenoid yields.

The chemical composition and concentration of nitrogen source in the medium might also be a means of physiological control and regulation of pigment metabolism in microorganisms. Several inorganic and organic nitrogen sources as well as flour extracts and protein hydrolysates have been studied for improvement of carotenoid production (Buzzini and Martini 1999, Čertík et al. 2005, Libkind and van Broock 2006). For example, a combination of $(NH_4)_2SO_4$ with yeast extract or soy protein hydrolysates favored astaxanthin production in a *Phaffia* strain. However, it seems that variation in carotene content in yeasts with regard to nitrogen source used in a medium and the rate of pigment production are influenced by the products of catabolism nitrogen source rather than the result of direct stimulation by the nitrogen compound itself.

25.4.1.2 Environmental Stress

Physiological regulation of the fermentation process by environmental stress is another possibility for overproduction of carotenoids demanded by selected cells. The alternations in environment induce adaptation mechanisms in the microbial cell, where physicochemical properties of cells are changed and new metabolic routes are triggered. These mechanisms, including active remodeling of pigment composition, are an essential feature of the adaptation response of microbial cells grown under unfavorable conditions.

There have been several reports on the enhancement of volumetric production (mg/L) as well as cellular accumulation (mg/g) of microbial carotenoid on supplementation of metal ions (copper, zinc, iron, calcium, cobalt, aluminum) in yeasts and molds (Govind et al. 1982, Bhosale 2004, Buzzini et al. 2005). Trace elements have been shown to exert a selective influence on the carotenoid profile in red yeasts. It may be explained by hypothesizing a possible activation or inhibition mechanism by selected metal ions on specific carotenogenic enzymes, in particular on specific desaturases involved in carotenoid biosynthesis, in agreement with previous studies reporting activation or inhibition by metal ions in microbial desaturases (Buzzini et al. 2005). The other explanation is based on observations that the presence of heavy metals results in formation of various active oxygen radicals, which in turn induces generation of protective carotenoid metabolites that reduce negative behavior of free radicals. Such strategy has been applied in several pigment-forming microorganisms to increase the yield of microbial pigments (Čertík and Breierová 2002, Breierová et al. 2005, Čertík et al. 2005, Rapta et al. 2005b). Schroeder and Johnson (1995) also reported that astaxanthin biosynthesis was regulated by singlet oxygen and peroxyl radicals in the *P. rhodozyma* cells.

During environmental stress response, many red yeasts exhibit cross-protective mechanisms (Sigler at al. 1999). Preincubation of yeast culture with a low concentration of one stress factor in inoculum media (e.g., salt, hydrogen peroxide) induced adaptation pathways that resulted in enhanced carotenoid production (Márová et al. 2005, Kočí et al. 2005). Further addition of a higher concentration of either the same or another stress factor can lead to a significant (5–10 times) increase of β-carotene production in *Rhodotorula glutinis* and *Sporobolomyces salmonicolor* (Márová et al. 2004, 2005). Simple preincubation of *R. glutinis* in the presence of 2% NaCl in inoculation medium followed by fermentation in inorganic production medium led to formation of 36 g biomass/L that corresponds to 5 mg carotenoids/L and 23 mg ergosterol/L (Márová et al. 2005). This combined environmental stress using a mild stress effect of salt or hydrogen peroxide could be industrially used for production of both carotene- and ergosterol-enriched biomass.

The addition of solvents such as ethanol, methanol, isopropanol, and ethylene glycol to the culture medium also stimulates microbial carotenogenesis. It should be noted that while ethanol supplementation (2% v/v) stimulated β-carotene and torulene formation in *Rhodotorula glutinis*, torularhodin formation was suppressed (Bhosale 2004). It was proposed that ethanol-mediated inhibition of torulene oxidation must be accompanied by an increase in β-carotene content, suggesting a shift in the metabolic pathway to favor ring closure. Gu et al. (1997) reported increased astaxanthin production on slight addition of ethanol (0.2%, v/v) to cultures of *Phaffia rhodozyma*. Detailed studies revealed that ethanol activates oxidative metabolism with induction of HMG-CoA reductase, which in turn enhances carotenoid production. However, stimulation of carotenoid accumulation by ethanol or H_2O_2 was more effective if stress factors were employed to the medium in the exponential growth phase than from the beginning of cultivation (Čertík et al. 2005, Márová et al. 2004).

25.4.1.3   Chemical Enhancers

To achieve rapid carotenoid overproduction, various stimulants can be added to the culture broth. One group of such enhancers is based on intermediates of the tricarboxylic acid cycle, which play an important role in metabolic reactions under aerobic conditions, forming a carbon skeleton for carotenoid and lipid biosynthesis in microbes. Because pigment increase is paralleled by decreased protein synthesis, restriction of protein synthesis is an important way to shift carbon flow to carotenoid synthesis (Flores-Cotera et al. 2001). It was also proposed that high respiratory and tricarboxylic acid cycle activity is associated with production of large quantities of reactive species, and these are known to enhance carotenoid production (An 2001). It should be emphasized that the degree of stimulation was dependent on the time of addition of the citric acid cycle intermediate to the culture medium. Some fungi showed that addition of organic acids to media elevated β-carotene content and concomitantly decreased γ-carotene level, with complete disappearance of lycopene (Bhosale 2004).

Chemical substances capable of inhibiting biosynthetic pathways have been applied to characterize metabolic pathways and elucidate reaction mechanisms. In general, compounds that inhibit biosynthesis can act through various mechanisms, such as inhibiting the active site directly by an allosteric effect (reversible or otherwise), altering the regulation of gene expression, and blocking the essential biochemical pathways or the availability of cofactors, among other possibilities. From this view, a number of chemical compounds, including terpenes, ionones, amines, alkaloids, antibiotics, pyridine, imidazole, and methylheptenone, have been studied for their effect on carotene synthesis (Govind et al. 1982, Britton et al. 1977, Bhosale 2004). For example, appropriate addition of 2-(4-chlorophenylthio) triethylamine (CPTA) as the cyclase inhibitor or N,N-diethylalkylamines to the culture broth reduced lycopene and γ-carotene accumulation with a subsequent increase in β-carotene (Hsu et al. 1990). Oppositely, there are several inhibitors, such as pyridine and imidazole, that enhance lycopene formation by inhibiting the enzymes responsible for cyclization in fungi (Feofilova et al. 1995).

To obtain commercially interesting carotenoid profiles, the effect of supplementation with diphenylamine (DPA) and nicotine in the culture media of *Rhodotorula rubra* and *R. glutinis* was investigated. DPA blocks the sequence of desaturation reactions by inhibiting phytoene synthase, leading to an accumulation of phytoene together with other saturated carotenoids, and nicotine inhibits lycopene cyclase and consequently the cyclization reactions (Howes and Balra 1970, Squina and Mercadante 2005). Cultivation of *Xanthophyllomyces dendrorhous* in the presence of diphenylamine and nicotine at 4°C was reported to trigger interconversion of β-carotene to astaxanthin (Ducrey Sanpietro and Kula 1998).

25.4.1.4 Aeration

It is widely accepted that availability of oxygen determines the amount of pigments and their profile since molecular oxygen participates in the desaturation mechanisms for carotene biosynthesis. It must be emphasized that many interactions between dissolved oxygen and temperature, medium composition, and oxygen requirement of the organisms have been investigated in a very limited manner; thus, all findings can only serve as guidelines for establishing the exact role of oxygen in microbial pigment biosynthesis. Studies on the effect of aeration on *Rhodotorula* spp. have revealed minor effects on the composition of carotenoids, although the overall synthesis was often stimulated by aeration (Simova et al. 2004). Also, production of astaxanthin was significantly enhanced when *Phaffia* strains were cultivated in media with a sufficient amount of oxygen (Čertík et al. 2005). Interestingly, the morphology of *Blakeslea trispora* depended on the aeration, and the zygospores were the morphological form responsible for β-carotene production (Mantzouridou et al. 2002).

25.4.1.5 Light

Generally, formation of carotenoid pigments by bacteria, yeast, fungi, and algae requires white-light irradiation, although it may vary according to the microorganism. In addition, the profile of carotenoid pigments depends on exposure time during the microbial life cycle. Among yeasts, *R. glutinis* has been very well studied for growth and carotenoid production under the influence of white light. It was shown that torularhodin biosynthesis increased by exposure to weak white light (Sakaki et al. 2001), whereas β-carotene production was significantly higher when *R. glutinis* was subjected to white light in the late exponential growth phase (Bhosale and Gadre 2002). On the other hand, *Mucor rouxii* accumulated about 10 times more β-carotene when grown continuously in the presence of light than corresponding cultures grown in the dark (Mosqueda-Cano and Gutierrez-Corona 1995). Production of astaxanthin was also proposed to be photoinducible in *Phaffia* strains (Meyer and Du Preez 1994, Vazquez 2001). Interestingly, supplementation of ergosterol increased the effect of photostimulation of carotenogenesis in yeasts. It was proposed that ergosterol may induce changes in the cell membrane that potentiates the response to the light stimulus (Bhosale 2004). Navarro et al. (2001) reported that *M. circinelloides* responds to blue light by activating carotenoid biosynthesis due to a rapid increase in the level of transcription of structural genes for carotenogenesis.

25.4.1.6 Temperature

Temperature was reported to control changes in enzyme activities that regulate carotenoid production in microorganisms. For example, *R. glutinis* biosynthesized β-carotene more efficiently at lower temperature, whereas increased torulene formation was accompanied by higher temperature (Bhosale and Gadre 2002). The reason might be found in γ-carotene, which acts as the branch point of carotenoid synthesis. Subsequent dehydrogenation and decarboxylation leading to torulene synthesis is known to be temperature dependent since the respective enzymes are less active at lower temperature compared to the activity of β-carotene synthase. This is a probable reason for an increase in the proportion of β-carotene at lower temperature in *R. glutinis*. The moderately psychrophilic yeast *Xanthophyllomyces dendrorhous* also displayed a 50% increase in total carotenoids at low temperatures with elevated levels of astaxanthin (Ducrey Sanpietro and Kula 1998). Oppositely, *M. rouxii* grown at 40°C under aerobic conditions synthesized three times more carotenoids compare to the amount obtained at the optimum growth temperature of 28°C (Mosqueda-Cano and Gutierrez-Corona 1995).

25.4.2 Gene Engineering

Progress in the genetic engineering of microorganisms has led to speculation that microbial pigment components could become marketable commodities if the genomes of traditional pigment-forming microbes would be modified appropriately. The ideal microorganism for commercial-scale application of pigments would produce a particular type of carotenoid that could be supplied constantly at a competitively low price. However, carotenoid biosynthesis requires the interaction of multiple gene products. To what degree must the carotenoid synthetic apparatus be modified to yield industrially applicable changes remains to be determined.

A major challenge in modifying the carotenoid composition is to regulate the key *crt* genes and gene clusters involved the metabolic pathway of pigment production. Alternation in metabolic networks using recombinant DNA technology has allowed noncarotenogenic microbes to produce a desired carotenoid. The availability of a large number of carotenogenic genes makes it possible to modify and engineer the carotenoid biosynthetic pathways in microorganisms. A number of genetically modified microbes (e.g., *Candida utilis, Escherichia coli, Saccharomyces cerevisiae, Zymomonas mobilis*, etc.) have been studied for carotenoid production (Misawa and Shimada 1998, Miura et al. 1998, Wang et al. 1999, 2000, Shimada et al. 1998, Schmidt-Dannert 2000, Lee and Schmidt-Dannert 2002, Sandmann 2001, 2003, Yamano et al. 1996). However, lack of sufficient precursors (such as IDP, DMADP, and GGDP) and limited carotenoid storage capability are main tasks for research how to exploit these organisms as commercial carotenoid producers. Therefore, effort has been focused on increasing the isoprenoid central flux and levels of carotenoid precursors. For example, overexpression of the IDP isomerase (*idi* catalyzes the isomerization of IDP to DMAP) together with an archaebacterial multifunctional GGDP synthase (*gps* converts IDP and DMADP directly to GGDP) resulted in a 50-fold increase of astaxanthin production in *E. coli* (Wang et al. 1999). *Neurospora crassa* was also genetically modified to overexpress 3-hydroxy-3-methylglutaryl-CoA, which led to a several-fold improvement in lycopene and neurosporaxanthin production (Wang and Keasling 2002).

By combination of genes from different organisms with different carotenoid biosynthetic branches, novel carotenoids not found in any other pathway can be synthesized. Most *Mucor* species accumulate β-carotene as the main carotenoid. The *crtW* and *crtZ* astaxanthin biosynthesis genes from *Agrobacterium aurantiacum* were placed under the control of *M. circinelloides* expression signals. Transformants that exhibited altered carotene production were isolated and analyzed. Studies revealed the presence of new carotenoid compounds and intermediates among the transformants (Papp et al. 2006). *Fusarium sporotrichioides* was genetically modified for lycopene production by redirection of the isoprenoid pathway toward the synthesis of carotenoids and introducing genes from the bacterium *Erwinia uredovora* (Leathers et al. 2004). The carotenoid biosynthetic pathway of astaxanthin producers of *Phaffia/Xanthophyllomyces* strains has also been engineered, and several genes, such as phytoene desaturase, isopentenyl diphosphate isomerase, and epoxide hydrolase, were isolated and expressed in *E. coli* (Verdoes et al. 2003, Lukacs et al. 2006).

25.5 BIOTECHNOLOGICAL PROCESSES FOR CAROTENOID PIGMENT PRODUCTION

The growing scientific evidence that carotenoid pigments may have potential benefits in human and animal health has increased commercial attention on the search for alternative natural sources. Biological sources of carotenoids receive major attention now because of the stringent rules and regulations applied to chemically synthesized/purified pigments. Because the availability of a variety of natural carotenoids is limited, there is currently renewed interest in microbial sources. Comparative success in microbial pigment production has led to a flourishing interest in the development of fermentation processes and has enabled several processes to attain commercial production levels. However, two main problems still exist: economic and marketing difficulties. On the other hand, the cost of microbial pigments may be secondary compared to the health (functional, dietetic, and therapeutical) benefits derived. Nevertheless, microbial carotenoids must yield to the best commercial varieties and have a broad production base to be competitive with other commodities. Although the manipulation of microbial pigment composition is a rapidly growing field of biotechnology, the supply of microbial carotenoids is still insufficient to meet industrial demand. Therefore, alternative strategies, such as mutation methods, molecular engineering techniques, and the use of inhibitors or activators of carotene biosynthesis, should be involved in developing fermentation processes for pigment overproduction.

The production biotechnological process proceeds essentially in two stages: fermentation and pigment recovery (Dufosse 2006). An important aspect of the fermentation process is the development of a suitable culture medium to obtain the maximum amount of desired product. In recent years, cheap raw materials and by-products of agroindustrial origin have been proposed as low-cost alternative carbohydrate sources for microbial metabolite production, with the view also of minimizing environmental and energetic problems related to residues and effluent disposal. For fermentation, seed cultures are produced from the original strain cultures and subsequently used in an aerobic submerged batch fermentation to produce a biomass rich in carotene pigment. In the second stage, the recovery process, the biomass is isolated and transformed into a form suitable for isolating carotene, which is isolated from the biomass with appropriate solvent, suitably purified and concentrated, and crystallized from the mother liquor. Thus, large-scale production of the carotenoids using microorganisms might be highly efficient because they are easily manipulated in the processing schemes. Several types of microbes, such as bacteria, algae, yeast, and fungi, have been reported to produce carotenoids, but only a few of them have been exploited commercially (Bhosale 2004).

Biotechnological production of β-carotene by several strains of the yeast *Rhodotorula* spp. is currently used industrially. This yeast is convenient for large-scale fermentation because of its unicellular nature and high growth rate. Because *R. glutinis* synthesizes β-carotene, torulene, and torularhodin, the rate of production of the individual carotenoid depends on the incubation conditions. Specially prepared mutants of *Rhodotorula* not only rapidly increased formation of torulene or thorularhodin, but also the amount of β-carotene reached the level of 70 mg/L (Sakaki et al. 2000). Mutation techniques as well as sexual stimulation of carotene biosynthesis improved the biosynthesis of β-carotene by *Phycomyces blakesleeanus* up to 35 mg/g (Cerda-Olmedo 2001). *Blakeslea trispora* is another important producer of β-carotene. The fungus is employed for two industrial processes, the first in Russia and the Ukraine and the second in Spain (Dufosse 2006). The process yielded up to 30 mg of β-carotene/gram dry mass or about 3 g β-carotene/liter. In addition, a yeastlike mutant of *Mucor circinelloides* was mentioned as a useful fungus for biotechnological production of β-carotene (Iturriaga et al. 2005).

Among the few astaxanthin-producing microorganisms, *Phaffia rhodozyma* (*Xanthophyllomyces dendrorhous*) is one of the best candidates for commercial production. Therefore, many academic laboratories and several companies have developed processes that could reach an industrial level. The genera *Phaffia/Xanthophyllomyces* has some advantageous properties that make it attractive for commercial astaxanthin production: (1) It synthesizes the natural form of astaxanthin ($3S,3'S$ configuration) as a principal carotenoid, (2) it does not require light for its growth and pigmentation, and (3) it can utilize many types of carbon and nitrogen sources (Lukacs et al. 2006, Dufosse 2006). Studies of physiological regulation of astaxanthin in flask cultivations was verified in bioreactors, and the astaxanthin amount reached 8.1 mg/L (Dufosse 2006). Enhanced production of the pigment was achieved during fed-batch fermentation with regulated additions of glucose, and optimized fermentation conditions finally yielded up to 20 mg astaxanthin/L (660 µg astaxanthin/g cell) (Čertík et al. 2005). Yamane et al. (1997) reported that a high carbon/nitrogen ratio induced an amount of astaxanthin, and C/N-regulated fed-batch fermentation of *P. rhodozyma* led to 16 mg astaxanthin/L. Thus, this strain can be considered a potential producer of astaxanthin. In addition, to avoid isolation of astaxanthin from cells, a two-stage batch fermentation technique was used by Fang and Wang (2002); *Bacillus circulans* with high cell wall lytic activity was added to the fermentation tank after the accumulation of astaxanthin in *P. rhodozyma* was completed.

Biotechnological overproduction of other carotenoid pigments has also been studied. Jones et al. (2004) reported elevated formation of lycopene by genetically modified fungus *Fusarium sporotrichioides* utilizing corn fiber as leftovers of the ethanol-making process. When carotenoid biosynthetic genes from the bacterium *Erwinia uredovora* was introduced to the fungus, the level of lycopene reached a maximum of 0.5 mg/g of dry mass. Dufosse (2006) reviewed that *Flavobacterium* sp. was able to produce up to 190 mg zeaxanthin/L, with a specific cell concentration of 16 mg/g dried cellular mass. Similarly, the photosynthetic bacterium *Bradyrhizobium* sp. and extremely halophilic bacterium *Halobacterium* sp. were described as potential producers of canthaxanthin. However, interest in canthaxanthin is still under debate due to its possible overdosage, leading to some health problems (Baker 2002).

25.6 PROPERTIES OF CAROTENOID PIGMENTS

Carotenoids are colored, fat-soluble pigments with functions ranging from light harvesting and photoprotective roles to protection of the eye and possible sexual attraction associated with the colors (Krinsky 1994, Stahl et al. 2002, Kiokias and Gordon 2004, El-Agamey et al. 2004). The important role of carotenoids in regulation of physical properties of biomembranes was reviewed (Gruszecki and Strzayka 2005). The properties of carotenoids are primarily dependent on their structure (Britton 1995). Most naturally occurring carotenoid pigments are C_{40} tetraterpenoids built from eight C_5 isoprenoid units, joined so that the sequence is reversed at the center (Britton et al. 2004). In several cases (xanthophylls), the hydrocarbon skeletons of carotenes contain oxygen functional groups such as hydroxy, keto, or epoxy groups. An extensive conjugated double-bond system serves as the light-absorbing chromophore responsible for the yellow, orange, or red color and the visible absorption spectrum. The conjugated C=C double-bond system is considered to be the most important factor in energy transfer reactions, such as those found in photosynthesis. The antioxidant activity of carotenoids arises primarily as a consequence of the ability of the conjugated double-bonded structure to delocalize unpaired electrons (Mortensen et al. 2001). The conjugated double-bond system constitutes a rigid, rod-like skeleton of carotenoid molecules. This feature seems to play a key role in the stabilization function of carotenoids, with respect to both lipid membranes and proteins (Gruszecki and Strzayka 2005).

Carotenoids are recognized as playing an important role in the prevention or inhibition of human diseases (e.g., certain types of cancer, artherosclerosis, age-related muscular degeneration) and maintaining good health (Rao and Rao 2007, Valko et al. 2006). They may influence the process of apoptosis in healthy cells, and β-carotene also exhibits a proapoptotic effect in colon and leukemic cancer cells (Sharoni et al. 2004). The functions of lycopene and astaxanthin include strong quenching of singlet oxygen, involvement in cancer prevention, and enhancement of immune responses (Heber and Lu 2002, Miki 1991). As isoprenoid membrane compounds, they serve as protection against photooxidative action of singlet oxygen and radicals that are formed during light (Sigler et al. 1999). Nonpolar pigments (β-carotene, lycopene, etc.) are localized at the inner part of hydrophobic membranes, where they show some mobility. On the other hand, xanthophylls are firmly anchored at polar lipids by their two polar groups (Gruszecki and Strzayka 2005). Therefore, they can more effectively quench reactive oxygen species (ROS) by either plunging into membranes or rising up from membranes (Rice-Evans et al. 1997, Liebler et al. 1997, Schroeder and Johnson 1995).

With very few exceptions, carotenoids are lipophilic. They are insoluble in water and soluble in organic solvents, such as acetone, alcohol, ethyl ether, chloroform, and ethyl acetate. The highly unsaturated carotenoid is prone to isomerization and oxidation. Isomerization reactions of metabolites of carotenoids are extremely important in biological systems such as the visual system. Carotenoids can undergo many reactions with a wide variety of chemical reagents, some of which might be similar to chemicals found in biological systems (Krinsky 1994, Rodriguez-Amaya 2001).

One of the most widely discussed roles of the carotenoids is the interaction with free radicals that initiate harmful reactions, such as lipid peroxidation (Edge et al. 1997, Bast et al. 1998). The hydrophobic core of biomembranes composed of polyunsaturated fatty acids is a potential target of attack of active oxygen species. For lipid peroxidation, carotenoids exhibit an antioxidant character, as evidenced by the reaction of carotenoids with free peroxyl radicals as an additional chain-breaking process. Generally, three mechanisms are proposed for the reaction of free radicals (ROO•, R•) with carotenoids: (1) radical addition, (2) hydrogen abstraction from the carotenoid, and (3) electron transfer reaction (El-Agamey et al. 2004). The key factors that determine the switch of carotenoids from antioxidants to prooxidants are the partial pressure of dioxygen (pO_2) and the carotenoid concentration (Rice-Evans et al. 1997). However, there are also some studies devoted to the appearance of prooxidant properties of carotenoids (Polyakov et al. 2001, Young and Lowe 2001). It was shown that β-carotene might act in the process of lipid peroxidation as a prooxidant at high oxygen pressure and high carotenoid concentration. It was suggested that β-carotene reacts with peroxyl radicals to give a carbon-centered carotenyl radical that, similar to lipid free radical, in the presence of oxygen produces β-carotene peroxyl radical, so that chain propagation may occur.

It is unlikely that carotenoids actually act as prooxidants in biological systems; rather, they exhibit a tendency to lose their effectiveness as antioxidants (Young and Lowe 2001).

A special case of antioxidant/prooxidant behavior of carotenoids emerges in the presence of metals (e.g., metal-induced lipid peroxidation). In this case, metal ions (Fe^{2+} or Cu^{2+}) react with hydroperoxides, via a Fenton-type reaction, to initiate free-radical chain processes. There are several studies that indicated that β-carotene offers protection against metal-induced lipid oxidation. However, Polyakov et al. (2001) have shown that the presence of carotenoid in the reaction system not only decreases the free-radical concentration but also the reduction of Fe^{3+} to Fe^{2+} by carotenoids may occur. Free-radical scavenging and antioxidant activities of metabolites produced by carotenogenic yeasts of *Rhodotorula* sp. and *Sporobolomyces* sp. grown under heavy metal presence were studied using various EPR (electron paramagnetic resonance) experiments (Rapta et al. 2005b). Since carotenogenic yeast differ from each other in resistance against the heavy metals due to their individual protective system, quenching properties and antioxidant activities of carotenogenic yeasts were modulated by metal ions variously. Thus, activated biosynthesis of carotenoids by yeasts exposed to heavy metals could be in part explained by their scavenger characters (Rapta et al. 2005a) as a protection against the harmful effect of the environment.

However, the function of carotenoids and their metabolites in the living organism is dependent on a wide range of factors. While *in vitro* studies provide an insight into these properties and into the interactions of carotenoids with ROS, coantioxidants, and other molecules, extrapolation of the results from such studies must be carried out carefully (Young and Lowe 2001).

25.7 FUTURE PERSPECTIVES FOR MICROBIAL PIGMENTS

Growing interest in pigment applications in various fields coupled with their significance in health and dietary requirements has encouraged "hunting" for more suitable sources of these compounds. Due to restrictions, there is no possibility to apply carotenoids prepared by chemical synthesis for food, pharmaceutical, and medical purposes. In addition, inadequacy of conventional agricultural pigments has put attention on developing new microbial technologies employing certain fungi, yeasts, microalgae, and bacteria. However, the success of microbial pigments depends on their acceptability in the market, regulatory approval, and the size of the capital investment required to bring the product to market. Therefore, the focus of biotechnology on highly valuable carotenoids requires knowledge of how microorganisms control and regulate the carotene biosynthetic machinery to obtain specific pigments in high yield and at low price.

From this view, attempts have been directed at the development and improvement of biotechnological processes for the production of carotenoids on an industrial scale. Current successes using mutation methods and molecular engineering techniques carried out over recent years have not only answered some fundamental questions related to pigment formation but also enabled the construction of new microbial varieties that can synthesize unusual carotene metabolites. Elucidation of these mechanisms represents a challenging and potentially rewarding subject for further research and may finally allow us to move from empirical technology to predictable carotenoid design. Thus, the manipulation and regulation of microbial carotenoid biosynthesis open a large number of possibilities for academic research, demonstrate the enormous potential in its application, and create new economic competitiveness and markets for microbial pigment compounds.

REFERENCES

An, G.H. 2001. Improved growth of the red yeast, *Phaffia rhodozyma* (*Xanthophyllomyces dendrorhous*), in the presence of tricarboxylic acid cycle intermediates. *Biotechnol Lett* 23:1005–1009.

Baker, R.T.M. 2002. Canthaxanthin in aquafeed applications: is there any risk? *Trends Food Sci Technol* 12:240–243.

Bast, A., Haenen, G.R., van den Berg, H. 1998. Antioxidant effect of carotenoids, *Int J Vitam Nutr Res* 68:399–403.

Bhosale, P. 2004. Environmental and cultural stimulants in the production of carotenoids from microorganisms. *Appl Microbiol Biotechnol* 63:351–361.

Bhosale, P., and Gadre, R.V. 2001. β-Carotene production in sugar cane molasses by a *Rhodotorula glutinis* mutant. *J Ind Microbiol Biotechnol* 26:327–332.

Bhosale, P., and Gadre, RV. 2002. Manipulation of temperature and illumination conditions for enhanced β-carotene production by mutant 32 of *Rhodotorula glutinis*. *Lett Appl Microbiol* 34:349–353.

Breierová, E., Márová, I., and Čertík, M. 2005. The role of the carotenoid pigments in yeast cells under stress conditions. *Chem Listy* 99:109–111.

Britton, G. 1995. Structure and properties of carotenoids in relation to function. *FASEB J* 9:1551–1558.

Britton, G., Liaaen-Jensen, S., and Pfander, H. 1998. *Carotenoids. Volume 3: Biosynthesis and Metabolism.* Birkhäuser, Basel.

Britton, G., Liaaen-Jensen, S., and Pfander, H. 2004. *Carotenoids Handbook.* Birkhauser, Basel.

Britton, G., Singh, R.K., Malhotra, H.C., Goodwin, T.W., and Ben-Aziz, A. 1977. Biosynthesis of 1,2-dihydrocarotenoids in *Rhodopseudomonas viridis*: experiments with inhibitors. *Phytochemistry* 16:1561–1566.

Bryant, D.A. 1996. *The Molecular Biology of Cyanobacteria.* Dordecht: Kluwer Academic.

Buzzini, P. 2000. An optimization study of carotenoid production by *Rhodotorula glutinis* DBVPGG 3853 from substrates containing concentrated rectified grape must as the sole carbohydrate source. *J Indust Microbiol Biotechnol* 4: 41–45.

Buzzini, P. 2001. Batch and fed-batch carotenoid production by *Rhodotorula glutinis–Debaryomyces castellii* co-cultures in corn syrup. *J Appl Microbiol* 90:843–847.

Buzzini, P., and Martini, A. 1999. Production of carotenoids by *Rhodotorula glutinis* cultured in raw materials of agro-industrial origin. *Bioresource Technol* 71:41–44.

Buzzini, P., Martini, A., Gaetani, M., Turchetti, B., Pagnoni, U.M., and Davoli, P. 2005. Optimization of carotenoid production by *Rhodotorula graminis* DBVPG 7021 as a function of trace element concentration by means of response surface analysis. *Enzyme Microb Technol* 36:687–692.

Cerda-Olmedo, E. 2001. *Phycomyces* and the biology of light and color. *FEMS Microbiol Rev* 25:503–512.

Čertík, M., and Breierová, E. 2002. Adaptation responses of yeasts to environmental stress. *Chem Listy* 96:147.

Čertík, M., Masrnová, S., Sitkey, V., Minárik, M., and Breierová, E. 2005. Biotechnological production of astaxanthin. *Chem Listy* 99:237–240.

Davoli, P., Mierau, V., and Weber, R.W.S. 2004. Carotenoids and fatty acids in red yeasts *Sporobolomyces roseus* and *Rhodotorula glutinis*. *Appl Biochem Microbiol* 40:392–397.

Dominguez-Bocanegra, A.R., and Torres-Munoz, J.A. 2004. Astaxanthin hyperproduction by *Phaffia rhodozyma* (now *Xanthophyllomyces dendrorhous*) with raw coconut milk as sole source of energy. *Appl Microbiol Biotechnol* 66:249–252.

Ducrey Sanpietro, L.M., and Kula, M.R. 1998. Studies of astaxanthin biosynthesis in *Xanthophyllomyces dendrorhous* (Phaffia rhodozyma). Effect of inhibitors and low temperature. *Yeast* 14:1007–1016.

Dufosse, L. 2006. Microbial production of food grade pigments. *Food Technol Biotechnol* 44:313–321.

Edge, R., McGarvey, D.J., and Truscott, T.G. 1997. The carotenoids as antioxidants. *J Photochem Photobiol B* 41:189–200.

El-Agamey, A., Lowe, G.M., McGarvey, D.J., Mortensen, A., Phillip, D.M., Truscott, T.G., and Young, A.J. 2004. Carotenoid radical chemistry and antioxidant/pro-oxidant properties. *Arch Biochem Biophys* 430:37–48.

Fang, T.J., and Wang, J.M. 2002. Extractibility of astaxanthin in a mixed culture of a carotenoid over-producing mutant of *Xanthophyllomyces dendrorhous* and *Bacillus circulans* in two-stage batch fermentation. *Process Biochem* 37:1235–1245.

Feofilova, E.P., Tereshina, V.M., and Memorskaya, A.S. 1995. Regulation of lycopene biosynthesis in mucorous fungus *Blakeslea trispora* by pyridine derivatives. *Mikrobiologiya* 64:734–740.

Flores-Cotera, L.B., Martin, R., and Sanchez, S. 2001. Citrate, a possible precursor of astaxanthin in *Phaffia rhodozyma*: influence of varying levels of ammonium, phosphate and citrate in a chemically defined medium. *Appl Microbiol Biotechnol* 55:341–347.

Fraser, P.D., and Bramley, P.M. 2004. The biosynthesis and nutritional uses of carotenoids. *Prog Lipid Res* 43:228–265.

Frengova, G., Simova, E., Parlova, K., Beshkova, D., and Grigorova, D. 1994. Formation of carotenoids by *Rhodotorula glutinis* in whey ultrafiltrate. *Biotechnol Bioeng* 44:888–894.

Govind, N.S., Amin, A.R., and Modi, V.V. 1982. Stimulation of carotenogenesis in *Blakeslea trispora* by cupric ions. *Phytochemistry* 21:1043–1044.

Gruszecki, W.I., and Strzayka, K. 2005. Carotenoids as modulators of lipid membrane physical properties. *Biochim Biophys Acta* 1740:108–115.

Gu, W.L., An, G.H., and Johnson, E.A. 1997. Ethanol increases carotenoid production in *Phaffia rhodozyma*. *J Ind Microbiol Biotechnol* 19:114–117.

Guerin, M., Huntley, M.E., and Olaizola, M.2003. *Haematococcus* astaxanthin: applications for human health and nutrition. *Trends Biotechnol* 21:210–216.

Heber, D., and Lu, Q.Y. 2002. Overview of mechanisms of action of lycopene, *Exp Biol Med* 227:920–923.

Higuera-Ciapara, I., Felix-Valenzuela, L., and Goycoolea, F.M. 2006. Astaxanthin. A review of its chemistry and applications. *Crit Rev Food Sci Nutr* 46:185–196.

Howes, C.D., and Batra, P.P. 1970. Accumulation of lycopene and inhibition of cyclic carotenoids in *Mycobacterium* in the presence of nicotine. *Biochim Biophys Acta* 222:174–179.

Hsu, W.J., Yokoyama, H., and DeBenedict, C. 1990. Chemical bioregulation of carotenogenesis in *Phycomyces blakesleeanus*. *Phytochemistry* 29:2447–2451.

Iturriaga, E.A., Papp, T., Breum, J., Arnau, J., and Eslava, A.P. 2005. Strain and culture conditions improvement for b-carotene production with *Mucor*. In: *Methods in Biotechnology: Microbial Processes and Products, Vol. 18*, J.L. Barredo (Ed.), Humana, Totowa, NJ, pp. 239–256.

Johnson, E.A., and Schroeder, W.A. 1995. Microbial carotenoids. *Adv Biochem Eng Biotechnol* 53:119–178.

Jones, J.D., Hohn, T.H., and Leathers, T.D. 2004. Genetically modified strains of *Fusarium sporotrichioides* for production of lycopene and β-carotene. Proceedings of Annual Meeting of Society of Industrial Microbiology, San Diego, CA, p. 91.

Kiokias, S., and Gordon, M.H. 2004. Antioxidant properties of carotenoids in vitro and in vivo. *Food Rev Int* 20:99–121.

Kočí, R., Drábková, M., and Márová, I. 2005: Production of industrial metabolites by red yeasts in stress conditions. *Chem Listy* 99:297–298.

Krinsky, N.I. 1994. The biological properties of carotenoids. *Pure Appl Chem* 66:1003–1010.

Leathers, T.D., Jones, J.D., and Hohn, T.M. 2004. System for the sequential, directional cloning of multiple DNA sequences. *U.S. patent* 6,696,282.

Lee, P.C., and Schmidt-Dannert, C. 2002. Metabolic engineering towards biotechnological production of carotenoids in microorganisms. *Appl Microbiol Biotechnol* 60:1–11.

Libkind, D., and van Broock, M. 2006. Biomass and carotenoid pigment production by Patagonian native yeasts. *World J Microbiol Biotechnol* 22:687–692.

Liebler, D.C., Stratton, S.P., and Kysen, K.L. 1997. Antioxidant actions of beta-carotene in liposomal and microsomal membranes: role of carotenoid-membrane incorporation and α-tocopherol. *Arch Biochem Biophys* 338:244–250.

Lukacs, G., Kovacs, N., Papp, T., and Vagvolgyi, C. 2005. The effect of vegetable oils on carotenoid production of *Phaffia rhodozyma*. *Acta Microbiol Immunol Hung* 52:267.

Lukacs, G., Linka, B., and Nyilasi, I. 2006. *Phaffia rhodozyma and Xanthophyllomuces dendrorhous*: astaxanthin-producing yeasts of biotechnological importance. *Acta Alimentaria* 5:99–107.

Maldonade, I.R., Rodriguez-Amaya, D.B., and Scamparini, A.R.P. 2008. Carotenoids of yeasts isolated from the Brazilian ecosystem. *Food Chem* 107:45–150.

Mantzouridou, F., Roukas, T., and Kotzekidou, P. 2002. Effect of the aeration rate and agitation speed on b-carotene production and morphology of *Blakeslea trispora* in a stirred tank reactor: mathematical modeling. *Biochem Eng J* 10:123–135.

Márová, I., Breierová, E., Kočí, R., Friedl, Z., Slovák, B., and Pokorná, J. 2004. Influence of exogenous stress factors on production of carotenoids by some strains of carotenogenic yeasts. *Ann Microbiol* 54:73–85.

Márová, I., Hrdličková, J., Kočí, R., Drábková, M., Kubešová, J., Vidláková, T., and Babák, L. 2005. Biotechnological production of carotenoids by transgenic bacteria and red yeasts. *Chem Listy* 99:322–323.

Metzler, D.E. 2003. *Biochemistry. The Chemical Reactions of Living Cells*. Vol. 2, Academic Press, Elsevier Science, New York.

Meyer, P.S., and Du Preez, J.C. 1994. Photo-regulated astaxanthin production by *Phaffia rhodozyma* mutants. *Syst Appl Microbiol* 17:24–31.

Miki, W. 1991. Biological functions and activities of animal carotenoids. *Pure Appl Chem* 63:141–146.

Misawa, N., and Shimada, H. 1998. Metabolic engineering for the production of carotenoids in non-carotenogenic bacteria and yeasts. *J Biotechnol* 59:169–181.

Miura, Y., Konda, K., Saito, T., Shimada, H., Fraser, P.D., and Misawa, N. 1998. Production of the carotenoids lycopene, β-carotene, and astaxanthin in the food yeast *Candida utilis*. *Appl Environ Microbiol* 64:1226–1229.

Mortensen, A., Skibsted, L.H., and Truscott, T.G. 2001. The interaction of dietary carotenoids with radical species. *Arch Biochem Biophys* 385:13–19.

Mosqueda-Cano, G., and Gutierrez-Corona, J.F. 1995. Environmental and developmental regulation of carotenogenesis in the dimorphic fungus *Mucor rouxii*. *Curr Microbiol* 31:141–145.

Navarro, E., Lorca-Pascual, J.M., Quiles-Rosillo, M.D., Nicolas, F.E., Garre, V., Torres-Martinez, S., and Ruiz-Vazquez, R.M. 2001. A negative regulator of light-inducible carotenogenesis in *Mucor circinelloides*. *Mol Genet Genom* 266:463–470.

Papp, T., Velayos, A., Bartók, T., Eslava, A.P., Vágvölgyi, C., and Iturriaga, E.A. 2006. Heterologous expression of astaxanthin biosynthesis genes in *Mucor circinelloides*. *Appl Microbiol Biotechnol* 69:526–531.

Polyakov, N.E., Leshina, T.V., Konovalova, T.A., and Kispert, L.D. 2001. Carotenoids as scavengers of free radicals in Fenton reaction: antioxidants or pro-oxidants? *Free Radic Biol Med* 31:398–404.

Rao, A.V., and Rao, L.G. 2007. Carotenoids and human health. *Pharmacol Res* 55:207–216.

Rapta, P., Polovka, M., Zalibera, M., Breierová, E., Žitňanová, I., Márová, I., and Čertík, M. 2005a. Scavenging and antioxidant properties of compounds synthesized by carotenogenic yeasts stressed by heavy metals—EPR spin trapping study. *Biophys Chem* 116:1–9.

Rapta, P., Zalibera, M., Čertík, M., and Breierová, E. 2005b. The influence of exogenous stress on pigment composition and antioxidant and radical scavenging activity of yeasts. *Chem Listy* 99:219–220.

Rice-Evans, C.A., Sampson, J., Bramley, P.M., and Holloway, D.E. 1997. Why do we expect carotenoids to be antioxidants in vivo? *Free Rad Res* 26:381–398.

Rodriguez-Amaya, D.B. 2001. *A Guide to Caroteniod Analysis in Foods*. International Life Sciences Institute, ILSI Press, Washington, DC.

Sakaki, H., Nakanishi, T., Satonaka, K.Y., Miki, W., Fujita, T., and Komemushi, S. 2000. Properties of a high-torularhodin mutant of *Rhodotorula glutinis* cultivated under oxidative stress. *J Biosci Bioeng* 89:203–205.

Sakaki, H., Nakanishi, T., Tada, A., Miki, W., and Komemushi, S. 2001. Activation of torularhodin production by *Rhodotorula glutinis* using weak white light irradiation. *J Biosci Bioeng* 92:294–297.

Sandmann, G. 2001. Carotenoid biosynthesis and biotechnological application. *Arch Biochem Biophys* 385:4–12.

Sandmann, G. 2003. Novel carotenoids genetically engineered in a heterologous host. *Chem Biol* 10:478–479.

Schmidt-Dannert, C. 2000. Engineering novel carotenoids in microorganisms. *Curr Opin Biotechnol* 11:255–261.

Schroeder, W.A., and Johnson, E.A. 1995. Singlet oxygen and peroxyl radicals regulate carotenoid biosynthesis in *Phaffia rhodozyma*. *J Biol Chem* 270:18374–18379.

Sharoni Y., Danilenko M., Dubi N., Ben-Dor A., and Levy J. 2004. Carotenoids and transcription. *Arch Biochem Biophys* 430:89–96.

Shimada, H., Kondo, K., Fraser, P., Miura, Y., Saito, T., and Misawa, N. 1998. Increased carotenoid production by the food yeast *Candida utilis* through metabolic engineering of the isoprenoid pathway. *Appl Environ Microbiol* 64:2676–2680.

Sigler, K., Chaloupka, J., Brozmanová, J., Stadler, N., and Höfer, M. 1999. Oxidative stress in microorganisms. I. Microbial versus higher cells—damage and defenses in relation to cell aging and death. *Folia Microbiologica* 44:587–624.

Simova, E.D., Frengova, G.I., and Beshkova, D.M. 2004. Synthesis of carotenoids by *Rhodotorula rubra* GED8 co-cultured with yogurt starter cultures in whey ultrafiltrate. *J Ind Microbiol Biotechnol* 31:115–121.

Squina, F.M., and Mercadante, A.Z. 2005. Influence of nicotine and diphenylamine on the carotenoid composition of *Rhodotorula* strains. *J Food Biochem* 29:638–652.

Squina, F.M., Yamashita, F., Pereira, J.L., and Mercadante, A.Z. 2002. Production of carotenoids by *Rhodotorula rubra* and *R. glutinis* in culture medium supplemented with sugar cane juice. *Food Biotechnol* 16:227–235.

Stahl, W., Ale-Agha, N., and Polidori, M.C. 2002. Non-antioxidant properties of carotenoids. *Biol Chem* 383:553–558.

Tinoi, J., Rakariyatham, N., and Deming, R.L. 2005. Simplex optimization of carotenoid production by *Rhodotorula glutinis* using hydrolyzed mung bean waste flour as substrate. *Process Biochem* 40:2551–2557.

Valko, M., Rhodes, C.J., Moncol, J., Izakovic, M., and Mazur, M. 2006. Free radicals, metals and antioxidants in oxidative stress-induced cancer. *Chem-Biol Interact* 160:1–40.

Vazquez, M. 2001. Effect of the light on carotenoid profiles of *Xanthophyllomyces dendrorhous* strains (formerly *Phaffia rhodozyma*). *Food Technol Biotechnol* 39:123–128.

Verdoes, J.C., Sandman, G., Visser, H., Diaz, M., Van Mossel, M., and Van Ooyen, A.J.J. 2003. Metabolic engineering of the carotenoid biosynthetic pathway in the yeast *Xanthophyllomyces dendrorhous* (*Phaffia rhodozyma*). *Appl Environ Microbiol* 69:3728–3738.

Wang, C., Oh, M.K., and Liao, J.C. 2000. Directed evolution of metabolically engineered *Escherichia coli* for carotenoid production. *Biotechnol Prog* 16:922–926.

Wang, C.W., Oh, M.K., and Liao, J.C. 1999. Engineered isoprenoid pathway enhances astaxanthin production in *Escherichia coli*. *Biotechnol Bioeng* 62:235–241.

Wang, G.Y., and Keasling, J.D. 2002. Amplification of HMG-CoA reductase production enhances carotenoid accumulation in *Neurospora crassa*. *Metabol Eng* 4:193–201.

Yamane, Y.-I., Higashida, K., Nakashimada, Y., Kakizono, T., and Nishio, N. 1997. Influence of oxygen and glucose on primary metabolism and astaxanthin production by *Phaffia rhodozyma* in batch and fed-batch cultures: kinetic and stoichiometric analysis. *Appl Environ Microbiol* 63:4471–4478.

Yamano, S., Ishii, T., Nakagawa, M., Ikenaga, H., and Misawa, N. 1994. Metabolic engineering for production of beta-carotene and lycopene in *Saccharomyces cerevisiae*. *Biosci Biotechnol Biochem* 58:1112–1114.

Young, A.J., and Lowe, G.M. 2001. Antioxidant and prooxidant properties of carotenoids. *Arch Biochem Biophys* 385:20–27.

Yuan, J.-P., Chen, F., Liu, X., and Li, X.-Z. 2002. Carotenoid composition in the green microalga *Chlorococcum*. *Food Chem* 76:319–325.

Zhang, D.H., and Lee, Y.K. 1999. Ketocarotenoid production by a mutant of *Chlorococcum* sp. in an outdoor tubular photobioreactor. *Biotechnol Lett* 21:7–10.

26 Marine Functional Ingredients and Advanced Technology for Health Food Development

Li-Cyuan Wu, Min-Hsiung Pan,
Jeng-Sheng Chang, Ke Liang B. Chang,
Bonnie Sun Pan, and Zwe-ling Kong

CONTENTS

Key Words: *Antrodia cinnamomea;* apoptosis; epidioxysterols; functional food; marine; *Meretrix lusoria*; nanoencapsulation.

26.1 INTRODUCTION

The human diet contains a great variety of natural carcinogens as well as many natural anticarcinogens. The physiological effects of food factors are of particular interest because they may be useful for human cancer prevention [1,2]. Characterizing and optimizing such defense systems may be an important part of the strategy for minimizing cancer and other age-related diseases. Marine resources are an important and potentially rewarding target for pharmacological research into antitumor activity and against the age-associated rise in malignancy, metabolic syndrome, and autoimmune disorders [3–5]. Bioactives from marine resources will be good candidates for both pharmacology and functional food development.

To clarify performance at the cellular level, human-derived tumor cell lines were cultured and treated in a serum-free system. Previous studies had shown that the precipitate of ammonium sulfate from hot water extract of fresh hard clam [6,7] not only promoted cell proliferation of hybridoma cell line (HB4C5) but also inhibited the growth of hepatoma cells (HuH-6KK and Hep G2). In this report, we clarify the partially purified fraction, including the structure and function for growth-promoting/-inhibiting function toward different cell lines. Through treatment with several kinds of general inhibitor, we elucidated its possible mechanism of apoptosis-inducing ability.

The rapid development of nanotechnology in recent years has created myriad engineered nanomaterials. Nanoscaled particulates [8–12] of metal (gold and silver nanoparticles), semiconductor (quantum dot), carbon (nanotube and buckyball), and oxide (iron oxide, titanium dioxide, and silica) are increasingly used in industrial production as well as scientific, biological, and medical research. Human exposure to crystalline silica dust occurs in the course of mining operation, foundry work, mineral processing, and construction. Precipitated silica is generally considered to be safe and approved for use as a food or animal feed ingredient by the U.S. Food and Drug Administration (FDA). The potential applications of amorphous silica in nanobiotechnology have encompassed areas such as diagnostics, bioanalysis and imaging, drug delivery, and gene transfer.

We demonstrated that a naturally derived polysaccharide, chitosan, is capable of forming composite nanoparticles with silica. For encapsulated particles, we used silicification and biosilicification to encapsulate curcumin and analyzed the physicochemical properties of curcumin nanoparticles. It proved that encapsulated curcumin nanoparticles enhanced stability toward ultraviolet (UV) irradiation, antioxidation and antitumor activity, enhanced/added function, solubility, bioactivities/bioavailability, and control release and overcame the immunobarrier. We present an in vitro study that examined the cytotoxicity of amorphous and composite silica nanoparticles to different cell lines. These bioactives include curcumin and *Antrodia cinnamomea*. It is hoped that by examining the response of multiple cell lines to silica nanoparticles more basic information regarding the cytotoxicity as well as potential functions of silica in future oncological applications could become available.

26.2 MATERIALS AND METHODS

26.2.1 CELL LINES

Human melanoma cell line A375, human lung carcinoma cell line A549, human cervical squamous carcinoma cell line HeLa, human hepatoma cell line Hep G2, human colon carcinoma cell line HT-29, human breast carcinoma cell line MCF-7, human gastric adenocarinoma cell line MKN-28, human skin fibroblast cell line CCD-966sk and WS1, and human lung fibroblast cell line MRC-5 were purchased from the American Type Culture Collection (Manassas, VA) and Dr. Murakami's research laboratory (Kyushu University). All cells were grown in medium supplemented with 10% fetal bovine serum (FBS) at 37°C in a humidified 95% air, 5% CO_2 environment. All cells were grown in Dulbecco's modified Eagle's medium (DMEM), but MKN-28 cells were grown in RPMI medium 1640. All media were purchased from Invitrogen Corporation (Carlsbad, CA). FBS was purchased from HyClone (Logan, UT)

26.2.2 PREPARATION OF NANOENCAPSULATED CURCUMIN

Chitosan samples purchased from a commercial supplier were analyzed for the degree of deacetylation (DD) and molecular weight (MW) according to previous reports [2]. The DD of chitosan samples were 90% with an MW of 20 kDa. Sodium silicate was dissolved in 100 mL 0.05M sodium acetate buffer to prepare 0.82% (w/w) solution. After agitation for 3 min on a magnetic stirrer, 10 mL chitosan solution (0.1% w/w) or 10 mL chitosan solution (0.1% w/w) and 10 mL curcumin (Kalsec International Inc., Kalamazoo, MI) were added. The solution containing curcumin, silicate, and chitosan was mixed completely with a magnetic stirrer. Nanoencapsulated curcumin was centrifuged after 4 days of synthesis and lyophilized for 3 days to obtain nanoparticles.

26.2.3 CHARACTERIZATION OF NANOENCAPSULATED CURCUMIN

Laser light scattering was performed to measure the size of nanoparticles with a Malvern 4700c submicron particle analyzer (Malvern Instruments, Malvern, Worcestershire, U.K.). Nanoencapsulated curcumin particles were suspended in 15 mL distilled, deionized water. Each 2-mL sample was placed in a quartz tube and measured for the intensity-averaged particle diameter. A Hitachi S-4800 scanning electron microscope (SEM) was used to observe the change in particle size and morphology of nanoparticles after their formation and aggregation. Lyophilized nanoparticles were adhered on carbon disks. A layer of gold was deposited, and the micrographs were taken at 15–20 kV in the SEM.

26.2.4 PREPARATION OF *ANTRODIA CINNAMOMEA* EXTRACT POLYSACCHARIDES

Powdered fruiting bodies of cultivated *Antrodia cinnamomea* (about 70 g) were soaked in 500 ml ethanol (70%) for 12 h. The suspensions were centrifuged at

5,520g for 30 min, and the insoluble matter was removed. The supernatant fraction solution was concentrated by evaporating to dryness under vacuum until the solution volume was 100 mL, then it was dialyzed (10 kDa MW cutoff) overnight. The soluble fraction solution was filtered with filter paper (pore size was 0.45 μm) and then freeze-dried to a powder form.

26.2.5 PREPARATION OF *ANTRODIA CINNAMOMEA* EXTRACT POLYSACCHARIDES ENCAPSULATED IN SILICA-CHITOSAN OR SILICA NANOPARTICLES

Chitosan samples purchased from a commercial supplier were analyzed for the degree of *N*-deacetylation (DD) and molecular weight (MW). The DD of chitosan samples were 81% with an MW of 200 kDa. *Antrodia cinnamomea* extract (ACE) polysaccharides encapsulated in silica-chitosan nanoparticles were prepared by synthesis from sodium silicate and chitosan solution. Sodium silicate was dissolved in 30 mL buffer (0.05M sodium acetate) to prepare 0.55% (w/w) solution (pH 6.0). Immediately after the dissolution of silicate on a magnetic stirrer, 6 mL ACE polysaccharide solution (0.1% w/w) and 3 mL chitosan solution (0.55% w/w) were added. The solution containing chitosan, silicate, and ACE polysaccharides was mixed completely. Afterward, it was moved from the stirrer to the benchtop. ACE polysaccharides encapsulated in silica-chitosan nanoparticles were centrifuged after 4 h of synthesis. Silica nanoparticles were synthesized by the same procedure for 12 h without the addition of chitosan solution.

26.2.6 CELL VIABILITY ASSAY

To determine the effect of curcumin and nanoencapsulated curcumin on cell viability, 3-[4,5-dimethylthiazol-2-yl]2,5-diphenyltetrazolium bromide (MTT) assay was used [13]. Cells were seeded into 96-well plates at a density of 2×10^5 cells/mL (100 μL/well) in medium supplemented with 2% FBS and incubated overnight. Cells were incubated for 4 h and were read at 570 nm on a microplate absorbance reader (Tecan Austria GmbH, Grödig/Salzburg, Austria). The mean and standard deviation for each treatment were determined and then converted to the percent relative to control, with solvent only equal to 100%. The concentration at which cell growth was inhibited by 50% (the 50% inhibitory concentration [IC$_{50}$]) was determined by linear interpolation [(50% – Low percentage)/(High percentage – Low percentage)] × [(High concentration – Low concentration) + Low concentration]. The Pearson product-moment correlation coefficient r was used to analyze correlations. One-way analysis of variance (ANOVA) and Duncan's test were used for significance testing, using a p value of .05 (SPSS 11, SPSS Inc., Chicago).

26.2.7 DNA FRAGMENTATION ASSAY

DNA fragmentation was used as an indicator of cell apoptosis [14]. Hep G2 cells were seeded into 10-cm dishes at a density of 2×10^5 cells/mL (10 mL/dish) in a medium supplemented with 2% FBS and incubated for 4 h. Solutions containing

ACE polysaccharides or ACE polysaccharides encapsulated in silica-chitosan or silica nanoparticles were added to the dishes. Cells were protected from light for the duration of sample treatment and were incubated for 48 h. After treating with ACE polysaccharides or their nanoparticles, we harvested the cells by centrifugation at 250g. Cells were lysed in a buffer containing 10 mM Tris-HCl (pH 7.4), 150 mM NaCl, 5 mM ethylenediamine tetraacetic acid (EDTA), and 0.5% Triton X-100 for 10 min on ice. Lysates were vortexed and centrifuged at 14,000g for 10 min. Fragmented DNA in the supernatant was extracted with an equal volume of a neutral mixture (phenol:chloroform:isoamyl alcohol mixture 25:24:1) and analyzed electrophoretically on 2% agarose gels. DNA fragments were stained with ethidium bromide and imaged with a FluoroImager (Pharmacia Biotech, D & R, Israel).

26.3 RESULTS AND DISCUSSION

26.3.1 Proliferating Inhibition of Partially Purified Hard Clam Extract

It was clarified that after being partially purified by Phenyl-Toyopearl, hydroxyapatite, FPLC Superdex 75, and Mono Q chromatography (Figure 26.1), we found that the growth-promoting and -inhibiting fractions toward HB4C5 and HuH-6KK were different. DNA fragmentation and PI stain provided evidence that the Superdex 75 second fraction (S75II) induced apoptosis in hepatoma (HuH-6KK), colon cancer

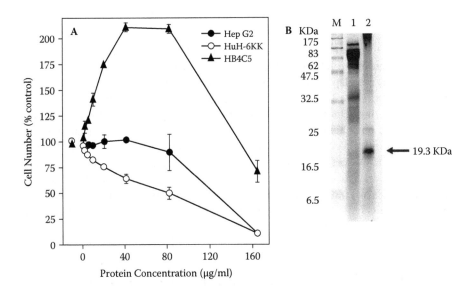

FIGURE 26.1 (A) Effect of three fractions separated by Phenyl-Toyopearl column chromatography of hard clam extract WG30-50 on the growth of HuH-6KK cells. The cells (2 × 10^4 cells) were precultured in 96 wells for 24 h then treated with different concentrations of PT fractions for 48 h. (B) The 10% sodium dodecyl sulfate polyacrylamide gel electrophoresis (SDS-PAGE) patterns of partially purified hard clam extract. M, standard protein marker; lane 1, crude extract WG30-50; lane 2, Superdex 75 second fraction (S75II).

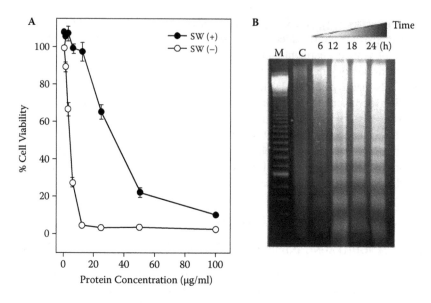

FIGURE 26.2 Effect of swainsonine on the growth of HuH-6KK treated with S75II. (A) The cells were primed with (●) or without (○) swainsonine (5 µg/mL) for 6 days, then treated with different concentrations of S75II for 48 h. Cell viability was detected by MTT assay. (B) DNA electrophoresis of HuH-6KK. The cells were primed with or without swainsonine (5 µg/mL) for 6 days, then treated with or without S75II (5 µg/mL).

(HT-29, COLO-201, and SK-CO-1), breast cancer (MCF-7), cervical cancer (HeLa), monocyte/macrophage-like cells (U937, J774.1, and CCRF-CEM), and mouse melanoma B16, but not Hep 3B or HL-60. HuH-6KK and HT-29 pretreated with lysosomal α-mannosidase II inhibitor (swainsonine 5, 2 µg/mL) for 6 days, the S75II apoptosis-inducing activity was blocked significantly (Figure 26.2). However, swainsonine showed no inhibitory activity in HuH-6KK with a higher concentration of S75II (>25 µg/mL). Neither Ca^{2+} chelator (ethylene glycol tetraacetic acid, EGTA) nor antioxidant (N-acetyl-L-cysteine) prevented cell death, excluding the participation of extracellular Ca^{2+} or intracellular reactive oxygen species (ROS). So, we suggested that S75II induced apoptosis by activating a preexisting apoptosis mechanism. Our results indicated that the principle of *Meretrix lusoria* may be a lectinlike component, and its apoptotic effect was through activation of proteases of caspase family-related apoptosis mediated by cell membrane glycosyl-moiety interaction.

26.3.2   EPIDIOXYSTEROL COMPOUNDS FROM ETHYL ACETATE FRACTION

However, we finally found that the lectinlike component, the potential candidate for the tumor apoptosis-inducing principle, was unstable in its bioactivity from the fraction recovery process. Here, Pan et al. [14,15] (Figure 26.3) conducted a modified method for extraction from *Meretrix lusoria* and found that some epidioxysterol compounds [6] (Figure 26.4) showed strong antitumor/apoptosis-inducing ability,

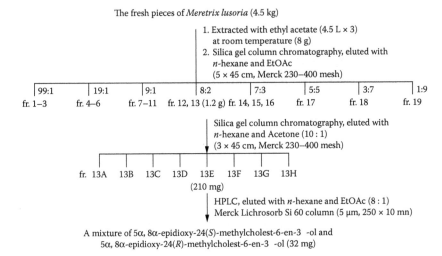

FIGURE 26.3 Separation scheme of the apoptosis-inducing substance. (Adapted from Pan, M.S., et al., 2007, with permission.)

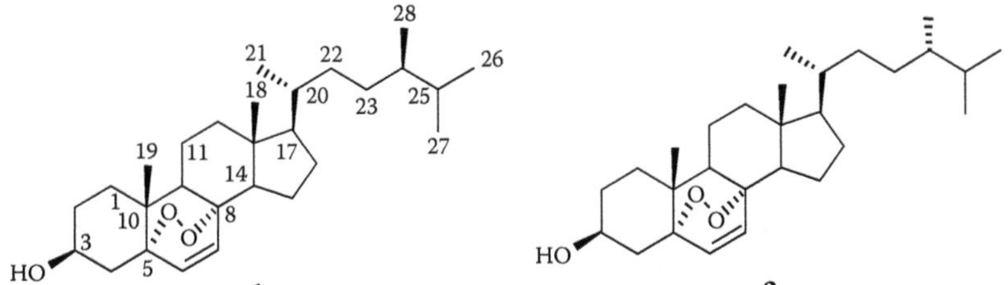

FIGURE 26.4 Structure of epidioxysterols (EDS) isolated from *Meretrix lusoria*. (Adapted from Pan, M.S., et al., 2007, with permission.)

which may suggest the real principle and reveal the contribution to the stability of the whole lectinlike component (Table 26.1 and Figure 26.5).

26.3.3 NANOPARTICLE PREPARATION

Average size of nanoparticles was determined by SEM by measuring the size of about 100 particles in the micrograph. The micrographs (Figure 26.6) show that freeze-dried silica nanoparticles were mostly spherical and aggregated. The aggregation results mainly from the freeze-drying and sample treatment procedures before SEM observations. The average size of individual silica nanoparticles prepared from sodium silicate were 21.58 ± 4.36 nm, and silica nanoparticles prepared from silica tetraethoxysilane (TEOS) were 80.21 ± 14.43 nm.

TABLE 26.1

Effect of HC-EA on the Growth of Various Human Cancer Cells and Human Polymorphonuclear Cells (PMNs)

Cell Line	HC-EA IC$_{50}$ (μg/mL)
COLO 205	448 ± 10
SK-Hep 1	580 ± 30
HL-60	198 ± 10
AGS	274 ± 9
THP-1	647 ± 40
PMNs	>2,500

Source: Adapted from Pan, M.H. et al., 2006, with permission.

HC-EA (400 μg/mL)

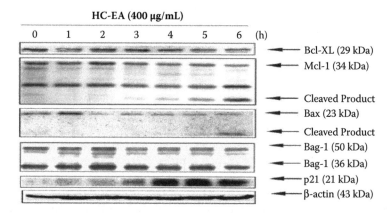

FIGURE 26.5 Effect of HC-EA on Bcl-2 family protein and p21Cip1/WAF1 expression in HL-60 cells. (Adapted from Pan, M.H. et al., 2006, with permission.)

The compounds from the ethyl acetate fraction showing strong apoptosis-inducing activity were identified by spectral methods as epidioxysterols. The *in vitro* results suggest that induction of apoptosis by epidioxysterols may provide a pivotal mechanism for its cancer chemoprevention.

26.3.3.1 Encapsulated Nanoparticle Preparation

The particle size of silica-chitosan composite nanoparticles increased slightly from about 10 to 14 nm with a reaction time of 15 to 360 min. The silica-chitosan nanoparticles prepared after different hours of reaction are designated as silica-chitosan 1H–6H (nanoparticles) in the rest of the discussion. The hydrodynamic particle sizes of silica from silicate were 188.3 ± 11.5 nm and of silica from TEOS were 236.3 ± 6.85 nm. The hydrodynamic particle size of silica-chitosan composite nanoparticles increased slightly from about 153 to 177 nm with a reaction time of 15 to 360 min.

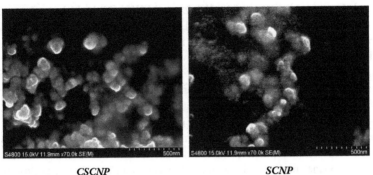

CSCNP

Particle Size: 112.88 ± 3.00 nm

SCNP

Particles Size: 111.05 ± 2.95 nm

FIGURE 26.6 Nanoparticle morphology. Malvern 4700 dynamic light-scattering apparatus; average particle size analyzed by SEM and light scattering.

By comparing the results from SEM and light-scattering measurements, it is likely that agglomerates of three to eight silica nanoparticles are present in the solution. No further aggregation was observed during the cytotoxicity experiment [16,17].

26.3.3.2 Antiproliferation of Encapsulated Nanoparticles

The results of the MTT assay, as a measure of metabolic competence of the cells following 48 h of contact with the silica nanoparticles or silica-chitosan 6H nanoparticles, are shown in Table 26.2. The cytotoxicity of silica nanoparticles increased with increasing particle mass concentration in all the cell cultures. At a concentration of 667 µg/mL silica nanoparticles, the cell viability decreased to a level between 60% and 80% of control. We use IC_{20} (20% inhibitory concentration, µg/mL) to represent the concentrations that cause cell proliferation rate to drop to 80%

TABLE 26.2

Antiproliferation of Encapsulated Nanoparticles: Cell Viability IC_{50} Toward Seven Tumor Cell Lines (µg/mL)

Cell Line	Curcumin	CSCNP	Difference[a]	SCNP	Difference[b]
A375	93 ± 3	53 ± 1	43%	65 ± 1	30%
A549	98 ± 0	56 ± 4	43%	81 ± 2	17%
HeLa	68 ± 4	42 ± 0	38%	56 ± 1	18%
Hep G2	90 ± 2	46 ± 3	49%	41 ± 10	54%
HT-29	106 ± 2	62 ± 0	42%	72 ± 3	32%
MCF-7	153 ± 7	80 ± 0	48%	112 ± 1	27%
MKN-28	166 ± 9	89 ± 1	46%	135 ± 1	19%

Note: Differences were shown by decreasing ratio of IC_{50} after nanoencapsulation.

[a] Formula = $100 - (CSCNP\ IC_{50}/Curcumin\ IC_{50}) \times 100$.

[b] Formula = $100 - (SCNP\ IC_{50}/Curcumin\ IC_{50}) \times 100$.

of control. Cancer cell lines (A549, HT-29, and MKN-28) had higher cell viability than lung and skin fibroblast cell lines (MRC-5, WS1, and CCD-966sk) when incubated with silica nanoparticle suspension in culture media. This indicates that cancer cells are more resistant to silica nanoparticles. These results correspond with the report that chitosan (MW = 213,000, DD = 88%) entities exhibited significant cytotoxicity only at concentrations higher than 0.741 mg/mL in A549 cell lines.

Our results revealed that silica nanoparticles may become slightly cytotoxic at high concentration even though they were nontoxic at low concentration. It has been reported that surface modification with aminosilanes increased the 50% lethal concentration of silica nanoparticles to Cos-1 cells to well above 1 mg/mL. By comparison, polylysine-modified silica showed cytotoxic effects (cell viability dropped to about 60% of control) to HNE1 and HeLa cells at concentrations of 500 µg/mL or above.

In view of the fact that silica-chitosan nanoparticles maintained the cell viability above 85% of control in most test conditions, it appears that the presence of chitosan is more effective than polylysine in reducing the cytotoxicity of silica. However, it may be more informative to relate the cytotoxicity data of different cell lines to their metabolic activity or their rate of proliferation. As a consequence, cell population doubling time could be an important indicator of the tolerance of cells exposed to silica or silica-chitosan nanoparticles. Most human tumor cell lines are capable of maintaining their cell proliferation after they are exposed to the high concentration of silica nanoparticles. However, the proliferation of A549 cells was inhibited by silica nanoparticles nearly as much as that of normal fibroblast cells. This result indicated that the exposure to a high concentration of silica nanoparticles might be toxic to lung cells no matter whether they are normal or tumorigenic. This suggests that chitosan is effective in reducing the cytotoxicity of silica to normal human cells that have much slower metabolic activity than tumor cells.

These results were the first investigation regarding the relationship between the cytotoxicity of silica and the population doubling time of different cell lines as well as the benefits of using chitosan to prepare composite silica nanoparticles. They would be relevant for future applications of silica nanoparticles in drug delivery and controlled release applications.

26.3.4   EFFECTS OF ACE NANOPARTICLES ON THE CELL GROWTH AND VIABILITY OF HEP G2 CELLS

Hepatocellular carcinoma (HCC) is one of the most common cancers worldwide. It is a leading cause of death by cancer, and its incidence is on the rise in developing countries. Apoptosis has an essential role in controlling cell number in many developmental and physiological settings. It is also an important phenomenon in chemotherapy-induced tumor cell killing. Recently, inducers of apoptosis have been used in cancer therapy. Several studies have attempted to induce apoptosis in cancer cells by triggering the core components of the cell death machinery, such as the BCL2 family of proteins, tumor necrosis factor (TNF)-related apoptosis-inducing ligand (TRAIL), and the caspases.

Antrodia cinnamomea is a medicinal fungus that has been used as a herbal medicine in Taiwan for the treatment of diarrhea, abdominal pain, and hypertension. It is also well known for its hepatoprotective and antitumor activities. Polysaccharides and triterpenoids are the major active components of medicinal fungi. For instance, the polysaccharides from *Antrodia cinnamomea* contribute to its antihepatitis B surface antigen effects and antitumor effects. Several studies have demonstrated that *Antrodia cinnamomea* induces apoptosis in cancer cells by triggering the mitochondrial pathway, Fas/Fas ligand pathway, the BCL2 family of proteins, Ca^{2+}-calpain pathway, the caspases, or the inhibition of NF-κB pathway.

Chitosan is a nontoxic, biocompatible, biodegradable, and bioadhesive polysaccharide. It is a linear polysaccharide composed of glucosamine and *N*-acetyl glucosamine linked by β-1, 4-glycosidic bonds. The applications of chitosan are under rapid development in diverse areas such as biology, medicine, and food industries. Chitosan nanoparticles have also been proposed as protein, drugs, and DNA carriers in previous studies. The purpose of this study was to evaluate the apoptotic activity of ACE polysaccharides and ACE polysaccharides encapsulated in silica-chitosan nanoparticles or silica nanoparticles in human liver cancer cell lines (Hep G2). We have assayed the death receptor and mitochondrial apoptotic pathway-related molecules (including cell cycle, the mitochondrial membrane potential, Fas/APO-1, caspase-8, caspase-9, and caspase-3 signaling molecules) that are strongly associated with the signal transduction pathway of apoptosis and are related to the sensitivity of tumor cells to anticancer compounds.

To investigate the potential effects of ACE polysaccharides and their nanoparticle encapsulated formulation on the proliferation and survival of Hep G2 cells, the cells were exposed to ACE polysaccharides or ACE polysaccharides encapsulated in silica or silica-chitosan nanoparticles for 24 and 48 h. The exposure to ACE polysaccharides encapsulated in silica-chitosan nanoparticles was associated with cell shrinkage, as detected in phase-contract micrographs (Figure 26.7), indicating that the ACE polysaccharides encapsulated in silica-chitosan nanoparticles induced the apoptosis of Hep G2 cells. These results are in agreement with those obtained from the MTT assay. Because the loss of membrane integrity is one of the physiological features of necrotic cell death, the LDH assay (data not shown) revealed that increasing the exposure to ACE polysaccharides or ACE polysaccharides encapsulated in silica or silica-chitosan nanoparticles induced more cells to die of necrosis.

ACKNOWLEDGMENTS

This research was partially supported by the National Science Council, Taiwan, Republic of China (grants NSC 94-2120-M-019-001 and NSC 93-2313-B-019-023); Center for Marine Bioscience and Biotechnology and Core Laboratory of Excellence in Fishery Functional Food, National Taiwan Ocean University; and university grant NTOU-AF94-04-03-01-01. We thank Chung-Yuan Mou, Che-Chen Chang, and Shi-Chang Tsai (Instrumentation Center, National Taiwan University) for their assistance in the x-ray photoelectron spectroscopy analysis.

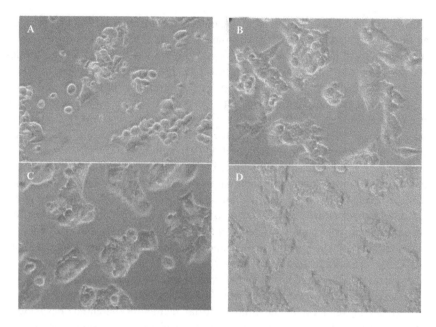

FIGURE 26.7 Phase-contrast microscopy of Hep G2 with ACE polysaccharides encapsulated in silica-chitosan nanoparticles (A) 22 µg/mL, (B) 11 µg/mL, (C) 5.5 µg/mL, (D) 2.75 µg/mL incubated for 48 h. All figures are of the same magnification (×200).

REFERENCES

1. Willett, W.C. Overview and perspective in human nutrition. *Asia Pac J Clin Nutr* 17 Suppl 1:1–4, 2008.
2. Dong, L.M., Kristal, A.R., Peters, U., Schenk, J.M., Sanchez, C.A., Rabinovitch, P.S., Blount, P.L., Odze, R.D., Ayub, K., Reid, B.J., and Vaughan, T.L. Dietary supplement use and risk of neoplastic progression in esophageal adenocarcinoma: a prospective study. *Nutr Cancer* 60:39–48, 2008.
3. Ikeda, I. Multifunctional effects of green tea catechins on prevention of the metabolic syndrome. *Asia Pac J Clin Nutr* 17 Suppl 1:273–274, 2008.
4. Yanagita, T., and Nagao, K. Functional lipids and the prevention of the metabolic syndrome. *Asia Pac J Clin Nutr* 17 Suppl 1:189–191, 2008.
5. Kim, W.Y., Kim, J.E., Choi, Y.J., and Huh, K.B. Nutritional risk and metabolic syndrome in Korean type 2 diabetes mellitus. *Asia Pac J Clin Nutr* 17 Suppl 1:47–51, 2008.
6. Pan, M.H., Huang, Y.T., Ho, C.T., Chang, C.I., Hsu, P.C., and Sun Pan, B. Induction of apoptosis by *Meretrix lusoria* through reactive oxygen species production, glutathione depletion, and caspase activation in human leukemia cells. *Life Sci* 79:1140–1152, 2006.
7. Kong, Z.L., Chiang, L.C., Fang, F., Shinohara, K., and Pan, P. Immune bioactivity in shellfish toward serum-free cultured human cell lines. *Biosci Biotechnol Biochem* 61:24–28, 1997.
8. Nahar, M., Mishra, D., Dubey, V., and Jain, N.K. Development, characterization, and toxicity evaluation of amphotericin B-loaded gelatin nanoparticles. *Nanomedicine* 4:252–261, 2008.

9. Kilian, O., Wenisch, S., Karnati, S., Baumgart-Vogt, E., Hild, A., Fuhrmann, R., Jonuleit, T., Dingeldein, E., Schnettler, R., and Franke, R.P. Observations on the microvasculature of bone defects filled with biodegradable nanoparticulate hydroxyapatite. *Biomaterials* 29:2429–3437, 2008.

10. Sadeghi, A.M., Dorkoosh, F.A., Avadi, M.R., Weinhold, M., Bayat, A., Delie, F., Gurny, R., Larijani, B., Rafiee-Tehrani, M., and Junginger, H.E. Permeation enhancer effect of chitosan and chitosan derivatives: comparison of formulations as soluble polymers and nanoparticulate systems on insulin absorption in Caco-2 cells. *Eur J Pharm Biopharm* 70:270–278, 2008.

11. Esposito, E., Drechsler, M., Mariani, P., Sivieri, E., Bozzini, R., Montesi, L., Menegatti, E., and Cortesi, R. Nanosystems for skin hydration: a comparative study. *Int J Cosmet Sci* 29:39–47, 2007.

12. Gaspar, M.M., Cruz, A., Fraga, A.G., Castro, A.G., Cruz, M.E., and Pedrosa, J. Developments on drug delivery systems for the treatment of mycobacterial infections. *Curr Top Med Chem* 8:579–591, 2008.

13. Scudiero, D.A., Shoemaker, R.H., Paull, K.D., Monks, A., Tierney, S., Nofziger, T.H., Currens, M.J., Seniff, D., and Boyd, M.R. Evaluation of a soluble tetrazolium/formazan assay for cell growth and drug sensitivity in culture using human and other tumor cell lines. *Cancer Res* 48:4827–4833, 1988.

14. Liu, H., Qin, C.Y., Han, G.Q., Xu, H.W., Meng, M., and Yang, Z. Mechanism of apoptotic effects induced selectively by ursodeoxycholic acid on human hepatoma cell lines. *World J Gastroenterol* 13:1652–1658, 2007.

15. Pan, M.H., Ghai, G., and Ho, C.T. Food bioactives, apoptosis, and cancer. *Mol Nutr Food Res* 52:43–52, 2008.

16. Chang, J.S., Kong, Z.L., Hwang, D.F., and Chang, K.L.B. Chitosan-catalyzed aggregation during the biomimetic synthesis of silica nanoparticles. *Chem Mater* 18:702–707, 2006.

17. Chang, J.S., Chang, K.L., Hwang, D.F., and Kong, Z.L. In vitro cytotoxicity of silica nanoparticles at high concentrations strongly depends on the metabolic activity type of the cell line. *Environ Sci Technol* 41:2064–2068, 2007.

27 A New Method for *In Vitro* Glycogen Synthesis and Immunostimulating Activity of Glycogen

*Hiroki Takata, Ryo Kakutani, Hideki Kajiura,
Takashi Furuyashiki, Tsunehisa Akiyama,
Yoshiyuki Adachi, Naohito Ohno, and
Takashi Kuriki*

CONTENTS

27.1 INTRODUCTION

In 1857, a French physiologist, Claude Bernard, first reported isolation of glycogen from dog liver and assigned its metabolic role as a storage form of sugar [1]. It is now well accepted that glycogen is the major storage polysaccharide in animals and micro-organisms. This polysaccharide is highly branched $(1\rightarrow4)(1\rightarrow6)$-linked α-D-glucans with a high molecular weight (10^6 to 10^9) (Figure 27.1) [2]. Based on the results of electron microscopy, it has been shown that glycogen consists of spherical particles with diameters of 20–40 nm (β-particles), and that the β-particles often associate into much larger α-particles (~200 nm). The molecular weight of an individual β-particle has been shown to be approximately 10^7 [2]. Glycogen is opalescent, milky white and slightly bluish, in aqueous solution, and gives a reddish-brown color with the addition of iodine. Amylopectin, a major component of starch, is also a highly branched $(1\rightarrow4)(1\rightarrow6)$-linked α-D-glucans. However, the degree of branching in glycogen is about twice that in amylopectin. Glycogen-type polymer, which is referred to as

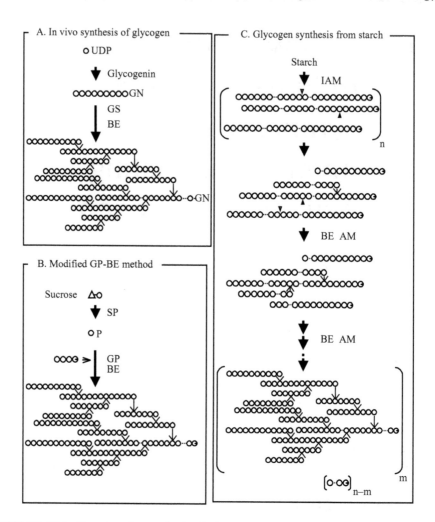

FIGURE 27.1 Reactions for synthesis of glycogen. Glycogen is synthesized *in vivo* (A) from UDP-glucose as a glucosyl donor via cooperative actions of glycogenin (GN), glycogen synthase (GS), and branching enzyme (BE). Glycogen can be synthesized from glucose-1-phosphate (G-1-P) and primer such as maltotetraose (B) by a combined action of α-glucan phophorylase (GP) and BE. G-1-P can be supplied from sucrose using SP. Glycogen can be synthesized from starch (C) using isoamylase (IAM), BE, and amylomaltase (AM). ○, glucosyl residue; ⊖, glucosyl residue with reducing end; Δ, fructosyl residue; —, α-1,4-linkage; ↓, α-1,6-linkage.

phytoglycogen, has also been isolated from several higher plants [3]. Glycogen has also been used as a raw material in the cosmetic industry and as a carrier to enhance the yield of DNA during precipitation with organic solvent [4]. From the viewpoint of chemistry, glycogen is a kind of hyperbranched polymer or dendritic polymer, and the hyperbranched polymer has received increasing attention in recent years [5].

27.2   PRODUCTION OF GLYCOGEN

In animal tissues, glycogen is synthesized from uridine diphosphate (UDP)-glucose via the cooperative action of glycogenin (EC 2.4.1.186), glycogen synthase (EC 2.4.1.11), and branching enzyme [BE; 1,4-α-D-glucan:1,4-α-D-glucan 6-α-D-(1,4-α-glucano)-transferase, EC 2.4.1.18] [6] (Figure 27.1A). Especially, shellfishes such as oysters and mussels have been considered to be sources of glycogen.

The *in vitro* synthesis of glycogen has been successfully achieved by the combined action of α-glucan phosphorylase (GP; EC 2.4.1.1) and BE from glucose-1-phosphate as a substrate (GP-BE method) (Figure 27.1B) [7–9]. We improved the method by supplying glucose-1-phosphate from sucrose using sucrose phosphorylase (SP; EC 2.4.1.7) (Figure 27.1B). The synthetic glycogen produced by the GP-BE method has properties similar to those of the native glycogen from natural resources [10].

In the GP-BE method, GP elongates a chain of α-(1→4)-glucan, and BE introduces branch points in the growing chains, mimicking the *in vivo* synthesis of glycogen by glycogen synthase and BE. Recently, we developed a novel enzymatic process for glycogen synthesis (isoamylase [IAM]-BE-AM [amylomaltase] method, Figure 27.1C) [11]. In this process, short-chain amylose is first prepared from starch or dextrin using IAM (EC 3.2.1.68), and then BE and AM (EC 2.4.1.25) act on the short-chain amylose. We tested eight types of BEs from various sources and found that *Aquifex aeolicus* BE [12] was the best to produce glucan that was similar in size to β-particles of glycogens from natural resource (NSGs). The molecular weight of enzymatically synthesized glycogens (ESGs) is controllable within the range of 3.0×10^6 to 3.0×10^7. On the other hand, BEs from *Bacillus cereus* [13] and kidney bean [14] synthesized glucan with only small molecular sizes (<1,000 k) in this process [11]. Our hypothesis regarding the mechanism of glycogen production from amylose by BE is illustrated in Figure 27.2A. In this model, a singly branched molecule produced in the first stage of the reaction is used as a recipient for subsequent reactions. Thus a small number of branched molecules are produced in the initial stage of the reaction and are used as a "nucleus" or "core" for the growth of a macromolecule. The number of cores is inversely proportional to the molecular size of the product. The substrate specificity of the BE plays an important role in determining the size of the product; if the activity of a BE for a linear molecule is much lower than that for a singly branched molecule produced in the first reaction, then the number of nuclei would be smaller and the molecular weight of the product would be higher. Oligosaccharides with very low molecular size (for example, DP < 8) can hardly be used by BE as substrates. AM can improve the yield of macromolecules in this process by elongating the oligosaccharide by its disproportionation reaction. Figure 27.2B illustrates an opposite case; when the activity of BE for a linear molecule is not so different from that for a singly branched molecule, relatively large numbers of cores are produced, and the molecular size of the final product would be relatively small. The structures and properties of ESGs closely resembled those of native glycogens from various natural resources (Table 27.1).

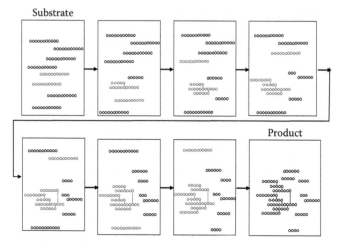

(A) Model of action of BE with high glycogen-synthetic ability

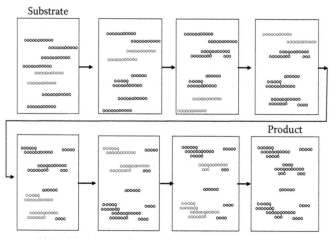

(B) Model of action of BE with low glycogen-synthetic ability

FIGURE 27.2   Model of BE action. Models of action of BE with high (A) and low (B) glycogen-synthetic abilities are shown. In each step, the acceptor and donor molecules for BE action are indicated by red and blue, respectively. (A) A singly branched molecule produced in the first stage of the reaction is used as a recipient for subsequent reactions. Very few branched molecules are produced in the initial stage of the reaction and are used as a "core" for the growth of a macromolecule. (B) Relatively large numbers of cores are produced, and the molecular size of the final product would be relatively small.

27.3   FUNCTION OF GLYCOGEN

Glycogen is exclusively known as energy and carbon reserves in animal cells and microorganisms. However, immunological activity of glycogen has long been suggested, although strong scientific evidence has not been obtained [15–17]. Indeed,

TABLE 27.1   Structural Parameters of Glycogens

Sample Name or Source of NSG	Remarks	M_w (k)[a]	CL	ECL	ICL	λmax (nm)
ESGs						
Product A	By IAM-BE-AM method	2,720	9.3	5.3	3.0	403
Product B		4,960	11.6	7.6	3.0	451
Product E		9,720	9.1	6.1	2.0	421
Product G		13,900	8.7	4.8	2.8	402
Product I		28,700	8.9	4.6	3.3	404
Gly A	By SP-GP-BE method	24,210	11.9	7.5	3.4	460
NSGs						
Oyster	Wako Pure Chemicals	6,010	10.2	6.6	2.7	444
Slipper limpet	Sigma	4,830	9.0	5.3	2.7	404
Bovine liver	Sigma	2,030	11.9	7.0	3.9	416
Rabbit liver	Sigma	15,700	12.5	7.7	3.9	466
Sweet corn	Q.P. Corp.	19,800	13.3	8.2	4.1	467

Note: CL, average chain length; ECL, external chain length; ICL, internal chain length.

[a] Weight-average molecular weight.

many scientists have been annoyed by the lack of reproducibility of their experimental results. Because their samples of glycogen have been extracted from natural resources, (1) the effect of a trace amount of the other materials cannot be ruled out; and (2) the important characteristics of each glycogen sample, such as the average molecular mass and the chain length, are quite different depending on the source and purification procedures. We have cleared up the immunological activity of glycogen by using a completely pure one with very uniform characteristics [18,19]. The results revealed that the molecular mass of glycogen was strongly related to its immuno-stimulating activity. The ESGs with molecular weight of 5,000 and 6,500 k strongly stimulated RAW264.7, a murine macrophage cell line, in the presence of interferon-γ, leading to augmented production of nitric oxide, tumor necrosis factor-α, and inter-leukin 6 (Figure 27.3). The peritoneal exudate cells (PECs) collected from C3H/HeJ mice, Toll-like receptor 4 mutant, were also activate by ESGs with similar profiles as RAW264.7 (data not shown). Furthermore, we demonstrated that biotinylated ESGs bound the macrophage cell line by flow cytometry (data not shown). These results strongly suggested that glycogen functions not only as a fuel reservoir but also as a signaling factor *in vivo*.

27.4   CONCLUSION AND FUTURE PROSPECTS

NSG binds to many kinds of protein, such as glycogen synthase and glycogen phos-phorylase, and contains minor constituents such as glucosamine and a phosphate group [6]. Furthermore, the fine structural properties of NSG can be influenced by the extraction method [20,21]. We can prepare ESG with high purity and controlled structure. While the major role of glycogen *in vivo* is to store glucose, other functions

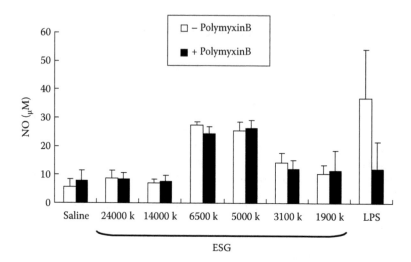

FIGURE 27.3 Relationship between molecular weight and macrophage-stimulating activity of ESG. RAW264.7 cells were cultured with ESGs or lypopolysaccharide (LPS) in the presence of interferon-γ. After cultivation for 48 h, NO in the culture supernatants was determined. Open and solid bars indicate the results without and with polymyxin B treatment, respectively.

have been proposed [22]. We have already demonstrated that glycogen can be recognized by some immune-competent cells and transmit a molecular signal to these cells. It has been demonstrated that some microorganisms produce glycogen-like polysaccharide in their extracellular spaces [23–25]. These findings imply that glycogen has a physiological role in addition to its well-known function as energy and carbon reserves. ESG with high purity and controlled structure should contribute to studies of the physiological role of glycogen.

REFERENCES

1. Young, F.G., Claude Bernard and the discovery of glycogen; a century of retrospect. *Br Med J* 1(5033): 1431–1437, 1957.
2. Geddes, R., Glycogen: a metabolic viewpoint. *Biosci Rep* 6(5): 415–428, 1986.
3. Manners, D.J., Recent developments in our understanding of glycogen structure. *Carbohydr Polym* 16: 37–82, 1991.
4. Heath, E.M., et al., Use of buccal cells collected in mouthwash as a source of DNA for clinical testing. *Arch Pathol Lab Med* 125(1): 127–133, 2001.
5. Satoh, T., et al., Synthesis, branched structure, and solution property of hyperbranched D-glucan and D-galactan. *Macromolecules* 38(10): 4202–4210, 2005.
6. Roach, P.J., Glycogen and its metabolism. *Curr Mol Med* 2(2): 101–120, 2002.
7. Cori, G.T., and C.F. Cori, Crystalline muscle phosphorylase. IV. Formation of glycogen. *J Biol Chem* 151: 57–63, 1943.
8. Fujii, K., et al., Bioengineering and application of novel glucose polymers. *Biocatal Biotransform* 21(4–5): 167–172, 2003.

9. Tolmasky, D.S., and C.R. Krisman, The degree of branching in (α1,4)-(α1,6)-linked glucopolysaccharides is dependent on intrinsic properties of the branching enzymes. *Eur J Biochem* 168: 393–397, 1987.

10. Kitahata, S., and S. Okada, Branching enzymes, in *Handbook of Amylase and Related Enzymes. Their Sources, Isolation Methods, Properties and Applications*, The Amylase Research Society of Japan, Ed. Pergamon Press, Oxford, U.K., 1988, pp. 143–154.

11. Kajiura, H., et al., A novel enzymatic process for glycogen production. *Biocatal Biotransform* 26(1 & 2): 133–140, 2008.

12. Takata, H., et al., Properties of branching enzyme from hyperthermophilic bacterium, *Aquifex aeolicus*, and its potential for production of highly-branched cyclic dextrin. *J Appl Glycosci* 50: 15–20, 2003.

13. Takata, H., et al., Cyclization reaction catalyzed by *Bacillus cereus* branching enzyme, and the structure of cyclic glucan produced by the enzyme from amylose. *J Appl Glycosci* 52: 359–365, 2005.

14. Hamada, S., et al., Two starch-branching-enzyme isoforms occur in different fractions of developing seeds of kidney bean. *Biochem J* 359(Pt 1): 23–34, 2001.

15. Mantovani, B., Phagocytosis of immune complexes mediated by IgM and C3 receptors by macrophages from mice treated with glycogen. *J Immunol* 126(1): 127–130, 1981.

16. Thorpe, B.D., and S. Marcus, Phagocytosis and intracellular fate of *Pasteurella tularensis: in vitro* effects of exudate stimulants and streptomycin on phagocytic cells. *J Reticuloendothel Soc* 4(1): 10–23, 1967.

17. Takaya, Y., et al., Antitumor glycogen from scallops and the interrelationship of structure and antitumor activity. *J Mar Biotechnol* 6: 208–213, 1998.

18. Kakutani, R., et al., Relationship between structure and immunostimulating activity of enzymatically synthesized glycogen. *Carbohydr Res* 342(16): 2371–2379, 2007.

19. Kakutani, R., et al., Stimulation of macrophage by enzymatically synthesized glycogen: the relationship between structure and biological activity. *Biocatal Biotransform* 26(1 & 2): 152–160, 2008.

20. Hata, K., et al., The structures of shellfish glycogens I. *J Jpn Soc Starch Sci* 30(1): 88–94, 1983.

21. Hata, K., et al., The structures of shellfish glycogens II. A comparative study of the structures of glycogens from oyster (*Crassostrea gigas*), scallop (*Patinopecten yessoensis*) and abalone (*Haliotis discus hannai*). *J Jpn Soc Starch Sci* 30(1): 95–101, 1983.

22. Greenberg, C.C., A.M. Danos, and M.J. Brady, Central role for protein targeting to glycogen in the maintenance of cellular glycogen stores in 3T3-L1 adipocytes. *Mol Cell Biol* 26(1): 334–342, 2006.

23. Arvindekar, A.U., and N.B. Patil, Glycogen—a covalently linked component of the cell wall in *Saccharomyces cerevisiae*. *Yeast* 19(2): 131–139, 2002.

24. Gunja-Smith, Z., N.B. Patil, and E.E. Smith, *Two pools of glycogen in Saccharomyces. J Bacteriol* 130(2): 818–825, 1977.

25. Dinadayala, P., et al., Revisiting the structure of the anti-neoplastic glucans of *Mycobacterium bovis* Bacille Calmette-Guerin. Structural analysis of the extracellular and boiling water extract-derived glucans of the vaccine substrains. *J Biol Chem* 279(13): 12369–12378, 2004.

28 Biological Synthesis of Metal Nanoparticles

Beom Soo Kim and Jae Yong Song

CONTENTS

28.1 INTRODUCTION

The term *nanoparticles* usually refers to particles with a size up to 100 nm [1]. Nanoparticles exhibit completely new or improved properties based on specific characteristics such as size, distribution, and morphology if compared with larger particles of their bulk material. Nanoparticles can be made of a wide range of materials, the most common being metal oxide ceramics, metals, silicates, and nonoxide ceramics. Even though other materials (e.g., polymer nanoparticles) exist, the former count for those used in most current applications [1].

Nanoparticles present a higher surface-to-volume ratio with decreasing size of nanoparticles. Specific surface area is relevant for catalytic reactivity and other related properties, such as antimicrobial activity in silver nanoparticles [1]. As specific surface area of nanoparticles increases, their biological effectiveness can increase due to the increase in surface energy. Since the surface area and volume of a sphere are $4\pi r^2$ and $(4/3)\pi r^3$, respectively, the surface area-to-volume ratio of a particle is $3/r$. Therefore, the ratio of surface area to volume for a spherical particle is inversely proportional to the particle radius, or size. For example, a particle of radius 1 mm has a surface area-to-volume ratio of 3000 m^{-1}, while a nanoparticle of radius 10 nm has a surface area-to-volume ratio of $3 \times 10^8\,m^{-1}$. The value of nanomaterials increases from that of the bulk materials due to changes of unique physical, mechanical, optical, and electromagnetic properties. For example, the sales costs of silver and gold increase from \$95/lb silver and \$6,650/lb gold for standard grades to \$415/lb silver and \$26,000/lb gold for nanoscale grades [2].

Noble metal nanoparticles such as gold, silver, and platinum nanoparticles are widely applied to human-contacting areas such as shampoo, soap, detergent, shoes, cosmetic products, and toothpaste as well as medical and pharmaceutical applications. Therefore, there is a growing need to develop environmentally friendly processes of nanoparticle synthesis that do not use toxic chemicals. Biological methods of nanoparticle synthesis using microorganisms, enzymes, and plants or plant extracts have been suggested as possible ecofriendly alternatives to chemical and physical methods. In this chapter, we summarize biological methods of metal nanoparticle production reported in the literature and mention the mechanism, advantages, and disadvantages of each biological nanoparticle synthesis method.

28.2 PRODUCTION METHODS AND APPLICATIONS OF NANOPARTICLES

There are various nanoparticle production methods reported. Most common approaches include solid-state methods (grinding and milling), vapor methods (physical vapor deposition and chemical vapor deposition), chemical synthesis/solution methods (sol-gel approach and colloidal chemistry), and gas-phase synthesis methods [1]. Chemical approaches are the most popular methods for the production of nanoparticles. Other novel production methods include microwave techniques, a supercritical fluid precipitation process, and biological techniques.

Current laboratory techniques utilized in the production of metal nanoparticles are often divided into wet and dry methods. Wet methods usually involve the chemical oxidation of salts via hazardous substances such as sodium borohydride, hydroxylamine, and polyvinylpyrrolidone [3–5]. Dry methods include ultraviolet (UV) irradiation, aerosol technology, and lithography [6,7].

Nanoparticle applications are clustered in six different fields: health care/medical, power/energy, engineering, consumer goods, environmental, and electronics [1]. Biologically synthesized nanoparticles may have special applications in health care/medical fields since chemical synthesis methods still lead to the presence of some toxic chemical species adsorbed on the surface of nanoparticles. Examples of health care/medical applications include targeted drug delivery, alternative drug and vaccine delivery (e.g., inhalation or oral delivery in place of injection), cancer treatments, biolabeling and detection using gold nanoparticles, and antibacterial creams and powders using silver nanoparticles [1].

28.3 BIOLOGICAL SYNTHESIS OF METAL NANOPARTICLES

Biological synthesis of metal nanoparticles is environmentally friendly processes that do not use toxic chemicals. Biological processes are classified into three methods: using microorganism such as bacteria and fungi, using enzyme, and using plant or plant extracts.

28.3.1 Nanoparticle Synthesis Using Microorganism

The initial report of microbial synthesis of nanoparticles was on intracellular silver nanoparticle formation in *Pseudomonas stutzeri* AG259, which was isolated from a silver mine [8]. Klaus et al. [9] reported on the biosynthesis of silver-based single crystals with well-defined shapes such as equilateral triangles and hexagons in *P. stutzeri* AG259. The crystals were up to 200 nm in size and were often located at the cell poles. Nair and Pradeep [10] reported the growth of gold, silver, and gold-silver alloy crystals assisted by most common *Lactobacillus* strains found in buttermilk when exposed to the precursor ions. Accumulation of metals was about 35% of the dry bacterial biomass harvested, and crystal growth did not affect the viability of the bacteria. Electron micrographic images showed that the crystals within the bacterial contour were much larger than the clusters outside.

Several strains of *Pseudomonas aeruginosa* were used for extracellular biosynthesis of gold nanoparticles in the range of 15–30 nm [11]. The extracellular reduction process would make the process simpler and easier for downstream processing. Shahverdi et al. [12] reported on the rapid synthesis of silver nanoparticles using the culture supernatants of *Enterobacteria*. Silver nanoparticles were formed within 5 min of silver ion coming in contact with the cell filtrate, but these particles were unstable after 5 min. Platinum nanoparticles were also deposited in the bacterium *Shewanella algae* [13]. Resting cells of *S. algae* were able to reduce aqueous $PtCl_6^{2-}$ ions into elemental platinum at room temperature and neutral pH within 60 min when lactate was provided as the electron donor. A transmission electron micrographic (TEM) image showed that platinum nanoparticles of about 5 nm were located in the periplasm.

The fungus *Verticillium* sp. resulted in the intracellular formation of gold and silver nanoparticles when treated with an aqueous solution of $AuCl_4^-$ or Ag^+ ions [14,15]. The extracellular synthesis of gold nanoparticles was observed by treatment of the fungus *Fusarium oxysporum* with aqueous $AuCl_4^-$ ions [16]. The reduction of the $AuCl_4^-$ ions occurred due to nicotinamide adenine dinucleotide (NADH)-dependent reductase released by the fungus into solution. Nanoparticulate magnetite was also produced extracellularly by challenging the fungi (*F. oxysporum* and *Verticillium* sp.) with mixtures of ferric and ferrous salts [17]. Extracellular hydrolysis of the anionic iron complexes by cationic proteins secreted by the fungi resulted in the synthesis of crystalline magnetic particles.

Besides noble metal nanoparticles such as gold, silver, and platinum, the syntheses of semiconductor nanocrystallites using microorganisms have been reported. Dameron et al. [18] showed that yeasts such as *Schizosaccharomyces pombe* and *Candida glabrata* produced intracellular CdS nanocrystallites when challenged with cadmium in solution. Kowshik et al. [19] reported that another yeast, *Torulopsis* sp., synthesized intracellular PbS nanocrystallites when challenged with lead. These findings suggest that microbial synthesis may offer a generic strategy for synthesis of a variety of metal and semiconductor nanoparticles.

28.3.2 Nanoparticle Synthesis Using Enzyme

Enzymes act as catalysts for the growth of metallic nanoparticles [20]. Numerous types of enzymes, such as oxidases, hydroxylases, hydrolytic proteins, or NAD(P)$^+$-dependent enzymes, may be employed as biocatalysts for the synthesis of metal nanoparticles and for the development of optical/electrochemical sensors for the respective substrates. The glucose oxidase-biocatalyzed oxidation of glucose leads to the formation of H_2O_2, which acts as a reducing agent for the catalytic deposition of gold on the gold nanoparticle seeds associated with the glass support [21]. This observation led to the development of an optical detection path for glucose oxidase activity and for the sensing of glucose.

Numerous redox enzymes utilize $NAD(P)^+/NAD(P)H$ cofactor systems. The NADH or NADPH (nicotinamide adenine dinucleotide phosphate) cofactors were found to enlarge gold nanoparticle seeds by the reduction of $AuCl_4^-$. The reduction of the salt was found to proceed in two steps (Eqs. 28.1 and 28.2). First, the gold III ion is rapidly reduced to gold I and then is reduced by NADH in the presence of gold nanoparticle seeds acting as catalyst to the enlarged particles [22].

$$AuCl_4^- + NADH \rightarrow Au(I) + 4Cl^- + NAD^+ + H^+ \qquad (28.1)$$

$$2Au(I) + NADH \rightarrow 2Au^0 + NAD^+ + H^+ \qquad (28.2)$$

The regeneration of NADH enabled the analysis of the enzyme activity (lactate dehydrogenase) and the quantitative detection of its substrate (lactate). When lactate is converted to pyruvate by lactate dehydrogenase, NADH is regenerated from NAD^+ and then used for reduction of $AuCl_4^-$ ion to gold nanoparticles. The reduced cofactor NADH was also employed to reduce Cu^{2+} ions to Cu^0 metal deposited on core gold nanoparticle seeds [23]. The amount of Cu^0 deposited on the gold nanoparticles was controlled by the amount of NADH. This enabled the quantitative analysis of NADH by following the enlargement of the nanoparticle seeds with copper.

28.3.3 Nanoparticle Synthesis Using Plants or Plant Extracts

Using plants for nanoparticle synthesis can be advantageous over other biological processes because it eliminates the elaborate process of maintaining cell cultures and can also be suitably scaled up for large-scale synthesis of nanoparticles [24]. Gardea-Torresdey et al. [25,26] demonstrated the synthesis of gold and silver nanoparticles within live alfalfa plants from solid media.

Extracellular nanoparticle synthesis using plant leaf extracts other than whole plants would be more economical due to the easier downstream processing. The pioneering works of nanoparticle synthesis using plant extracts have been carried out by Sastry's group [24,27–32]. They reported that the rates of nanoparticle synthesis using plant extracts are comparable to those of chemical methods. The shape of nanoparticles plays a crucial role in modulating their optical properties. Gold nanotriangles were formed when lemongrass (*Cymbopogon flexuosus*) leaf extract was reacted with aqueous $AuCl_4^-$ ions [29]. The role of halide ions and temperature on the morphology of biogenic gold nanotriangles was also studied by the same group [30]. They suggested that the presence of Cl^- ions during the synthesis promoted the growth of nanotriangles, whereas the presence of I^- ions distorted the nanotriangle morphology and induced the formation of aggregated spherical nanoparticles.

Shankar et al. [24] reported on the synthesis of pure metallic nanoparticles of silver and gold by the reduction of Ag^+ and Au^{3+} ions using neem (*Azadirachta indica*) leaf broth. The times required for more than 90% reduction of Ag+ and Au3+ ions using neem leaf broth were about 4 and 2 h, respectively. If biological synthesis of nanoparticles can compete with chemical methods, there is a need to achieve faster synthesis rates. We carried out engineering approaches such as rapid nanoparticle synthesis using plant extracts and size control of the synthesized nanoparticles [33,34]. Several plant leaf extracts were screened and compared for their synthesis of gold, silver, or platinum nanoparticles by monitoring the conversion using UV-visible spectroscopy. The effects of reaction conditions such as temperature and composition of the reaction mixture were also investigated to control the size of nanoparticles.

Figure 28.1 shows the time courses of silver nanoparticle production with different reaction temperatures obtained with magnolia (*Magnolia kobus*) leaf broth. As the reaction temperature increased, both synthesis rate and final conversion to silver nanoparticles increased. The final conversion at 25°C was 65%, and it reached almost 100% at more than 65°C. Figure 28.2 is a scanning electron micrographic (SEM) image of silver nanoparticles obtained with 5% persimmon (*Pinus desiflora*) leaf broth and 1 m*M* AgNO3 solution. It is shown that relatively spherical and uniform nanoparticles were formed. Figure 28.3 is a TEM image of gold nanoparticles obtained with 1 m*M* HAuCl4 and 5% magnolia leaf broth. A nanoparticle mixture of triangles, pentagons, hexagons, and spheres is shown with a size range of 5 to 100 nm. It is known that nanoparticles synthesized using plant extracts are stable in solution for 4 weeks after their synthesis, possibly due to the capping material on the surface of nanoparticles [24]. Our energy-dispersive x-ray spectroscopy (EDS) profile of gold and silver nanoparticles synthesized using plant extracts also showed strong gold or silver signal along with a weak oxygen and carbon peak, which may originate from the biomolecules that are bound to the surface of nanoparticles.

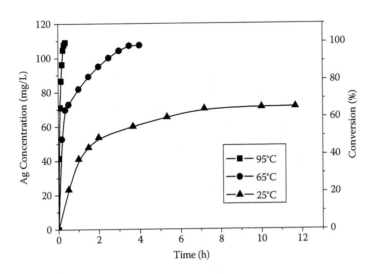

FIGURE 28.1 Time courses of silver nanoparticle formation obtained with 1 m*M* AgNO3 and 5% *Magnolia kobus* leaf broth with different reaction temperatures.

FIGURE 28.2 SEM image of silver nanoparticles formed with 1 mM AgNO$_3$ and 5% *Pinus desiflora* leaf broth.

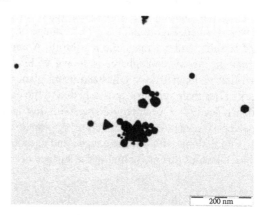

FIGURE 28.3 TEM image of gold nanoparticles formed with 1 mM HAuCl$_4$ and 5% *Magnolia kobus* leaf broth.

Table 28.1 summarizes the synthesis rates of gold, silver, and platinum nanoparticles obtained with several plant leaf extracts (pine, persimmon, ginkgo, magnolia, and platanus). The synthesis rate was highest with magnolia leaf broth. Only 3 and 11 min were required for more than 90% conversion at 95°C to gold and silver nanoparticles, respectively. Although rapid synthesis of silver nanoparticles within 5 min was recently reported using culture supernatants of *Enterobacteria*, the silver nanoparticles synthesized were unstable after 5 min [12]. Using plant extracts for nanoparticle synthesis is another advantage over using bacteria because the nanoparticles are stable for a long time.

Currently, the mechanism of biological nanoparticle synthesis is not fully understood. For gold nanoparticles synthesized extracellularly by the fungus *F. oxysporum*, it was reported that the reduction occurs due to NADH-dependent reductase released into the solution [16]. With neem leaf broth, it was reported that terpenoids are believed to be the surface-active molecules stabilizing the nanoparticles, and reaction of the metal ions is possibly facilitated by reducing sugars or terpenoids

TABLE 28.1
Summary of Nanoparticle Production Using Plant Extracts

Reducing Agent	Nanoparticles	Time for > 90% Conversion
Magnolia leaf broth	Silver	11 min
Persimmon leaf broth	Silver	50 min
Platanus leaf broth	Silver	90 min
Ginkgo leaf broth	Silver	100 min
Pine leaf broth	Silver	150 min
Magnolia leaf broth	Gold	3 min
Magnolia leaf broth	Platinum	4 h

present in the neem leaf broth [24]. Terpenoids (isoprenoids) are a large and diverse class of naturally occurring organic chemicals derived from five-carbon isoprene units assembled and modified in thousands of ways. These lipids can be found in all classes of living things and are the largest group of natural products. Well-known terpenoids include citral, menthol, and camphor. We tested citral as a reducing agent by treatment with $AuCl_4^-$ ions and observed the formation of gold nanoparticles. More elaborate studies are required to elucidate the mechanism of biological nanoparticle synthesis.

28.4 CONCLUSION

Biological methods of metal nanoparticle synthesis provide a new possibility of conveniently synthesizing nanoparticles using natural products. As possible ecofriendly alternatives to chemical and physical methods, biological synthesis of nanoparticles using microorganisms, enzymes, and plants or plant extracts has been suggested. Microbial synthesis using bacteria, yeast, and fungi offers a generic strategy for synthesis of a variety of metal and semiconductor nanoparticles. Numerous types of enzymes have been employed as biocatalysts for the synthesis of metal nanoparticles and for the development of optical/electrochemical sensors for the respective substrates. Using plants for nanoparticle synthesis can be advantageous over other biological processes because it eliminates the elaborate process of maintaining cell cultures and can be suitably scaled up for large-scale synthesis of nanoparticles. Another advantage of using plant extracts for nanoparticle synthesis over using bacteria is stable nanoparticle formation for a long time due to the capping material on the surface of nanoparticles. By using screened plant extracts with high production capability and increasing the reaction temperature, we obtained synthesis rates faster or comparable to those of chemical methods. The particle size and shape can also be controlled by changing the plant type or temperature and composition of the reaction mixture. This environmentally friendly method of biological metal nanoparticle production can potentially be used in various human-contacting areas such as cosmetics, foods, and medical applications.

REFERENCES

1. Willems and van den Wildenberg, Roadmap report on nanoparticles, W&W Espana, Barcelona, 2005.
2. Yoo, J.N., Chemical engineering industry and nanotechnology. Paper presented at the fall meeting of the Korean Institute of Chemical Engineers, Seoul, 2006.
3. Vorobyova, S.A., Sobal, N.S., and Lesnikovich, A.I., Colloidal gold, prepared by interphase reduction, *Colloids Surf.*, 176, 273–277, 2001.
4. Koel, B.E., Meltzer, S., Resch, R., Thompson, M.E., Madhukar, A., Requicha, A.A.G., and Will, P., Fabrication of nanostructures by hydroxylamine seeding of gold nanoparticle templates, *Langmuir*, 17, 1713–1718, 2001.
5. Han, M.Y., Quek, C.H., Huang, W., Chew, C.H., and Gan, L.M., A simple and effective chemical route for the preparation of uniform nonaqueous gold colloids, *Chem. Mater.*, 11, 1144–1147, 1999.
6. Chen, Z.Y., Zhou, Y., Wang, C.Y., and Zhu, Y.R., A novel ultraviolet irradiation technique for shape-controlled synthesis of gold nanoparticle at room temperature, *Chem. Mater.*, 11, 2310–2312, 1999.
7. Magnusson, M.H., Deppert, K., Maim, J., Bovin, J., and Samuelson, L., Size-selected gold nanoparticles by aerosol technology, *Nanostruc. Mater.*, 12, 45–48, 1999.
8. Gadd, G.M., Laurence, O.S., Briscoe, P.A., and Trevors, J.T., Silver accumulation in *Pseudomonas stutzeri* AG259, *BioMetals*, 2, 168–173, 1989.
9. Klaus, T., Joerger, R., Olsson, E., and Granqvist, C.G., Silver-based crystalline nanoparticles, microbially fabricated, *Proc. Natl. Acad. Sci. U.S.A.*, 96, 13611–13614, 1999.
10. Nair, B., and Pradeep, T., Coalescense of nanoclusters and formation of submicron crystallites assisted by *Lactobacillus* strains. *Cryst. Growth Des.*, 2, 293–298, 2002.
11. Husseiny, M.I., Abd El-Aziz, M., Badr, Y., and Mahmoud, M.A., Biosynthesis of gold nanoparticles using *Pseudomonas aeruginosa*, *Spectrochim. Acta Part A*, 67, 1003–1006, 2007.
12. Shahverdi, A., Minaeian, S., Shahverdi, H.R., Jamalifar, H., and Nohi, A.-A., Rapid synthesis of silver nanoparticles using culture supernatants of *Enterobacteria*: a novel biological approach, *Proc. Biochem.*, 42, 919–923, 2007.
13. Konishi, Y., Ohno, K., Saitoh, N., Nomura, T., Nagamine, S., Hishida, H., Takahashi, Y., and Uruga, T., Bioreductive deposition of platinum nanoparticles on the bacterium *Shewanella algae*, *J. Biotechnol.*, 128, 648–653, 2007.
14. Mukherjee, P., Ahmad, A., Mandal, D., Senapati, S., Sainkar, S.R., Khan, M.I., Ramani, R., Parischa, R., Ajayakumar, P.V., Alam, M., Sastry, M., and Kumar, R., Bioreduction of AuCl$_4^-$ ions by the fungus, *Verticillium* sp. and surface trapping of the gold nanoparticles formed, *Angew. Chem. Int. Ed.*, 40, 3585–3588, 2001.
15. Mukherjee, P., Ahmad, A., Mandal, D., Senapati, S., Sainkar, S.R., Khan, M.I., Parischa, R., Ajayakumar, P.V., Alam, M., Kumar, R., and Sastry, M., Fungus-mediated synthesis of silver nanoparticles and their immobilization in the mycelial matrix: a novel biological approach to nanoparticle synthesis, *Nano Lett.*, 1, 515–519, 2001.
16. Mukherjee, P., Senapati, S., Mandal, D., Ahmad, A., Khan, M.I., Kumar, R., and Sastri, M., Extracellular synthesis of gold nanoparticles by the fungus *Fusarium oxysporum*, *Chem. Biochem.*, 5, 461–463, 2002.
17. Bharde, A., Rautaray, D., Bansal, V., Ahmad, A., Sarkar, I., Yusuf, S.M., Sanyal, M., and Sastri, M., Extracellular biosynthesis of magnetite using fungi, *Small*, 2, 135–141, 2006.
18. Dameron, C.T., Reese, R.N., Mehra, R.K., Kortan, A.R., Carroll, P.J., Steigerwald, M.L., Brus, L.E., and Winge, D.R., Biosynthesis of cadmium sulphide quantum semiconductor crystallites, *Nature*, 338, 596–597, 1989.

19. Kowshik, M., Vogel, W., Urban, J., Kulkarni, S.K., and Paknikar, K.M., Microbial synthesis of semiconductor PbS nanocrystallites, *Adv. Mater.*, 14, 815–818, 2002.
20. Willner, I., Baron, R., and Willner, B., Growing metal nanoparticles by enzymes, *Adv. Mater.*, 18, 1109–1120, 2006.
21. Zayats, M., Baron, R., Popov, I., and Willner, I., Biocatalytic growth of Au nanoparticles: from mechanistic aspects to biosensors design, *Nano Lett.*, 5, 21–25, 2005.
22. Xiao, Y., Pavlov, V., Levine, S., Niazov, T., Markovich, G., and Willner, I., Catalytic growth of Au-nanoparticles by NAD(P)H cofactors: optical sensors for NAD(P)⁺-dependent biocatalyzed transformations, *Angew. Chem. Int. Ed.*, 43, 4519–4522, 2004.
23. Shlyahovsky, B., Katz, E., Xiao, Y., Pavlov, V., and Willner, I., Optical and electrochemical detection of NADH and of NAD⁺-dependent biocatalyzed processes by the catalytic deposition of copper on gold nanoparticles, *Small*, 213–216, 2005.
24. Shankar, S.S., Rai, A., Ahmad, A., and Sastry, M., Rapid synthesis of Au, Ag, and bimetallic Au core Ag shell nanoparticles using Neem (*Azadirachta indica*) leaf broth, *J. Colloid Interface Sci.*, 275, 496–502, 2004.
25. Gardea-Torresdey, J.L., Parsons, J.G., Gomez, E., Peralta-Videa, J., Troiani, H.E., Santiago, P., and Jose-Yacaman, M., Formation and growth of Au nanoparticles inside live alfalfa plants, *Nano Lett.*, 2, 397–401, 2002.
26. Gardea-Torresdey, J.L., Gomez, E., Peralta-Videa, J., Parsons, J.G., Troiani, H.E., Santiago, P., and Jose-Yacaman, M., Alfalfa sprouts: a natural source for the synthesis of silver nanoparticles, *Langmuir*, 19, 1357–1361, 2003.
27. Shankar, S.S., Ahmad, A., Pasricha, R., and Sastry, M., Bioreduction of chloroaurate ions by geranium leaves and its endophytic fungus yields gold nanoparticles of different shapes, *J. Mater. Chem.*, 13, 1822–1826, 2003.
28. Shankar, S.S., Ahmad, A., and Sastry, M., Geranium leaf assisted biosynthesis of silver nanoparticles, *Biotechnol. Prog.*, 19, 1627–1631, 2003.
29. Shankar, S.S., Rai, A., Ankamwar, B., Singh, A., Ahmad, A., and Sastry, M., Biological synthesis of triangular gold nanoprisms, *Nat. Mater.*, 3, 482–488, 2004.
30. Rai, A., Singh, A., Ahmad, A., and Sastry, M., Role of halide ions and temperature on the morphology of biologically synthesized gold nanotriangles, *Langmuir*, 22, 736–741, 2006.
31. Rai, A., Chaudhary, M., Ahmad, A., Bhargava, S., and Sastry, M., Synthesis of triangular Au core-Ag shell nanoparticles, *Mater. Res. Bull.*, 42, 1212–1220, 2007.
32. Chandran, S.P., Chaudhary, M., Pasricha, R., Ahmad, A., and Sastry, M., Synthesis of gold nanotriangles and silver nanoparticles using *Aloe vera* plant extract, *Biotechnol. Prog.*, 22, 577–583, 2006.
33. Song, J.Y., and Kim, B.S., Rapid biological synthesis of silver nanoparticles using plant leaf extract, *Bioprocess Biosyst. Eng.*, in press.
34. Song, J.Y., and Kim, B.S., Biological synthesis of bimetallic Au/Ag nanoparticles using Persimmon (*Diopyros kaki*) leaf extract, *Korean J. Chem. Eng.*, 25, 808–811, 2008.

Index

Milton Keynes UK
Ingram Content Group UK Ltd.
UKHW021841071024
449327UK00021B/1527

9 780367 385699